Understanding Basic Statistics

THIRD EDITION

3

Understanding Basic Statistics

Charles Henry Brase
Regis University

Corrinne Pellillo Brase
Arapahoe Community College

HOUGHTON MIFFLIN COMPANY
Boston   New York

Publisher: Jack Shira

Sponsoring Editor: Lauren Schultz

Assistant Editor: Jennifer King

Editorial Assistant: Kasey McGarrigle

Senior Project Editor: Nancy Blodget

Editorial Assistant: Celeste Ng

Senior Production/Design Coordinator: Jennifer Meyer Dare

Senior Manufacturing Coordinator: Marie Barnes

Senior Marketing Manager: Ben Rivera

Marketing Associate: Alexandra Shaw

Cover photographer: Harold Burch, NYC
Viewpoint artist: Lauren Arnest

A complete list of photo credits appears in the back of the book, immediately following the Appendix.

ComputerStat displays for Windows reprinted with permission of Houghton Mifflin Company.
TI-83 is a registered trademark of Texas Instruments, Inc.
Minitab is a registered trademark of Minitab, Inc.
Microsoft Excel screen shots reprinted by permission from Microsoft Corporation. Excel, Microsoft, and Windows are either registered trademarks or trademarks of Microsoft Corporation in the United States and/or other countries.

Printed in the U.S.A.

Library of Congress Control Number: 2002109371

ISBN
Student Text: 0-618-31553-5
High School Edition: 0-618-33359-2

1 2 3 4 5 6 7 8 9 — VHP — 07 06 05 04 03

Contents

1 Getting Started   2

2 Organizing Data   34

3 Averages and Variation   78

Appendix: *Tables*   *A1*

Preface

Welcome to the exciting world of statistics! We have written this text to make statistics accessible to everyone, including those with a limited mathematics background. Statistics affects all aspects of our lives. Whether we are testing new medical devices or determining what will entertain us, applications of statistics are so numerous that, in a sense, we are limited only by our own imagination.

Overview

The third edition of *Understanding Basic Statistics* continues to emphasize concepts of statistics. Statistical methods are carefully presented with a focus on understanding both the *suitability of the method* and the *meaning of the result*. Statistical methods and measurements are developed in the context of applications.

We have retained and expanded features that made the first two editions of the text readable. Examples, exercises, and problems touch on applications appropriate to a broad range of interests.

Technology-based components give both students and professors additional resources. New with the third edition is the HM StatPass CD-ROM packaged with every text. This CD-ROM contains over 100 data sets (in Excel, Minitab, and TI-83Plus/ASCII file formats), interactive tutorial exercises for each chapter section, and the text-specific computer software package ComputerStat. The Web site for the text provides a wealth of material including course management tools, PowerPoint slides, a quizzing and tutorial program, and links to other relevant Web sites. Instructional videos covering every section of the text provide even more learning support.

Content Changes in the Third Edition

With each new edition we reevaluate the scope, appropriateness, and effectiveness of the text's presentation and reflect on extensive user feedback. Revisions have been made throughout the text to clarify explanations of important concepts and to update problems.

Sections and Organization

- The second edition's Chapter 1, Organizing Data, has been divided into two chapters in the third edition. The new Chapter 1, Getting Started, includes material about samples and populations, the concept of random samples, and a new section that introduces experimental design. Material on graphs, histograms, and stem-and-leaf displays now appears in the new Chapter 2.

- Section 12.4, Inferences Concerning Regression Parameters, replaces Section 11.4 of the second edition. This section includes testing the correlation coefficient as well as the slope of the least-squares line. The Student's *t* distribution is used for both tests.

Table for Standard Normal Distribution (cumulative area to the left of *z*)

New with the third edition is a two-page table that gives the cumulative area to the left of a specified *z* value. This type of table is easy to use and is also compatible with cumulative distribution results given on computers and calculators. The *Standard Normal Distribution Table* giving the area between 0 and *z* that was used in previous editions is available on the Web site and CDs for the text.

New Topics

Back-to-back stem plots are now briefly discussed. In Section 5.4, Trees and Counting Techniques, trees are used to show the probability of various events and to further illustrate conditional probability.

Features in the Third Edition

Chapter and Section Lead-ins

- NEW! *Preview Questions* at the beginning of each chapter are keyed to the sections.
- NEW! *Focus Problems* at the beginning of each section demonstrate some of the types of questions students can answer once they master the concepts and skills presented in the chapter.
- NEW! *Focus Points* at the beginning of each section describe the primary learning objectives of the section.

Carefully Developed Pedagogy

- *Examples* show students how to select and use appropriate procedures.
- *Guided exercises* within the sections give students an opportunity to work with a new concept. Completely worked-out solutions appear beside each exercise to give immediate reinforcement.
- NEW! *Labels* for each example or guided exercise highlight the technique, concept, or process illustrated by the example or guided exercise. In addition, new labels for section and chapter problems describe the field of application and show the wide variety of subjects in which statistics are used.
- *Section and chapter problems* require the student to use all the new concepts mastered in the section or chapter. The problem sets include a variety of real-world applications with data or setting from identifiable sources. Key steps and solutions to odd-numbered problems appear at the end of the book.
- *Data Highlights* and *Linking Concepts* provide group projects and writing projects.
- *Viewpoints* are brief essays presenting diverse situations in which statistics are used.
- *Enhanced design and photos* provide improved readability.

Technology within the Text

- NEW! *Tech Notes* within sections provide brief point-of-use instructions for the TI-83Plus calculator, Excel, and Minitab. These Notes replace the Calculator Notes of the second edition.
- *Using Technology* sections show the use of Excel as well as the TI-83Plus calculator, Minitab, and ComputerStat.

Supplements for Students

- NEW! *HM StatPass CD-ROM.* This state-of-the-art statistics CD-ROM contains:
  - *Data Sets* for additional experimentation with Minitab, Excel, and the TI-83Plus.
  - *ComputerStat* software designed specifically to accompany the text.
  - *Algorithmically Generated Exercises* for each chapter section, complete with optional hints, solution steps, and lessons.
- NEW! *Companion Web Site for Students* features a number of student resources, including data sets, tutorials, quizzes, glossary, and Web links. Go to http://math.college.hmco.com/students and follow the statistics links to the Brase/Brase, *Understanding Basic Statistics, 3e* site.
- NEW! SMARTHINKING™ *Live On-line Tutoring.* Houghton Mifflin has partnered with SMARTHINKING to provide an easy-to-use and effective on-line tutoring service. A dynamic Whiteboard and Graphing Calculator function enables students and e-structors to collaborate easily. SMARTHINKING offers three levels of service:
  - *Text-Specific Tutoring* provides real-time, one-on-one instruction with a specially qualified e-structor.
  - *Questions Any Time* allows students to submit questions to the tutor outside the scheduled hours and receive a reply within 24 hours.
  - *Independent Study Resources* connect students with around-the-clock access to additional educational services, including interactive Web sites, diagnostic tests, and Frequently Asked Questions posed to SMARTHINKING e-structors.
- *Technology Guide* contains information and examples for the TI-83Plus graphing calculator, Minitab software, and the tutorial program ComputerStat.
- *Excel Guide* contains information and examples for using Excel.
- *Student Solutions Manual* provides the solutions to the odd-numbered exercises in the student textbook.
- *Houghton Mifflin Mathematics Instructional Videos and DVDs.* A lecture series featuring Dana Mosely in which he provides careful explanations of key concepts, examples, exercises, and applications. Ideal if students miss a lecture or need more step-by-step explanations, and a great tool for reviewing course material before an upcoming class, quiz, or test.
- *Student Version of MINITAB (Release 12) CD-ROM* manipulates and interprets data to produce textual, graphical, and tabular results. Minitab may be packaged with the textbook. Please visit the Houghton Mifflin Web site or e-mail a representative at college_math@hmco.com.

Supplements for Instructors

- *Instructor's Resource Guide with Complete Solutions* provides the complete solutions to all exercises in the text, sample tests for each chapter, Teaching Hints, and Transparency Masters for the tables and frequently used formulas found in the third edition. Also included is the Test Item File, a printed version of the computerized test bank.

- NEW! *HM ClassPrep with HM Testing CD-ROM* is designed to produce an unlimited number of tests for each chapter of the text, including cumulative tests and final exams. This computerized test generator hybrid CD-ROM, which works on both Microsoft Windows and Macintosh platforms, contains numerous algorithms as well as on-line testing and gradebook functions. ClassPrep provides a multitude of text-specific resources to enhance the classroom experience and includes files for every printed ancillary as well as Digital Figures and Tables from the book. Resources can be easily accessed from the CD-ROM by chapter or resource type.

- NEW! *Companion Web Site for Instructors.* In addition to the resources found on the student Web site, instructors can access Digital Lessons and classroom management features by going to http://math.college.hmco.com/instructors and following the statistics links to the Brase/Brase, *Understanding Basic Statistics, 3e* site.

Alternate Routes Through the Text

Understanding Basic Statistics, Third Edition, is designed to be flexible. It offers the professor a choice of teaching possibilities. In most one-semester courses, it is not practical to cover all the material in depth. However, depending on the emphasis of the course, the professor may choose to cover various topics. For help in topic selection, refer to the *Table of Prerequisite Material* on page 1.

- *Linear Regression.* Chapter 4, Regression and Correlation, may be delayed until after Chapter 10. The descriptive topics of linear regression may then be followed immediately by the inferential topics of linear regression presented in Chapter 12.

- *Probability.* For courses requiring minimal probability, Section 5.1, What Is Probability? and the first part of Section 5.2, Some Probability Rules—Compound Events, will be sufficient.

Acknowledgments

It is our pleasure to acknowledge the reviewers of this text. All of their insights and comments have been very valuable to us. Reviewers of this and previous editions include:

Delores Anderson, Truett-McConnell College
Jonathan Baker, Columbus State Community College
Lynda L. Ballou, Kansas State University
Randall Boan, Aims Community College
Brian Bradie, Christopher Newport University
John Bray, Broward Community College
Barbara Burrows, Santa Fe Community College
Don Campbell, Western Illinois University
Bradford Crain, Portland State University
Reid Davis, University of Tennessee at Knoxville
Jennifer M. Dollar, Grand Rapids Community College
Mark Ecker, University of Northern Iowa
Michael Ecker, Pennsylvania State University
C. Wayne Ehler, Anne Arundel Community College
Andrew Ellett, Indiana University

Mary Fine, Moberly Area Community College
Tom Fox, Cleveland State Community College
Douglas Frank, Indiana University of Pennsylvania
Rene Garcia, Miami-Dade Community College
Larry Green, Lake Tahoe Community College
David Gurney, Southeastern Louisiana University
Susan Hoy, Bristol Community College
Vijay Joshi, Virginia Intermont College
Jane Keller, Metropolitan Community College
Raja Khoury, Collin County Community College
Dennis Kimzey, Rogue Community College
Rita Kolb, Community College of Baltimore County—Cantonsville
Charles Laws, Cleveland State Community College
Lawrence Mark Lesser, University of Northern Colorado
Michael R. Lloyd, Henderson State University
Beth Long, Pellissippi State Technical and Community College
Darcy P. Mays, Virginia Commonwealth University
Charles C. Okeke, College of Southern Nevada, Las Vegas
Peg Pankowski, Community College of Allegheny County
Geetha Ramachandran, California State University—Sacramento
Michael L. Russo, Suffolk County Community College
Janel Schultz, Saint Mary's University of Minnesota
Aileen Solomon, Trident Technical College
Richard Stockbridge, University of Kentucky
Winson Taam, Oakland University
Jennifer L. Taggart, Rockford College
William Truman, University of North Carolina at Pembroke
Delmar Tunnell, Illinois Valley Community College
Jim Wienckowski, State University of New York at Buffalo
Stephen M. Wilkerson, Susquehanna University
Hongkai Zhang, East Central University
Shunpu Zhang, University of Alaska Fairbanks

We would especially like to thank Helen Medley for her careful accuracy review of this text, and Lauren Arnest for her creative illustrations that accompany the Viewpoint essays. Finally, we acknowledge the cooperation of Minitab, Inc., Texas Instruments, and Microsoft Excel.

Charles Henry Brase
Corrinne Pellillo Brase

A User's Guide to Features

The third edition of *Understanding Basic Statistics* includes a variety of features designed to enhance a student's understanding by providing overviews and summaries of concepts and methods, interesting real-world problems using real data sets, and information about using technology. The newly enhanced design highlights important features and provides visual interest.

Key features of the text are described on the following pages. New features have been added to the third edition, while other hallmark features have been retained.

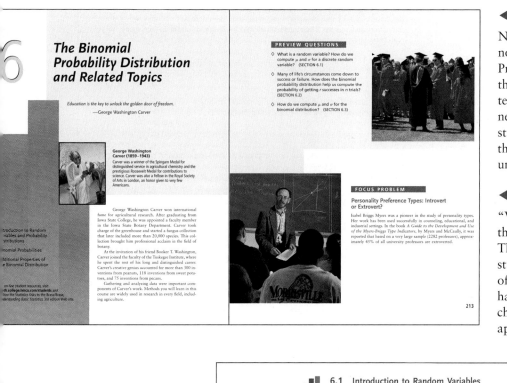

◀ Preview Questions

Newly designed chapter openers now contain a helpful list of Preview Questions that focus on the main objectives of the chapter. Section references appear next to each question to show students where they will learn the answers to develop an understanding of these topics.

◀ Focus Problems

"What kinds of problems will this chapter help me solve?" The Focus Problems motivate students by showing examples of work they can do once they have mastered the skills in the chapter. A Focus Problem appears in each chapter opener.

Focus Points ▶

Located at the beginning of each section, Focus Points briefly list the main objectives of the section for easy review and reference.

6.1 Introduction to Random Variables and Probability Distributions

FOCUS POINTS
✓ Distinguish between discrete and continuous random variables.
✓ Graph discrete probability distributions.
✓ Compute μ and σ for a discrete probability distribution.

For our purposes, we will say that a *statistical experiment* or *observation* is any process by which measurements are obtained. Examples are

1. Counting the number of eggs in a robin's nest
2. Measuring daily rainfall in inches
3. Counting the number of defective light bulbs in a case of bulbs
4. Measuring the weight in kilograms of a polar bear cub

Let x represent a quantitative variable that is measured or observed in an experiment. We are interested in the numerical values that x can take on. So x = number of eggs in a robin's nest and x = weight in kilograms of a polar bear cub would be examples of such quantitative variables. Furthermore, we say that the quantitative variable x is a *random variable* because the value that x takes on in a given experiment is a chance or random outcome. We will study two types of random variables: *discrete random variables* and *continuous random variables*.

Discrete random variable

Definition

When the observations of a quantitative random variable can take on only a finite number of values or a countable number of values, we say that the variable is a *discrete random variable*.

Guided Exercises

These unique exercises appear after selected examples in the text. Each Guided Exercise encourages students to examine and analyze a problem similar to the preceding example. Completely worked-out solutions appear beside each exercise to give immediate reinforcement in the learning process. In this way, students have a chance to work through and learn a new concept before being presented with additional concepts.

GUIDED EXERCISE 2

Using a sample space

Professor Gutierrez is making up a final exam for a course in literature of the Southwest. He wants the last three questions to be of the true–false type. To guarantee that the answers do not follow his favorite pattern, he lists all possible true–false combinations for three questions on slips of paper and then picks one at random from a hat.

(a) Finish listing the outcomes in the given sample space.

| TTT | FTT | TFT | _____ |
| TTF | FTF | TFF | _____ |

➡ The missing outcomes are FFT and FFF.

(b) What is the probability that all three items will be false? Use the formula

$$P(\text{all F}) = \frac{\text{No. of favorable outcomes}}{\text{Total no. of outcomes}}$$

➡ There is only one outcome, FFF, favorable to all false, so

$$P(\text{all F}) = \frac{1}{8}$$

(c) What is the probability that exactly two items will be true?

➡ There are three outcomes that have exactly two true items: TTF, TFT, and FTT. Thus,

$$P(\text{two T}) = \frac{\text{No. of favorable outcomes}}{\text{Total no. of outcomes}} = \frac{3}{8}$$

---

where $SS_x = \Sigma(x - \mu)^2 = \Sigma x^2 - \dfrac{(\Sigma x)^2}{N}$ ◇

... that the standard deviation (sample or population) is a measure of data spread. We will use the standard deviation extensively in later chapters. In Chapter 7 we will use it to study standard z values and areas under normal curves. In Chapters 9, 10, and 11 we will use it to study the inferential statistics topics of estimation and testing. The standard deviation will appear again in our study of regression and correlation.

TECH NOTE Most scientific or business calculators have a statistics mode, and provide the mean and sample standard deviation directly. The TI-83Plus, Excel, and Minitab provide the median and several other measures as well.

Many technologies display only the sample standard deviation s. You can quickly compute σ if you know s by using the formula

$$\sigma = s\sqrt{\frac{n-1}{n}}$$

The mean given in displays can be interpreted as the sample mean $\bar{x}$ or the population mean μ as appropriate.

The following three displays show output for the hybrid rose data of Guided Exercise 3.

TI-83Plus Display
Press **STAT** ➤ **CALC** ➤ **1:1-Var Stats**. S_x is the sample standard deviation. σ_x is the population standard deviation.

```
1-Var Stats
x̄=6
Σx=48
Σx2=296
Sx=1.069044968
σx=1
↓n=8
```

Tech Notes ▶

Tech Notes provide optional information about using the TI-83Plus graphing calculator, Excel, Minitab, and the text-specific program ComputerStat.

SECTION 3.1 PROBLEMS

1. *Agriculture: Growing Season* The average length of the growing season is often measured in average number of frost-free days. The front range of Colorado (Fort Collins, Boulder, Denver, Colorado Springs, Pueblo) was studied by J. F. Benci and T. B. McKee, from the Department of Atmospheric Science at Colorado State University. Based on data from their Climatology Report No. 77-3, different locations in the Colorado front range had the following average number of frost-free days per year:

| 156 | 161 | 152 | 162 | 144 | 153 |
| 148 | 157 | 168 | 157 | 161 | 157 |

Compute the mean, median, and mode. Write a brief description of the meaning of these numbers from the point of view of a gardener.

2. *Baseball: Home Runs* Babe Ruth was the American League Home Run Champion 12 times (during the period from 1918 to 1931). The number of home runs he hit to earn the 12 titles were

| 11 | 29 | 54 | 59 | 41 | 46 |
| 47 | 60 | 54 | 46 | 49 | 46 |

Find the mean, median, and mode of the number of home runs.

3. *Environmental Studies: Death Valley* How hot does it get in Death Valley? The following data are taken from a study conducted by the National Park System, of which Death Valley is a unit. The ground temperatures (°F) were taken from May to November in the vicinity of Furnace Creek.

| 146 | 152 | 168 | 174 | 180 | 178 | 179 |
| 180 | 178 | 178 | 168 | 165 | 152 | 144 |

Compute the mean, median, and mode for these ground temperatures.

4. *Leisure: Maui Vacation* How expensive is Maui? If you want a vacation rental condominium (for up to four people), visit the Brase/Brase *Understanding Basic Statistics* site at http://math.college.hmco.com/students, find the link to Maui, and then search for accommodations. *The Maui News* gave the following cost in dollars per day for a random sample of condominiums located throughout the island of Maui.

| 89 | 50 | 68 | 60 | 375 | 55 | 500 | 71 | 40 | 350 |
| 60 | 50 | 250 | 45 | 45 | 125 | 235 | 65 | 60 | 130 |

(a) Compute the mean, median, and mode for the data.
(b) Compute a 5% trimmed mean for the data, and compare it with the mean computed in part (a). Does the trimmed mean more accurately reflect the general level of the daily rental costs?
(c) If you were a travel agent, and a client asked about the daily cost of renting a condominium on Maui, what average would you use? Explain. Is there any other information about the costs that you think might be useful, such as the spread of the costs?

5. *Medical: Injuries* The Grand Canyon and the Colorado River are beautiful, rugged, and sometimes dangerous. Thomas Myers is a physician at the park clinic in Grand Canyon Village. Dr. Myers has recorded (for a 5-year period) the number of visitor injuries at

Real-World Exercises

Numerous application problems utilizing real data and real-world situations are included in the text. These exercises use identifiable sources, including some Web sites, and also cover a wide range of fields such as natural science, business, economics, medicine, social science, archaeology, and consumer interest.

Viewpoints ▶

These brief illustrated essays show the broad scope of statistical applications to a variety of human experiences and endeavors. In many cases, Internet Web site URLs are provided for students interested in a further explanation of topics. The Viewpoint feature is located immediately before most section problem sets and chapter problem sets.

> ## VIEWPOINT
>
> ### *The First Measured Century*
>
> The 20th century saw measurements of aspects of American life that had never been systematically studied before. Social conditions involving crime, sex, food, fun, religion, and work have been numerically investigated. The measurements and survey responses taken over the entire century reveal unsuspected statistical trends. *The First Measured Century* is a book by Caplow, Hicks, and Wattenberg. It is also a PBS documentary available on video. For more information, visit the Brase/Brase *Understanding Basic Statistics* site at http://math.college.hmco.com/students and find the link to the PBS first measured century documentary.

▼ End-of-Chapter Material

The following features are included at the end of each chapter: a brief chapter **Summary**, a list of **Important Words & Symbols** grouped by section for easy review, and **Chapter Review Problems**.

28   Chapter 1   Getting Started

other volunteers receive skin patches with no drugs. At the end of two months, each subject is surveyed regarding his or her current smoking habits.

6. *General: Randomized Two-Treatment Experiment* How would you use a randomized two-treatment experiment in each of the following settings? Is this a double-blind experiment or not? Be specific and give details.

(a) A new high-temperature bonding process is used to manufacture automobile tires. A group of 43 people volunteered to test the new tires on their cars. Out of this group, 25 cars are given the new high-temperature-bond tires. The other cars are given new tires without the high-temperature bond. Only the serial numbers on the tires tell which type is being used, and this information is held confidential. After 35,000 miles, all tires are thoroughly examined by the manufacturer.

(b) The FBI wants to test the Miami Airport security system. The FBI has inserted weapons into 10 out of 250 carry-on bags that agents take through security checks. The agent carrying the bag knows whether or not the bag contains a weapon. Those bags that get through the security check and those that do not are observed by a remote security camera.

(c) The medical school is investigating new eye drops as a treatment for glaucoma. Out of 63 volunteers, 35 will get the new eye drops. The others will get the currently used (not new) eye drops. The new eye drops come in a grey plastic bottle and the old drops come in a red plastic bottle. Neither the patient nor the doctor knows which color contains which eye drops. After six months, eye pressure on each patient is measured and a sealed report revealing medication is opened.

SUMMARY

In this chapter, we've seen that statistics is the study of how to collect, organize, analyze, and interpret numerical information from populations or samples. First, we looked at classifying data as quantitative or qualitative. Next, we examined levels of measurement. The technique of drawing a simple random sample was discussed along with other sampling methods. Finally, we looked at methods of collecting data and some of the issues involved in experimental design and surveys.

IMPORTANT WORDS & SYMBOLS

Section 1.1*
Statistics
Individual
Variable
Quantitative variable
Qualitative variable
Population data
Sample data
Levels of measurement
  Nominal
  Ordinal

Interval
Ratio
Descriptive statistics
Inferential statistics

Section 1.2
Simple random sample
Random-number table
Simulation
Stratified sampling
Systematic sampling

*Indicates section of first appearance.

Chapter Review Problems   29

Cluster sampling
Convenience sampling

Section 1.3
Census
Observational study
Experiment
Placebo
Control group
Treatment group

Randomized two-treatment experiment
Double-blind
Lurking variable
Confounding variable
Randomization
Replication
Survey
Nonresponse
Voluntary response
Hidden bias

> ## VIEWPOINT
>
> ### *Is Chocolate Good for Your Heart?*
>
> A study of 7,841 Harvard alumni showed the death rate was 30% lower in those who ate candy compared with those who abstained. It turns out that candy, especially chocolate, contains antioxidants that help slow the aging process. Also, chocolate, like aspirin, reduces the activity of blood platelets that contribute to plaque and blood clotting. Furthermore, chocolate seems to raise high-density lipoprotein (HDL), the good cholesterol. However, these results are all preliminary. The investigation is far from complete. A wealth of information on this topic was published in the August 2000 issue of the *Journal of Nutrition*. Statistical studies and reliable experimental design are indispensable in this type of research.

CHAPTER REVIEW PROBLEMS

1. *General: Samples and Variables* Find a newspaper or Web site article that uses statistics. Are the data from the entire population or just from a sample? What are the variables?

2. *Radio Talk Show: Sample Bias* A radio talk show asked listeners to respond either yes or no to the question, Is the candidate who spends the most on a campaign the most likely to win? Fifteen people called in and nine said yes. What is the implied population? What is the variable? Can you detect any bias in the selection of the sample?

3. *Essay: Levels of Measurement* In your own words, give a complete and careful description of the four levels of measurement. Which level is the highest? Which is the lowest? What are the different suitable uses for each level of measurement?

4. *Personal Data: Levels of Measurement* Write a brief description of yourself in which you list your name, age, year of birth, height, Social Security number, color of your hair and eyes, address, phone number, place of birth, letter grade on test, intended college major (if decided), distance you live from college, and so forth. In one column, list each item in the descriptions of yourself. In a second column to the right of the first, list the level of measurement corresponding to each item in the description.

5. *Colorado Lotto: Simulation* Lotto is the name of the Colorado lottery. The Lotto boards consist of 42 numbers (from 1 to 42). To play, you select six distinct

Data Highlights: Group Projects ▶

These in-depth projects give students an additional opportunity to practice their skills by asking them to solve problems using appropriate methods from the chapter. Newspapers, magazines, and journals are the sources for these projects.

Linking Concepts: Writing Projects ▶

These questions help students extend and integrate their thinking to develop a broader conceptual understanding of statistics. Students are asked to discuss and write about key concepts from the chapter and related topics from prior chapters.

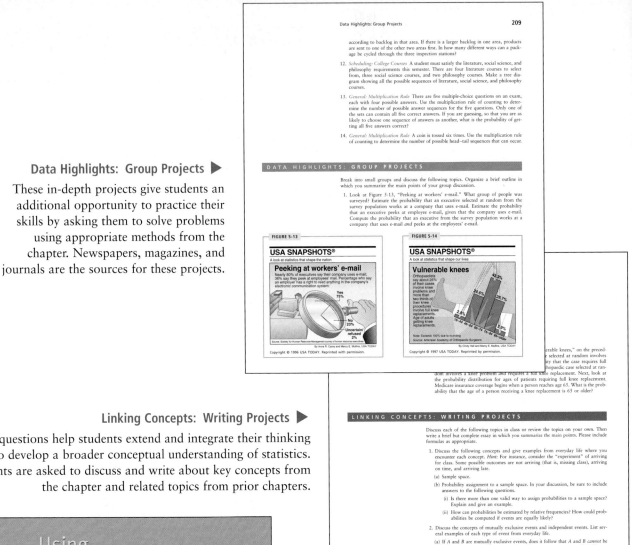

◀ Using Technology

These features have been revised for this edition to include information on using the TI-83Plus graphing calculator, Excel, Minitab, and ComputerStat to solve statistical problems.

Table of Prerequisite Material

Chapter	Prerequisite Sections
1 Getting Started	None
2 Organizing Data	1.1, 1.2
3 Averages and Variation	1.1, 1.2, 2.2
4 Regression and Correlation	1.1, 1.2, 3.1, 3.2
5 Elementary Probability Theory	1.1, 1.2, 2.2
6 The Binomial Probability Distribution and Related Topics	1.1, 1.2, 2.2, 3.1, 3.2, 5.1, 5.2 5.3 useful but not essential
7 Normal Distributions (omit 7.4) (include 7.4)	1.1, 1.2, 2.2, 3.1, 3.2, 5.1, 5.2, 6.1 also 6.2, 6.3
8 Introduction to Sampling Distributions	1.1, 1.2, 2.2, 3.1, 3.2, 5.1, 5.2, 6.1, 7.1, 7.2, 7.3
9 Estimation (omit 9.3 and parts of 9.4) (include 9.3 and all of 9.4)	1.1, 1.2, 2.2, 3.1, 3.2, 5.1, 5.2, 6.1, 7.1, 7.2, 7.3, 8.1, 8.2 also 6.2, 6.3, 7.4
10 Hypothesis Testing (omit 10.5) (include 10.5)	1.1, 1.2, 2.2, 3.1, 3.2, 5.1, 5.2, 6.1, 7.1, 7.2, 7.3, 8.1, 8.2 also 6.2, 6.3, 7.4
11 Inferences About Differences (omit 11.4) (include 11.4)	1.1, 1.2, 2.2, 3.1, 3.2, 5.1, 5.2, 6.1, 7.1, 7.2, 7.3, 8.1, 8.2, 9.1, 9.2, 10.1, 10.2, 10.3, 10.4 6.2, 6.3, 7.4, 10.5 also
12 Additional Topics Using Inference (Part I: 12.1, 12.2, 12.3) (Part II: 12.4, 12.5)	1.1, 1.2, 2.2, 3.1, 3.2, 5.1, 5.2, 6.1, 7.1, 7.2, 7.3, 8.1, 8.2, 10.1 Chapter 4, 9.1, 9.2 also

1

Getting Started

To guess is cheap,
To guess wrongly is expensive.

Tell me, I'll forget.
Show me, I may remember.
But involve me and I'll understand.

—Old Chinese Proverbs

Dragon Gate, Chinatown San Francisco

Dragon Gate is one of the many landmarks in San Francisco that exemplifies the city's ethnic diversity.

For on-line student resources, visit **math.college.hmco.com/students** and follow the Statistics links to the Brase/Brase, *Understanding Basic Statistics*, 3rd edition Web site.

Most of the important decisions in life involve incomplete information. Such decisions often involve so many complicated factors that a complete analysis is not practical or even possible. We are often forced into the position of making a guess based on limited information. However, as the first proverb implies, a blind guess is not the best solution. Statistical methods, such as you will learn in this book, can help you make the best "educated guess."

The authors of this book want you to understand and enjoy statistics. The reading material will *tell you* about the subject. The examples will *show you* how it works. To understand, however, the second proverb says you must *get involved*. Guided exercises, calculator and computer applications, section and chapter problems, and writing exercises are all designed to get you involved in the subject. As you grow in your understanding of statistics, we believe you will enjoy learning a subject that has a world full of interesting and relevant applications.

PREVIEW QUESTIONS

◇ Why is statistics important?   (SECTION 1.1)

◇ What is the nature of data?   (SECTION 1.1)

◇ How can we draw a random sample?
(SECTION 1.2)

◇ What are other sampling techniques?
(SECTION 1.2)

◇ How can we design ways to collect data?
(SECTION 1.3)

Adapted from The Ohio State University Firefly Files logo

FOCUS PROBLEM

Where Have All the Fireflies Gone?

A feature article in the *Wall Street Journal* discusses the alarming disappearance of fireflies. In the article, Professor Sara Lewis (of Tufts University) and other scholars express concern about the decline in the worldwide population of fireflies. So far the evidence is anecdotal: stories from around the world about the alarming decline of the fireflies. Japan has a tradition that nighttime fireflies are a beautiful reminder of the souls of friends and family who have passed away. For reasons such as this, and possibly for the simple joy of watching fireflies, Japan has a number of special sanctuaries to protect the declining population of fireflies.

There are a number of possible explanations for the world decline in the firefly population. Artificial nighttime lighting might be interfering with the Morse code-like mating ritual of the fireflies. Woodlands, wetlands, and open fields have given way to cities and subdivisions. Chemicals and pesticides may be killing the fireflies.

Fireflies are commercially valuable because they contain two rare chemicals used in research on cancer, multiple sclerosis, cystic fibrosis, and heart disease. The chemicals are also used on spacecraft in electronic detectors designed to look for earth-life forms in outer space. Near St. Louis there is a bounty on fireflies: A chemical

3

company pays 1 cent for each firefly. One family was so successful in collecting fireflies that they could send three children to college from the proceeds!

What does any of this have to do with statistics?

The truth, at this time, is that no one really knows (a) how much the world firefly population has declined or (b) how to explain the decline. The population of all fireflies is simply too large to study in its entirety. In any study of fireflies, we must rely on incomplete information from samples. Furthermore, from these samples we must draw realistic conclusions that have statistical integrity.

You may be sure that a number of very fine scholars in universities across the world are beginning to examine the possible world decline in fireflies. You also may be sure that these scholars will use statistical methods when they collect, analyze, and investigate their data about fireflies.

The Ohio State University Department of Entomology does research on anthropods and maintains extensive information about fireflies. For more information about fireflies, visit the Brase/Brase *Understanding Basic Statistics* site at http://math. college.hmco.com/students and find the link to The Ohio State University Department of Entomology. Then search for firefly in the index of The Ohio State site.

1.1
What Is Statistics?

FOCUS POINTS
✓ Identify variables in a statistical study.
✓ Distinguish between quantitative and qualitative variables.
✓ Identify populations and samples.
✓ Determine the level of measurement.
✓ Compare descriptive and inferential statistics.

Introduction

Decision making is an important aspect of our lives. We make decisions based on the information we have, our attitudes, and our values. Statistical methods help examine information. Moreover, statistics can be used for making decisions when we are faced with uncertainties. For instance, if we wish to estimate the proportion of people who will have a severe reaction to a flu shot without giving the shot to everyone who wants it, statistics provides appropriate methods. Statistical methods enable us to look at information from a small collection of people or items and make inferences about a larger collection of people or items.

Procedures for analyzing data, together with rules of inference, are central topics in the study of statistics.

Statistics

> **Statistics** is the study of how to collect, organize, analyze, and interpret numerical information from data.

The statistical procedures you will learn in this book should supplement your built-in system of inference—that is, the results from statistical procedures and good sense should dovetail. Of course, statistical methods themselves have no power to work miracles. These methods can help us make some decisions, but not all conceivable decisions. Remember, a properly applied statistical procedure is no more accurate than the data, or facts, on which it is based. Finally, statistical results should be interpreted by one who understands not only the methods, but also the subject matter to which they have been applied.

The general prerequisite for statistical decision making is the gathering of data. First, we need to identify the individuals or objects to be included in the study and the characteristics or features of the individuals that are of interest.

Individual
Variable

> **Individuals** are the people or objects included in the study.
> A **variable** is the characteristic of the individual to be measured or observed.

For instance, if we want to do a study about the people who have climbed Mt. Everest, then the individuals in the study are all people who have actually made it to the summit. One variable might be the height of such individuals. Other variables might be age, weight, gender, nationality, income, and so on. Regardless of the variables we use, we would not include measurements or observations from people who have not climbed the mountain.

The variables in a study may be *quantitative* or *qualitative* in nature.

Quantitative variable
Qualitative variable

> A **quantitative variable** has a value or numerical measurement for which operations such as addition or averaging make sense. A **qualitative variable** describes an individual by placing the individual into a category or group such as male or female.

For the Mt. Everest climbers, variables such as height, weight, age, or income are *quantitative* variables. *Qualitative variables* involve nonnumerical observations such as gender or nationality. Sometimes qualitative variables are referred to as *categorical*.

Another important issue regarding data is their source. Do the data comprise information from *all* individuals of interest, or from just *some* of the individuals?

Population data
Sample data

> In **population data,** the variable is from *every* individual of interest.
> In **sample data,** the variable is from *only some* of the individuals of interest.

For instance, if we have data from *all* the individuals who have climbed Mt. Everest, then we have population data. On the other hand, if our data come from just some of the climbers, we have sample data.

It is interesting to note that the population is defined in terms of our *desire for knowledge*. The population can be thought of as measurements or observations for the entire group of objects or individuals about which information is desired. In this sense, a population can be an existing set of data, or it can be a set of data that is clear in our understanding but is not yet complete. The ages of all U.S. presidents at the time of inauguration—from George Washington to George W. Bush—can be thought of as an existing set of data. However, the set of ages of all presidents at the time of inauguration from George Washington into the future is not complete. Nevertheless, we have a clear understanding of how this population is to be constructed. In a way, we can think of such an incomplete population as being open-ended.

EXAMPLE 1

Using basic terminology

The Hawaii Department of Tropical Agriculture is doing a study of ready-to-harvest pineapples in an experimental field.

(a) The pineapples are the *objects* (individuals) of the study. If the researchers are interested in the individual weights of pineapples in the field, then the *variable* consists of weights. At this point, it is important to specify units of measurement and degree of accuracy of measurement. The weights could be to the nearest ounce or gram. Weight is a *quantitative* variable since it is a numerical measure. If weights of *all* the ready-to-harvest pineapples in the field are included in the data, then we have a *population*.

(b) Suppose the researchers also want data on taste. A panel of tasters rates the pineapples according to the categories poor, acceptable, good. Only some of the pineapples are included in the taste test. In this case, the *variable* is taste. This is a *qualitative* or *categorical* variable. Because only some of the pineapples in the field are included in the study, we have a *sample*.   ◇

Throughout this text, you will encounter *guided exercises* embedded in the reading material. These exercises are included to give you an opportunity to work immediately with new ideas. The questions guide you through appropriate analysis. Cover the answers on the right side (an index card will fit this purpose). After you have thought about or written down *your own response,* check the answers. If there are several parts to an exercise, check each part before you continue. You should be able to answer most of these exercise questions, but don't skip them—they are important.

GUIDED EXERCISE 1

Using basic terminology

Television station QUE wants to know the proportion of TV owners in Virginia who watch the station's new program at least once a week. The station asked a group of 1000 TV owners in Virginia if they watch the program at least once a week.

(a) Identify the individuals of the study and the variable.

➡ The individuals are the 1000 TV owners surveyed. The variable is the response does, or does not, watch the new program at least once a week.

(b) Do the data comprise a sample? If so, what is the underlying population?

➡ The data comprise a sample of the population of responses from all TV owners in Virginia.

(c) Is the variable qualitative or quantitative?

➡ Qualitative—the categories are the two possible responses, does or does not watch the program.

(d) Identify a quantitative variable that might be of interest.

➡ Age or income might be of interest.

Levels of Measurement

We have categorized data as either qualitative or quantitative. Another way to classify data is according to one of the following four *levels of measurement:*

Nominal	Ordinal	Interval	Ratio

Nominal level

Since the nominal level is the lowest level, let's examine it first. A dictionary meaning of the word *nominal* is "in name only." This is an easy way to remember the meaning of the nominal level of measurement. Data at this level of measurement consist of "names only," or qualities, with no implied criteria by which the data can be identified as greater than or less than other data items.

EXAMPLE 2
Nominal level

The following are examples of data at the nominal level of measurement.

(a) Aspen, Vail, and Breckenridge are names of three ski resorts from the population of names of all ski resorts in Colorado.

(b) Taos, Acoma, Zuni, and Cochiti are names of four Native American pueblos from the population of all names of Native American pueblos in Arizona and New Mexico.   ◊

It is clear that the nominal data in Example 2 are not intended for numerical calculation. The specific names or qualities do not contain any implied ordering or numerical significance.

Ordinal level

The next level of measurement is the *ordinal level*. Data at the ordinal level may be arranged in some order, but actual differences between data values either cannot be determined or are meaningless.

EXAMPLE 3
Ordinal level

The following are examples of data at the ordinal level of measurement.

(a) In a fishing tackle catalogue, 17 fishing reels are advertised. Of these reels, 6 were rated as good quality, 4 were rated as better quality, and 7 were rated as best quality.

(b) In a high school graduating class of 319 students, Jim ranked 25th, June ranked 19th, Walter ranked 10th, and Julia ranked 4th.   ◊

In Example 3, we should not try to determine specific quantitative differences between "good," "better," and "best." The difference between June's and Jim's rank was 6, and this is the same difference that exists between Walter's and Julia's ranks. However, this difference doesn't really mean anything significant. For instance, if you look at grade point average, Walter and Julia may have had a big gap between them, whereas June and Jim may have been closer together. In any ranking system, it is only the relative standing that matters. Differences in ranks can be meaningless.

In general, the ordinal level of measurement provides information about relative comparisons, but exact differences are not computed.

Interval level

The *interval level of measurement* is like the ordinal level, but it has the additional property that meaningful differences between data values can be computed. However, interval-level data may not have an intrinsic zero or starting point. Consequently, *differences* are meaningful, but *ratios* of data values are not.

EXAMPLE 4

Interval level

The following are examples of data at the interval level of measurement.

(a) Years in which Democrats won presidential elections.

(b) Body temperatures (in degrees Celsius) of trout in the Yellowstone River.   ◇

Temperature readings in Celsius (or Fahrenheit) are examples of data at the interval level of measurement. Such values are certainly ordered, and we can compute meaningful differences. However, for Celsius-scale temperatures, there is not an inherent starting point. The value 0°C may seem to be a starting point, but the value of 0°C does not indicate the state of "no heat." Furthermore, it is not correct to say 20°C is twice as hot as 10°C. Calendar times are also interval measurements, since the date 0 A.D. does not signify "no time." However, a time lapse is at a higher level of measurement and is, in fact, an example of our top level of measurement, the *ratio level*.

Ratio level

The *ratio level of measurement* is the highest level. The ratio level is similar to the interval level, but it includes an inherent zero as a starting point for all measurements. Consequently, at this level, both differences *and* ratios are meaningful.

EXAMPLE 5

Ratio level

The following are examples of data at the ratio level of measurement.

(a) The core temperatures of stars in the Milky Way when measured in degrees Kelvin. Notice that in the Kelvin scale of measurement 0°K means "no heat." This is a special temperature scale used primarily by scientists.

(b) Length of trout swimming in the Yellowstone River. A trout 18 inches long is three times as long as a 6-inch trout. Observe that we can divide 6 into 18 to determine the *ratio* of the trout lengths.   ◇

In summary, there are four levels of measurement. The nominal is considered the lowest, and in ascending order we have the ordinal, interval, and ratio levels. In general, calculations based on one level of measurement may not be appropriate for a lower level.

Level of Measurement	Suitable Calculation
Nominal	We can put the data into categories.
Ordinal	We can order the data from smallest to largest or "worst" to "best." Each data value can be *compared* with another data value.
Interval	We can order the data and also take the differences between data values. At this level, it makes sense to compare the differences between data values. For instance, we can say that one data value is 5 more than another or 12 less than another data value.
Ratio	We can order the data, take differences, and also find the ratio between data values. For instance, it makes sense to say that one data value is twice as large as another.

GUIDED EXERCISE 2

Levels of measurement

The following describe different data associated with a state senator. For each data entry, indicate the corresponding *level of measurement*.

(a) The senator's name is Sam Wilson.   ⟹ Nominal level

(b) The senator is 58 years old.   ⟹ Ratio level

(c) The years in which the senator was elected to   ⟹ Interval level
the senate are 1980, 1986, 1992, and 1998.

(d) His total taxable income last year was   ⟹ Ratio level
$878,314.19.

(e) The senator sponsored a bill to protect water   ⟹ The opinions about the bill are at the ordinal level.
rights. Out of 1100 voters in his district, 400
said they strongly favored the bill, 300 said they
favored the bill, 200 said they were neutral, 150
said they did not favor the bill, and 50 said they
strongly did not favor the bill.

(f) The senator is married now.   ⟹ Nominal level

(g) A leading news magazine claims the senator is   ⟹ Ordinal level
ranked seventh for his voting record on bills
regarding public education.

Looking Ahead

The purpose of collecting and analyzing data is to obtain information. Statistical methods provide us tools to obtain information from data. These methods break into two branches.

Descriptive statistics

Inferential statistics

> **Descriptive statistics** involves methods of organizing, picturing, and summarizing information from samples or populations.
> **Inferential statistics** involves methods of using information from a sample to draw conclusions regarding the population.

We will look at methods of descriptive statistics in the next four chapters. These methods may be applied to data from samples or populations.

Sometimes we do not have access to the entire population. At other times, the difficulties or expense of working with the entire population are prohibitive. In such cases, we will use inferential statistics together with probability. These are the topics of Chapters 5 through 12.

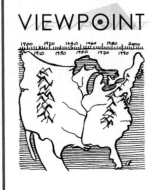

VIEWPOINT   *The First Measured Century*

The 20th century saw measurements of aspects of American life that had never been systematically studied before. Social conditions involving crime, sex, food, fun, religion, and work have been numerically investigated. The measurements and survey responses taken over the entire century reveal unsuspected statistical trends. *The First Measured Century* is a book by Caplow, Hicks, and Wattenberg. It is also a PBS documentary available on video. For more information, visit the Brase/Brase *Understanding Basic Statistics* site at http://math.college.hmco.com/students and find the link to the PBS first measured century documentary.

SECTION 1.1   PROBLEMS

1. *Marketing: Fast Food* USA Today reported that 44.9% of those surveyed (1261 adults) ate in fast-food restaurants from one to three times each week.
   (a) Identify the variable.
   (b) Is the variable quantitative or qualitative?
   (c) What is the implied population?

2. *Advertising: Auto Mileage* What is the average miles per gallon (mpg) for all new 2003 cars? Using *Consumer Reports,* a random sample of 35 new 2003 cars gave an average of 21.1 mpg.
   (a) Identify the variable.
   (b) Is the variable quantitative or qualitative?
   (c) What is the implied population?

3. *Academic: Student Fees* The students at Eastmore College are concerned about the level of student fees. They took a random sample of 30 colleges and universities throughout the nation and obtained information about the student fees at these institutions. From this information they concluded that their student fees are higher than those of most colleges in the nation.
   (a) Identify the variable.
   (b) Is the variable quantitative or qualitative?
   (c) What is the implied population?

4. *Quality Control: Shelf Life* The quality-control department at Healthy Crunch, Inc., wants to estimate the shelf life of all Healthy Crunch granola bars. A random sample of 10 of these bars was tested, and the shelf life was determined. From the sample results, a shelf life for all Healthy Crunch granola bars was estimated.
   (a) Identify the variable.
   (b) Is the variable quantitative or qualitative?
   (c) What is the implied population?

5. *Insurance: Claim Payment* An insurance company wants to determine the time interval between the arrival of an insurance payment check and the time that the check clears. A central payment office processes the payments for a five-state region. A random sample of 32 payment checks from this five-state region was received and processed. The time interval between receipt and check clearance was determined for each check. From this information the company estimated the time interval necessary for all checks sent to this office to clear.

(a) Identify the variable.

(b) Is the variable quantitative or qualitative?

(c) What is the implied population?

6. *Education: Teacher Evaluation* If you were going to apply *statistical methods* to analyze teacher evaluations, which question form, A or B, would be better?
Form A: In your own words, tell how this teacher compares with other teachers you have had.
Form B: Use the following scale to rank your teacher as compared with other teachers you have had.

1	2	3	4	5
worst	below average	average	above average	best

7. *Student Life: Levels of Measurement* Categorize these measurements associated with student life according to level: nominal, ordinal, interval, or ratio.

(a) Length of time to complete an exam

(b) Time of first class

(c) Major field of study

(d) Course evaluation scale: poor, acceptable, good

(e) Score on last exam (based on 100 possible points)

(f) Age of student

8. *Business: Levels of Measurement* Categorize these measurements associated with a robotics company according to level: nominal, ordinal, interval, or ratio.

(a) Salesperson's performance: below average, average, above average

(b) Price of company's stock

(c) Names of new products

(d) Room temperature (°F) in CEO's private office

(e) Gross income for each of past 5 years

(f) Color of packaging

9. *Fishing: Levels of Measurement* Categorize these measurements associated with fishing according to level: nominal, ordinal, interval, or ratio.

(a) Species of fish caught: perch, bass, pike, trout

(b) Cost of rod and reel

(c) Time of return home

(d) Guidebook rating of fishing area: poor, fair, good

(e) Number of fish caught

(f) Temperature of water

1.2
Random Samples

FOCUS POINTS

✓ Explain the importance of random samples.

✓ Construct a simple random sample using random numbers.

✓ Simulate a random process.

✓ Describe stratified sampling, cluster sampling, systematic sampling, and convenience sampling.

Simple Random Samples

Eat lamb—20,000 coyotes can't be wrong!

This slogan is sometimes found on bumper stickers in the western United States. The slogan indicates the trouble that ranchers have experienced in protecting their flocks from predators. Based on their experience with this sample of the coyote population, the ranchers concluded that *all* coyotes are dangerous to their flocks and should be eliminated! The ranchers used a special poison bait to get rid of the coyotes. Not only was this poison distributed on ranch land, but with government cooperation it also was distributed widely on public lands.

The ranchers found that the results of the widespread poisoning were not very satisfactory. The sheep-eating coyotes continued to thrive while the general population of coyotes and other predators declined. What was the problem? The sheep-eating coyotes the ranchers observed were not a representative sample of all coyotes. Modern methods of predator control target the sheep-eating coyotes. To a certain extent the new methods have come about through a closer examination of the sampling techniques used.

In this section, we will examine several widely used sampling techniques. One of the most important sampling techniques is a *simple random sample*.

Simple random sample

> A **simple random sample** of *n* measurements from a population is a subset of the population selected in a manner such that
>
> (a) every sample of size *n* from the population has an equal chance of being selected and
>
> (b) every member of the population has an equal chance of being included in the sample.

Notice that two criteria are necessary for a simple random sample of a specified size. Not only must every member of the population have an equal chance of being included, but every sample of the specified size must have an equal chance of being included as well.

GUIDED EXERCISE 3

Simple random sample

Is open space around metropolitan areas important? Players of the Colorado Lottery might think so because some of the proceeds of the game go to fund open space and outdoor recreational space. To play the game, you pay one dollar and choose any 6 different numbers from the group of numbers 1 through 42. If your group of 6 numbers matches the winning group of 6 numbers selected by simple random sampling, then you are a winner of a grand prize of at least 1.5 million dollars.

(a) Is the number 25 as likely to be selected in the winning group of six numbers as the number 5?

⟹ Yes, since the winning numbers constitute a simple random sample, each number from 1 through 42 has an equal chance of being selected.

(b) Could all the winning numbers be even?

⟹ Yes, since 6 even numbers is one of the possible groups of six numbers.

(c) Your friend always plays these numbers

1 2 3 4 5 6

Could she ever win?

⟹ Yes, in a simple random sample the listed group of six numbers is *as likely as any* of the 5,245,786 groups of six numbers to be selected as the winner. (See Section 5.3 to learn how to compute the number of groups of 6 numbers selected from 42 numbers.)

How do we get random samples? Suppose you need to know if the emission system of the latest shipment of Toyotas satisfies pollution-control standards. You want to pick a random sample of 30 cars from this shipment of 500 cars and test them. One way to pick a random sample is to number the cars 1 through 500. Write these numbers on cards, mix up the cards, and then draw 30 numbers. The sample will consist of the cars with the chosen numbers. If you mix the cards sufficiently, this procedure produces a random sample.

Random-number table

An easier way to select the numbers is to use a *random-number table*. You can make one yourself by writing the digits 0 through 9 on separate cards and mixing up these cards in a hat. Then draw a card, record the digit, return the card, and mix up the cards again. Draw another card, record the digit, and so on. However, Table 1 in the Appendix is a ready-made random-number table (adapted from Rand Corporation, *A Million Random Digits with 100,000 Normal Deviates*). Let's see how to pick our random sample of 30 Toyotas by using this random-number table.

EXAMPLE 6
Random-number table

Use a random-number table to pick a random sample of 30 cars from a population of 500 cars.

SOLUTION: Again, we assign each car a different number between 1 and 500, inclusive. Then we use the random-number table to choose the sample. Table 1 in the Appendix has 50 rows and 10 blocks of five digits each; it can be thought of as a solid mass of digits that has been broken up into rows and blocks for user convenience.

You read the digits by beginning anywhere in the table. We dropped a pin on the table, and the head of the pin landed in row 15, block 5. We'll begin there and list all the digits in that row. If we need more digits, we'll move on to row 16, and so on. The digits we begin with are

99281   59640   15221   96079   09961   05371

Since the highest number assigned to a car is 500, and this number has three digits, we regroup our digits into blocks of 3:

992   815   964   015   221   960   790   996   105   371

To construct our random sample, we use the first 30 car numbers we encounter in the random-number table when we start at row 15, block 5. We skip the first three groups—992, 815, and 964—because these numbers are all too large. The next group of three digits is 015, which corresponds to 15. Car number 15 is the first car included in our sample, and the next is car number 221. We skip the next three groups and then include car numbers 105 and 371. To get the rest of the cars in the sample, we continue to the next line and use the random-number table in this fashion. If we encounter a number we've used before, we'll skip it. ◇

◇ **COMMENT** When we use the term *(simple) random sample*, we have very specific criteria in mind for selecting the sample. One proper method for selecting a simple random sample is to use a computer-based or calculator-based random-number generator or to use a table of random numbers as we have done in the example. The term *random* should not be confused with *haphazard!* ◇

Simulation

Another important use of random-number tables is in *simulation*. We use the word *simulation* to refer to the process of providing mathematical imitations of "real" phenomena. Simulation methods have been productive in studying a diverse array of subjects such as nuclear reactors, cloud formation, cardiology (and medical science in general), highway design, production control, shipbuilding, airplane design, war games, economics, and electronics. A complete list would probably include something from every aspect of modern life. In Example 7 and Guided Exercise 4 we'll perform a brief simulation.

EXAMPLE 7

Simulation

A well-known theory in stock market analysis is the "random walk" theory (see *A Random Walk Down Wall Street,* 6th Edition, Burton Malkiel, W. W. Norton & Co.). The term *random walk,* as applied to stock prices, means that short-term changes in stock prices cannot be predicted but rather are random. In particular, according to the random walk theory, the next move in the price of a stock (up or down) is completely unpredictable on the basis of what price changes happened before.

Let's use a very simplified model to *simulate* the stock price changes of a hypothetical company, Fun Boards (maker of skate boards, surf boards, snow boards, and in-line skates). Suppose the initial price of Fun Boards stock is $50 per share. Use the random-number table to simulate daily price changes for the next 15 trading days in the following way. Notice we are interested in price *changes*. Days during which the stock does not change price will be ignored.

Trading floor, New York Stock Exchange

SOLUTION: The daily stock price change will be dictated by a number from the random-number table. When you encounter an even digit (0, 2, 4, 6, 8), increase the stock price by $1. When you encounter an odd digit (1, 3, 5, 7, 9), decrease the stock price by $1. Do this for a sequence of 15 trading days during which the price changed.

Beginning with line 6, block 2 in Table 1 of the Appendix, we see the 15 random digits

51709   94456   48396

Since the first random digit is odd, we will decrease the price by $1 on the first day. In fact, the next two digits are also odd, so we will decrease the price again by $1 on day 2 and then again on day 3. The next digit is even, so we will increase the price by $1 on day 4. Table 1-1 shows the simulated price changes for the 15-trading-day period.   ◇

◇ **COMMENT** Recall that the random samples we have been constructing so far are called *simple random samples.* Throughout this text we will use the term *random sample* to mean simple random sample. All the statistical methods in this text assume that a simple random sampling has been used to collect the data.   ◇

TABLE 1-1   Simulated Price Moves of Fun Boards Stock

Price	50	49	48	47	48	47	46	47	48	47	48	49	50	49	48	49
Day	Initial	1	2	3	4	5	6	7	8	9	10	11	12	13	14	15
Digit		5	1	7	0	9	9	4	4	5	6	4	8	3	9	6

GUIDED EXERCISE 4

Simulation

Use a random-number table to simulate the outcomes of tossing a balanced (that is, fair) penny 10 times.

(a) How many outcomes are possible when you toss a coin once?

⟹ Two; heads or tails

(b) There are several ways to assign numbers to the two outcomes. Because we assume a fair coin, assign an even digit to the outcome heads and an odd digit to the outcome tails. Then, starting at block 3 of row 2 of Table 1 in the Appendix, list the first 10 single digits.

⟹ 7 1 5 4 9 4 4 8 4 3

(c) What are the outcomes associated with the 10 digits?

⟹ T T T H T H H H H T

(d) If you start in a different block and row of Table 1 in the Appendix, will you get the same sequence of outcomes?

⟹ It is possible, but not very likely. (In Section 5.3 you will learn how to determine that there are 1024 possible sequences of outcomes for 10 tosses of a coin.)

Sampling with replacement

TECH NOTE Most statistical software packages, spreadsheet programs, and statistical calculators generate random numbers. In general, these devices sample *with replacement*. Sampling with replacement means that although a number is selected for the sample, it is *not removed* from the population. Therefore, the same number may be selected for the sample more than once. If you need to sample without replacement, generate more items than you need for the sample. Then sort the sample and remove duplicate values. Specific procedures for generating random samples on the TI-83Plus calculator, Excel, Minitab, and ComputerStat are shown in Using Technology at the end of this chapter. More details are given in the separate Technology Guides for each of these technologies.

Other Sampling Techniques

Although we will always assume that (simple) random samples are used throughout this text, other methods of sampling are also widely used. Appropriate statistical techniques exist for these sampling methods, but they are beyond the scope of this text.

Stratified sampling

One of these sampling methods is called *stratified sampling*. Groups or classes inside a population that share a common characteristic are called *strata* (plural of *stratum*). For example, in the population of all undergraduate college students, some strata might be freshmen, sophomores, juniors, or seniors. Other strata might be

men or women or in-state students or out-of-state students, and so on. In the method of stratified sampling, the population is divided into at least two distinct strata. Then a (simple) random sample of a certain size is drawn from each stratum, and the information obtained is carefully adjusted or weighted in all resulting calculations.

The groups or strata are often sampled in proportion to their actual percentages of occurrence in the overall population. However, other (more sophisticated) ways to determine the optimal sample size in each stratum may give the best results. In general, statistical analysis and tests based on data obtained from stratified samples are somewhat different from techniques discussed in an introductory course in statistics. Such methods for stratified sampling will not be discussed in this text.

Systematic sampling

Another popular method of sampling is called *systematic sampling.* In this method, it is assumed that the elements of the population are arranged in some natural sequential order. Then we select a (random) starting point and select every kth element for our sample. For example, people lining up to buy rock concert tickets are "in order." To generate a systematic sample of these people (and ask questions regarding topics such as age, smoking habits, income level, etc.), we could include every 5th person in line. The "starting" person could be selected at random from the first five.

The advantage of a systematic sample is that it is easy to get. However, there are dangers in using systematic sampling. When the population is repetitive or cyclic in nature, systematic sampling should not be used. For example, consider a fabric mill that produces dress material. Suppose the loom that produces the material makes a mistake every 17th yard, but we check only every 16th yard with an automated electronic scanner. In this case, a random starting point may or may not result in detection of fabric flaws before a large amount of fabric is produced.

Cluster sampling

Cluster sampling is a method used extensively by government agencies and certain private research organizations. In cluster sampling we begin by dividing the demographic area into sections. Then we randomly select sections or clusters. Every member of the cluster is included in the sample. For example, in conducting a survey of school children in a large city, we could first randomly select 5 schools and then include all the children from each selected school.

Convenience sampling

Convenience sampling simply uses results or data that are conveniently and readily obtained. In some cases, this may be all that is available, and in many cases, it is better than no information at all. However, convenience sampling does run the risk of being severely biased. For instance, consider a newsperson who wishes to get the "opinions of the people" about a proposed seat tax to be imposed on tickets to all sporting events. The revenues from the seat tax will then be used to support the local symphony. The newsperson stands in front of a classical music store at noon and surveys the first five people coming out of the store who will cooperate. This method of choosing a sample will produce some opinions, and perhaps some human interest stories, but it certainly has bias. It is hoped that the city council will not use these opinions as the sole basis for a decision about the proposed tax. It is good advice to be very cautious indeed when the data come from the method of convenience sampling.

Our discussion of sampling methods is intended to be brief and general. This is not the place for an extensive treatment of theory and practice of sampling. However, the interested reader is referred to the book *A Sampler on Sampling,* by Bill Williams, John Wiley and Sons, Inc.

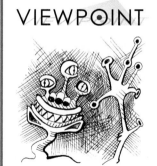

VIEWP⦿INT   *Extraterrestrial Life?*

Do you believe intelligent life exists on other planets? Using methods of random sampling, a Fox News opinion poll found that about 54% of all U.S. men do believe in intelligent life on other planets, whereas only 47% of women believe there is such life. How could you conduct a random survey of students on your campus regarding belief in extraterrestrial life?

SECTION 1.2   PROBLEMS

1. *Sampling: Random*
   (a) In your own words, explain the meaning of the terms *random numbers* and *random samples*.
   (b) Why are random samples so important in statistics?

2. *Sampling: Random* Use a random-number table to get a list of eight random numbers from 1 to 976. Explain your work.

3. *Sampling: Random* Use a random-number table to get a list of six random numbers from 1 to 8615. Explain your work.

4. *Sampling: Random* Use a random-number table to get a list of five random numbers from 43 to 719. Explain your work.

5. *Psychology: Random Selection* How do colds affect analytical thinking performance? Results of a study conducted by McGraw and Schleser were reported in *Psychology Today*. The study showed that under certain conditions, persons with colds do better than their healthy colleagues. The study considered 62 subjects: 40 healthy men and women and 22 suffering from colds or flu. A key component in this study was the formation of two groups of equal size from the 62 participants, with each group containing both healthy and sick participants.
   (a) Describe how you could take the 40 healthy subjects and randomly divide them into two groups of equal size using the random-number table.
   (b) Repeat part (a) for the 22 sick subjects.
   (c) How would you combine the groups of healthy and sick subjects found in parts (a) and (b) to form two groups of equal size so that each group contained both healthy and sick subjects?

6. *Simulation: Coin Toss* Use a random-number table to simulate the outcomes of tossing a quarter 25 times. Assume that the quarter is balanced (i.e., fair).

7. *Computer Simulation: Roll of a Die* A die is a cube with dots on each face. The faces have 1, 2, 3, 4, 5, or 6 dots. The table below is a computer simulation (from the software package Minitab) of the results of rolling a fair die 20 times.

DATA DISPLAY

ROW	C1	C2	C3	C4	C5	C6	C7	C8	C9	C10
1	5	2	2	2	5	3	2	3	1	4
2	3	2	4	5	4	5	3	5	3	4

(a) Assume that each number corresponds to the number of dots on the face of the die. Is it appropriate that the same number appear more than once? Why? What is the outcome of the 4th roll?

(b) If we simulate more "rolls of the die," do you expect to get the same sequence of outcomes? Why or why not?

8. *Simulation: Birthday Problem* Suppose there are 30 people at a party. Do you think any two share the same birthday? Let's use the random-number table to simulate the birthdays of the 30 people at the party. Ignoring leap year, let's assume that the year has 365 days. Number the days, with 1 representing January 1, 2 representing January 2, and so forth, with 365 representing December 31. Draw a random sample of 30 days (with replacement). These days represent the birthdays of the people at the party. Were any two of the birthdays the same? Compare your results with those obtained by other students in the class. Would you expect the results to be the same or different?

9. *Sampling: Students in Class* Suppose you are given the number 1, and each of the other students in your statistics class calls out consecutive numbers until each person in class has his or her own number. Explain how you could get a random sample of four students from your statistics class.

(a) Explain why the first four students walking into the classroom would not necessarily form a random sample.

(b) Explain why four students coming in late would not necessarily form a random sample.

(c) Explain why four students sitting in the back row would not necessarily form a random sample.

(d) Explain why the four tallest students may not necessarily form a random sample.

10. *Sampling: Quality Control* Products are inspected during production, and equipment is adjusted to correct defects. It is usually not possible to examine every product, so a random sample is examined. For each of the following, give a detailed explanation of how you could get the requested random sample. Be sure to include the random numbers you use to get the random sample.

(a) How would you draw a random sample of 6 of the next 500 stereo headsets coming off an assembly line?

(b) How would you obtain a random sample of 10 men's dress shirts coming off an assembly line from 8 A.M. to 12 noon?

(c) Serial numbers are placed on radios as they come off an assembly line. How could you get a random sample of nine radios with serial numbers from 21942 to 98756?

(d) A truck has just delivered 800 cartons of eggs to a supermarket. How would you get a random sample of 12 cartons to check for broken eggs?

11. *Sampling: General* For each of the following, give a detailed explanation of how you would get the requested random sample. Be sure to include the random numbers you use to get the random sample.

(a) You are a veterinarian. How would you get a random sample of 15 sheep to check for ticks on a farm that has 250 sheep?

(b) You are a medical records technician. Patients from 1992 to the present were given file numbers starting at 1024 and ending with 8342. How would you get a random sample of 10 of these patients?

(c) You are a security agent for an airline. How could you get a random sample of five pieces of luggage that are moving on a conveyor belt in the next 25 minutes?

(d) You are conducting a survey for your sociology class project. How can you get a random sample of 12 adults walking past the information booth at a shopping center between 6 and 7 P.M.?

12. *Education: Test Construction* Professor Gill uses true–false questions. She wishes to place 20 such questions on the next test. To decide whether to place a true statement or false statement in each of the 20 questions, she uses a random-number table. She selects 20 digits from the table. An even digit tells her to use a true statement. An odd digit tells her to use a false statement. Use a random-number table to pick a sequence of 20 digits, and describe the corresponding sequence of 20 true–false questions. What would the test key for your sequence look like?

13. *Education: Test Construction* Professor Gill is designing a multiple-choice test. There are to be 10 questions. Each question is to have five choices for answers. The choices are to be designated by the letters *a, b, c, d,* and *e.* Professor Gill wishes to use a random-number table to determine which letter choice should contain the correct answer for a question. Using the number correspondence 1 for *a,* 2 for *b,* 3 for *c,* 4 for *d,* and 5 for *e,* use a random-number table to determine the letter choice for the correct answer in each of the 10 questions.

14. *Sampling Methods: Health Care* Modern Managed Hospitals (MMH) is a national for-profit chain of hospitals. Management wants to survey patients discharged this past year to obtain patient satisfaction profiles. They wish to use a sample of such patients. Several sampling techniques are described below. Categorize each technique as *simple random sample, stratified sample, systematic sample, cluster sample,* or *convenience sample.*
    (a) Obtain a list of patients discharged from all MMH facilities. Divide the patients according to length of hospital stay (2 days or less, 3–7 days, 8–14 days, more than 14 days). Draw simple random samples from each group.
    (b) Obtain lists of patients discharged from all MMH facilities. Number these patients, and then use a random-number table to obtain the sample.
    (c) Randomly select some MMH facilities from each of five geographic regions, and then include all the patients on the discharge lists of the selected hospitals.
    (d) At the beginning of the year, instruct each MMH facility to survey every 500th patient discharged.
    (e) Instruct each MMH facility to survey 10 discharged patients this week and send in the results.

15. *Sampling Methods: Benefits Package* An important part of employee compensation is a benefits package that might include health insurance, life insurance, child care, vacation days, retirement plan, parental leave, bonuses, etc. Suppose you want to conduct a survey of benefit packages available in private businesses in Hawaii. You want a sample size of 100. Some sampling techniques are described below. Categorize each technique as *simple random sample, stratified sample, systematic sample, cluster sample,* or *convenience sample.*
    (a) Assign each business in the Island Business Directory a number, and then use a random-number table to select the businesses to be included in the sample.
    (b) Use the postal ZIP Codes to divide the state into regions. Pick a random sample of 10 ZIP Code areas and then include all the businesses in each selected ZIP Code area.
    (c) Send a team of five research assistants to Bishop Street in downtown Honolulu. Let each assistant select a block or building and interview an employee from each business found. Each researcher can have the rest of the day off after getting responses from 20 different businesses.
    (d) Use the Island Business Directory. Number all the businesses. Select a starting place at random, and then use every 50th business listed until you have 100 businesses.
    (e) Group the businesses according to type: medical, shipping, retail, manufacturing, financial, construction, restaurant, hotel, tourism, other. Then select a random sample of 10 businesses from each business type.

1.3
Introduction to Experimental Design

FOCUS POINTS
✓ Discuss what it means to take a census.
✓ Describe simulations, observational studies, and experiments.
✓ Identify control groups, placebo effects, and randomized two-treatment design.
✓ Discuss potential pitfalls that might make your data unreliable.

Planning a Statistical Study

Planning a statistical study and gathering data are essential components for obtaining reliable information. Depending on the nature of the statistical study, a great deal of expertise and resources may be required during the planning stage. In this section, we look at some of the basics for planning a statistical study.

> **Basic guidelines for planning a statistical study**
>
> 1. First, identify the individuals or objects of interest.
>
> 2. Specify the variables as well as protocols or procedures for taking measurements or making observations.
>
> 3. Determine if you will use an entire population or a representative sample. If using a sample, decide on a viable sampling method.
>
> 4. Collect the data.
>
> 5. Use appropriate descriptive statistics methods (Chapters 2, 3, 4) and make decisions using appropriate inferential statistics methods (Chapters 9–12).
>
> 6. Finally, note any concerns you might have about your data collection methods and list any recommendations for future studies.

Census

One issue to consider is whether to use the entire population in a study or a representative sample. If we use data from the entire population, we have a *census*.

> In a **census,** measurements or observations from the *entire* population are used.

When the population is small and easily accessible, a census is very useful because it gives complete information about the population. However, obtaining a census can be both expensive and difficult. Every 10 years, the U.S. Department of Commerce, Bureau of Census is required to conduct a census of the United States. However, contacting some members of the population—such as the homeless—is almost impossible. Sometimes members of the population will not respond. In such cases, statistical estimates for the missing responses are often supplied.

Sample

If we use data from only part of the population of interest, we have a *sample*.

> In a **sample,** measurements or observations from a *representative part* of the population should be used.

In the previous section, we examined several sampling strategies: simple random sampling, stratified sampling, cluster sampling, systematic sampling, and convenience sampling. In this text, we will study methods of inferential statistics based on simple random samples.

Simulation

As discussed in Section 1.2, *simulation* is a numerical facsimile of real-world phenomena. Sometimes simulation is called a "dry lab" approach, in the sense that it is a mathematical imitation of a real situation. Advantages of simulation are that mathematical and statistical simulations can fit real-world problems extremely well. The researcher can explore procedures in simulation that might be very dangerous in real life. In the real world, you might not want to introduce a high level of a drug into a diabetic person's bloodstream. However, you might want to simulate the injection statistically and study the results. You will harm no one, and the information gained may be of real medical value. Similarly, you might test the effect of wind shear on an airplane wing in a simulated environment rather than with an actual airplane in flight.

Experiments and Observation

When gathering data for a statistical study, we want to distinguish between observational studies and experiments.

Observational study

Experiment

> In an **observational study,** observations and measurements of individuals are conducted in a way that doesn't change the response or the variable being measured.
>
> In an **experiment,** a *treatment* is deliberately imposed on the individuals in order to observe a possible change in the response or variable being measured.

EXAMPLE 8

Experiment

In 1778, Captain James Cook landed in what we now call the Hawaiian Islands. He gave the islanders a present of several goats, and over the years these animals multiplied into wild herds totaling several thousand. They eat almost anything, including the famous silver sword plant, which was once unique to Hawaii. At one time, the silver sword grew abundantly on the island of Maui (in Haleakala, a national park on that island, the silver sword can still be found), but each year there seemed to be fewer and fewer plants. Biologists suspected that the goats were partially responsible for the decline in the plants and conducted a statistical study that verified their theory.

(a) To test the theory, park biologists set up stations in remote areas of Haleakala. At each station two plots of land similar in soil conditions, climate, and plant count were selected. One plot was fenced to keep out the goats, while the other was not. At regular intervals a plant count was made in each plot. This study involves an *experiment* because a *treatment* (the fence) was imposed on one plot.

(b) The experiment involved two plots at each station. The plot that was not fenced represents the *control* plot. This is the plot where a treatment was specifically not imposed, but it was similar to the fenced plot in every other way.   ◊

Silver sword plant, Haleakala National Park

Placebo

Control group

Treatment group

Randomized two-treatment experiment

Statistical experiments are commonly used to determine the effect of a treatment. However, the design of the experiment needs to *control* for other possible causes of the effect. For instance, in medical experiments the *placebo* effect is the improvement or change that is the result of patients just believing in the treatment whether the treatment itself is effective or not. To account for the placebo effect, patients are divided into two groups. One group receives the prescribed treatment. The other group, called the *control group*, receives a dummy or placebo treatment that is disguised to look like the real treatment. Finally, after the treatment cycle, the medical condition of the patients in the *treatment group* is compared to that of the patients in the control group.

It is difficult to account for all variables that might influence a patient's response to a treatment. Sometimes patients assigned to a treatment group and a control group are carefully matched by age, gender, level of medical condition, smoker, etc. However, a more common way to assign patients to the treatment and control groups is by using a random process. This is the essence of a *randomized two-treatment experiment*.

EXAMPLE 9

Randomized two-treatment experiment

Can chest pain be relieved by drilling holes in the heart? Since 1980, surgeons have been using a laser procedure to drill holes in the heart. Many patients report a lasting and dramatic decrease in angina (chest pain) symptoms. Is the relief due to the procedure, or is it a placebo effect? A recent research project at Lenox Hill Hospital in New York City provided some information about this issue by using a randomized two-treatment experiment. The laser treatment was applied through a less invasive (catheter laser) process. A group of 298 volunteers with severe, untreatable chest pain were randomly assigned to get the laser or not. The patients were sedated but awake. They could hear the doctors discuss the laser process. Each patient thought he or she was receiving the treatment.

The experimental design can be pictured as

The laser patients did well. But shockingly, the placebo group showed more improvement in pain relief. The medical impacts of this study are still being investigated. ◇

Double-blind experiment

Many experiments are also *double-blind*. This means that neither the individuals in the study nor the observers know which subjects are receiving the treatment. Double-blind experiments help control for subtle biases that a doctor might pass on to a patient.

Experimental design is an essential component of good statistical research. The design of experiments can be quite complicated, particularly if more than two populations are to be compared. If the experiment is complex, the services of a professional statistician may be required.

The study cited in Example 9 has many features of good experimental design.

Control group

There is a **control group.** This group received a dummy treatment enabling the researchers to control for the placebo effect. In general, a control group is used to account for the influence of other known or unknown variables that might be an underlying cause of a change in response in the experimental group. Such variables are called **lurking** or **confounding variables.**

Lurking variable
Confounding variable

Randomization

Randomization is used to assign individuals to the two treatment groups. This helps prevent bias in selecting members for each group.

Replication

Replication of the experiment on many patients reduces the possibility that the differences in pain relief for the two groups occurred by chance alone.

GUIDED EXERCISE 5

Collecting data

Which technique (sampling, experiment, simulation, or census) for gathering data do you think might be the most appropriate for the following studies?

(a) Study of the effect of stopping the cooling process of a nuclear reactor.

⟹ Probably simulation, since you may not want to risk a nuclear meltdown.

(b) Study of the amount of time college students taking a full course load spend watching television.

⟹ Sampling and using an observational study would work well. Notice that obtaining the information from a student will probably not change the amount of time a student spends watching television.

(c) Study of the effect of a calcium supplement given to young girls on bone mass.

⟹ Experimentation. A study by Tom Lloyd reported in the *Journal of the American Medical Association* utilized 94 young girls. Half were randomly selected and given a placebo. The other half were given calcium supplements to bring their daily calcium intake up to about 1400 milligrams per day. The group getting the experimental treatment of calcium gained 1.3% more bone mass in a year than the girls getting the placebo.

(d) Study of the credit hours load of *each* student enrolled at your college at the end of the drop/add period this semester.

⟹ Census. The registrar can obtain records for *every* student.

Potential pitfalls

Nonresponse

Voluntary response

Hidden bias

Other variables

Surveys

Once you decide you are going to use sampling, census, observation, or experiments, a common means to gather data about people is to ask them questions. This process is the essence of *surveying*. Sometimes the possible responses are simply yes or no. Other times the respondents choose a number on a scale that represents their feelings from, say, strongly disagree to strongly agree. In the case of an open-ended, discussion-type response, the researcher must determine a way to convert the response to a category or number.

A number of issues can arise when using a survey. Are the questions asked in a neutral way, or is conscious or unconscious bias built into the wording? In the case of an interviewer, is he or she giving the respondents subtle feedback that might influence the responses? How can you be sure the respondents are answering truthfully? Is your sample representative of the population? For instance, when conducting election polls, some studies use only registered voters because these are the only people eligible to vote. Other polls use only "likely" voters. They first inquire if the respondent is planning to vote.

Other problems arise if the selected respondent cannot be contacted or refuses to respond. This is known as *nonresponse*. If the nonresponse is sufficiently high, the study may be biased and some sort of adjustment may need to be made.

Voluntary response samples often overrepresent people with strong opinions. A Colorado newspaper with statewide distribution asked the question, "Should grazing fees for use of public lands be increased?" Of the people who called in, about 85% said, "No!" However, the results were misleading as an indicator of the opinions held by the population of *all* people in Colorado. The sample who responded were self-selected people. The sample consisted mainly of ranchers who felt strongly enough to call the newspaper. These people did not want their grazing fees on public land increased.

Data from voluntary responses can be useful and interesting. However, the information given is anecdotal in nature. It would be questionable to generalize the results to the entire population of interest. Surveys must be designed and administered carefully for the results to generalize.

Whenever you gather data, whether by sampling a population, by results of experiment, or by simulation, you should view the data with a critical eye. The way the data are gathered may produce a *hidden bias*. This means that in reality you are not actually measuring what you hoped to measure. For instance, if a uniformed police officer conducts a survey regarding opinions about and use of illegal drugs, the responses might be different than if a casually dressed person asks the same questions. Asking students if parking lots are large enough and safe before asking if they would approve an increase in parking fees might bias responses. Care must be taken to deal with all units or subjects in the same way so that no (conscious or unconscious) preferential treatment or selection can occur.

Sometimes our goal is to understand the cause-and-effect relationships between two variables (as might occur in regression and correlation of two variables presented in Chapter 4). However, the effect of one variable on another can be hidden by other variables for which no data have been obtained. These variables are called *lurking* or *confounding* variables. For instance, a study of ticket price and attendance at a sporting event might show that higher ticket prices and higher attendance are related, since events with higher ticket prices seem to have greater attendance. One might be led to conclude that if you want to increase attendance at an event, you should raise the price of the tickets. What is missing from this analysis is the lurking variable of *event popularity*. A Super Bowl football game is so special that people are willing to pay higher prices just to be there. A preseason football game, however, does not have the drawing power and may need lower ticket prices to ensure a reasonable crowd at the game. Other variables—such as the location of the game, weather, and records of the teams involved—also might influence attendance.

The problem of other variables that influence one or both of the original variables often can be overcome if the researcher not only is familiar with statistics but also is well versed in the field of investigation.

Generalizing results

Some researchers want to generalize their findings to a situation wider than that of the actual data setting. The true scope of a new discovery must be determined by repeated studies in various real-world settings. Statistical experiments showing that a drug had a certain effect on a collection of laboratory rats do not guarantee that the drug will have a similar effect on a herd of wild horses in Montana.

GUIDED EXERCISE 6

Cautions about data

Comment on the usefulness of the data collected as described.

(a) A uniformed police officer interviews a group of 20 college freshmen. She asks each one his or her name and then if he or she has used an illegal drug in the last month.

⇒ Respondents may not answer truthfully. Some may refuse to participate.

(b) Jessica saw some data that show that cities with more low-income housing have more homeless people. Does building low-income housing cause homelessness?

⇒ There may be some other confounding or lurking variables such as the size of the city. Larger cities may have more low-income housing and more homeless.

(c) A survey about food in the student cafeteria was conducted by having forms available for customers to pick up at the cash register. A drop box for completed forms was available outside the cafeteria.

⇒ The voluntary response will likely produce more negative comments.

(d) Extensive studies on coronary problems were conducted using men over age 50 as the subjects.

⇒ Conclusions for men over age 50 may or may not generalize to other age and gender groups. These results may be useful for women or younger people, but studies specifically involving these groups may need to be performed.

VIEWPOINT

Is the Placebo Effect a Myth?

Henry Beecher, Chief of Anesthesiology at Massachusetts General Hospital, published a paper in the *Journal of the American Medical Association* (1955) in which he claimed that the placebo effect is so powerful that about 35% of patients would improve simply if they believed a dummy treatment (placebo) was real. However, two Danish medical researchers refute this widely accepted claim in the *New England Journal of Medicine* (May 2001). They say the placebo effect is nothing more than a "regression effect," referring to a well-known statistical observation that patients who feel especially bad one day will almost always feel better the next day no matter what is done for them. However, other respected statisticians question the findings of the Danish researchers. Regardless of the new controversy surrounding the placebo effect, medical researchers agree that placebos are still needed in clinical research. Double-blind research using placebos prevents the researchers from inadvertently biasing results.

SECTION 1.3   PROBLEMS

1. *Ecology: Gathering Data* Which technique (observational study or experiment) for gathering data do you think was used in the following studies?
   (a) The Colorado Division of Wildlife netted and released 774 fish at Quincy Reservoir. There were 219 perch, 315 blue gill, 83 pike, and 157 rainbow trout.
   (b) The Colorado Division of Wildlife caught 41 bighorn sheep on Mt. Evans and gave each one an injection to prevent heartworm. A year later, 38 of these sheep did not have heartworm, while the other three did.
   (c) The Colorado Division of Wildlife imposed special fishing regulations on the Deckers section of the South Platte River. All trout under 15 inches must be released. A study of trout before and after the regulation went into effect showed the average length increased by 4.2 inches after the new regulation.
   (d) An ecology class used binoculars to watch 23 turtles at Lowell Ponds. It was found that 18 were box turtles, and five were snapping turtles.

2. *General: Gathering Data* Which technique (sampling, experiment, simulation, or census) for gathering data do you think was used in the following studies?
   (a) One way to find information on the Super Bowl football game is to look at the NFL web site. Visit the Brase/Brase *Understanding Basic Statistics* site at http://math.college.hmco.com/students and find the link to the NFL site. Explore the Super Bowl results. There the winning scores for all the Super Bowl games played to date are given. Using the data for all the games, we find that as of Super Bowl XXXIV, the average score for the winning teams was 31.
   (b) A sample of 82 healthy female and male subjects was recruited to participate in a study on pain (*Physical Therapy*, Vol. 70, No. 1). The subjects were divided into two groups. The experimental group received laser stimulation, and the control group received sham stimulation. Tests of pain tolerance were then conducted on each group.
   (c) Computer imaging of runners shows the effect of stride length on running efficiency.

(d) Do the Chinese like chocolate? Gallup Chinese is conducting surveys in China to answer the question for the U.S. Chocolate Manufacturers Association. Gallup Chinese is surveying a portion of the Chinese population to determine whether there is a market for chocolate in China (*Wall Street Journal*).

3. *General: Gathering Data* Which technique (sampling, experiment, simulation, or census) for gathering data do you think was used in the following studies?
   (a) An analysis of a sample of 31,000 patients from New York hospitals suggests that the poor and the elderly sue for malpractice at one-fifth the rate of wealthier patients (*Journal of the American Medical Association*).
   (b) The effects of wind shear on airplanes during both landing and takeoff are studied by using complex computer programs that mimic actual flight.
   (c) A study of football scores attained through touchdowns and field goals in all NFL season games was conducted by the National Football League to determine whether field goals account for more scoring events than touchdowns (*USA Today*).
   (d) An Australian study included 588 men and women who already had some precancerous skin lesions. Half got a skin cream containing a sunscreen with a sun protection factor of 17; half got an inactive cream. After 7 months, those using the sunscreen with the sun protection had fewer new precancerous skin lesions (*New England Journal of Medicine*).

4. *Surveys: Manipulation* The *New York Times* did a special report on polling that was carried in papers across the nation. The article points out how readily the results of a survey can be manipulated. Some features that can influence the results of a poll include the following: the number of possible responses, the phrasing of the question, the sampling techniques used (voluntary response or sample designed to be representative), the fact that words may mean different things to different people, the questions that precede the question of interest, and finally, the fact that respondents can offer opinions on issues that they know nothing about.
   (a) Consider the expression "over the last few years." Do you think that this expression means the same time span to everyone? What would be a more precise phrase?
   (b) Consider this question: "Do you think fines for running stop signs should be doubled?" Do you think the response would be different if the question "Have you ever run a stop sign?" preceded the question about fines?
   (c) Consider this question: "Do you watch too much television?" What do you think the responses would be if the only responses possible were yes or no? What do you think the responses would be if the possible responses were rarely, sometimes, or frequently?

5. *General: Randomized Two-Treatment Experiment* How would you use a randomized two-treatment experiment in each of the following settings? Is a placebo being used or not? Be specific and give details.
   (a) A veterinarian wants to test a strain of antibiotic on calves to determine their resistance to common infection. In a pasture are 22 newborn calves. There is enough vaccine for 10 calves. However, blood tests to determine resistance to infection can be done on all calves.
   (b) The Denver Police Department wants to improve its image with teenagers. A uniformed officer is sent to a school one day a week for ten weeks. Each day the officer visits with students, eats lunch with students, attends pep rallies, and so on. There are 18 schools, but the police department can visit only half of these schools this semester. A survey regarding how teenagers view police is sent to all 18 schools at the end of the semester.
   (c) A skin patch contains a new drug to help people quit smoking. A group of 75 cigarette smokers have volunteered as subjects to test the new skin patch. For one month 40 of the volunteers receive skin patches with the new drug. The

numbers. Every week a drawing machine randomly selects six numbered Ping-Pong balls. If one of your boards contains all six winning numbers, in any order, you've hit the jackpot! You can pick your numbers any way you wish. However, suppose you want to use a random-number table to pick your six numbers. Describe how you would do so, and list your selected numbers. (To play, you must pay $1 to have your selections entered into a computer for a specified week's drawing.)

6. *General: Type of Sampling* Categorize the type (simple random, stratified, systematic, cluster, or convenience) of sampling used in each of the following situations.
   (a) To conduct a preelection opinion poll on a proposed amendment to the state constitution, a random sample of 10 telephone prefixes (first three digits of the phone number) was selected, and all households from the phone prefixes selected were called.
   (b) To conduct a study on depression among the elderly, a sample of 30 patients in one nursing home was used.
   (c) To maintain quality control in a brewery, every 20th bottle of beer coming off the production line was opened and tested.
   (d) Subscribers to the magazine *Sound Alive* were assigned numbers. Then a sample of 30 subscribers was selected by using a random-number table. The subscribers in the sample were invited to rate new compact disc players for a "What the Subscribers Think" column.
   (e) To judge the appeal of a proposed television sitcom, a random sample of 10 people from each of three different age categories was selected and those chosen were asked to rate a pilot show.

7. *General: Gathering Data* Which technique (observational study or experiment) for gathering data do you think was used in the following studies? Explain your answer.
   (a) The U.S. Census Bureau tracks population age. In 1900, the percentage of the population that was nineteen years old or younger was 44.4%. In 1930, the percentage was 38.8%; in 1970, the percentage was 37.9%; and in 2000, the percentage in the age group was down to 28.5% (*The First Measured Century*, T. Caplow, L. Hicks, B. J. Wattenberg).
   (b) After receiving the same lessons, a class of 100 students was randomly divided into two groups of 50 each. One group was given a multiple-choice exam over the material in the lessons. The other group was given an essay exam. The average test scores for the two groups were then compared.

8. *General: Randomized Two-Treatment Experiment* How would you use a randomized two-treatment experiment in each of the following settings? Is a placebo being used or not? Be specific and give details.
   (a) A charitable nonprofit organization wants to test two methods of fund raising. From a list of 1,000 past donors, half will be sent literature about the successful activities of the charity and asked to make another donation. The other 500 donors will be contacted by phone and asked to make another donation. The percentage of people from each group who make a new donation will be compared.
   (b) A tooth-whitening gel is to be tested for effectiveness. A group of 85 adults have volunteered to participate in the study. Of these, 43 are to be given a gel that contains the tooth-whitening chemicals. The remaining 42 are to be given a similar-looking package of gel that does not contain the tooth-whitening chemicals. A standard method will be used to evaluate the whiteness of teeth for all participants. Then the results for the two groups will be compared. How could this experiment be designed to be double-blind?

9. *Student Life: Data Collection Project* Make a statistical profile of your own statistics class. Items of interest might be
   (a) Height, age, gender, pulse, number of siblings, marital status

(b) Number of college credit hours completed (as of beginning of term); grade point average
(c) Major; number of credit hours enrolled in this term
(d) Number of scheduled hours working per week
(e) Distance from residence to first class; time it takes to travel from residence to first class
(f) Year, make, and color of car usually driven

What directions would you give to people answering these questions? For instance, how accurate should the measurements be? Should age be recorded as of last birthday?

10. *Census: Web Site Census and You,* a publication of the Bureau of the Census, indicates that "Wherever your Web journey ends up, it should start at the Census Bureau's site." Visit the Brase/Brase *Understanding Basic Statistics* site at http://math.college.hmco.com/students and find a link to the Census Bureau's site. The site touts itself as the source of "official statistics." But it is willing to share the spotlight. The web site now has links to other "official" sources: other federal agencies, foreign statistical agencies, and state data centers. If you have access to the Internet, try the Census Bureau's site.

DATA HIGHLIGHTS: GROUP PROJECTS

1. Use a random-number table or random-number generator to simulate tossing a fair coin 10 times. Generate 20 such simulations of 10 coin tosses. Compare the simulations. Are there any strings of 10 heads? of 4 heads? Does it seem that in most of the simulations half the outcomes are heads? half are tails? In Chapter 6, we will study the probability of getting from 0 to 10 heads in such a simulation.

2. Use a random-number table or random-number generator to get a random sample of 30 distinct values from the set of integers from 1 to 100. Instructions for doing this, using the TI-83, Excel, Minitab, or ComputerStat, are in Using Technology at the end of this chapter. Generate five such samples. How many of the samples include the number 1? the number 100? Comment about the differences among the samples. How well do the samples seem to represent the numbers between 1 and 100?

LINKING CONCEPTS: WRITING PROJECTS

Discuss each of the following topics in class or review the topics on your own. Then write a brief but complete essay in which you summarize the main points. Please include formulas and graphs as appropriate.

1. What does it mean to say we are going to use a sample to draw an inference about the population? Why is a random sample so important for this process? If we wanted a random sample of students in the cafeteria, why couldn't we just take the students who order Diet Pepsi with their lunch? Comment on the statement, "A random sample is kind of a miniature population, whereas samples that are not random are likely to be biased." Why would the students who order Diet Pepsi with lunch not be a random sample of students in the cafeteria?

2. In your own words, explain the differences among the following sampling techniques: simple random sample, stratified sample, systematic sample, cluster sample, and convenience sample. Describe situations in which each type might be useful.

Using TECHNOLOGY

TI-83PLUS   •   EXCEL   •   MINITAB   •   COMPUTERSTAT

General spreadsheet programs such as Microsoft's Excel, specific statistical software packages such as Minitab, and graphing calculators such as the TI-83Plus all offer computing support for statistical methods. Applications in this section may be completed using software or calculators with statistical functions. Select keystroke or menu choices are shown for the TI-83Plus, Excel, Minitab, and ComputerStat in the Technology Hints portion of this section. More details can be found in the software-specific Technology Guides that accompany this text.

APPLICATIONS

Most software packages sample *with replacement*. That is, the same number may be used more than once in the sample. If your applications require sampling without replacement, draw more items than you need. Then use sort commands in the software to put the data in order, and delete repeated data.

1. Simulate the results of tossing a fair die 18 times. Repeat the simulation. Are the results the same? Do you expect them to be the same? Why or why not? Do there appear to be equal numbers of outcomes 1 through 6 in each simulation? In Chapter 5, we will see the law of large numbers, which tells us that we would expect equal numbers of outcomes only when the simulation is very large.

2. A college has 5,000 students, and the registrar wishes to use a random sample of 50 students to examine credit hour enrollment for this semester. Write a brief description of how a random sample can be drawn. Draw a random sample of 50 students. Are you sampling with or without replacement?

Technology Hints: Random Numbers

TI-83Plus

To select a random set of integers between two specified values, press the **MATH** key and highlight **PRB** with **5:randInt** (low value, high value, sample size). Press Enter and fill in the low value, high value, and sample size. To store the sample in list L1, press the **STO➡** key and then L1. The screen display shows two random samples of size 5 drawn from the integers between 1 and 100.

```
randInt(1,100,5)
{63 89 13 46 47}
randInt(1,100,5)
{29 82 99 50 41}
```

Excel

To select a random number between two specified values, type the command=**Randbetween**(low value, high value) in the formula bar. Alternatively, access a dialogue box for the command by clicking on the **Paste Function** f_x key on the menu bar. Then choose **All** in the left drop-down menu and **RANDBETWEEN** in the right menu. Fill in the dialogue box. This command requires the Analysis ToolPack add-in.

A5		=	=RANDBETWEEN(1,100)		
	A	B	C	D	E
1	99				
2	58				
3	69				
4	86				
5	13				

Minitab

To generate random integers between specified values, use the menu selection **Calc ➤ Random Data ➤ Integer.** Fill in the dialogue box to get 5 random numbers between 1 and 100.

Worksheet 2 ***	C1
1	8
2	35
3	33
4	9
5	15

ComputerStat

In ComputerStat, select the program Random Samples found in the Descriptive Statistics menu. This program lets you select the range of integers from which you wish to draw random samples. It also gives you the option of sampling with or without replacement.

2

Organizing Data

In dwelling upon the vital importance of sound observation, it must never be lost sight of what observation is for. It is not for the sake of piling up miscellaneous information or curious facts, but for the sake of saving life and increasing health and comfort.

—Florence Nightingale, *Notes on Nursing*

**Florence Nightingale
(1820–1910)**

Florence Nightingale has been described as a "passionate statistician" and a "relevant statistician." She viewed statistics as a science that allows one to transcend his or her narrow individual experience and aspire to the broader service of humanity. She was one of the first nurses to use graphic representation of statistics, illustrating with charts and diagrams how improved sanitation decreased the rate of mortality. Her statistical reports about the appalling sanitary conditions at Scutari (the main British hospital during the Crimean War) were taken very seriously by the English Secretary at War, Sidney Herbert. When sanitary reforms recommended by Nightingale were instituted in military hospitals, the mortality rate dropped from an incredible 42.7% to only 2.2%.

For on-line student resources, visit **math.college.hmco.com/students** and follow the Statistics links to the Brase/Brase, *Understanding Basic Statistics,* 3rd edition Web site.

◊ How can we select graphs appropriate for given data sets? (SECTION 2.1)

◊ What are histograms and relative-frequency diagrams? Where are they used? (SECTION 2.2)

◊ What are common distribution shapes? (SECTION 2.2)

◊ How can we quickly order data and, at the same time, reveal the distribution shape? (SECTION 2.3)

FOCUS PROBLEM

Say It with Pictures

Edward R. Tufte, in his book *The Visual Display of Quantitative Information*, has a number of guidelines for producing good graphics. According to the criteria, a graphical display should

• show the data;

• induce the viewer to think about the substance of the graphic rather than about the methodology, the design, the technology, or other production devices;

• avoid distorting what the data have to say.

As an example of a graph that violates some of the criteria, Tufte includes a graphic that appeared in a well-known newspaper. Figure 2-1(a) shows a facsimile of the problem graphic, whereas part (b) of the figure shows a better rendition of the data display. Look at both graphics, and identify some of the problems in part (a). Examine part (b). How is this graphic different from part (a), and what features better highlight the information displayed?

FIGURE 2-1

Fuel Economy Standards for Autos

(a)

Fuel Economy Standards for Autos

Set by Congress and supplemented by the Transportation Department. In miles per gallon.

Source: Copyright © 1978 by The New York Times Company. Reprinted by permission.

(b)

REQUIRED FUEL ECONOMY STANDARDS: NEW CARS BUILT FROM 1978 TO 1985

19.1 mpg, expected average for all cars, 1985

13.7 mpg, average for all cars, 1978

Source: The Visual Display of Quantitative Information by Edward R. Tufte, p. 57. Copyright © 1983. Reprinted by permission of Graphics Press.

2.1

Bar Graphs, Circle Graphs, and Time Plots

FOCUS POINTS

✓ Determine types of graphs appropriate for specific data.

✓ Construct bar graphs, Pareto charts, circle graphs, and time plots.

✓ Interpret information displayed in graphs.

Open almost any magazine, newspaper, or large Web site, and you will find graphs. In fact, because graphs display information so effectively, most word processors and spreadsheet programs include extensive chart- and graph-making tools. In this section, we identify several common types of graphs, show how to construct them, and show how to interpret information presented in graphical form.

Let's start with *bar graphs*. These are graphs that can be used to display quantitative or qualitative data.

Features of a bar graph

1. Bars can be vertical or horizontal.

2. Bars are of uniform width and uniformly spaced.

3. The lengths of the bars represent values of the variable being displayed, the frequency of occurrence, or the percentage of occurrence. The same measurement scale is used for the length of each bar.

4. The graph is well annotated with title, label for each bar, and vertical scale or actual value for the length of each bar.

Bar graphs

EXAMPLE 1

Bar graph

What's a square foot worth? In retail, annual sales per square foot is an important industry measurement. A recent article in *The Wall Street Journal* gave average annual sales per square foot for nationwide department stores: $147 for Dillard's

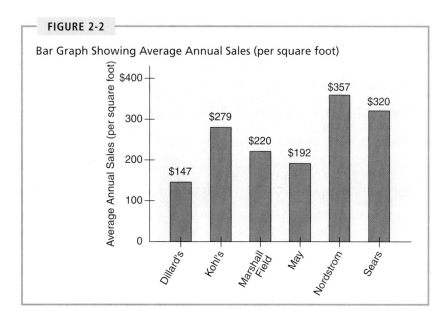

FIGURE 2-2

Bar Graph Showing Average Annual Sales (per square foot)

Inc., $279 for Kohl's Corp., $220 for Marshall Field, $192 for May Department Stores, $357 for Nordstrom, Inc., and $320 for Sears. To make a bar graph of this information, we let the length of the bars in Figure 2-2 represent the sales (in dollars) per square foot. Each store has its own bar.

Usually, it is not necessary to display both the vertical scale and the individual height of each bar. However, the height label gives the reader exact information while the scale display gives the viewer a sense of the range of data values. In addition, the graph is titled and the bars are labeled by store.

From the graph, it is readily apparent that Nordstrom and Sears have the highest sales per square foot. *The Wall Street Journal* reports that Nordstrom's upscale merchandise and Sears' appliance sales push their sales figures up. Among the other four stores that carry similar merchandise, we see that Kohl's has the highest average sales per square foot.   ◇

Changing scales

There are several variations of bar graphs. Figure 2-3 shows life expectancy for men and women born in specific years. Notice the bars for men and women are clustered over the year of birth, and a legend identifies which bars are for men and which are for women.

An important feature illustrated in Figure 2-3(b) is that of a *changing scale*. Notice that the scale between 0 and 65 is compressed. The changing scale amplifies the apparent difference between life spans for men and women, as well as the increase in life spans from those born in 1970 to the projected span of those born in 2010. Whenever you use a change in scale, warn the viewer by using a squiggle ∿ on the changed axis. Sometimes, if a single bar is unusually long, the bar length is compressed with a squiggle in the bar itself.

Quality control is an important aspect of today's production and service industries. Dr. W. Edwards Deming was one of the developers of total quality management (TQM). In his book *Out of Crisis*, he outlines many strategies for monitoring and improving service and production industries. In particular, Dr. Deming recommends the use of some statistical methods to organize and analyze data

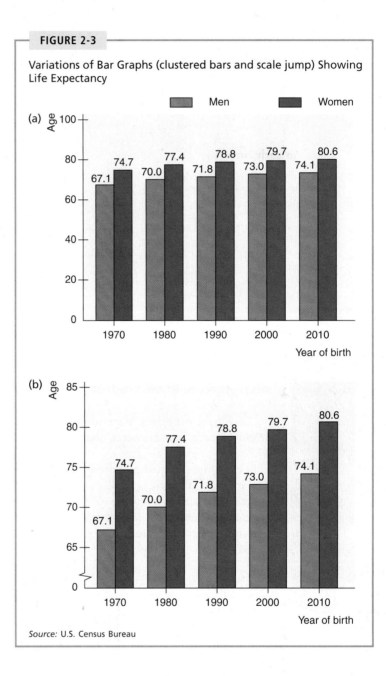

FIGURE 2-3

Variations of Bar Graphs (clustered bars and scale jump) Showing Life Expectancy

Source: U.S. Census Bureau

Pareto charts

from industries so that sources of problems can be identified and then corrected. *Pareto* (pronounced "Pah-rāy-tō) *charts* are among the many techniques used in quality-control programs.

A **Pareto chart** is a bar graph in which the bar height represents frequency of an event. In addition, the bars are arranged from left to right according to decreasing height.

GUIDED EXERCISE 1

Pareto charts

This exercise is adapted from *The Deming Management Method* by Mary Walton. Suppose you want to arrive at the college 15 minutes before your first class so that you can feel relaxed when you walk into class. An early arrival time also allows room for unexpected delays. However, you always find yourself arriving "just in time" or slightly late. What causes you to be late? Charlotte made a list of possible causes and then kept a checklist for 2 months (Table 2-1). On some days more than one item was checked because several events occurred that caused her to be late.

TABLE 2-1 Causes for Lateness
(September–October)

Cause	Frequency
Snoozing after alarm goes off	15
Car trouble	5
Too long over breakfast	13
Last-minute studying	20
Finding something to wear	8
Talking too long with roommate	9
Other	3

(a) Make a Pareto chart showing the causes for lateness. Be sure to label the causes, and draw the bars using the same vertical scale.

FIGURE 2-4 Pareto Chart: Conditions That Might Cause Lateness

(b) Looking at the Pareto chart, what recommendations do you have for Charlotte?

According to the chart, rearranging study time, or getting up earlier to allow for studying, would cure her most frequent cause for lateness. Repairing the car might be important, but for getting to campus early, it would not be as effective as adjusting study time.

Circle graphs or pie charts

Another popular pictorial representation of data is the *circle graph*, or *pie chart*. It is relatively safe from misinterpretation and is especially useful for showing the division of a total quantity into its component parts. The total quantity, or 100%, is represented by the entire circle. Each wedge of the circle represents a component part of the total. These proportional segments are usually labeled with corresponding percentages of the total. Guided Exercise 2 shows how to make a circle graph.

GUIDED EXERCISE 2

Circle graph

How long do we spend talking on the telephone after hours (at home after 5 P.M.)? The results from a recent survey of 500 people (as reported in *USA Today*) are shown in Table 2-2. We'll make a circle graph to display these data.

TABLE 2-2   Time Spent on Home Telephone After 5 P.M.

Time	Number	Fractional Part	Percentage	Number of Degrees
Less than ½ hour	296	296/500	59.2	59.2% × 360° ≈ 213°
½ hour to 1 hour	83	83/500	16.6	16.6% × 360° ≈ 60°
More than 1 hour	121	_____	_____	_____
Total	_____		_____	_____

(a) Fill in the missing parts in Table 2-2 for "More than 1 hour." Remember that the central angle of a circle is 360°. Round to the nearest degree.

⇨ For "More than 1 hour," Fractional Part = 121/500; Percentage = 24.2%; Number of Degrees = 24.2% × 360° ≈ 87°. The symbol ≈ means approximately equal.

(b) Fill in the totals. What is the total number of responses? Do the percentages total 100% (within rounding error)? Do the numbers of degrees total 360° (within rounding error)?

⇨ The total number of responses is 500. The percentages total 100%. You must have such a total in order to create a circle graph. The numbers of degrees total 360°.

(c) Draw a circle graph. Divide the circle into pieces with the designated numbers of degrees. Label each piece, and show the percentage corresponding to each piece. The numbers of degrees are usually omitted from pie charts shown in newspapers, magazines, journals, and reports.

⇨ **FIGURE 2-5** Hours on Home Telephone After 5 P.M.

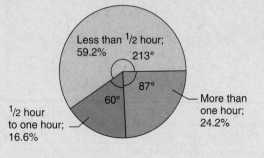

Suppose you begin an exercise program that involves walking or jogging for 30 minutes. You exercise several times a week but monitor yourself by logging the distance you cover in 30 minutes each Saturday. How do you display these data in a meaningful way? Making a bar chart showing the frequency of distances you cover might be interesting, but it does not really show how the distance you cover in 30 minutes has changed over time. A graph showing the distance covered on each date will let you track your performance over time.

Time plot

We will use a *time plot*. A time plot is a graph showing data measurements in chronological order. To make a time plot, we put time on the horizontal scale and the variable being measured on the vertical scale. In a basic time plot, we connect the data points by line segments.

EXAMPLE 2

Time plot

Suppose you have been in the walking/jogging exercise program for 20 weeks, and for each week you have recorded the distance you covered in 30 minutes. Your data log is shown in Table 2-3.

TABLE 2-3   Distance in Miles Walked/Jogged in 30 Minutes

Week	1	2	3	4	5	6	7	8	9	10
Distance	1.5	1.4	1.7	1.6	1.9	2.0	1.8	2.0	1.9	2.0
Week	11	12	13	14	15	16	17	18	19	20
Distance	2.1	2.1	2.3	2.3	2.2	2.4	2.5	2.6	2.4	2.7

(a) Make a time plot.

SOLUTION: The data are appropriate for a time plot because they represent the same measurement (distance covered in a 30-minute period) taken at different times. The measurements are also recorded at equal time intervals (every week). To make our time plot, we list the weeks in order on the horizontal scale. Above each week, plot the distance covered that week on the vertical scale. Then connect the dots. Figure 2-6 shows the time plot. Be sure the scales are labeled.

FIGURE 2-6

Time Plot of Distance (in miles) Jogged in 30 Minutes

(b) From looking at Figure 2-6, can you detect any patterns?

> **SOLUTION:** There seems to be an upward trend in distance covered. The distances covered in the last few weeks are about a mile farther than those for the first few weeks. However, we cannot conclude that this trend will continue. Perhaps you have reached your goal for this training activity and now wish to maintain a distance of about 2.5 miles in 30 minutes.   ◇

Time series

Data sets composed of similar measurements taken at regular intervals over time are called *time series*. Time series are often used in economics, finance, sociology, medicine, and any situation wherein we want to study or monitor a similar measure over a period of time. A time plot can reveal some of the main features of time series.

We've seen several styles of graphs. Which kinds are suitable for a specific data collection?

> **Bar graphs** are useful for quantitative or qualitative data. With qualitative data, the frequency or percentage of occurrence can be displayed. With quantitative data, the measurement itself can be displayed as was done in the bar graph showing life expectancy. Watch that the measurement scale is consistent or that a jump scale squiggle is used.

> **Pareto charts** identify the frequency of events or categories in decreasing order of frequency of occurrence.

> **Circle graphs** display how a *total* is dispersed into several categories. The circle graph is very appropriate for qualitative data, or any data where percentage of occurrence makes sense. Circle graphs are most effective when the number of segments are ten or fewer.

> **Time plots** display how data change over time. It is best if the units of time are consistent in a given plot. For instance, measurements taken every day should not be mixed on the same plot with data taken every week.

> **For any graph:** Provide a title, label axes, identify units of measure. As Edward Tufte suggests in his book *The Visual Display of Quantitative Information,* don't let artwork or skewed perspective cloud the clarity of the information displayed.

TECH NOTES   *Bar graphs, circle graphs, time plot*

TI-83Plus Only time plot. Place consecutive values 1 through the number of time segments in list L1 and data in L2. Press **Stat Plot** and highlight an *xy* line plot.

Excel Use the chart wizard on the toolbar . Select the desired option and follow the instructions in the dialogue boxes.

Minitab Use the menu selection **Graph.** Select the desired option and follow the instructions in the dialogue boxes.

VIEWPOINT

Do Ethical Standards Vary by the Situation?

Lutheran Brotherhood did a national survey and found that nearly 60% of all U.S. adults claim that ethics vary by the situation; 33% claim that there is only one ethical standard; and 7% were not sure. How could you draw a circle graph to make a visual impression of Americans' views on ethical standards?

SECTION 2.1   PROBLEMS

1. *Education: Does College Pay Off?* It is costly in both time and money to go to college. Does it pay off? According to the Bureau of Census, the answer is yes. The average annual income (in thousands of dollars) of a *household* headed by a person with the stated education level is as follows: 16.1 if 9th grade is the highest level achieved, 34.3 for high school graduates, 48.6 for those holding associate degrees, 62.1 for those with bachelor's degrees, 71.0 for those with master's degrees, and 84.1 for those with doctoral degrees. Make a bar graph showing household income for each education level.

2. *Accidents: Child Deaths* How safe is the world for kids? Unfortunately, some children between the ages of 1 and 14 die of injuries every year. United Nations data show that by nation, the annual number of deaths from injuries per 100,000 children are as follows: Australia, 9.5; Canada, 9.7; Denmark, 8.1; France, 9.1; Germany, 8.3; Hungary, 10.8; Ireland, 8.3; Italy, 6.1; Japan, 8.4; Korea, 25.6; Mexico, 19.8; New Zealand, 13.7; Netherlands, 6.6; Poland, 13.4; Portugal, 17.8; Spain, 8.1; Sweden, 5.2; Switzerland, 9.6; U.K., 6.1; United States, 14.1. Display these data in a Pareto chart.

3. *Airlines: Fear of Flying* In an article entitled "How to Cure the Fear of Flying," *Fortune* magazine gave the following information: The number of people who died in a calendar year while on a commercial flight, 33; in the bathtub, 318; by poisoning, 5900; as pedestrians, 6500; murdered by spouses, 7000; from falls, 12,400; in motor vehicle accidents, 33,800. Make a Pareto chart showing activities causing death.

4. *Business: Company Failures* Suppose you are interested in investing in a business corporation. What are some pitfalls that might cause a business corporation to fail? To answer this question, we can look at history. Buccino & Associates surveyed more than 1300 business professionals and asked them about leading causes of business failures. According to the report in *USA Today,* 88% of those surveyed cited internal problems, not external factors, as the leading cause of business failure. Of the internal problems listed, 13% of those interviewed cited insufficient management experience as the leading cause of business failure, 13% cited poor business planning, 18% cited inadequate leadership, and 29% cited excessive debt.
   (a) Make a Pareto chart for the causes of business failure.
   (b) Does your chart show *all* the causes for business failure cited by business professionals interviewed? Which internal cause was cited most frequently?

5. *Lifestyle: Hide the Mess!* A survey of 1000 adults (reported in *USA Today*) uncovered some interesting housekeeping secrets. When unexpected company comes, where do we hide the mess? The survey showed that 68% of the respondents toss their mess in the closet, 23% shove things under the bed, 6% put things in the bathtub, and 3% put the mess in the freezer. Make a circle graph to display this information.

6. *Lifestyle: Fast Food* What meal are we most likely to eat in a fast-food restaurant? A survey of 1261 adults (reported in *USA Today*) revealed that 48.9% of the respondents are most likely to eat lunch at a fast-food restaurant; 7.7%, breakfast; 31.6%, dinner; 10%, a snack; and 1.8% answered "don't know." Display this information in a circle graph.

7. *Education: College Professors' Time* How do college professors spend their time? *The National Education Association Almanac of Higher Education* gives the following average distribution of professional time allocation: teaching, 51%; research,

16%; professional growth, 5%; community service, 11%; service to the college, 11%; and consulting outside the college, 6%. Make a pie chart showing the allocation of professional time for college professors.

8. *Education: College Professors' Ages* How old are college professors? The NEA reference in Problem 7 gave the following age distribution in years: under 35 years, 8%; 35–44 years, 29%; 45–54 years, 37%; 55–59 years, 13%; 60–64 years, 9%; and 65 years and over, 4%. Make a pie chart showing the age distribution of college professors.

9. *Merchandise: Consumer Electronics* Telephone gadgets fill our home. Consumer Electronics Manufacturers Association reported that 94.1% of U.S. households have corded phones, 78% have cordless phones, 74% have a telephone answering device, 51% have a cellular phone, 28% have pagers, and 51% have a modem or fax machine. Display this information in a bar chart showing percentage of households owning each gadget. Could the information as reported be put into a circle graph? Explain.

10. *Driving: Bad Habits* Driving would be more pleasant if we didn't have to put up with bad habits of other drivers. *USA Today* reported the results of a Valvoline Oil Company survey of 500 drivers in which the drivers marked their complaints about other drivers. The top complaints turned out to be tailgating, marked by 22% of the respondents; not using turn signals, marked by 19%; 16% marked being cut off; 11% complained about other drivers driving too slowly; and 8% complained about other drivers being inconsiderate. Make a Pareto chart showing percentage of drivers listing each stated complaint. Could this information as reported be put in a circle graph? Why or why not?

11. *Ecology: Lakes* Pyramid Lake, Nevada, is described as the pride of the Paiute Indian Nation. It is a beautiful desert lake famous for very large trout. The elevation of the lake surface (feet above sea level) varies according to the annual flow of the Truckee River from Lake Tahoe. The U.S. Geological Survey provided the following data:

Year	Elevation	Year	Elevation	Year	Elevation
1986	3817	1992	3798	1998	3811
1987	3815	1993	3797	1999	3816
1988	3810	1994	3795	2000	3817
1989	3812	1995	3797		
1990	3808	1996	3802		
1991	3803	1997	3807		

Make a time plot displaying the data. For more information, visit the Brase/Brase *Understanding Basic Statistics* site at http://math.college.hmco.com/students and find the link to the Pyramid Lake Fisheries.

12. *Vital Statistics: Height* How does average height for boys change as the boy gets older? According to *Physician's Handbook*, the heights at different ages are as follows:

Age (years)	0.5	1	2	3	4	5	6	7
Height (inches)	26	29	33	36	39	42	45	47

Age (years)	8	9	10	11	12	13	14
Height (inches)	50	52	54	56	58	60	62

Make a time plot for average height for age 0.5 through 14 years.

2.2
Frequency Distributions and Histograms

Histograms

Healthy Crunch cereal is about to take over sponsorship of the TV program *Space Voyage*. The advertising manager has requested a report on the age distribution of the viewers so the spot ads can be tailored to appeal to the age groups with the most viewers. Figure 2-7 is a *histogram* showing the *frequency distribution* of ages.

Figure 2-7 looks like (and in fact is) a bar graph. However, the graph also meets several specialized criteria that make it a histogram.

Histograms are bar graphs in which

(a) The bars have the same width and always touch.

(b) The width of a bar represents a quantitative value, such as age, rather than a category.

(c) The height of each bar indicates *frequency*.

Information is presented in condensed form in a histogram. The original data for the viewer age report included the *exact* number of 21-year-olds in the sample. In the histogram this number was grouped with the ages 14.5–24.5. We can tell how many viewers are in that age group, but we cannot tell exactly how many are 21 years old. However, the condensed information in the histogram can be assimilated more quickly than the same information in more detailed form.

FIGURE 2-7

Histogram Showing Viewer Age for *Space Voyage*

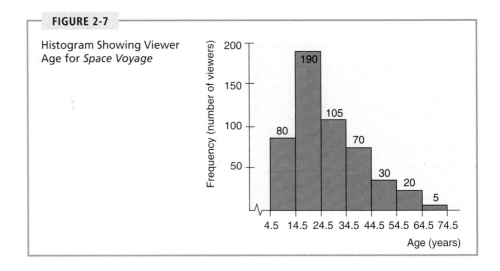

If you are given many pieces of data, how do you condense the information to make a histogram? The task force to encourage car pools did a study of one-way commuting distances for workers in the downtown Dallas area. A random sample of 60 of these workers was taken. The commuting distances of the workers in the sample are given in Table 2-4.

TABLE 2-4   One-Way Commuting Distances in Miles for 60 Workers in Downtown Dallas

13	47	10	3	16	20	17	40	4	2
7	25	8	21	19	15	3	17	14	6
12	45	1	8	4	16	11	18	23	12
6	2	14	13	7	15	46	12	9	18
34	13	41	28	36	17	24	27	29	9
14	26	10	24	37	31	8	16	12	16

The first thing to do is to decide *how many bars or classes* you want in the histogram. Five to 15 classes are usually used. If you use fewer than 5 classes, you risk losing too much information, but if you use more than 15 classes, the clarity of the diagram might be sacrificed for detail. Let the spread of the data and the purpose of the histogram be your guide when selecting the number of classes.

Procedure to make histogram for integer data

Class width

Next, find a convenient *class width*. To do this, find the difference between the largest and smallest data values and divide by the number of classes. If you want the class width to be a whole number, always *increase* the result to the next whole number so that the classes cover the data.

> **Class width**
>
> 1. Compute $$\dfrac{\text{Largest data value} - \text{Smallest data value}}{\text{Desired number of classes}}$$
>
> 2. Increase the value computed to the next highest whole number.

Note: To ensure that the classes cover the data, we need to increase the result of Step 1 to the *next* whole number, even if Step 1 produced a whole number. For instance, if the calculation in Step 1 produces the value 5, we make the class width 6.

In this case, let's use 6 classes. The largest distance commuted is 47 miles and the smallest is 1 mile.

$$\text{Class width} = \frac{47 - 1}{6} \approx 7.7 \quad \text{increase to} \quad 8$$

Lower class limit
Upper class limit

The lowest and highest values that can fit into a class are called the *lower class limit* and *upper class limit,* respectively. The *class width* is the difference between the lower class limit of one class and the lower class limit of the next class. We follow the convention of using the lowest data value for the lower class limit of the first class. Then add the class width to this value to determine the lower class limit for the second class. Continue in this manner to generate all the lower class limits. The upper class limits are then the highest possible data value for each respective class. Note that the difference between consecutive upper class limits is also the class width.

TABLE 2-5   Frequency Table of One-Way Commuting Distances for 60 Downtown Dallas Workers (data in miles)

Class Limits Lower–Upper	Class Boundaries Lower–Upper	Tally*	Frequency	Class Midpoint
1–8	0.5–8.5	ﬀﬀﬀ ﬀﬀﬀ IIII	14	4.5
9–16	8.5–16.5	ﬀﬀﬀ ﬀﬀﬀ ﬀﬀﬀ ﬀﬀﬀ I	21	12.5
17–24	16.5–24.5	ﬀﬀﬀ ﬀﬀﬀ I	11	20.5
25–32	24.5–32.5	ﬀﬀﬀ I	6	28.5
33–40	32.5–40.5	IIII	4	36.5
41–48	40.5–48.5	IIII	4	44.5

*The tally column is included simply as an aid for determining the frequencies. It is not a necessary part of a frequency table.

Class midpoint
Class mark

The center of the class is called the *midpoint* (or *class mark*). This is found by adding the lower and upper class limits of one class and dividing by 2.

$$\text{Midpoint} = \frac{\text{Lower class limit } + \text{ Upper class limit}}{2}$$

The midpoint is often used as a representative value of the entire class.

Frequency table

Now we can organize the commuting distance data into a *frequency table* (Table 2-5). Such a table shows the limits of each class, the frequency with which the data fall into a class, and the class midpoint. A tally will help us find the frequencies.

We're almost ready to make a histogram. But in a histogram we want the bars to touch. There is a space between the upper limit of one class and the lower limit

Class boundaries

of the next class. The halfway points of these intervals are called *class boundaries*.

FIGURE 2-8

One-Way Commuting Distances in Miles Driven by Downtown Dallas Workers

We use the class boundaries as the endpoints of the bars in the histogram. Then there is no space between the bars. Figure 2-8 is the histogram of commuter distances. The class boundaries are shown. (Sometimes only the class midpoints are labeled.)

Relative-frequency Histograms

Other useful tools for organizing data are *relative-frequency tables* and *relative-frequency histograms*. Once we have made a frequency table, it is easy to construct a relative-frequency table. The relative frequency for a particular class is found by dividing the class frequency by the total of all frequencies (sample size).

$$\text{Relative frequency} = \frac{f}{n} = \frac{\text{Class frequency}}{\text{Total of all frequencies}}$$

Table 2-6 shows the relative frequencies for the commuter data of Table 2-4. Since we already have the frequency table (Table 2-5), the relative-frequency table is obtained easily. The sample size is $n = 60$. Notice that the sample size is the total of all the frequencies. Therefore, the relative frequency for the first class (the class from 1 to 8) is

$$\text{Relative frequency} = \frac{f}{n} = \frac{14}{60} \approx 0.23$$

The symbol $\approx$ means "approximately equal to." We use the symbol because we rounded the relative frequency. Relative frequencies for the other classes are computed in a similar way.

The total of the relative frequencies should be 1. However, rounded results may make the total slightly higher or lower than 1.

Using Table 2-6 and Figure 2-8, we can quickly make a relative-frequency histogram (Figure 2-9 on the following page). The horizontal scale will be the same, but the vertical scale will be marked with *relative frequencies f/n* instead of the actual frequencies *f*. The basic shape of the two graphs otherwise will be the same.

In Guided Exercise 3 we will ask you to make a frequency table, histogram, relative-frequency table, and relative-frequency histogram. There are a number of steps to follow. The individual steps are not difficult, but you need to keep them all in mind. The steps are listed for your convenience.

Relative frequencies

TABLE 2-6   Relative Frequencies of One-Way Commuting Distances

Class	Frequency f	Relative Frequency f/n
1–8	14	14/60 ≈ 0.23
9–16	21	21/60 ≈ 0.35
17–24	11	11/60 ≈ 0.18
25–32	6	6/60 ≈ 0.10
33–40	4	4/60 ≈ 0.07
41–48	4	4/60 ≈ 0.07

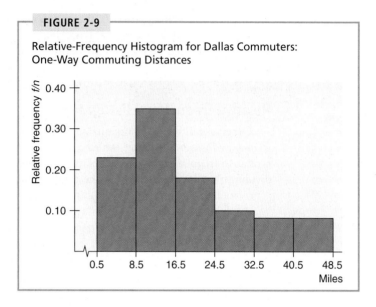

FIGURE 2-9

Relative-Frequency Histogram for Dallas Commuters:
One-Way Commuting Distances

To make a frequency table

1. Determine the class width.

2. Create the distinct classes. We use the convention that the *lower class limit* of the first class is the smallest data value. Add the class width to this number to get the *lower class limit* of the next class.

3. Tally the data into classes. Each data value should fall into exactly one class. Total the tallies to obtain each *class frequency*.

4. Compute the *midpoint* (class mark) for each class.

5. Determine the *class boundaries*.

To make a relative-frequency table

6. For each class, compute the *relative frequency f/n*, where f is the class frequency and n is the total sample size.

◇ **COMMENT** The use of class boundaries in histograms assures us that the bars of the histogram touch and that no data fall on the boundaries. Both these features are important. But a histogram displaying class boundaries may look awkward. For instance, the age range of 14.5 to 24.5 years shown in Figure 2-7 isn't as natural a choice as an age range of 15 to 25 years. For this reason, many magazines and newspapers do not use class boundaries as labels on a histogram. Instead, some use lower class limits as labels, with the convention that *a data value falling on the class limit is included in the next higher class (class to the right of the limit)*. Another convention is to label midpoints instead of class boundaries. Determine the convention being used before creating frequency tables and histograms on a computer.   ◇

GUIDED EXERCISE 3

Histogram and relative-frequency histogram

One irate customer called Dollar Day Mail Order Company 40 times during the last two weeks to see why his order had not arrived. Each time he called, he recorded the length of time he was put "on hold" before being allowed to talk to a customer representative.

TABLE 2-7 Length of Time on Hold, in Minutes

1	5	5	6	7	4	8	7	6	5
5	6	7	6	6	5	8	9	9	10
7	8	11	2	4	6	5	12	13	6
3	7	8	8	9	9	10	9	8	9

(a) What are the largest and smallest values in Table 2-7? If we want five classes, what should the class width be?

The largest value is 13; the smallest value is 1. The class width is

$$\frac{13 - 1}{5} = 2.4 \approx 3$$

Note that we *increase* the value to 3.

(b) Complete the following frequency table.

TABLE 2-8 Time on Hold

Class Limits Lower–Upper		Tally	Frequency	Midpoint
1	– 3	3	3	2
4	– 6	15	15	5
7	– 9		7	8
10	– 12		4	11
13	– 15		1	14

TABLE 2-9 Completion of Table 2-8

Class Limits Lower–Upper	Tally	Frequency	Midpoint
1–3	III	3	2
4–6	︳HI HI HI	15	5
7–9	HI HI HI II	17	8
10–12	IIII	4	11
13–15	I	1	14

(c) Recall that the class boundary is halfway between the upper limit of one class and the lower limit of the next. Use this fact to find the class boundaries and to complete the partial histogram in Figure 2-10.

TABLE 2-10 Class Boundaries

Class Limits	Class Boundaries
1–3	0.5–3.5
4–6	3.5–6.5
7–9	6.5–9.5
10–12	9.5–12.5
13–15	12.5–15.5

TABLE 2-11 Completion of Table 2-10

Class Limits	Class Boundaries
1–3	0.5–3.5
4–6	3.5–6.5
7–9	6.5–9.5
10–12	9.5–12.5
13–15	12.5–15.5

Continued

FIGURE 2-10

➡ **FIGURE 2-11** Completion of Figure 2-10

(d) Compute the relative class frequency *f/n* for each class and complete the partial relative-frequency histogram in Figure 2-12.

TABLE 2-12 Relative Class Frequency

Class	f/n
1–3	3/40 = 0.075
4–6	15/40 = 0.375
7–9	17/40= 0.43
10–12	4/40 = 0.10
13–15	1/40= .003

➡ **TABLE 2-13** Completion of Table 2-12

Class	f/n
1–3	0.075
4–6	0.375
7–9	0.425
10–12	0.100
13–15	0.025

FIGURE 2-12

➡ **FIGURE 2-13** Completion of Figure 2-12

We will see relative-frequency distributions again when we study probability in Chapter 5. There we will see that if a random sample is large enough, then we can estimate the probability of an event by the relative frequency of the event. The relative-frequency distribution then can be interpreted as a *probability distribution*. Such distributions will form the basis of our work in inferential statistics.

Distribution Shapes

Histograms are valuable and useful tools. If the raw data came from a random sample of population values, the histogram constructed from your sample values should have a distribution shape that is reasonably similar to that of the population.

Several terms are commonly used to describe histograms and their associated population distribution.

(a) *Symmetrical:* This term refers to a histogram in which both sides are (more or less) the same when the graph is folded vertically down the middle. Figure 2-14(a) shows a typical histogram with a symmetrical shape.

(b) *Uniform* or *rectangular:* These terms refer to a histogram in which every class has equal frequency. From one point of view, a uniform distribution is symmetrical with the added property that the bars are of the same height. Figure 2-14(b) illustrates a typical histogram with a uniform shape.

(c) *Skewed left* or *skewed right:* These terms refer to a histogram in which one tail is stretched out longer than the other. The direction of skewness is on the side of the *longer* tail. So if the longer tail is on the left, we say the histogram is skewed to the left. Figure 2-14(c) shows a typical histogram skewed to the left and another skewed to the right.

(d) *Bimodal:* This term refers to a histogram in which the two classes with largest frequencies are separated by at least one class. The top two frequencies of these classes may have slightly different values. This type of situation sometimes indicates we are sampling from two different populations. Figure 2-14(d) illustrates a typical histogram with a bimodal shape.

FIGURE 2-14

Types of Histograms

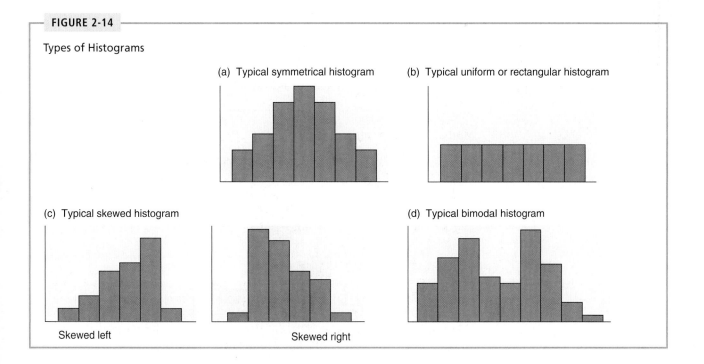

(a) Typical symmetrical histogram   (b) Typical uniform or rectangular histogram

(c) Typical skewed histogram   (d) Typical bimodal histogram

Skewed left   Skewed right

TECH NOTE The TI-83Plus, Excel, and Minitab all create histograms. However, each technology automatically selects the number of classes to use. In Using Technology at the end of this chapter, you will see instructions for specifying the number of classes yourself and for generating histograms such as we create "by hand."

The displays show the default histograms for the commuting data of Table 2-4. For each technology, enter the data and then follow the designated menu/key choices.

TI-83Plus Stat Plot ➤ On;Histogram Type ➤ Zoom ➤ 9:ZoomStat. Use **Trace** to see the number in a class.

Excel Tools ➤ Data Analysis ➤ Histogram. Select data range for input and check Chart. In Excel, classes are called bins.

Minitab Graph ➤ Histogram

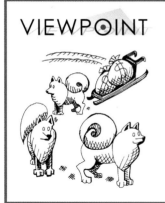

VIEWPOINT   *Mush, You Huskies!*

In 1925, the village of Nome, Alaska, had a terrible diphtheria epidemic. Serum was available in Anchorage but had to be brought to Nome by dogsled over the 1161-mile Iditarod Trail. Since 1973, the Iditarod Dog Sled Race from Anchorage to Nome has been an annual sporting event with a current winner's purse of more than $350,000. Winning times range from more than 20 days to a little over 9 days.

To collect data on winning times, visit the Brase/Brase *Understanding Basic Statistics* site at http://math.college.hmco.com/students and find the link to the Iditarod. Make a frequency distribution for these times.

SECTION 2.2   PROBLEMS

For Problems 1–6, use the specified number of classes to do the following:
(a) Find the class width.
(b) Make a frequency table showing class limits, class boundaries, midpoints, frequencies, and relative frequencies.
(c) Draw a histogram.
(d) Draw a relative-frequency histogram.

1. *Sports: Dog Sled Racing* How long does it take to finish the 1161-mile Iditarod Dog Sled Race from Anchorage to Nome, Alaska (*see* Viewpoint)? Finish times (to the nearest hour) for 57 dog-sled teams are shown below.

261	271	236	244	279	296	284	299	288	288	247	256
338	360	341	333	261	266	287	296	313	311	307	307
299	303	277	283	304	305	288	290	288	289	297	299
332	330	309	328	307	328	285	291	295	298	306	315
310	318	318	320	333	321	323	324	327			

    Use five classes.

2. *Sports: Skiing* Recreational skiing is big business in some of the western and New England states. Many recreational skiers are beyond the beginning level and want intermediate or "more difficult" terrain (but not the most difficult). *Snow Country* described the 35 top-rated ski areas and gave the percentage of the skiing terrain that was at the more difficult level. The percentages follow:

36	54	51	49	30	43	40	46	35	40	52	28
57	51	40	40	45	40	60	20	25	50	40	65
58	43	59	49	55	30	33	60	30	46	65	

    Use five classes.

3. *Sociology: Denver Neighborhoods* A random sample of 50 Denver neighborhoods gave the following information regarding children as a percent of total population (*Source:* The Piton Foundation). For more information, visit the Brase/Brase *Understanding Basic Statistics* site at http://math.college.hmco.com/students and find the link to the Piton Foundation.

30 39 32 35 31 24 19 28 5   10 8   30
12 14 18 21 19 33 38 37 22 20 27 29
39 33 27 42 16 39 20 19 20 30 24 36
24 12 37 15 40 32 29 7   35 14 33 27
53 21

Use seven classes.

4. *Business: Fast-Food* Franchise and Business Opportunities *Annual Report* contains information about franchise opportunities in the United States and Canada. A franchise fee is one of the expenses associated with owning a franchise. Other expenses include startup, advertising, royalties, and so on. A large category of franchises is the fast-food business, which includes franchises such as baked goods, donuts, hamburgers, chicken, and hot dogs. Franchise fees (in thousands of dollars) for the fast-food category are as follows:

21	25	25	18	44	20	25	15	19	24	10
28	30	25	25	25	25	10	25	25	20	20
20	15	10	20	25	25	13	30	15	28	15
15	35	24	40	15	20	35	5	50	30	15
25	40	15	25	15	75	33	23	30	10	15
8	25	10	8	20	25	20	30			

Use five classes.

5. *Nursing: Workload* Nurses on the eighth floor of Community Hospital believe they need extra staffing at night. To estimate the night workload, a random sample of 35 nights was used. For each night, the total number of room calls to the nurses' station on the eighth floor was recorded as follows:

68	60	69	70	83	58	90	86	71	71	92
95	70	74	46	18	84	82	75	63	101	77
102	80	86	85	73	86	62	100	90	37	88
70	87									

Use five classes.

6. *Advertising: Readability* "Readability Levels of Magazine Ads," by F. K. Shuptrine and D. D. McVicker, is an article in the *Journal of Advertising Research*. (For more information, visit the Brase/Brase *Understanding Basic Statistics* site at http://math.college.hmco.com/students and find the link to DASL, the Carnegie Mellon University Data and Story Library. Look in Data Subjects under Consumer and then Magazine Ads Readability file.) The following is a list of the number of three-syllable (or longer) words in advertising copy of randomly selected magazine advertisements:

34	21	37	31	10	24	39	10	17	18	32
17	3	10	6	5	6	6	13	22	25	3
5	2	9	3	0	4	29	26	5	5	24
15	3	8	16	9	10	3	12	10	10	10
11	12	13	1	9	43	13	14	32	24	15

Use eight classes.

[handwritten in left margin: width (Max−min)/#classes +1]

7. *Business: Profits and Sales* *Fortune* published statistics on its Fortune 500 companies. Forty-eight of the companies were categorized as food companies (including companies such as RJR Nabisco, Pillsbury, General Mills, Quaker Oats, Tyson Foods, Hershey Foods). For 39 of these companies, profits were given as a percentage of sales (with a negative number indicating a loss). The data follow:

2	8	3	2	7	4	5	1	6	5	7	6	5
6	11	6	2	0	8	3	1	4	2	3	0	2
4	2	−1	2	2	10	−3	6	4	1	0	0	1

Also listed in the Fortune 500 companies were 45 electronics companies (including companies such as General Electric, Westinghouse Electronic, Texas Instruments). For 44 of these companies, profit as a percentage of sales was also given. The data follow:

7	7	5	6	−6	4	8	6	1	2	9	5	10	16	12
3	5	1	8	1	7	2	5	5	6	5	2	2	11	−3
2	3	5	6	4	−2	2	1	12	5	3	11	6	12	

(a) Make a frequency table and histogram for the food companies, and then make another frequency table and histogram for the electronics companies. Use five classes for each.

(b) By looking at the two histograms, can we determine which category (food or electronics) has the greatest profits as a percentage of sales? Discuss some of the problems involved in comparing these two distributions.

8. *Football: Weights of Professional Players* How do football teams stack up weight-wise? Let's look at two teams in the National Football League. *Lindy's Pro Football* gave the preseason player statistics for all the NFL teams. For the 70 players with the Miami Dolphins, the weights (in pounds) came in as follows:

242	220	224	200	220	222	222	185	223	192
225	196	208	193	202	232	200	225	226	220
198	185	193	190	190	210	192	240	238	227
235	228	237	264	228	235	221	275	275	289
280	275	280	275	295	290	282	295	290	282
295	277	248	182	175	184	234	190	180	238
240	260	270	275	255	280	265	273	280	249

For the 72 players with the San Diego Chargers, the weights were as follows:

213	196	185	119	208	205	203	222	220	214
198	195	180	192	170	200	185	202	198	184
207	223	212	188	206	240	255	242	242	236.
246	230	230	230	275	278	280	310	260	280
293	310	295	275	270	275	230	265	305	248
248	165	210	195	188	277	184	300	259	291
237	245	267	271	209	170	285	279	233	195
200	205								

(a) Make frequency tables and histograms for each of the two teams, using six classes.

FIGURE 2-15

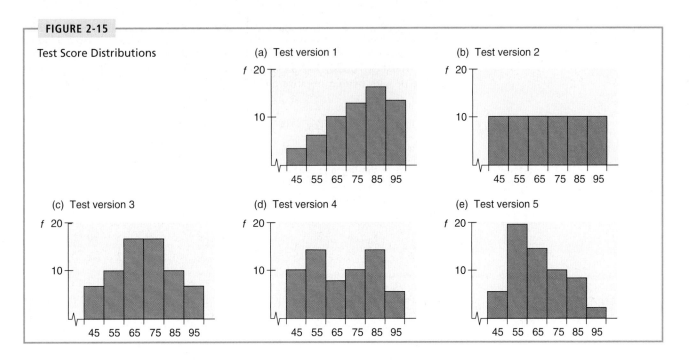

Test Score Distributions

(a) Test version 1

(b) Test version 2

(c) Test version 3

(d) Test version 4

(e) Test version 5

(b) From the histograms, can you tell if either team seems to have the heavier distribution of players' weights? Discuss some of the problems involved in comparing these distributions.

9. *Education: Testing* Professor Silva teaches anatomy and physiology. He has developed five different versions of a test on the same material. On giving each version to a different sample of 60 students, he discovered that the test score distributions looked like those shown in Figure 2-15.
   (a) Categorize the distribution shapes as uniform, symmetric, bimodal, skewed left, or skewed right.
   (b) Comment on some advantages or problems with each test version. As a student, which version might you prefer? Which version would you like the least?

10. *Consumer: Warranty Cards* Many products come with owner registration or warranty cards. Usually, the consumer is asked a few questions about his or her family and household income. Random samples of warranty or registration cards for the indicated product revealed the household income distribution shown in Figure 2-16.
    (a) Categorize the distribution shapes as uniform, symmetric, bimodal, skewed left, or skewed right.
    (b) If you were in charge of advertising, how would you use income-distribution information of present customers to target ads for the indicated product?
    (c) How valid do you think income information is on warranty cards?

Histogram for data with decimals

11. *Agriculture: Wheat Harvest* The following data represent tonnes of wheat harvested each year (1894–1925) from Plot 19 at the Rothamsted Agricultural Experiment Stations, England.

2.71	1.62	2.60	1.64	2.20	2.02	1.67	1.99	2.34	1.26	1.31
1.80	2.82	2.15	2.07	1.62	1.47	2.19	0.59	1.48	0.77	2.04
1.32	0.89	1.35	0.95	0.94	1.39	1.19	1.18	0.46	0.70	

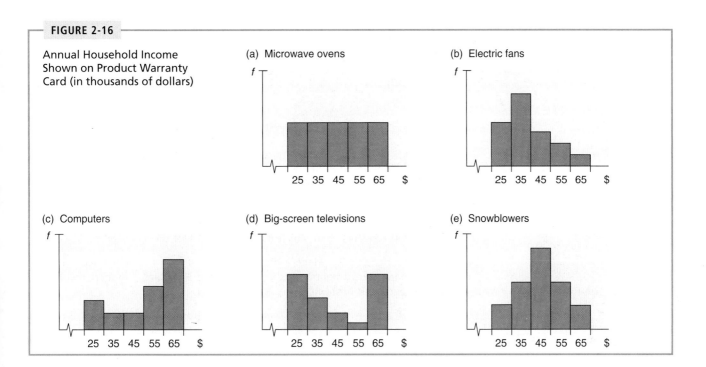

FIGURE 2-16

Annual Household Income Shown on Product Warranty Card (in thousands of dollars)

(a) Microwave ovens

(b) Electric fans

(c) Computers

(d) Big-screen televisions

(e) Snowblowers

(a) Multiply each data value by 100 to "clear" the decimals.
(b) Use the standard procedures of this section to make a frequency table and histogram with your whole-number data. Use six classes.
(c) Divide class limits, class boundaries, and class midpoints by 100 to get back to your original data values.

Histogram for data with decimals

12. *Baseball: Batting Averages* The following data represent baseball batting averages for a random sample of National League players near the end of the baseball season. The data are from the baseball statistics section of *The Denver Post*.

0.194	0.258	0.190	0.291	0.158	0.295	0.261	0.250	0.181
0.125	0.107	0.260	0.309	0.309	0.276	0.287	0.317	0.252
0.215	0.250	0.246	0.260	0.265	0.182	0.113	0.200	

(a) Multiply each data value by 1000 to "clear" the decimals.
(b) Use the standard procedures of this section to make a frequency table and histogram with your whole-number data. Use five classes.
(c) Divide class limits, class boundaries, and class midpoints by 1000 to get back to your original data.

Dotplot

13. *Dotplots: Minitab* Another display technique that is somewhat similar to a histogram is a *dotplot*. In a dotplot, the data values are displayed along the horizontal axis. A dot is then plotted over each data value in the data set. The display on the next page shows a dotplot generated by Minitab (➤ **Graph** ➤ **Dotplot**) for the number of licensed drivers per 1000 residents by state, including the District of Columbia (*Source:* U.S. Department of Transportation).

Dotplot for Licensed Drivers per 1000 Residents

Licensed drivers

(a) From the dotplot, how many states have 600 or fewer licensed drivers per 1000 residents?

(b) About what percentage of the states (out of 51) seem to have close to 800 licensed drivers per 1000 residents?

(c) Consider the intervals 550 to 650, 650 to 750, and 750 to 850 licensed drivers per 1000 residents. Into which interval do most of the states fall?

Dotplot

14. *Sports: Dog Sled Racing* Make a dotplot for the data in Problem 1 regarding the finish time (number of hours) for the Iditarod Dog Sled Race. Compare the dotplot to the histogram of Problem 1.

Dotplot

15. *Sociology: Denver Neighborhoods* Make a dotplot for the data in Problem 3 regarding the percentage of children in the population of Denver neighborhoods. Compare the dotplot to the histogram of Problem 3.

2.3
Stem-and-Leaf Displays

Exploratory Data Analysis

FOCUS POINTS

✓ Construct a stem-and-leaf display from raw data.

✓ Use a stem-and-leaf display to visualize data distribution.

✓ Compare a stem-and-leaf display to a histogram.

EDA

Together with histograms and other graphics techniques, the stem-and-leaf display is one of many useful ways of studying data in a field called *exploratory data analysis* (often abbreviated as *EDA*). John W. Tukey wrote one of the definitive books on the subject, *Exploratory Data Analysis* (Addison-Wesley). Another very useful reference for EDA techniques is the book *Applications, Basics, and Computing of Exploratory Data Analysis*, by Paul F. Velleman and David C. Hoaglin (Duxbury Press). Exploratory data analysis techniques are particularly useful for detecting patterns and extreme data values. They are designed to help us explore a data set, to ask questions we had not thought of before, or to pursue leads in many directions.

EDA techniques are similar to those of an explorer. An explorer has a general idea of destination but is always alert to the unexpected. An explorer needs to assess situations quickly and often simplify and clarify them. An explorer makes pictures—that is, maps showing the relationships of landscape features. The aspects of rapid implementation, visual displays such as graphs and charts, data simplification, and robustness (that is, analysis that is not influenced much by extreme data values) are key ingredients of EDA techniques. In addition, these techniques are good for exploration because they require very few prior assumptions about the data.

EDA methods are especially useful when our data have been gathered for general interest and observation of subjects. For example, we may have data regarding the age of applicants to graduate programs. We don't have a specific question in mind. We want to see what the data reveal. Are the ages fairly uniform or spread out? Are there exceptionally young or old applicants? If there are, we might look at other characteristics of these applicants, such as field of study. EDA methods help us quickly absorb some aspects of the data and then may lead us to ask specific questions for which we might apply methods of traditional statistics.

In contrast, when we design an experiment to produce data to answer a specific question, we focus on particular aspects of the data that are useful to us. If we want to determine the average highway gas mileage of a specific sports car, we use that model car in well-designed tests. We don't need to worry about unexpected road conditions, poorly trained drivers, different fuel grades, sudden stops and starts, etc. Our experiment is designed to control outside factors. Consequently, we do not need to "explore" our data as much. We can often make valid assumptions about the data. Methods of traditional statistics will be very useful to analyze such data and answer our specific questions.

Stem-and-Leaf Display

Stem-and-leaf display

In this text, we will introduce two EDA techniques: stem-and-leaf displays and, in Section 3.3, box-and-whisker plots. Let's first look at a stem-and-leaf display. We know that frequency distributions and histograms provide a useful organization and summary of data. However, in a histogram, we lose most of the specific data values. A *stem-and-leaf display* is a device that organizes and groups data but allows us to recover the original data if desired. In the next example, we will make a stem-and-leaf display.

EXAMPLE 3

Stem-and-leaf display

Many airline passengers seem weighted down with their carry-on luggage. Just how much weight are they carrying? The carry-on luggage weights for a random sample of 40 passengers returning from a vacation to Hawaii were recorded in pounds (see Table 2-14).

Stem
Leaf

To make a stem-and-leaf display, we break the digits of each data value into *two* parts. The left group of digits is called a *stem*, and the remaining group of digits on the right is called a *leaf*. We are free to choose the number of digits to be included in the stem.

The weights in our example consist of two-digit numbers. For a two-digit number, the stem selection is obviously the left digit. In our case, the tens digits will form the stems, and the units digits will form the leaves. For example, for the weight 12, the stem is 1, and the leaf is 2. For the weight 18, the stem is again 1, but the leaf is 8. In the stem-and-leaf display, we list each possible stem once on the left and all its leaves in the same row on the right as in Figure 2-17(a) on the next page. Finally, we order the leaves as shown in Figure 2-17(b).

Figure 2-17 shows a stem-and-leaf display for the weights of carry-on luggage. We see that two bags weighed 27 lb, one weighed 3 lb, one weighed 51 lb, and so

TABLE 2-14 Weights of Carry-on Luggage in Pounds

30	27	12	42	35	47	38	36	27	35
22	17	29	3	21	0	38	32	41	33
26	45	18	43	18	32	31	32	19	21
33	31	28	29	51	12	32	18	21	26

FIGURE 2-17

Stem-and-Leaf Displays of Airline Carry-on Luggage Weights

(a) Leaves not ordered

```
    3 | 2   represents 32 lb
 Stem | Leaves
    0 | 3  0
    1 | 2  7  8  8  9  2  8
    2 | 7  7  2  9  1  6  1  8  9  1  6
    3 | 0  5  8  6  5  8  2  3  2  1  2  3  1  2
    4 | 2  7  1  5  3
    5 | 1
```

(b) Final display with leaves ordered

```
    3 | 2   represents 32 lb
 Stem | Leaves
    0 | 0  3
    1 | 2  2  7  8  8  8  9
    2 | 1  1  1  2  6  6  7  7  8  9  9
    3 | 0  1  1  2  2  2  2  3  3  5  5  6  8  8
    4 | 1  2  3  5  7
    5 | 1
```

on. We see that most of the weights were in the 30-lb range, only two were less than 10 lb, and six were over 40 lb. Note that the length of line containing the leaves gives the visual impression that a sideways histogram would present.

As a final step, we need to indicate the scale. This is usually done by indicating the value represented by the stem and one leaf.   ◇

There are no firm rules for selecting the group of digits for the stem. But whichever group you select, you must list all the possible stems from smallest to largest in the data collection.

To make a stem-and-leaf display

1. Supply 0's as appropriate so that all data values have the same number of digits. Divide the digits of each data value into two parts. The leftmost part is called the *stem* while the rightmost part is called the *leaf*.

2. Align all the stems in a vertical column from smallest to largest. Draw a vertical line to the right of all the stems.

3. Place all the leaves with the same stem on the same row as the stem, and arrange the leaves in increasing order.

4. Use a label to indicate the magnitude of the numbers in the display. We include the decimal position in the label rather than with the stems or leaves.

In the example that follows, we show a stem-and-leaf display for data with three digits.

EXAMPLE 4

Selecting the stem

What does it take to win at sports? If you're talking about basketball, one sports writer gave the answer. He listed the winning scores of the conference championship games over the last 35 years. The scores for those games follow below.

> **FIGURE 2-18**
>
> Winning Scores of
> Conference Basketball
> Championship Games
>
> ```
> 08 | 3  represents 083 or 83 points
>
> 08 | 3
> 09 | 2 7 8 9
> 10 | 1 2 3 4 5 5 6 6 9
> 11 | 0 1 2 2 2 2 6 7 7 8 9
> 12 | 0 4 5 5 6 8
> 13 | 1 2 5
> 14 | 3
> ```

132	118	124	109	104	101	125	83	99
131	98	125	97	106	112	92	120	103
111	117	135	143	112	112	116	106	117
119	110	105	128	112	126	105	102	

To make a stem-and-leaf display, we'll use the first *two* digits as the stems (see Figure 2-18). Notice that the distribution of scores is fairly symmetrical.   ◊

GUIDED EXERCISE 4

Stem-and-leaf display

Tel-a-Message is experimenting with computer-delivered telephone advertisements. Of primary concern is how much of the 4-minute advertisement is heard. A study was done to see how long the advertisement ran before the listeners hung up. A random sample of 30 calls gave the information in Table 2-15.

TABLE 2-15   Time Spent Listening to Advertisement (in minutes)

1.3	0.7	2.1	0.5	0.2	0.9	1.1	3.2	4.0	3.8
1.4	3.1	2.5	0.6	0.5	2.1	4.0	4.0	0.3	1.2
1.0	1.5	0.4	4.0	2.3	2.7	4.0	0.7	0.5	4.0

(a) We'll make a stem-and-leaf display using the first digit as the stem and the second as a leaf. What is the leaf unit?

⟹ The trailing digit is in the tenths position so

$$1 \text{ leaf unit} = 0.1 \text{ min}$$

(b) List all the stem values.

⟹ Stem: 0
1
2
3
4

Continued

35	23	90	72	85	42	59	50	49
48	38	60	46	87	50	89	49	67

Make a stem-and-leaf display of these data. Be sure to indicate the scale. How are the percentages distributed? Is the distribution skewed? Are there any gaps?

State	No. of Hospitals	Average Length of Stay	State	No. of Hospitals	Average Length of Stay	State	No. of Hospitals	Average Length of Stay
Alabama	119	7.0	Kentucky	107	6.9	N. Dakota	47	11.1
Alaska	16	5.7	Louisiana	136	6.7	Ohio	193	6.6
Arizona	61	5.5	Maine	38	7.2	Oklahoma	113	6.7
Arkansas	88	7.0	Maryland	51	6.8	Oregon	66	5.3
California	440	6.0	Massachusetts	101	7.0	Pennsylvania	236	7.5
Colorado	71	6.8	Michigan	175	7.3	Rhode Island	12	6.9
Connecticut	35	7.4	Minnesota	148	8.7	S. Carolina	68	7.1
Delaware	8	6.8	Mississippi	102	7.2	S. Dakota	52	10.3
Dist. of Columbia	11	7.5	Missouri	133	7.4	Tennessee	122	6.8
Florida	227	7.0	Montana	53	10.0	Texas	421	6.2
Georgia	162	7.2	Nebraska	90	9.6	Utah	42	5.2
Hawaii	19	9.4	Nevada	21	6.4	Vermont	15	7.6
Idaho	41	7.1	New Hampshire	27	7.0	Virginia	98	7.0
Illinois	209	7.3	New Jersey	96	7.6	Washington	92	5.6
Indiana	113	6.6	New Mexico	37	5.5	W. Virginia	59	7.1
Iowa	123	8.4	New York	231	9.9	Wisconsin	129	7.3
Kansas	133	7.8	N. Carolina	117	7.3	Wyoming	27	8.5

3. *Health Care: Hospitals* The American Medical Association Center for Health Policy Research included data, by state, on the number of community hospitals and the average patient stay (in days) in its publication *State Health Care Data: Utilization, Spending, and Characteristics*. The data (by state) are shown in the table above. Make a stem-and-leaf display of the data for the average length of stay in days. Comment about the general shape of the distribution.

4. *Health Care: Hospitals* Using the number of hospitals per state listed in the table, make a stem-and-leaf display for the number of community hospitals per state. Which states have an unusually high number of hospitals?

Stem-and-leaf diagram with two lines per stem

5. *Boston Marathon: Winning Times* The Boston Marathon is the oldest and best-known U.S. marathon. It covers a route from Hopkinton, Massachusetts, to downtown Boston. The distance is approximately 26 miles. Visit the Brase/Brase *Understanding Basic Statistics* site at http://math.college.hmco.com/students and find the link to Boston Marathon. Search the site for the marathon to find a wealth of information about the history of the race. In particular, it gives the winning times for the Boston Marathon. They are all over 2 hours. The following data are the minutes over 2 hours for the winning male runner:

1961–1980

23	23	18	19	16	17	15	22	13	10
18	15	16	13	9	20	14	10	9	12

1981–2000

9	8	9	10	14	7	11	8	9	8
11	8	9	7	9	9	10	7	9	9

(a) Make a stem-and-leaf diagram for the minutes over 2 hours of the winning times for the years 1961 to 1980. Use two lines per stem. For instance, to use two lines for the stem value 0, we designate the first stem as 0* and put leaf digits 0 through 4 on this line. The second line for stem 0 is designated with stem 0˙, and we put leaf digits 5 through 9 on this line.

(b) Make a stem-and-leaf diagram for the minutes over 2 hours of the winning times for the years 1981 to 2000. Use two lines per stem.

(c) Compare the two distributions. How many times under 15 minutes are in each distribution?

Stem-and-leaf diagram with two lines per stem

6. *Golf: Scores* The U.S. Open Golf Tournament was played at Congressional Country Club, Bethesda, Maryland, with prizes ranging from $465,000 for first place to $5000. Par for the course is 70. The tournament consists of four rounds played on different days. The scores for each round of the 32 players who placed in the money (more than $17,000) were given on a Web site. For more information, visit the Brase/Brase *Understanding Basic Statistics* site at http://math.college.hmco.com/students and find the link to golf. The scores for the first round were as follows:

71	65	67	73	74	73	71	71	74	73	71
70	75	71	72	71	75	75	71	71	74	75
66	75	75	75	71	72	72	73	71	67	

The scores for the fourth round for these players were as follows:

69	69	73	74	72	72	70	71	71	70	72
73	73	72	71	71	71	69	70	71	72	73
74	72	71	68	69	70	69	71	73	74	

(a) Make a stem-and-leaf diagram for the first-round scores. Use two lines per stem. (*See* Problem 5.)

(b) Make a stem-and-leaf diagram for the fourth-round scores. Use two lines per stem.

(c) Compare the two distributions. How do the highest scores compare? How do the lowest scores compare?

Are cigarettes bad for people? Cigarette smoking involves tar, carbon monoxide, and nicotine. The first two are definitely not good for a person's health, and the last ingredient can cause addiction. Problems 7, 8, and 9 refer to Table 2-16 on the following page, which was taken from the Web site maintained by the *Journal of Statistics Education*. For more information, visit the Brase/Brase *Understanding Basic Statistics* site at http://math.college. hmco.com/students and find the link to the *Journal of Statistics Education*. Follow the links to the cigarette data.

7. *Health: Cigarette Smoke* Use the data in Table 2-16 to make a stem-and-leaf display for milligrams of tar per cigarette smoked.

8. *Health: Cigarette Smoke* Use the data in Table 2-16 to make a stem-and-leaf display for milligrams of carbon monoxide per cigarette smoked.

9. *Health: Cigarette Smoke* Use the data in Table 2-16 to make a stem-and-leaf display for milligrams of nicotine per cigarette smoked. Truncate the measurement to the tenth position (i.e., simply ignore the digit in the hundredth position). Use two lines per stem (*see* Problem 5, part a).

TABLE 2-16   Milligrams of Tar, Nicotine, and Carbon Monoxide (CO) Content for One Cigarette

Brand	Tar	Nicotine	CO	Brand	Tar	Nicotine	CO
Alpine	14.1	0.86	13.6	MultiFilter	11.4	0.78	10.2
Benson & Hedges	16.0	1.06	16.6	Newport Lights	9.0	0.74	9.5
Bull Durham	29.8	2.03	23.5	Now	1.0	0.13	1.5
Camel Lights	8.0	0.67	10.2	Old Gold	17.0	1.26	18.5
Carlton	4.1	0.40	5.4	Pall Mall Lights	12.8	1.08	12.6
Chesterfield	15.0	1.04	15.0	Raleigh	15.8	0.96	17.5
Golden Lights	8.8	0.76	9.0	Salem Ultra	4.5	0.42	4.9
Kent	12.4	0.95	12.3	Tareyton	14.5	1.01	15.9
Kool	16.6	1.12	16.3	True	7.3	0.61	8.5
L&M	14.9	1.02	15.4	Viceroy Rich Light	8.6	0.69	10.6
Lark Lights	13.7	1.01	13.0	Virginia Slim	15.2	1.02	13.9
Marlboro	15.1	0.90	14.4	Winston Lights	12.0	0.82	14.9
Merit	7.8	0.57	10.0				

Source: Journal of Statistics Education Web Site http://www.stat.nesu.edu/info/jse/homepage.html. Reprinted with permission.

Back-to-back stem plot

10. *Archaeology: Depth of Artifacts* In archaeology, the depth (below surface grade) at which artifacts are found is very important. Greater depths sometimes indicate older artifacts, perhaps from a different archaeological period. Figure 2-20 is a *back-to-back stem plot* showing the depths of artifact locations at two different archaeological sites. These sites are from similar geographic locations. Notice that the stems are in the center of the diagram. The leaves for Site I artifact depths are shown to the left of the stem, while the leaves for Site II are to the right of the stem (*see Mimbres Mogollon Archaeology* by A. I. Woosley and A. J. McIntyre, University of New Mexico Press).

(a) What are the least and greatest depths of artifact finds at Site I? at Site II?

(b) Describe the data distribution of depths of artifact finds at Site I and at Site II.

FIGURE 2-20

Depth (in cm) of Artifact Location

```
                  5 | 2 | 0    = 25 cm at Site I and 20 cm at Site II
        Site I                   Site II

                    5 | 2 | 0 5 5
                  5 0 | 3 | 0 0 0 0 5 5
              5 5 5 5 | 4 | 0 0 0 5
                  5 0 | 5 | 0 0 5 5
          5 5 5 5 5 0 | 6 | 0 0 0 5 5 5
      5 5 5 5 5 5 0 0 | 7 |
          5 5 0 0 0 | 8 |
          5 5 0 0 | 9 |
              5 5 |10 |
                0 |11 | 0 0 5 5 5 5
                  |12 | 0 0 0 0 5
```

FIGURE 2-21

Back-to-Back Stem Plot
Showing Annual Salaries
(in thousands of dollars)
for Full Professors in
California and in New
York

9 | 4 | 5   represents $49,000 for California and $45,000 for New York

California		New York
9	4	5 6 8
9 6 3 0	5	0 1 1 5 6
9 3 0 0 0	6	0 1 1 3 4 5 6 9 9 9 9
7 7 7 6 6 5 5 5 5 0	7	2 2 4 5 5 6
9 7 5 5 3	8	2 5 6
5 4 4 3	9	3 5 7
4 2	10	0 3
3 2	11	
6 2	12	0 0

(c) At Site II, there is a gap in the depths at which artifacts were found. Does the Site II data distribution suggest that there might have been a period of no occupation?

11. *Education: Professors' Salaries* Figure 2-21 is a *back-to-back stem plot* showing the average annual salaries (in thousands of dollars) of full professors in California universities compared to New York universities (*Source: The Chronicle of Higher Education*). For more information, visit the Brase/Brase *Understanding Basic Statistics* site at http://math.college.hmco.com/students and find the link to the *Chronicle of Higher Education*.
   (a) What are the low and high average salaries for professors in California? in New York?
   (b) Which state has a greater number of average salaries in the $60,000 range? in the $70,000 range?
   (c) In general, which state, California or New York, would you say had higher average salaries for professors in colleges and universities?

SUMMARY

Organizing and presenting data are the main purposes of that part of statistics called descriptive statistics. In this chapter, we have studied bar graphs, Pareto charts, circle graphs, time plots, histograms, relative-frequency histograms, and stem-and-leaf displays. From the viewpoint of future applications, histograms are the most important because the area under a bar can represent the likelihood of data values falling into that class. Histograms and stem-and-leaf displays both reveal distribution properties such as uniformity, symmetry, or skewness.

IMPORTANT WORDS & SYMBOLS

Section 2.1
Bar graph
Pareto chart

Circle graph or pie chart
Time plot
Time series

Section 2.2
Histogram
Frequency distribution
Frequency
Class width
Class, lower limit, upper limit
Class frequency
Class midpoint
Class mark
Frequency table
Class boundaries
Relative-frequency table
Relative-frequency histogram
Relative frequencies

Symmetric distribution
Uniform distribution
Skewed left
Skewed right
Bimodal distribution
Dotplot

Section 2.3
EDA
Stem-and-leaf display
Stem
Leaf
Back-to-back stem plot

VIEWPOINT

"This land is your land, This land is my land"＊
—*Woody Guthrie*

But who actually owns the forest? On many maps, forest land (including national forests) is colored green. Such maps give the impression that vast areas of the western United States are public land. This is not the case! *USA Today* gave the following information about ownership of U.S. timber lands: state/local, 17%; federal, 10%; forest industry, 14%; and private nonindustry, 59%. Organize these data for better visual presentation, using a Pareto chart and a circle graph.

*Words and Music by Woody Guthrie TRO © Copyright 1956(Renewed) 1958(Renewed) Ludlow Music, Inc. New York, New York. Used by permission.

CHAPTER REVIEW PROBLEMS

1. *Cars: Fuel Economy* The Focus Problem at the beginning of Chapter 2 shows you two graphics [Figure 2-1(a) and (b)] displaying the same information regarding fuel economy standards for autos. Review the criteria given in the Focus Problem for producing good graphs (from Edward R. Tufte, *The Visual Display of Quantitative Information*). Look at the graph in Figure 2-1(a). Is it essentially a bar graph? Explain. What are some of the flaws of Figure 2-1(a) as a bar graph? Next, examine Figure 2-1(b), which shows the same information. Is it essentially a time plot? Explain. In what ways does the second graph seem to display the information in a clearer manner?

2. *Criminal Justice: Prisoners* The time plot in Figure 2-22 gives the number of state and federal prisoners per 100,000 population (*Source: Statistical Abstract of the United States*, 120th Edition).
   (a) Estimate the number of prisoners per 100,000 people for 1980 and for 1997.

FIGURE 2-22

Number of State and Federal Prisoners per 100,000 Population

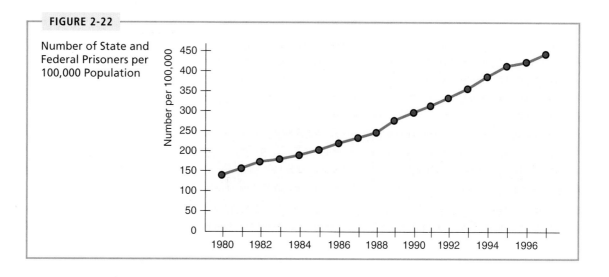

(b) During the time period shown, there was increased prosecution of drug offenses, longer sentences for common crimes, and reduced access to parole. What does the time plot say about the prison population change per 100,000 people?

(c) In 1997, the U.S. population was approximately 266,574,000 people. At the rate of 444 prisoners per 100,000 population, about how many prisoners were in the system? The projected U.S. population for the year 2020 is 323,724,000. If the rate of prisoners per 100,000 stays the same as in 1997, about how many prisoners do we expect to have in the system? To obtain the most recent information, visit the Brase/Brase *Understanding Basic Statistics* site at http://math.college.hmco.com/students and find the link to the Census Bureau.

3. *IRS: Tax Returns* Almost everyone files (or will sometime file) a federal income tax return. A research poll for Turbo Tax (a computer software package to aid in tax-return preparation) asked what aspect of filing a return people thought to be the most difficult. The results showed that 43% of the people said understanding the IRS jargon, 28% said knowing deductions, 10% said getting the right form, 8% said calculating the numbers, and 10% didn't know. Make a circle graph to display this information. *Note:* Percentages will not total 100% because of rounding.

4. *Law Enforcement: DUI* Driving under the influence of alcohol (DUI) is a serious offense. The following data give the ages of a random sample of 50 drivers arrested while driving under the influence of alcohol. This distribution is based on the age distribution of DUI arrests given in the *Statistical Abstract of the United States* (112th Edition).

46	16	41	26	22	33	30	22	36	34
63	21	26	18	27	24	31	38	26	55
31	47	27	43	35	22	64	40	58	20
49	37	53	25	29	32	23	49	39	40
24	56	30	51	21	45	27	34	47	35

(a) Make a stem-and-leaf display of the age distribution.
(b) Make a frequency table using seven classes.
(c) Make a histogram showing class boundaries.

5. *Medical: Glucose Blood Test* A fasting glucose blood test was given to 53 (non-pregnant) women at Boston City Hospital. The results of the blood tests (milligrams of glucose per 100 milliliters of blood) are shown below. (*Source: American Journal of Clinical Nutrition*, Vol. 19, pp. 345–351.)

66	59	69	84	85	96	85	82	80	90
75	80	73	69	85	82	91	84	77	86
75	83	87	74	80	83	77	69	78	93
93	87	70	85	76	87	87	79	91	86
82	82	84	85	90	79	83	65	87	70
89	81	74							

(a) Make a frequency table with seven classes showing class limits, class boundaries, midpoints, frequencies, and relative frequencies.
(b) Draw a histogram.
(c) Draw a relative-frequency histogram.

6. *Law: Corporation Lawsuits* Many people say the civil justice system is overburdened. Many cases center on suits involving businesses. The following data are based on a *Wall Street Journal* report. Researchers conducted a study of lawsuits involving 1908 businesses ranked in the Fortune 1000 over a 20-year period. They found the following distribution of civil justice caseloads brought before the federal courts involving the businesses:

Case Type	Number of Filings (in thousands)
Contracts	107
General torts (personal injury)	191
Asbestos liability	49
Other product liability	38
All other	21

Note: Contracts cases involve disputes over contracts between businesses.
(a) Make a Pareto chart of the caseloads. Which type of cases occur most frequently?
(b) Make a circle chart showing the percentage of cases in each type.

7. *Archaeology: Tree-Ring Data* The Sand Canyon Archaeological Project, edited by W. D. Lipe and published by Crow Canyon Archaeological Center, contains the stem-and-leaf diagram in Figure 2-23. The study uses tree rings to accurately determine the year in which a tree was cut. The figure gives the tree-ring-cutting dates for samples of timbers found in the architectural units at Sand Canyon Pueblo. The text referring to the figure says, "The three-digit numbers in the left column represent centuries and decades A.D. The numbers to the right represent individual years, with each number derived from an individual sample. Thus, **124 2 2 2** represents three samples dated to A.D. 1242." Use Figure 2-23 and the verbal description to answer the following questions.
(a) Which decade contained the most samples?
(b) How many samples had a tree-ring-cutting date between (and including) 1200 A.D. and 1239 A.D.?

FIGURE 2-23

Tree-Ring-Cutting
Dates from Archi-
tectural Units at Sand
Canyon Pueblo: *The
Sand Canyon Archaeo-
logical Project*

```
119 | 5 6
120 | 0 0 1 2 3 3 3 3 3 3 3 3 3 3 3 3 3 3 3 3 3 3 3 3 3 3 3 3 3 3
120 |
121 | 2
121 | 5 5
122 | 0 0 1 1 1 1 2 2 3 4 4 4 4 4 4 4
122 | 5 8 9
123 | 0 1 2 3 3 4
123 | 5 5 5 5 5 5 5 5 5 5 5 5 5 5 5 6 8 8 9
124 | 1 2 2 2 2 2 2 2 2 2 2 2 2 2 2 2 2 2 2 2 2 2 2 3 4 4
124 | 5 6 8 9 9 9 9 9 9 9 9 9
125 | 0 0 0 0 0 0 0 0 0 0 0 0 0 0 0 1 1 1 1 1 1 1 2 2 2
125 |
126 | 0 0 0 1 2 2 2 2 2 2 2 2 2 2 2 2 2 2 4 4 4 4 4 4 4
126 | 5 5 5 6 6 7
127 | 0 1 1 1 1 4 4
```

(c) What are the dates of the longest interval in which no tree-cutting samples occurred? What might this indicate about new construction or renovation of the pueblo structures during this period?

8. *Education: Grades* The following relative-frequency histogram is based on information from the *Statistical Abstract of the United States* (112th Edition). In Figure 2-24, x = high-school grade point average for a person who has received a bachelor's degree. We use a 4-point scale for grading, so A = 4, B+ = 3.5, B = 3, C+ = 2.5, C = 2, D+ = 1.5, and D = 1. The vertical axis gives the relative frequency for the high-school grade point average of students who graduated from college.

(a) Comment on the shape of the distribution. Is it skewed? Is it symmetrical? Is it uniform?

FIGURE 2-24

High-School Grade
Point Average of
Students Who
Graduated from
College

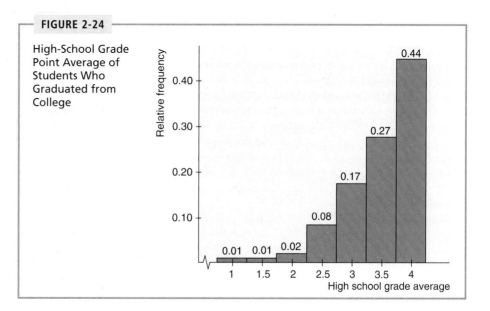

(b) The class boundaries are halfway between the class midpoints. Find the class boundaries. *Hint:* Use the fact that the classes are all the same width to find the lower boundary of the first class and the upper boundary of the last class.

(c) Convert the relative frequencies into percentages. Then find the percentage of college graduates who had high-school grade point averages of less than 3.25 (low B). What percentage had high-school grade point averages of less than 3.75 (high B)?

DATA HIGHLIGHTS: GROUP PROJECTS

Break into small groups and discuss the following topics. Organize a brief outline in which you summarize the main points of your group discussion.

1. Examine Figure 2-25, "Slobs make worst roommates." This is a double bar graph because two percentages are given for each response category: responses from men and responses from women. Comment about how the artistic rendition has slightly changed the format of a bar graph. Do the bars seem to have lengths that accurately reflect the relative percentages of the responses? In your own opinion, does the artistic rendition enhance or confuse the information? Explain. Which characteristic of "worst roommates" does the graphic seem to illustrate? Can this graph be considered a Pareto chart for men? For women? Why or why not? From the information given in the figure, do you think the survey listed only the four given annoying characteristics? Do you think a respondent could choose more than one characteristic? Explain your answer in terms of the percentages given and in terms of the explanation given in the graphic. Could this information also be displayed in one circle graph for men and another for women? Explain.

2. Examine Figure 2-26, "Global teen worries." How many countries were contained in the sample? The graph contains bars and a circle. Which bar is the longest? Which

FIGURE 2-25

USA SNAPSHOTS®
A look at statistics that shape our lives

Genders agree:
Slobs make worst roommates
What men and women say bothers them the most when living with another person:

Sloppiness — 41% / 35%
Uneven sharing of chores — 15% / 32%
Irritating personal habits — 22% / 24%
Invasions of privacy — 22% / 9%

Men / Women

Source: Advantage Business Research for Mattel *Compatibility*
By Anne R. Carey and Marcia Staimer, USA TODAY

Copyright © 1996 USA TODAY. Reprinted with permission.

FIGURE 2-26

USA SNAPSHOTS®
A look at statistics that shape our lives

Global teen worries
Leading sources of major concern for teenagers, ages 15–18, surveyed in 41 countries:

Getting a good job **70%**
Parents' health **63%**
Losing someone they love **57%**
Finishing education **54%**
Personal health **54%**
Getting university degree **52%**

Note: May name more than one

Source: BrainWaves Group's New World Teen Study
By Anne R. Carey and Genevieve Lynn, USA TODAY

Copyright © 1996 USA TODAY. Reprinted with permission.

bar represents the greatest percentage? Is this a bar graph or not? If not, what changes would need to be made to put the information into a bar graph? Could it be made into a Pareto chart? Could it be made into a circle graph? Explain.

LINKING CONCEPTS: WRITING PROJECTS

Discuss each of the following topics in class or review the topics on your own. Then write a brief but complete essay in which you summarize the main points. Please include formulas and graphs as appropriate.

1. In your own words, explain the differences among bar graphs, circle graphs, time plots, Pareto charts, histograms, relative-frequency histograms, and stem-and-leaf plots. If you have nominal data, which graphic displays might be useful? What if you have ordinal, interval, or ratio data?

2. What do we mean when we say a histogram is skewed to the left? to the right? What is a bimodal histogram? Discuss the following statement: "A bimodal histogram usually results if we draw a sample from two populations at once." Suppose you took a sample of weights of college football players and with this sample you included weights of cheerleaders. Do you think a histogram made from the combined weights would be bimodal? Explain.

3. Discuss the statement that stem-and-leaf displays are quick and easy to construct. How can we use a stem-and-leaf display to make the construction of a frequency table easier? How does a stem-and-leaf display help you spot extreme values fast?

4. Go to the library and pick up a current issue of *The Wall Street Journal, Newsweek, Time, USA Today,* or other news media. Examine each newspaper or magazine for graphs of the type discussed in this chapter. List the variables used, method of data collection, and general type of conclusion drawn from the graphs. Another source for information is the Internet. Explore several Web sites, and categorize the graphs you find as you did for the print media. For interesting Web sites, visit the Brase/Brase *Understanding Basic Statistics* site at http://math.college.hmco.com/students and find links to the Social Statistics Briefing Room, to law enforcement, and to golf.

Using TECHNOLOGY

TI-83PLUS • MINITAB • EXCEL • COMPUTERSTAT

APPLICATIONS

The following tables show the first-round winning scores of the NCAA men's and women's basketball teams.

TABLE 2-17 Men's Winning First-Round NCAA Tournament Scores

95	70	79	99	83	72	79	101
69	82	86	70	79	69	69	70
95	70	77	61	69	68	69	72
89	66	84	77	50	83	63	58

TABLE 2-18 Women's Winning First-Round NCAA Tournament Scores

80	68	51	80	83	75	77	100
96	68	89	80	67	84	76	70
98	81	79	89	98	83	72	100
101	83	66	76	77	84	71	77

1. Use the software or method of your choice to construct separate histograms for the men's and women's winning scores. Try 5, 7, and 10 classes for each. Which number of classes seems to be the best choice? Why?

2. Use the same class boundaries for histograms of men's and of women's scores. How do the scores for the two groups compare? What general shape do the histograms follow?

3. Use the software or method of your choice to make stem-and-leaf diagrams for each set of scores. If your software does not make stem-and-leaf displays, sort the data first and then make a back-to-back display by hand. Do there seem to be any extreme values in either set? How do the data sets compare?

Technology Hints: Creating Histograms

The default histograms produced by the TI-83Plus, Minitab, and Excel all determine the number of classes to use automatically. To control the number of classes the technology uses, follow the key steps as indicated. The display screens are generated for data found in Table 2-4—Commuting Distances of Dallas Workers—using five classes.

TI-83Plus

Determine the class width for the number of classes you want and the lower class boundary for the first class. Enter the data in list L1.

Press **STATPLOT** and highlight On and the histogram plot.

Press **WINDOW** and set Xmin = lowest class boundary, Xscl = class width. Use appropriate values for the other settings.

```
WINDOW
  Xmin=.5
  Xmax=51
  Xscl=10
  Ymin=-5
  Ymax=30
  Yscl=1
  Xres=1
```

Press **GRAPH. TRACE** gives boundaries and frequency.

```
P1:L1

min=20.5
max<30.5      n=9
```

Excel

Determine the upper class boundaries for the five classes. Enter the data. In a separate column, enter the upper class boundaries. Use the menu selection **Tools ➤ Data Analysis ➤ Histogram.**

Put the data range in the Input Range. Put the upper class boundaries range for the Bin Range.

To make bars touch, right click on a bar, select **Format Data Series ➤ Options tab.** Set **gap width** to 0.

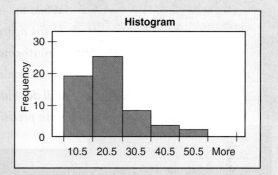

Minitab

Determine the class boundaries. Enter the data. Use the menu selection **Graph ➤ Histogram.**

In the Dialogue Box, press **Options.** Then select cutpoints, and enter the class boundaries as cutpoint positions.

Select **Frame** to adjust scales. Otherwise, press **OK.**

the data from USA SNAPSHOT, "College affordable for most," we estimate mid-points for the cost intervals. Say 51% are about $3,500; 21% are about $6,000; 6% are about $10,000; 8% are about $14,000; 7% are about $18,000; and 7% are about $25,000. Compute the weighted average of college tuition charged at all colleges.

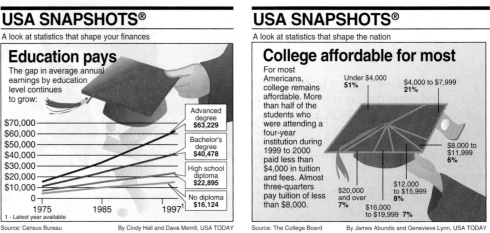

Copyright © 1999 USA TODAY. Reprinted with permission.   Copyright © 1999 USA TODAY. Reprinted with permission.

3.1   Measures of Central Tendency: Mode, Median, and Mean

FOCUS POINTS

✓ Compute mean, median, and mode from raw data.

✓ Interpret what mean, median, and mode tell us.

✓ Explain how mean, median, and mode can be affected by extreme data values.

✓ Why do we trim data? Compute a trimmed mean.

The average price of an ounce of gold is $295. The Zippy car averages 39 miles per gallon on the highway. A survey showed the average shoe size for women is size 8.

In each of the preceding statements, *one* number is used to describe the entire sample or population. Such a number is called an *average*. There are many ways to compute averages, but we will study only three of the major ones.

The easiest average to compute is the *mode*.

The **mode** of a data set is the value that occurs most frequently.

EXAMPLE 1

Mode

Count the letters in each word of this sentence and give the mode. The number of letters in the words of the sentence are

5   3   7   2   4   4   2   4   8   3   4   3   4

Scanning the data, we see that 4 is the mode because more words have 4 letters than any other number. For larger data sets, it is useful to order—or sort—the data before scanning them for the mode.   ◊

Not every data set has a mode. For example, if Professor Fair gives equal numbers of As, Bs, Cs, Ds, and Fs, then there is no modal grade. In addition, the mode is not very stable. Changing just one number in a data set can change the mode dramatically. However, the mode is a useful average when we want to know the most frequently occurring data value, such as the most frequently requested shoe size.

Median

Another average that is useful is the *median*, or central value, of an ordered distribution. When you are given the median, you know there are an equal number of data values in the ordered distribution that are above it and below it.

The **median** is the central value of an ordered distribution. To find it,

(a) Order the data from smallest to largest.

(b) For an *odd* number of data values in the distribution,

Median = Middle data value

(c) For an *even* number of data values in the distribution,

$$\text{Median} = \frac{\text{Sum of middle two values}}{2}$$

EXAMPLE 2

Median

What do barbecue-flavored potato chips cost? According to *Consumer Reports*, Volume 66, No. 5, the prices per ounce in cents of the rated chips are

19   19   27   28   18   35

(a) To find the median, we first order the data, and then note that there are an even number of entries, so the median is constructed by using the two middle values.

18   19   19   27   28   35
↑   ↑
middle values

$$\text{Median} = \frac{19 + 27}{2} = 23 \text{ cents}$$

(b) According to *Consumer Reports*, the brand with the lowest overall taste rating costs 35 cents per ounce. Eliminate that brand, and find the median price per ounce for the remaining barbecue-flavored chips. Again order the data. Note that there are an odd number of entries, so the median is simply the middle value.

18   19   19   27   28
↑
middle value

Median = Middle value = 19 cents

(c) One ounce of potato chips is considered a small serving. Is it reasonable to budget about $10.45 to serve the barbecue-flavored chips to 55 people?

Yes, since the median price of the better tasting chips is 19 cents per small serving. This budget for chips assumes that there is plenty of other food!   ◊

The median uses the *position* rather than the specific value of each data entry. If the extreme values of a data set change, the median usually does not change. This is why the median is often used as the average for house prices. If one mansion costing several million dollars sells in a community of much lower-priced homes, the median selling price for houses in the community would be affected very little, if at all.

GUIDED EXERCISE 1

Median and mode

Belleview College must make a report to the budget committee about the average credit hour load a full-time student takes. (A 12-credit-hour load is the minimum requirement for full-time status. For the same tuition, students may take up to 20 credit hours.) A random sample of 40 students yielded the following information (in credit hours):

17	12	14	17	13	16	18	20	13	12
12	17	16	15	14	12	12	13	17	14
15	12	15	16	12	18	20	19	12	15
18	14	16	17	15	19	12	13	12	15

(a) Organize the data from smallest to largest number of credit hours.

⟹ 12 12 12 12 12 12 12 12 12 12
13 13 13 13 14 14 14 14 15 (15)
(15) 15 15 15 16 16 16 16 17 17
17 17 17 18 18 18 19 19 20 20

(b) Since there are an _____ (odd, even) number of values, we add the two middle values and divide by 2 to get the median. What is the median credit hour load?

⟹ There are an even number of entries. The two middle values are circled in part (a).

$$\text{Median} = \frac{15 + 15}{2} = 15$$

(c) What is the mode of this distribution? Is it different from the median? If the budget committee is going to fund the school according to the average student credit hour load (more money for higher loads), which of these two averages do you think the college will use?

⟹ The mode is 12. It is different from the median. Since the median is higher, the school will probably use it and indicate that the average being used is the median.

Mean

An average that uses the exact value of each entry is the *mean* (sometimes called the *arithmetic mean*). To compute the mean, we add the values of all the entries and then divide by the number of entries.

$$\text{Mean} = \frac{\text{Sum of all the entries}}{\text{Number of entries}}$$

The mean is the average usually used to compute a test average.

EXAMPLE 3

Mean

To graduate, Linda needs at least a B in biology. She did not do very well on her first three tests; however, she did well on the last four. Here are her scores:

58   67   60   84   93   98   100

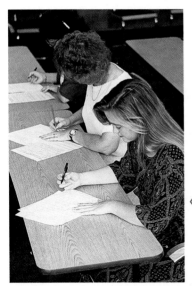

Compute the mean and determine if Linda's grade will be a B (80 to 89 average) or a C (70 to 79 average).

SOLUTION:

$$\text{Mean} = \frac{\text{Sum of scores}}{\text{Number of scores}} = \frac{58 + 67 + 60 + 84 + 93 + 98 + 100}{7}$$

$$= \frac{560}{7} = 80$$

Since the average is 80, Linda will get the needed B.   ◇

◇ **COMMENT** When we compute the mean, we sum the given data. There is a convenient notation to indicate the sum. Let x represent any value in the data set. Then the notation

Σx (read "the sum of all given x values")

means that we are to sum all the data values. In other words, we are to sum all the entries in the distribution. The *summation symbol* Σ means *sum the following* and is capital sigma, the S of the Greek alphabet.   ◇

Summation symbol

Formulas for the mean

The symbol for the mean of a *sample* distribution of x values is denoted by $\bar{x}$ (read "x bar"). If your data comprise the entire *population*, we use the symbol μ (lowercase Greek letter mu, pronounced "mew") to represent the mean. The procedure to compute the mean is the same regardless of whether we have population or sample data. If we let n represent the number of entries in a *sample* data set and N represent the number of entries in a *population* data set, the formulas are

Sample mean
Population mean

$$\text{Sample mean} = \bar{x} = \frac{\Sigma x}{n} \qquad\qquad \text{Population mean} = \mu = \frac{\Sigma x}{N} \qquad (1)$$

● **CALCULATOR NOTE** It is very easy to compute the mean on *any* calculator: Simply add the data values and divide the total by the number of data items. However, on calculators with a statistics mode, you place the calculator into that mode, *enter* the data, and then press the key for the mean. The key is usually designated $\bar{x}$. Because the formula for the population mean is the same as that of the sample mean, the same key gives the value for μ.

We have seen three averages: the mode, the median, and the mean. For later work, the mean is the most important. A disadvantage of the mean, however, is that it can be affected by exceptional values.

Resistant measures

A *resistant measure* is one that is not influenced by extremely high or low data values. The mean is not a resistant measure of center because we can make the mean as large as we want by increasing the size of only one data value. The median, on the other hand, is more resistant. However, a disadvantage of the median is that it is not sensitive to the specific size of a data value.

Trimmed mean

A measure of center that is more resistant than the mean but still sensitive to specific data values is the *trimmed mean*. A trimmed mean is the mean of the data values left after "trimming" a specified percentage of the smallest and largest data values from the data set. Usually a 5% trimmed mean is used. This implies that we trim the lowest 5% of the data as well as the highest 5% of the data. A similar procedure is used for a 10% trimmed mean.

To compute a 5% trimmed mean

1. Order the data from smallest to largest.

2. Delete the bottom 5% of the data and the top 5% of the data. *Note:* If the calculation of 5% of the number of data values does not produce a whole number, *round* to the nearest integer.

3. Compute the mean of the remaining 90% of the data.

GUIDED EXERCISE 2

Mean and trimmed mean

Barron's Profiles of American Colleges, 19th Edition, lists average class size for introductory lecture courses at each of the profiled institutions. A sample of 20 colleges and universities in California showed class size for introductory lecture courses to be

⑭	20	20	20	20	23	25	30	30	30
35	35	35	40	40	42	50	50	80	⑧⓪

(a) Compute the mean for the entire sample. ⟹ Add all the values and divide by 20:

$$\bar{x} = \frac{\Sigma x}{n} = \frac{719}{20} \approx 36.0$$

(b) Compute a 5% trimmed mean for the sample. ⟹ The data are already ordered. Since 5% of 20 is 1, we eliminate one data value from the bottom of the list and one from the top. These values are circled in the data set. Then take the mean of the remaining 18 entries.

$$5\% \text{ trimmed mean} = \frac{\Sigma x}{n} = \frac{625}{18} \approx 34.7$$

(c) Find the median for the original data set. ⟹ Note the data are already ordered.

$$\text{Median} = \frac{30 + 35}{2} = 32.5$$

(d) Find the median of the 5% trimmed data set. Does the median change because of trimming the data? ⟹ The median is still 32.5. Notice that trimming the same number of entries from both ends leaves the middle position of the data set unchanged.

(e) Is the trimmed mean or the original mean closer to the median? ⟹ The trimmed mean is closer to the median.

TECH NOTE   Minitab, Excel, and the TI-83Plus all provide the mean and median of a data set. Minitab and Excel also provide the mode. The TI-83Plus sorts data, so you can easily scan the sorted data for the mode. Minitab provides the 5% trimmed mean, as does Excel.

All this technology is a wonderful aid for analyzing data. However, *a measurement has no meaning if you do not know what it represents or how a change in data values might affect the measurement.* The defining formulas and procedures for computing the measures tell you a great deal about the measures. Even if you use a calculator to evaluate all the statistical measures, pay attention to the information the formulas and procedures give you about the components or features of the measurement.

◇ **COMMENT**   In Chapter 1, we examined four levels of data: nominal, ordinal, interval, and ratio. The mode (if it exists) can be used with all four levels, including nominal. For instance, the modal color of all passenger cars sold last year might be blue. The median may be used with data at the ordinal level or above. If we ranked the passenger cars in order of customer satisfaction level, we could identify the median satisfaction level. For the mean, our data need to be at the interval or ratio level (although there are exceptions in which the mean of ordinal-level data is computed). We can certainly find the mean model year of used passenger cars sold or the mean price of new passenger cars.   ◇

VIEWPOINT

What's Wrong with Pitching Today?

One way to answer this question is to look at *averages*. Batting averages and average hits per game are shown for select years 1901 to 2000 (*Source: The Wall Street Journal*).

Year	1901	1920	1930	1941	1951	1961	1968	1976	1986	2000
B.A.	0.277	0.284	0.288	0.267	0.263	0.256	0.231	0.256	0.262	0.276
Hits	19.2	19.2	20.0	18.4	17.9	17.3	15.2	17.3	17.8	19.1

A quick scan of the averages shows that batting averages and average hits per game are virtually the same as almost 100 years ago. It seems there is *nothing* wrong with today's pitching! So what's changed? For one thing, the rules have changed! The strike zone is considerably smaller than it once was, and the pitching mound is lower. Both give the hitter an advantage over the pitcher. Even so, pitchers don't give up hits with any greater frequency than they did a century ago (look at the averages). However, modern hits go much farther, which is something a pitcher can't control.

SECTION 3.1   PROBLEMS

1. *Agriculture: Growing Season* The average length of the growing season is often measured in average number of frost-free days. The front range of Colorado (Fort Collins, Boulder, Denver, Colorado Springs, Pueblo) was studied by J. F. Benci and T. B. McKee, from the Department of Atmospheric Science at Colorado State University. Based on data from their Climatology Report No. 77-3, different locations in the Colorado front range had the following average number of frost-free days per year:

156	161	152	162	144	153
148	157	168	157	161	157

    Compute the mean, median, and mode. Write a brief description of the meaning of these numbers from the point of view of a gardener.

2. *Baseball: Home Runs* Babe Ruth was the American League Home Run Champion 12 times (during the period from 1918 to 1931). The number of home runs he hit to earn the 12 titles were

11	29	54	59	41	46
47	60	54	46	49	46

    Find the mean, median, and mode of the number of home runs.

3. *Environmental Studies: Death Valley* How hot does it get in Death Valley? The following data are taken from a study conducted by the National Park System, of which Death Valley is a unit. The ground temperatures (°F) were taken from May to November in the vicinity of Furnace Creek.

146	152	168	174	180	178	179
180	178	178	168	165	152	144

    Compute the mean, median, and mode for these ground temperatures.

4. *Leisure: Maui Vacation* How expensive is Maui? If you want a vacation rental condominium (for up to four people), visit the Brase/Brase *Understanding Basic Statistics* site at http://math.college.hmco.com/students, find the link to Maui, and then search for accommodations. *The Maui News* gave the following cost in dollars per day for a random sample of condominiums located throughout the island of Maui.

89	50	68	60	375	55	500	71	40	350
60	50	250	45	45	125	235	65	60	130

    (a) Compute the mean, median, and mode for the data.
    (b) Compute a 5% trimmed mean for the data, and compare it with the mean computed in part (a). Does the trimmed mean more accurately reflect the general level of the daily rental costs?
    (c) If you were a travel agent, and a client asked about the daily cost of renting a condominium on Maui, what average would you use? Explain. Is there any other information about the costs that you think might be useful, such as the spread of the costs?

5. *Medical: Injuries* The Grand Canyon and the Colorado River are beautiful, rugged, and sometimes dangerous. Thomas Myers is a physician at the park clinic in Grand Canyon Village. Dr. Myers has recorded (for a 5-year period) the number of visitor injuries at

Now obtain the standard deviation by taking the square root of the variance.

$$s = \sqrt{s^2} = \sqrt{9.43} \approx 3.07$$

(Use a calculator to compute a square root. Because of rounding, we use the approximately equal symbol, $\approx$.)   ◊

GUIDED EXERCISE 3

Sample standard deviation

Big Blossom Greenhouse gathered another random sample of mature peak blooms from Hybrid B. The eight blossoms had these widths (in inches):

5   5   5   6   6   6   7   8

(a) Again, we will construct a table so we can find the mean, variance, and standard deviation more easily. In this case, what is the value of n? Find the sum of Column I in Table 3-2, and compute the mean.

➡ $n = 8$. The sum of Column I is $\Sigma x = 48$, so the mean is

$$\bar{x} = \frac{48}{8} = 6 \text{ in.}$$

TABLE 3-2   Complete Columns II and III

I x	II $x - \bar{x}$	III $(x - \bar{x})^2$
5	$5-6=-1$	$(-1)^2=1$
5	$5-6=-1$	$(-1)^2=1$
5	$5-6=-1$	$(-1)^2=1$
6	$6-6=0$	$(0)^2=0$
6	$6-6=0$	$(0)^2=0$
6	$6-6=0$	$(0)^2=0$
7	$7-6=1$	$(1)^2=1$
8	$8-6=2$	$(2)^2=4$
$\Sigma x = 48$		$\Sigma(x-\bar{x})^2 = 8$

TABLE 3-3   Completion of Table 3-2

I x	II $x - \bar{x}$	III $(x - \bar{x})^2$
5	-1	1
5	-1	1
5	-1	1
6	0	0
6	0	0
6	0	0
7	1	1
8	2	4
$\Sigma x = 48$		$\Sigma(x-\bar{x})^2 = 8$

(b) What is the value of $n - 1$? Divide the total sum of Column III by $n - 1$ to find the variance.

➡ $n - 1 = 7$

$$\text{Variance} = s^2 = \frac{\Sigma(x - \bar{x})^2}{n - 1} = \frac{8}{7} \approx 1.14$$

(c) Use a calculator to find the square root of the variance. Is this the standard deviation?

➡ $\sqrt{\text{variance}} = \sqrt{s^2} = \sqrt{1.14} \approx 1.07$ in. The square root of the variance is the standard deviation.

Let's summarize and compare the results of Guided Exercise 3 and Example 5. The greenhouse found the following blossom diameters for Hybrid A and Hybrid B:

Hybrid A: mean, 6.0 in.; standard deviation, 3.07 in.

Hybrid B: mean, 6.0 in.; standard deviation, 1.07 in.

In both cases, the means are the same: 6 in. But the first hybrid has a larger standard deviation. This means that the blossoms of Hybrid A are less consistent than those of Hybrid B. If you want a rosebush that occasionally has 10-in. blooms and 2-in. blooms, use the first hybrid. But if you want a bush that consistently produces roses close to 6 in. across, use Hybrid B.

Sum of squares

◇ **COMMENT** The value $\Sigma(x - \overline{x})^2$ is used so commonly in other measures in statistics that it has its own name, the *sum of squares,* and symbol SS_x. A computationally simpler (requires fewer subtractions) version of the expression that is often used is

$$\text{Sum of squares} = SS_x = \Sigma(x - \overline{x})^2 = \Sigma x^2 - \frac{(\Sigma x)^2}{n} \tag{4}$$

Then $s = \sqrt{\dfrac{SS_x}{n - 1}}$. We will see the sum of squares SS_x when we study linear regression. ◇

● **ROUNDING NOTE** Rounding errors cannot be completely eliminated, even if a computer or calculator does all the computations. However, software and calculator routines are designed to minimize the error. If the mean is rounded, the value of the standard deviation will change slightly depending on how much the mean is rounded. If you do your calculations "by hand" or reenter intermediate values into a calculator, try to carry one or two more digits than occur in the original data. If your resulting answers vary slightly from those in this text, do not be overly concerned. The text answers are computer- or calculator-generated.

Population mean and standard deviation

In almost all applications of statistics, we work with a random sample of data rather than the entire population of *all* possible data values. However, if we do in fact have data for the entire population, we can compute the *population mean* μ (lowercase Greek letter mu, pronounced "mew") *and population standard deviation* σ (lowercase Greek letter sigma) by using the following formulas:

$$\mu = \frac{\Sigma x}{N} \quad \text{population mean}$$

$$\sigma = \sqrt{\frac{\Sigma(x - \mu)^2}{N}} \quad \text{population standard deviation}$$

where N is the number of data values in the population, and x represents the individual data values of the population. We note that the formula for μ is the same as the formula for $\overline{x}$ (the sample mean) and the formula for σ is the same as the formula for s (the sample standard deviation), except that the population size N is used instead of $n - 1$. Also, μ is used instead of $\overline{x}$ in the formula for σ.

Population size

In the formulas for s and σ, we use $n - 1$ to compute s, and N to compute σ. Why? The reason is that N (capital letter) represents the *population size*, while n (lowercase letter) represents the sample size. Since a random sample usually will not contain extreme data values (large or small), we divide by $n - 1$ in the formula for s to make s a little larger than it would have been had we divided by n. Courses in advanced theoretical statistics show that this procedure will give us the best possible estimate for the standard deviation σ. In fact, s is called the *unbiased estimate* for σ. If we have the population of all data values, then extreme data values are, of course, present, so we divide by N instead of $N - 1$.

◇ **COMMENT** The computation formula for the population standard deviation is

$$\sigma = \sqrt{\frac{SS_x}{N}} \quad \text{where} \quad SS_x = \Sigma(x - \mu)^2 = \Sigma x^2 - \frac{(\Sigma x)^2}{N} \qquad \diamond$$

We've seen that the standard deviation (sample or population) is a measure of data spread. We will use the standard deviation extensively in later chapters. In Chapter 7 we will use it to study standard z values and areas under normal curves. In Chapters 9, 10, and 11 we will use it to study the inferential statistics topics of estimation and testing. The standard deviation will appear again in our study of regression and correlation.

TECH NOTE Most scientific or business calculators have a statistics mode, and provide the mean and sample standard deviation directly. The TI-83Plus, Excel, and Minitab provide the median and several other measures as well.

Many technologies display only the sample standard deviation s. You can quickly compute σ if you know s by using the formula

$$\sigma = s\sqrt{\frac{n - 1}{n}}$$

The mean given in displays can be interpreted as the sample mean $\bar{x}$ or the population mean μ as appropriate.

The following three displays show output for the hybrid rose data of Guided Exercise 3.

TI-83Plus Display
Press **STAT** ➤ **CALC** ➤ **1:1-Var Stats**. S_x is the sample standard deviation. σ_x is the population standard deviation.

```
1-Var Stats
 x̄=6
 Σx=48
 Σx²=296
 Sx=1.069044968
 σx=1
↓n=8
```

Excel Display

Menu choices: **Tools ➤ Data Analysis ➤ Descriptive Statistics.** The standard deviation is the sample standard deviation.

Mean	6
Standard Error	0.377964
Median	6
Mode	5
Standard Deviation	1.069045
Sample Variance	1.142857
Kurtosis	0.35
Skewness	0.935414
Range	3
Minimum	5
Maximum	8
Sum	48
Count	8

Minitab Display

Menu choices: **Stat ➤ Basic Statistics ➤ Display Descriptive Statistics.** StDev is the sample standard deviation. TrMean is a 5% trimmed mean.

```
Descriptive Statistics
  N          Mean      Median      TrMean      StDev     SE Mean
  8          6.000      6.000       6.000       1.069      0.378
Minimum         Maximum         Q1          Q3
  5.000           8.000         5.000       6.750
```

Now let's look at two immediate applications of the standard deviation. The first is the coefficient of variation, and the second is Chebyshev's theorem.

Coefficient of Variation

A disadvantage of the standard deviation as a comparative measure of variation is that it depends on the units of measurement. This means that it is difficult to use the standard deviation to compare measurements from different populations. For this reason, statisticians have defined the *coefficient of variation*, which expresses the standard deviation as a percentage of the sample or population mean.

Coefficient of variation

If $\bar{x}$ and s represent the sample mean and sample standard deviation, then the sample **coefficient of variation** CV is defined to be

$$CV = \frac{s}{\bar{x}} \cdot 100$$

If μ and σ represent the population mean and standard deviation, then the population coefficient of variation CV is defined to be

$$CV = \frac{\sigma}{\mu} \cdot 100$$

Notice that the numerator and denominator in the definition of CV have the same units, so CV itself has no units of measurement. This gives us the advantage of being able to directly compare the variability of two different populations by using the coefficient of variation.

In the next example and guided exercise, we will compute the CV of a population and of a sample and then compare the results.

EXAMPLE 6

Coefficient of variation

The Trading Post on Grand Mesa is a small, family-run store in a remote part of Colorado. The Grand Mesa region contains many good fishing lakes, so the Trading Post sells spinners (a type of fishing lure). The store has a very limited selection of spinners, however. In fact, the Trading Post has only eight different types of spinners for sale. The prices (in dollars) are

| 2.10 | 1.95 | 2.60 | 2.00 | 1.85 | 2.25 | 2.15 | 2.25 |

Since the Trading Post has only eight different kinds of spinners for sale, we consider the eight data values to be the *population*.

(a) Use a calculator with appropriate statistics keys to verify that for the Trading Post data, $\mu = \$2.14$ and $\sigma = \$0.22$.

   SOLUTION: Since the computation formula for $\bar{x}$ and μ are identical, most calculators provide the value of $\bar{x}$ only. Use the output of this key for μ. The computation formulas for the sample standard deviation s and the population standard deviation σ are slightly different. Be sure that you use the key for σ (sometimes designated as σ_n or σ_x).

(b) Compute the CV of prices for the Trading Post and comment on the meaning of the result.

   SOLUTION:

   $$CV = \frac{\sigma}{\mu} \times 100 = \frac{0.22}{2.14} \times 100 = 10.28\%$$

   The coefficient of variation can be thought of as a measure of the spread of the data relative to the average of the data. Since the Trading Post is very small, it carries a small selection of spinners that are all priced similarly. The CV tells us that the standard deviation of the spinner prices is only 10.28% of the mean.   ◊

GUIDED EXERCISE 4

Coefficient of variation

Cabela's in Sidney, Nebraska, is a very large outfitter that carries a broad selection of fishing tackle. It markets its products nationwide through a catalog service. A random sample of 10 spinners from Cabela's extensive spring catalog gave the following prices (in dollars):

| 1.69 | 1.49 | 3.09 | 1.79 | 1.39 |
| 2.89 | 1.49 | 1.39 | 1.49 | 1.99 |

Continued

GUIDED EXERCISE 4 continued

(a) Use a calculator with sample mean and sample standard deviation keys to compute $\bar{x}$ and s.

⟹ $\bar{x} = \$1.87$ and $s = \$0.62$.

(b) Compute the *CV* for the spinner prices at Cabela's.

⟹ $CV = \dfrac{s}{\bar{x}} \times 100 = \dfrac{0.62}{1.87} \times 100 = 33.16\%$.

(c) Compare the mean, standard deviation, and *CV* for the spinner prices at the Grand Mesa Trading Post (Example 6) and Cabela's. Comment on the differences.

⟹ The *CV*'s for Cabela's and the Trading Post are pure numbers (no units), so a direct comparison is possible. The *CV* for Cabela's is more than three times the *CV* for the Trading Post. Why? First, because of the remote location, the Trading Post tends to have somewhat higher prices (larger μ). Second, the Trading Post is very small, so it has a rather limited selection of spinners with a smaller variation in price. For Cabela's, however, the average price $\bar{x}$ is lower and the variety larger (larger s). It makes sense that the *CV* for Cabela's is larger than the *CV* for the Trading Post.

Chebyshev's Theorem

From our earlier discussion about standard deviation, we recall that the spread or dispersion of a set of data about the mean will be small if the standard deviation is small, and it will be large if the standard deviation is large. If we are dealing with a symmetrical bell-shaped distribution, then we can make very definite statements about the proportion of the data that must lie within a certain number of standard deviations on either side of the mean. This will be discussed in detail in Chapter 7 when we talk about normal distributions.

However, the concept of data spread about the mean can be expressed quite generally for *all data distributions* (skewed, symmetric, or other shape) by using the remarkable theorem of Chebyshev:

Chebyshev's theorem

Chebyshev's theorem

For *any* set of data (either population or sample) and for any constant k greater than 1, the proportion of the data that must lie within k standard deviations on either side of the mean is *at least*

$$1 - \dfrac{1}{k^2}$$

Results of Chebyshev's theorem

For *any* set of data:

- at *least* 75% of the data fall into the interval from $\mu - 2\sigma$ to $\mu + 2\sigma$

- at *least* 88.9% of the data fall into the interval from $\mu - 3\sigma$ to $\mu + 3\sigma$

- at *least* 93.8% of the data fall into the interval from $\mu - 4\sigma$ to $\mu + 4\sigma$

The results of Chebyshev's theorem can be derived by using the theorem and a little arithmetic. For instance, if we create an interval $k = 2$ standard deviations on either side of the mean, Chebyshev's theorem tells us that

$$1 - \frac{1}{2^2} = 1 - \frac{1}{4} = \frac{3}{4} \text{ or } 75\%$$

is the minimum percentage of data in the $\mu - 2\sigma$ to $\mu + 2\sigma$ interval.

Notice that Chebyshev's theorem refers to the *minimum* percentage of data that must fall within the specified number of standard deviations of the mean. If the distribution is mound-shaped, an even *greater* percentage of data will fall into the specified intervals (see the Empirical Rule in Section 7.1).

EXAMPLE 7

Chebyshev's theorem

Students Who Care is a student volunteer program in which college students donate work time to various community projects such as planting trees. Professor Gill is the faculty sponsor for this student volunteer program. For several years, Dr. Gill has kept a careful record of x = total number of work hours volunteered by a student in the program each semester. For a random sample of students in the program, the mean number of hours was $\bar{x} = 29.1$ hours each semester, with a standard deviation $s = 1.7$ hours each semester. Find an interval A to B for the number of hours volunteered into which at least 75% of the students in this program would fit.

SOLUTION: According to results of Chebyshev's theorem, at least 75% of the data must fall within 2 standard deviations of the mean. Because the mean $\bar{x} = 29.1$ and the standard deviation $s = 1.7$, the interval is

$$\bar{x} - 2s \text{ to } \bar{x} + 2s$$
$$29.1 - 2(1.7) \text{ to } 29.1 + 2(1.7)$$
$$25.7 \text{ to } 32.5$$

At least 75% of the students would fit into the group that volunteered from 25.7 to 32.5 hours each semester.   ◇

GUIDED EXERCISE 5

Chebyshev interval

The *East Coast Independent News* periodically runs ads in its own classified section offering a month's free subscription to those who respond. In this way, management can get a sense about the number of subscribers who read the classified section each day. Over a period of 2 years, careful records have been kept. The mean number of responses per ad is $\bar{x} = 525$ with standard deviation $s = 30$.

Determine a Chebyshev interval about the mean in which at least 88.9% of the data fall.

⟹

By Chebyshev's theorem, at least 88.9% of the data fall into the interval

$\bar{x} - 3s$ to $\bar{x} + 3s$

Because $\bar{x} = 525$ and $s = 30$, the interval is

$525 - 3(30)$ to $525 + 3(30)$

or from 435 to 615 responses per ad.

VIEWPOINT

Socially Responsible Investing

Make a difference *and* make money! Socially responsible mutual funds tend to screen out corporations that sell tobacco, weapons, and alcohol, as well as companies that are environmentally unfriendly. In addition, these funds screen out companies that use child labor in sweatshops. There are 68 socially responsible funds tracked by the Social Investment Forum. For more information, visit the Brase/Brase *Understanding Basic Statistics* site at http://math.college.hmco.com/students and find the link to social investing.

How do these funds rate compared to other funds? One way to answer this question is to study the annual percent returns of the funds using both the *mean and standard deviation*. (See Problem 5 of this section.)

SECTION 3.2   PROBLEMS

1. *Ecology: Deer* By sampling different landscapes in Mesa Verde National Park over a 2-year period, the number of deer per square kilometer was determined (*The Mule Deer of Mesa Verde National Park*, by G. W. Mieran and J. L. Schmidt, published by Mesa Verde Museum Association, 1981). The results were (deer per square kilometer)

30	20	5	29	58	7
20	18	4	29	22	9

(a) Compute the range, sample mean, sample variance, and sample standard deviation.

(b) Compute the coefficient of variation. Is the coefficient of variation a fairly large number? Would you say there was a considerable variation in the distribution of deer from one section of the park to another? Explain.

2. *Wildlife Management: Canada Geese* Hatching success of game birds is a topic discussed in the book *Wildlife Management Techniques,* edited by R. H. Giles (The Wildlife Society Press, Washington, D.C.). What percentage of Canada goose nests are successful (at least one gosling survives)? Studies in regions of Montana, Illinois, Wyoming, Utah, and California gave the following percentage of successful nests:

   ~~23.9~~   ~~52.5~~   60.0   68.5   78.6   71.0   ~~17.8~~   57.5   ~~59.0~~   ~~52.0~~

   (a) Compute the range and mean.
   (b) Compute the sample variance and standard deviation.
   (c) Compute the coefficient of variation. Write a brief explanation of the meaning of this number.

3. *Wildlife Management: Mallard Ducks* Hatching success for mallard duck nests is also given in the same reference as in Problem 2. A successful nest means at least one duckling survives. The following are percentages of successful mallard duck nests:

   55.7   85.2   51.5   12.7   38.5   90.3   70.0   39.8   65.0   48.0

   (a) Compute the range and mean.
   (b) Compute the sample variance and standard deviation.
   (c) Compute the coefficient of variation. Write a brief explanation of the meaning of this number. Compare this value with the coefficient of variation for Canada geese in Problem 2. What does this mean about relative variability of nesting success rates for Canada geese compared with mallard ducks?

4. *Business Administration: Profits* Jobs and productivity! How productive are American workers? One way to answer this question is to examine annual profits per employee. The results may change, however, from one industry group to another. This problem is based on information taken from *Forbes Top Companies* (edited by J. T. Davis, John Wiley Publisher).

   (a) A random sample of seven automotive parts companies gave the following information about *profits* per employee (in thousands of dollars per employee):

   14.1   12.4   7.7   6.9   9.0   6.8   6.8

   Compute the range, mean, sample variance, sample standard deviation, and coefficient of variation.

   (b) For a random sample of seven computer companies, the profits per employee in the same units were

   29.0   24.5   31.0   29.8   21.8   27.7   19.1

   Compute the range, mean, sample variance, sample standard deviation, and coefficient of variation.

   (c) Examine your answers for parts (a) and (b). Although the standard deviation for part (a) is the smaller of the two, the mean for part (b) is the larger of the two. The *CV* for part (b) is almost half that of part (a) In this case, why would a small *CV* and high mean indicate more reliable productivity from the point of view of annual profits per employee? Explain.

5. *Investing: Socially Responsible Mutual Funds* Pax World Balanced is a highly respected, socially responsible mutual fund of stocks and bonds (see Viewpoint). Vanguard Balanced Index is another highly regarded fund that represents the entire U.S. stock and bond market (an index fund). The mean and standard deviation of annualized percent returns are shown below. The annualized mean and standard deviation are based on 36 months of data prior to May 2001 (*Source:* Morningstar).

   Pax World Balanced: $\bar{x} = 11.69\%$; $s = 11.56\%$

   Vanguard Balanced Index: $\bar{x} = 5.61\%$; $s = 12.50\%$

   (a) Compute the coefficient of variation for each fund. If $\bar{x}$ represents return and s represents risk, then explain why the coefficient of variation can be taken to represent risk per unit of return. From this point of view, which fund appears to be better? Explain.

   (b) Compute a 75% Chebyshev interval around the mean for each fund. Use the intervals to compare the two funds. As usual, past performance does not guarantee future performance.

6. *Meteorology: Tornados* The U.S. Weather Bureau has provided the following information about the total annual number of reported tornados in the United States for the years 1956 to 1975:

504	856	564	604	616	697	657	464	704	906
585	926	660	608	653	888	741	1102	947	918

   (a) Use a calculator with mean and standard deviation keys to verify that the mean number of tornados per year is 730, with a sample standard deviation of 172 tornados.

   (b) Use Chebyshev's theorem to find an interval centered about the mean for the annual number of tornados in which you would expect at least 75% of the years to fall.

   (c) Use Chebyshev's theorem to find an interval centered about the mean in which you would expect at least 88.9% of the years to fall.

7. *History: Billy the Kid* In 1881, Billy the Kid killed two deputies and escaped from a jail cell in the Lincoln County Courthouse, New Mexico Territory. Famous frontier personalities such as Kit Carson, Jesse Chisum, and Sheriff Pat Garrett (who eventually shot Billy the Kid) also were involved in the notorious Lincoln County Cattle Baron Wars. Before the 1985 renovation of the now famous courthouse, the Museum of New Mexico commissioned anthropologist/historian Yvonne Oakes to do a thorough analysis of the area. The distribution of artifacts (of the 1880s' period) for seven excavation sites were

851	596	444	956	576	219	326

   (*Source: Museum of New Mexico: Laboratory of Anthropology Notes No. 357.*)

   (a) Compute the range and mean.

   (b) Compute the sample variance and standard deviation.

   (c) Compute the coefficient of variation. Write a brief explanation of the meaning of this number in the context of this problem.

   (d) Use Chebyshev's theorem to find an interval centered about the mean in which at least 75% of the artifact counts for all such excavation sites would fall.

8. *Medical: Physician Visits* In some reports, the mean and coefficient of variation are given. For instance, in *Statistical Abstract of the United States,* 116th Edition, one report gives the average number of physician visits by males per year. The

average reported is 2.2, and the reported coefficient of variation is 1.5%. Use this information to determine the standard deviation of the number of visits to physicians made by males.

9. *Financial: Wal-Mart and Disney Stock* The majority of U.S. households now own stocks. Two widely held stocks are Wal-Mart and Disney. Let's look at daily closing price per share (in dollars) for the two stocks during the period May 11 to June 9, 2001.

Wal-Mart, $: 54.10 54.35 52.00 51.65 51.76 52.04 53.20 53.52 52.52 52.89
51.20 51.21 51.47 51.75 51.72 50.99 51.30 50.75 51.10 51.02

Disney, $: 31.27 30.95 31.10 32.06 32.51 32.60 34.28 34.50 33.11 33.33
32.64 32.62 31.78 31.62 31.72 31.62 31.35 31.60 32.15 31.85

(a) Use a calculator to verify the following means and standard deviations:
Wal-Mart: $\bar{x} \approx \$52.03$, $s \approx \$1.06$; Disney: $\bar{x} \approx \$32.23$, $s \approx \$0.98$.

(b) Compute the coefficient of variation for Wal-Mart and for Disney. If $\bar{x}$ represents return and s represents risk, then explain why the coefficient of variation can be taken to represent risk per unit of return. From this point of view, do the stocks appear to be about equally attractive (in this period)? Explain.

(c) Compute an 88.9% Chebyshev interval for Wal-Mart and for Disney. Many stock brokers refer to *support* and *resistance* as the trading range of a stock. The *support* is a value the broker is "confident" the stock will not sink below. The *resistance* is a value the broker is "confident" the stock will not rise above. There are many different ways to compute supports and resistances. An 88.9% Chebyshev interval is one way, with the support being the lower end of the interval, and the resistance being the upper end. Compute the support and resistance for Wal-Mart and for Disney by using this method.

10. *Lifestyle: Gathering Data* For the next 2 weeks, record how long it takes you to go from your residence (or job) to your first class. Record the values to the nearest minute. Compute the range, mean, sample standard deviation, and coefficient of variation. Use Chebyshev's theorem to compute an interval in which at least 88.9% of the travel times fall. Were there any unusual data values in your data set? How would you explain them?

Grouped data

Approximating $\bar{x}$ and s from grouped data

Grouped data

When data are grouped, such as in a frequency table or histogram, we can estimate the mean and standard deviation by using the following formulas. Notice that all data values in a given class are treated as though each of them equals the midpoint x of the class.

Sample mean for a frequency distribution

$$\bar{x} = \frac{\Sigma xf}{n} \tag{5}$$

Sample standard deviation for a frequency distribution

$$s = \sqrt{\frac{\Sigma(x - \bar{x})^2 f}{n - 1}} \tag{6}$$

Computation formula for the standard deviation

$$s = \sqrt{\frac{SS_x}{n-1}}$$ (7)

where $SS_x = \Sigma x^2 f - \dfrac{(\Sigma xf)^2}{n}$

x is the midpoint of a class,
f is the number of entries in that class,
n is the total number of entries in the distribution, and $n = \Sigma f$
The summation Σ is over all classes in the distribution.

Use formulas (5) and (6) or (5) and (7) to solve Problems 11–15. To use formulas (5) and (6) to evaluate the sample mean and standard deviation, use the following column heads:

| Midpoint x | Frequency f | xf | $(x - \bar{x})$ | $(x - \bar{x})^2$ | $(x - \bar{x})^2 f$ |

For formulas (5) and (7), use these column heads:

| Midpoint x | Frequency f | xf | x^2 | $x^2 f$ |

Note: On the TI-83 calculator, enter the midpoints in column L_1 and the frequencies in column L_2. Then use **1-VarStats** L_1, L_2.

11. *Crime: Shoplifting* What is the age distribution of adult shoplifters (21 years of age or older) in supermarkets? The following is based on information taken from the National Retail Federation. A random sample of 895 incidents of shoplifting gave the following age distribution:

Age range (years)	21–30	31–40	41 and over
Number of shoplifters	260	348	287

Estimate the mean age, sample variance, and sample standard deviation for the shoplifters. For the class 41 and over, use 45.5 as the class midpoint.

12. *Anthropology: Navajo Reservation* What was the age distribution of prehistoric Native Americans? Extensive anthropologic studies in the southwestern United States gave the following information about a prehistoric extended family group of 80 members on what is now the Navajo Reservation in northwestern New Mexico. (*Source:* Based on information taken from *Prehistory in the Navajo Reservation District,* by F. W. Eddy, Museum of New Mexico Press.)

Age range (years)	1–10*	11–20	21–30	31 and over
Number of individuals	34	18	17	11

*Includes infants.

Estimate the mean age expressed in years for this community, the sample variance, and the sample standard deviation. For the class 31 and over, use 35.5 as the class midpoint.

13. *Business Administration: Profits/Assets* What are the big corporations doing with their wealth? One way to answer this question is to examine profits as percentage of assets. A random sample of *50 Fortune 500* companies gave the following information. (*Source:* Based on information from *Fortune 500*, Vol. 135, No. 8.)

Profit as percentage of assets	8.6–12.5	12.6–16.5	16.6–20.5	20.6–24.5	24.6–28.5
Number of companies	15	20	5	7	3

Estimate the sample mean, sample variance, and sample standard deviation for profit as percentage of assets.

14. *Sales: Catalog Shoppers* Based on data from *USA Today*, the ages of a random sample of 300 adults who shop by catalog are

Age	18–24	25–34	35–44	45–54	55–64	65–80
Number	78	75	48	33	33	33

Estimate the mean age of the adults who shop by catalog. Estimate the standard deviation of the age of the shoppers and the coefficient of variation.

15. *Medical: Hours of Sleep per Day* Alexander Borbely is a professor at the University of Zurich Medical School, where he is director of the sleep laboratory. The histogram in Figure 3-1 is based on information in his book *Secrets of Sleep*. The histogram displays hours of sleep per day for a random sample of 200 subjects. Estimate the mean hours of sleep, standard deviation of hours of sleep, and coefficient of variation.

FIGURE 3-1

Hours of Sleep Each Day (24-hour period)

3.3
Percentiles and Box-and-Whisker Plots

Percentile

We've seen measures of central tendency and spread for a set of data. The arithmetic mean $\bar{x}$ and the standard deviation s will be very useful in later work. However, because they each utilize every data value, they can be heavily influenced by one or two extreme data values. In cases where our data distributions are heavily skewed or even bimodal, we often get a better summary of the distribution by utilizing relative position of data rather than exact values.

Recall that the median is an average computed by using relative position of the data. If we are told that 81 is the median score on a biology test, we know that after the data have been ordered, 50% of the data fall at or below the median value of 81. The median is an example of a *percentile*; in fact, it is the 50th percentile. The general definition of the Pth percentile follows.

> For whole numbers P (where $1 \leq P \leq 99$), the Pth **percentile** of a distribution is a value such that $P\%$ of the data fall at or below it and $(100 - P)\%$ of the data fall at or above it.

In Figure 3-2, we see the 60th percentile marked on a histogram. We see that 60% of the data lie below the mark and 40% lie above it.

FIGURE 3-2

A Histogram with the 60th Percentile Shown

FIGURE 3-3

Percentiles

GUIDED EXERCISE 6

Percentiles

You took the English achievement test to obtain college credit in freshman English by examination.

(a) If your score was at the 89th percentile, what percentage of scores are at or below yours?

⟹ The percentile means that 89% of the scores were at or below yours.

(b) If the scores ranged from 1 to 100 and your raw score is 95, does this necessarily mean that your score is at the 95th percentile?

⟹ No, the percentile gives an indication of relative position of the scores. The determination of your percentile has to do with the number of scores at or below yours. If everyone did very well and only 80% of the scores fell at or below yours, you would be at the 80th percentile even though you got 95 out of 100 points on the exam.

There are 99 percentiles, and in an ideal situation, the 99 percentiles divide the data set into 100 equal parts. (See Figure 3-3.)

However, if the number of data elements is not exactly divisible by 100, the percentiles will not divide the data into equal parts.

There are several widely used conventions for finding percentiles. They lead to slightly different values for different situations, but these values are close together. For all conventions, the data are first *ranked* or ordered from smallest to largest. A natural way to find the *P*th percentile is to then find a value so that *P*% of the data fall at or below it. This will not always be possible, so we take the nearest value satisfying the criterion. It is at this point that there are a variety of processes to determine the exact value of the percentile.

We will not be very concerned about exact procedures for evaluating percentiles in general. However, *quartiles* are special percentiles used so frequently that we want to adopt a specific procedure for their computation.

Quartiles

Quartiles are those percentiles that divide the data into fourths. The *first quartile* Q_1 is the 25th percentile, the *second quartile* Q_2 is the median, and the *third quartile* Q_3 is the 75th percentile. (See Figure 3-4.)

FIGURE 3-4

Quartiles

	25%	25%	25%	25%	
Lowest	Q_1	Q_2	Q_3	Highest

Median
50th percentile

Again, several conventions are used for computing quartiles, but the following one utilizes the median and is widely adopted.

Procedure to compute quartiles

1. Order the data from smallest to largest.

2. Find the median. This is the 2nd quartile.

3. The first quartile Q_1 is then the median of the lower half of the data; that is, it is the median of the data falling *below* the Q_2 position (and not including Q_2).

4. The third quartile Q_3 is the median of the upper half of the data; that is, it is the median of the data falling *above* the Q_2 position (and not including Q_2).

In short, all we do to find the quartiles is to find three medians.

The median, or second quartile, is a popular measure of the center utilizing relative position. A useful measure of data spread utilizing relative position is the *interquartile range (IQR)*. It is simply the difference between the third and first quartiles.

Interquartile range

$$\text{Interquartile range} = Q_3 - Q_1$$

The interquartile range tells us the spread of the middle half of the data. Now let's look at an example to see how to compute all these quantities.

EXAMPLE 8

Quartiles

In a hurry? On the run? Hungry as well? How about an ice cream bar as a snack? Ice cream bars are popular among all age groups. *Consumer Reports* did a study of ice cream bars. Twenty-seven bars with taste ratings of at least "fair" were listed, and cost per bar was included in the report. Just how much will an ice cream bar cost? The data, expressed in dollars, appear in Table 3-4. As you can see, the cost varies quite a bit, partly because the bars are not of uniform size.

(a) Find the quartiles.

SOLUTION: We first order the data from smallest to largest.
Next, we find the median. Since the number of data values is 27, there are an

TABLE 3-4   Cost of Ice Cream Bars (in dollars)

0.99	1.07	1.00	0.50	0.37	1.03	1.07	1.07
0.97	0.63	0.33	0.50	0.97	1.08	0.47	0.84
1.23	0.25	0.50	0.40	0.33	0.35	0.17	0.38
0.20	0.18	0.16					

TABLE 3-5   Ordered Cost of Ice Cream Bars (in dollars)

0.16	0.17	0.18	0.20	0.25	0.33	0.33	0.35
0.37	0.38	0.40	0.47	0.50	0.50	0.50	0.63
0.84	0.97	0.97	0.99	1.00	1.03	1.07	1.07
1.07	1.08	1.23					

odd number of data, and the median is simply the center or 14th value. The value is shown boxed in Table 3-5.

$$\text{Median} = Q_2 = 0.50$$

There are 13 values below the median position, and Q_1 is the median of these values. It is the middle or 7th value and is shaded in Table 3-5.

$$\text{First quartile} = Q_1 = 0.33$$

There are also 13 values above the median position. The median of these is the 7th value from the right end. This value is also shaded in Table 3-5.

$$\text{Third quartile} = Q_3 = 1.00$$

(b) Find the interquartile range.

SOLUTION:

$$
\begin{aligned}
IQR &= Q_3 - Q_1 \\
&= 1.00 - 0.33 \\
&= 0.67
\end{aligned}
$$

This means that the middle half of the data has a cost spread of 67¢.   ◇

GUIDED EXERCISE **7**

Quartiles

Many people consider the number of calories in an ice cream bar as important as, if not more important than, the cost. The *Consumer Reports* article also included the calorie count of the rated ice cream bars (Table 3-6). There were 22 vanilla-flavored bars rated. Again, the bars varied in size, and some of the smaller bars had fewer calories. The calorie counts for the vanilla bars follow.

TABLE 3-6   Calories in Vanilla-Flavored Ice Cream Bars

342	377	319	353	295
234	294	286	377	182
310	439	111	201	182
197	209	147	190	151
131	151			

Continued

GUIDED EXERCISE 7 continued

(a) Our first step is to order the data. Do so.

➡ **TABLE 3-7   Ordered Data**

111	131	147	151	151	182
182	190	197	201	209	234
286	294	295	310	319	342
353	377	377	439		

(b) There are 22 data values. Find the median.

➡ Average the 11th and 12th data values boxed together in Table 3-7.

$$\text{Median} = \frac{209 + 234}{2}$$
$$= 221.5$$

(c) How many values are below the median position? Find Q_1.

➡ Since the median lies halfway between the 11th and 12th values, there are 11 values below the median position. Q_1 is the median of these values.

$Q_1 = 182$

(d) There are the same number of data above as below the median. Use this fact to find Q_3.

➡ Q_3 is the median of the upper half of the data. There are 11 values in the upper portion.

$Q_3 = 319$

(e) Find the interquartile range and comment on its meaning.

➡ $IQR = Q_3 - Q_1$
$= 319 - 182$
$= 137$

The middle portion of the data has a spread of 137 calories.

Box-and-Whisker Plots

Five-number summary

The quartiles together with the low and high data values give us a very useful *five-number summary* of the data and their spread.

> **Five-number summary**
>
> Lowest value, Q_1, median, Q_3, highest value

Box-and-whisker plot

We will use these five numbers to create a graphic sketch of the data called a *box-and-whisker plot*. Box-and-whisker plots provide another useful technique from exploratory data analysis (EDA) for describing data.

To make a box-and-whisker plot

1. Draw a vertical scale to include the lowest and highest data values.

2. To the right of the scale draw a box from Q_1 to Q_3.

3. Include a solid line through the box at the median level.

4. Draw solid lines, called *whiskers*, from Q_1 to the lowest value and from Q_3 to the highest value.

FIGURE 3-5

Box-and-Whisker Plot

The next example demonstrates the process of making a box-and-whisker plot.

EXAMPLE 9
Box-and-whisker plot

Using the data shown in Guided Exercise 7, make a box-and-whisker plot showing the calories in vanilla-flavored ice cream bars. Use the plot to make observations about the distribution of calories.

(a) In Guided Exercise 7, we ordered the data (see Table 3-7) and found the values of the median, Q_1 and Q_3. From this previous work we have the five-number summary:

low value = 111; Q_1 = 182; median = 221.5; Q_3 = 319; high value = 439

(b) We select an appropriate vertical scale and make the plot (Figure 3-6).

(c) A quick glance at the box-and-whisker plot reveals the following:

(i) The box tells us where the middle half of the data lies, so we see that half of the ice cream bars have between 182 and 319 calories, with an interquartile range of 137 calories.

FIGURE 3-6

Box-and-Whisker Plot for Calories in Vanilla-Flavored Ice Cream Bars

(ii) The median is slightly closer to the lower part of the box. This means that the lower calorie counts are more concentrated. The calorie counts above the median are more spread out, indicating that the distribution is slightly skewed toward the higher values.

(iii) The upper whisker is also longer than the lower, which again emphasizes skewness toward the higher values.   ◊

◊ **COMMENT** In exploratory data analysis, *hinges* rather than quartiles are used to create the box. Hinges are computed in a manner similar to the way we compute quartiles. However, in the case of an odd number of data values, include the median itself in both the lower and upper halves of the data (see *Applications, Basics, and Computing of Exploratory Data Analysis*, by Paul Velleman and David Hoaglin, Duxbury Press). This has the effect of shrinking the box and moving the ends of the box slightly toward the median. For an even number of data, the quartiles as we computed them equal the hinges.   ◊

GUIDED EXERCISE 8

Box-and-whisker plot

The Renata College Development Office sent salary surveys to alumni who graduated two and five years ago. The voluntary responses received are summarized in the box-and-whisker plots shown in Figure 3-7.

(a) From Figure 3-7, estimate the median and extreme values of salaries of alumni graduating two years ago. In what range are the middle half of the salaries?

⇒ The median seems to be about $44,000. The extremes are about $33,000 and $54,000. The middle half of the salaries fall between $40,000 and $47,000.

FIGURE 3-7 Box-and-Whisker Plots for Alumni Salaries (in thousands of dollars)

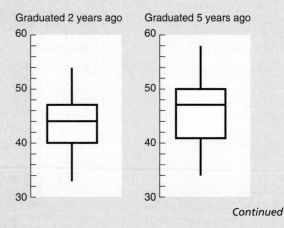

Continued

GUIDED EXERCISE 8 continued

(b) From Figure 3-7, estimate the median and the extreme values of salaries of alumni graduating five years ago. What is the location of the middle half of the salaries?

⟹ The median seems to be $47,000. The extremes are $34,000 and $58,000. The middle half of the data is enclosed by the box with low side at $41,000 and high side at $50,000.

(c) Compare the two box plots and make comments about the salaries of alumni graduating two and five years ago.

⟹ The salaries of the alumni graduating five years ago have a larger spread but begin slightly higher and extend to levels about $4000 above those graduating two years ago. The middle half of the data is also more spread out with higher boundaries and a higher median.

Outlier

We have developed the skeletal box-and-whisker plot. Other variations may include *fences,* which are marks placed on either side of the box to represent various portions of the data. Values that lie outside the fences are called *outliers* (see Problem 10 of this section). These values seem to stand by themselves, away from most of the data. They might be exceptional values and deserve closer study. Or they may be the result of data entry error. For a more complete discussion of outliers and variations of box plots, see *Applications, Basics, and Computing of Exploratory Data Analysis,* by Velleman and Hoaglin.

TECH NOTE *Box-and-Whisker Plot* Both Minitab and the TI-83Plus support box-and-whisker plots. On the TI-83Plus, the quartiles Q_1 and Q_3 are calculated as we calculate them in this text. In Minitab and Excel, they are calculated by using a slightly different process.

TI-83Plus
STATPLOT ➤ On. Highlight box plot. Use **Trace** and arrow keys to display values of five-number summary. The display shows the plot for calories in ice cream bars.

Excel
Does not produce box-and-whisker plot. **Paste Function** (f_x) ➤ **Statistics** ➤ **Quartiles** gives five-number summary.

Minitab
Graph ➤ Boxplot. In dialogue box set Display to IQRange Box.

VIEWPOINT

Is Shorter Higher?

Can you estimate a person's *height* from the *pitch* of his or her voice? Is a soprano shorter than an alto? Is a bass taller than a tenor? A statistical study of singers in the New York Choral Society provided information. For more information, visit the Brase/Brase *Understanding Basic Statistics* site at http://math.college.hmco.com/students and find the link to DASL, the Carnegie Mellon University Data and Story Library. From the Data Subjects, select music and then singers. Methods of this chapter can be used with new methods we will learn in Chapters 9 and 10 to examine such questions from a statistical point of view.

SECTION 3.3   PROBLEMS

1. *Education: Aptitude Test* Angela took a general aptitude test and scored in the 82nd percentile for aptitude in accounting. What percentage of the scores were at or below her score? What percentage were above?

2. *Education: College Admission* One standard for admission to Redfield College is that the student must rank in the upper quartile of his or her graduating high school class. What is the minimal percentile rank of a successful applicant?

3. *Education: Competency Exam* The town of Butler, Nebraska, decided to give a teacher-competency exam and defined the passing scores to be those in the 70th percentile or higher. The raw test scores ranged from 0 to 100. Was a raw score of 82 necessarily a passing score? Explain.

4. *Education: Test Scores* Clayton and Timothy took different sections of Introduction to Economics. Each section had a different final exam. Timothy scored 83 out of 100 and had a percentile rank in his class of 72. Clayton scored 85 out of 100 but his percentile rank in his class was 70. Who performed better with respect to the rest of the students in the class: Clayton or Timothy? Explain your answer.

5. *Health Care: Nurses* At Center Hospital, there is some concern about the high turnover of nurses. A survey was done to determine how long (in months) nurses had been in their current positions. The responses of 20 nurses were

   Make a box-and-whisker plot of the data. Find the interquartile range.

6. *Health Care: Staff* Another survey was done at Center Hospital to determine how long (in months) clerical staff had been in their current positions. The responses of 20 clerical staff members were

| 25 | 22 | 7 | 24 | 26 | 31 | 18 | 14 | 17 | 20 |
| 31 | 42 | 6 | 25 | 22 | 3 | 29 | 32 | 15 | 72 |

(a) Make a box-and-whisker plot. Find the interquartile range.

(b) Compare this plot with the one in Problem 5. Discuss the location of the medians, the location of the middle half of the data banks, and the distance from Q_1 and Q_3 to the extreme values.

7. *Sociology: College Graduates* What percentage of the general U.S. population have bachelor's degrees? The *Statistical Abstract of the United States*, 120th Edition, gives the percentage of bachelor's degrees by state.

22	26	24	17	27	38	34	24	22	22	26	21	26
18	22	26	20	21	23	35	31	21	32	19	23	24
20	20	27	31	25	27	24	22	26	24	27	24	27
21	26	18	24	28	28	32	29	18	24	22		

(a) Make a box-and-whisker plot and find the interquartile range.

(b) Illinois has a bachelor's degree percentage rate of about 26%. Into what quartile does this rate fall?

8. *Sociology: High-School Dropouts* What percentage of the general U.S. population are high-school dropouts? The *Statistical Abstract of the United States*, 120th Edition, gives the percentage of high-school dropouts by state.

13	11	14	14	10	9	10	14	14	14	8	10
11	11	7	9	13	13	8	11	9	10	6	12
11	8	7	15	9	10	12	10	13	5	9	10
12	9	11	12	8	13	13	9	8	10	11	11
7	7										

(a) Make a box-and-whisker plot and find the interquartile range.

(b) Wyoming has a dropout rate of about 7%. Into what quartile does this rate fall?

9. *Auto Insurance: Interpret Graphs* *Consumer Reports* rated automobile insurance companies and gave annual premiums for top-rated companies in several states. Figure 3-8 shows box plots for annual premiums for urban customers (married couple with one 17-year-old son) in three states. The box plots in Figure 3-8 are all drawn on the same scale on a TI-83Plus calculator.

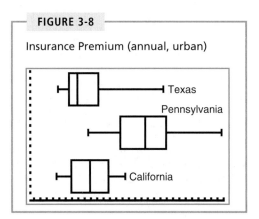

FIGURE 3-8

Insurance Premium (annual, urban)

(a) Which state has the lowest premium? the highest?
(b) Which state has the highest median premium?
(c) Which state has the smallest range of premiums? the smallest interquartile range?
(d) Figure 3-9 gives the five-number summaries generated on the TI-83 for the box plots of Figure 3-8. Match the five-number summaries to the appropriate box plot.

FIGURE 3-9

Five-Number Summaries for Insurance Premiums

(a)

```
1-Var Stats
↑n=10
  minX=2382
  Q1=2758
  Med=2991
  Q3=3652
  maxX=5715
```

(b)

```
1-Var Stats
↑n=10
  minX=3314
  Q1=4326
  Med=5116.5
  Q3=5801
  maxX=7527
```

(c)

```
1-Var Stats
↑n=10
  minX=2323
  Q1=2801
  Med=3377.5
  Q3=3966
  maxX=4482
```

---

10. *General: Outliers* Some data sets include values so high or so low that they seem to stand apart from the rest of the data. These data are called *outliers*. Outliers may be from data collection errors, data entry errors, or simply valid but unusual data values. Regardless of the reason, it is important to identify outliers in the data set and examine the outliers carefully to determine if they are in error. One way to detect outliers is to use a box-and-whisker plot. Data values that fall beyond the limits,

Lower Limit: $Q_1 - 1.5 \times (IQR)$

Upper Limit: $Q_3 + 1.5 \times (IQR)$

where *IQR* is the interquartile range, are suspected outliers. In the computer software package Minitab, values beyond these limits are plotted with asterisks (*).

Students from a statistics class were asked to record their heights in inches. The heights (as recorded) were

65	72	68	64	60	55	73	71	52	63	61	74
69	67	74	50	4	75	67	62	66	80	64	65

(a) Make a box-and-whisker plot of the data.
(b) Find the value of the interquartile range (*IQR*).
(c) Multiply the *IQR* by 1.5 and find the lower and upper limits.
(d) Are there any data values below the lower limit? above the upper limit? List any suspected outliers. What might be some explanations for the outliers?

11. *Salary Increases: Interpret Graphs* The following Minitab display shows box-and-whisker plots for the percentage increases in annual salaries for the university and college faculty in Tennessee (by rank). *Source: Academe: Bulletin of the American Association of University Professors*, Volume 85, Number 2.

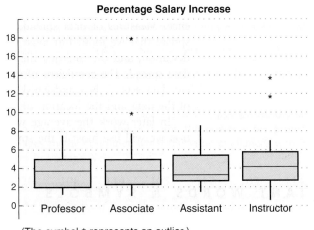

Percentage Salary Increase

(The symbol * represents an outlier.)

Minitab menu selections: ➤ **Graph** ➤ **Boxplot.**

Note: Minitab uses a procedure to find Q_1 and Q_3 that is slightly different from the method shown in this text. As a result, Minitab values for Q_1 and Q_3 will differ slightly from text or TI-83 values.

(a) Which faculty rank had the smallest median percentage salary increase? Which faculty rank had the single highest percentage salary increase?

(b) Which faculty rank had the largest spread between the first and third quartiles?

(c) Which faculty rank had the smallest spread for the lower 50% of the percentage salary increases?

(d) Which faculty rank had the most symmetric percentage salary increases? If the outliers for the associate professors were omitted, would that distribution appear to be symmetric?

(e) Look at the following descriptive statistics from Minitab.

VARIABLE	N	MEAN	MEDIAN	TRMEAN	STDEV	SE MEAN
Professo	20	3.685	3.700	3.611	1.766	0.395
Associat	24	4.362	3.750	3.927	3.592	0.733
Assistan	23	4.330	3.500	4.262	2.007	0.419
Instruct	21	4.781	4.100	4.568	3.132	0.684

Minitab Display: Descriptive Statistics

VARIABLE	MINIMUM	MAXIMUM	Q_1	Q_3
Professo	1.200	7.500	2.100	4.975
Associat	0.600	17.700	2.350	5.075
Assistan	1.500	8.600	2.900	5.500
Instruct	0.200	13.400	2.850	5.800

For the ranks of associate professor and instructor, compute the upper limit for outliers (see Problem 10). Do these ranks show outliers?

SUMMARY

To characterize numerical data, we use measures both of center and of variation. An average is an attempt to summarize the data into just one number. We studied several important averages: the mode, the median, and the mean. We also looked at trimmed means, which are more resistant to the effects of extreme values. However, rows and quadrants. The soil water content (percent of water by volume of soil) was determined for each plot. An important first step in such a research project is to give a statistical description of the data.

Soil Water Content for Bell Pepper Study

```
15  14  14  14  13  12  11  11  11  11  10  11  13  16  10
 9  15  12   9  10   7  14  13  14   8   9   8  11  13  13
15  12   9  10   9   9  16  16  12  10  11  11  12  15   6
10  10  10  11   9
```

(a) Make a box-and-whisker plot of the data. Find the interquartile range.
(b) *Grouped Data* Make a frequency table using 4 classes. Then estimate the mean and sample standard deviation by using the frequency table. Compute a 75% Chebyshev interval centered about the mean.
(c) If you have a statistical calculator or computer, use it to find the actual sample mean and sample standard deviation.

5. *Fishing: Line Strength* A certain brand of nylon monofilament fishing line is known to deteriorate in very cold temperatures. A spool of 10-pound test monofilament line was left out overnight at Fairbanks, Alaska, when temperatures dropped to $-35°F$. A random sample of six pieces of line gave the following breaking strengths (in pounds):

   10.1   6.2   9.8   5.3   9.9   5.7

(a) Compute the sample mean, sample standard deviation, coefficient of variation, and range.

A second spool of this line that had not been subjected to extreme cold temperatures gave the following breaking strengths (in pounds) for a random sample of six pieces of line:

   10.2   9.7   9.8   10.3   9.6   10.1

(b) Compute the sample mean, sample standard deviation, coefficient of variation, and range for these values.
(c) Compare your answers for parts (a) and (b) and comment on the observed differences. Which line had the more consistent performance? How was this reflected in the sample standard deviations and in the coefficients of variation? In the ranges?

6. *Medical: Glucose Blood Level* The data in Table 3-8 represent glucose blood level (mg/100 ml) after a 12-hour fast for a random sample of 70 women (based on infor-

TABLE 3-8   Glucose Blood Level (mg/100 ml)

45	66	83	71	76	64	59	59	76	82	80
81	85	77	82	90	87	72	79	69	83	71
87	69	81	76	96	83	67	94	101	94	89
94	73	99	93	85	83	80	78	80	85	83
84	74	81	70	65	89	70	80	84	77	65
46	80	70	75	45	101	71	109	73	73	80
72	81	63	74							

mation from the *American Journal of Clinical Nutrition,* Vol. 19, 345–351). Compute the five-number summary and the interquartile range, and make a box-and-whisker plot.

7. *General: Average Weight* An elevator is loaded with 16 people and is at its load limit of 2500 lb. What is the mean weight of these people?

8. *General: Create Examples*
   (a) Is it possible that the range and standard deviation can be equal? If your answer is yes, give an example of a data set in which they are equal.
   (b) Is it possible that the mean, median, and mode can all be equal? Is it possible that they can all be different? If your answer is yes to either question, give an example to illustrate your answer.

DATA HIGHLIGHTS: GROUP PROJECTS

Old Faithful Geyser, Yellowstone National Park

Break into small groups and discuss the following topics. Organize a brief outline in which you summarize the main points of your group discussion.

1. *The Story of Old Faithful* is a short book written by George Marler and published by the Yellowstone Association. Chapter 7 of this interesting book talks about the effect of the 1959 earthquake on eruption intervals for Old Faithful Geyser. Dr. John Rinehart (a senior research scientist with the National Oceanic and Atmospheric Administration) has done extensive studies of the eruption intervals before and after the 1959 earthquake. Examine Figure 3-10. Notice the general shape. Is the graph more or less symmetrical? Does it have a single mode frequency? The mean interval between eruptions has remained steady at about 65 minutes for the past 100 years. Therefore, the 1959 earthquake did not significantly change the mean, but it did change the distribution of eruption intervals. Examine Figure 3-11. Would you say there are really two frequency modes? One shorter and the other longer?

FIGURE 3-10

Typical Behavior of Old Faithful Geyser Before 1959 Quake

Why is the median a midway point? Why is the mode the most common data point? List three areas of daily life in which you think either the mean, median, or mode would be the best choice to describe an "average."

2. Why do we need to study the variation of a collection of data? Why isn't the average by itself adequate? We have studied three ways to measure variation. The range, standard deviation, and to a large extent, a box-and-whisker plot all indicate the variation within a data collection. Discuss similarities and differences among these ways to measure data variation. Why would it seem reasonable to pair the median with a box-and-whisker plot and to pair the mean with the standard deviation? What are the advantages and disadvantages of each method of describing data spread? Comment on statements such as the following: (a) The range is easy to compute, but it doesn't give much information; (b) although the standard deviation is more complicated to compute, it has some significant applications; (c) the box-and-whisker plot is fairly easy to construct, and it gives a lot of information at a glance.

3. Why is the coefficient of variation important? What do we mean when we say the coefficient of variation has no units? What advantage can there be in having no units? Why is *relative size* important?

   Consider robin eggs; the mean weight of a collection of robin eggs is 0.72 ounce and the standard deviation is 0.12 ounce. Now consider elephants; the mean weight of elephants in the zoo is 6.42 tons, with a standard deviation 1.07 tons. The units of measurement are different and there is a great deal of difference between the size of an elephant and a robin's egg. Yet the coefficient of variation is about the same for both. Comment on this from the viewpoint of the size of the standard deviation relative to the mean.

4. What is Chebyshev's theorem? Suppose you have a friend who knows very little about statistics. Write a paragraph or two in which you describe Chebyshev's theorem for your friend. Keep the discussion as simple as possible, but be sure to get the main ideas across to your friend. Suppose he or she asks, "What is this stuff good for?" and suppose you respond (a little sarcastically) that Chebyshev's theorem applies to everything from butterflies to the orbits of the planets! Would you be correct? Explain.

Using TECHNOLOGY

TI-83PLUS • EXCEL • MINITAB • COMPUTERSTAT

Raw Data

APPLICATION

Using the software or calculator available to you, do the following.

Trade winds are one of the beautiful features of island life in Hawaii. The following data represent total air movement in miles each day over a weather station in Hawaii as determined by a continuous anemometer recorder. The period of observation is January 1 to February 15, 1971.

26	14	18	14	113	50	13	22
27	57	28	50	72	52	105	138
16	33	18	16	32	26	11	16
17	14	57	100	35	20	21	34
18	13	18	28	21	13	25	19
11	19	22	19	15	20		

Source: United States Department of Commerce, National Oceanic and Atmospheric Administration, Environmental Data Service. *Climatological Data, Annual Summary, Hawaii,* Vol. 67, No. 13. Asheville: National Climatic Center, 1971, pp. 11, 24.

(a) Enter the data.
(b) Use the computer to find the sample mean, median, and (if it exists) mode.
(c) Use the computer to find the range, sample variance, and sample standard deviation.
(d) We studied the box-and-whisker plot as a topic in exploratory data analysis (EDA). Use the five-number summary provided by the computer to make a box-and-whisker plot of total air movement over the weather station.
(e) Four data values are exceptionally high: 113, 105, 138, and 100. The strong winds of January 5 (113 reading) brought in a cold front that dropped snow on Haleakala National Park (at the 8000 ft elevation). The residents were so

excited that they drove up to see the snow and caused such a massive traffic jam that the Park Service had to close the road. The winds of January 15 and 16 (readings 105 and 138) brought in a storm that created more damaging funnel clouds than any other storm to that date in the recorded history of Hawaii. The strong winds of January 28 (reading 100) accompanied a storm with funnel clouds that did much damage. Eliminate these values (i.e., 100, 105, 113, 138) from the data bank and redo parts (a) through (d). Compare your results with those previously obtained. Which average is most affected? What happens to the standard deviation? How do the two box-and-whisker plots compare?

(f) What is "normal" weather in a region? This is a relevant question since climate is an important factor in agriculture, commerce, transportation, and tourism. The National Climate Data Center keeps temperature, moisture, wind, and other climate records. It establishes "climatic normals." A "climatic normal" is an arithmetic average of a meteorological element over 30 years. For more information, visit the Brase/Brase *Understanding Basic Statistics* site at http://math.college.hmco.com/students and find the links to NOAA (National Oceanic and Atmospheric Administration) and *USA Today.* Once at the NOAA site, use the index to find "climatic normals." At the *USA Today* site, select weather.

Technology Hints: Raw Data

The Tech Note of Section 3.2 gives brief instructions for finding summary statistics for raw data using the TI-83, Excel, and Minitab. The Tech Note of Section 3.3 gives brief instructions for constructing box plots on the TI-83 and in Minitab. In ComputerStat, select Descriptive Statistics, then Frequency Distribution.

123

4

Regression and Correlation

When it is not in our power to determine what is true, we ought to follow what is most probable.

—René Descartes

René Descartes (1596–1650)
French mathematician and philosopher.

It is important to realize that statistics and probability do not deal in the realm of certainty. If there is any realm of human knowledge where genuine certainty exists, you may be sure that our statistical methods are not needed there. In most human endeavors, and in almost all of the natural world around us, the element of chance happenings cannot be avoided. When we cannot expect something with true certainty, we must rely on probability to be our guide. In this chapter, we will study regression, correlation, and forecasting. One of the tools we use is a scatter plot. René Decartes was the first mathematician to systematically use rectangular coordinate plots. For this reason, such a coordinate axis is called a Cartesian axis.

For on-line student resources, visit **math.college.hmco.com/students** and follow the Statistics links to the Brase/Brase, *Understanding Basic Statistics,* 3rd edition Web site.

◇ How can we use a scatter plot to visually estimate the degree of linear correlation of random variables? (SECTION 4.1)

◇ How is the least-squares line mathematically determined? (SECTION 4.2)

◇ What is the mathematical definition of the correlation coefficient, and how is it computed? What does it tell us? (SECTION 4.3)

Changing Populations and Crime Rate

Is the crime rate higher in neighborhoods where people might not know each other very well? Is there a relationship between crime rate and population change? If so, can we make predictions based on such a relationship? Is the relationship statistically significant? Is it possible to predict crime rate from population changes?

Denver is a city that has had a lot of growth and consequently a lot of population change in recent years. Sociologists studying population changes and crime rate could find a wealth of information in Denver statistics. Let x be a random variable representing percentage change in neighborhood population in the past few years, and let y be a random variable representing crime rate (crimes per 1,000 population). A random sample of six Denver neighborhoods gave the following information (*Source: Neighborhood Facts*, The Piton Foundation). To find out more about the Piton Foundation, visit the Brase/Brase *Understanding Basic Statistics* site at http://math.college.hmco.com/students and find the link to the Piton Foundation.

x	29	2	11	17	7	6
y	173	35	132	127	69	53

Using information presented in this chapter, you will be able to make the following analysis:

(a) Draw a scatter diagram to display the data.

(b) Compute the least-squares line and graph it on the scatter diagram.

(c) If a neighborhood had a 12% change in population in the past few years, could you predict the crime rate from the least-squares line?

(d) What is the sample correlation coefficient?

(e) Does a high correlation guarantee a "cause-and-effect" situation, or is it simply an indication of a mathematical relationship between variables?

FOCUS POINTS

✓ Make a scatter diagram.
✓ Visually estimate the location of the "best-fitting" line for a scatter diagram.
✓ Visually estimate the degree of linear correlation.

Paired data values

4.1
Introduction to Paired Data and Scatter Diagrams

The study of correlation and regression of two variables usually begins with a table and/or a graph of *paired data values*. Example 1 shows what we mean.

EXAMPLE 1

Scatter diagram

For her botany class project, Jan wonders if there is a connection between the average weight of watermelons a vine produces and the root depth of the vine. Jan suspects that vines with deeper roots have a better water supply and also larger average melons. From a large watermelon field, 15 vines are chosen at random. At the end of 8 weeks, the watermelons are removed from each vine, and the average weight of watermelons from each vine is determined. Each plant is carefully dug up and its root depth (or length) is measured. Table 4-1 shows the results.

Scatter diagram

(a) For each plant, there is an ordered pair (x, y) of data values. If we plot the pairs (x, y) as points on a coordinate system, we obtain the graph of Figure 4-1. Figure 4-1 is called a *scatter diagram* for the paired data values of Table 4-1.

(b) By inspecting Figure 4-1, we see that to some extent, larger values of x tend to be associated with larger y values, and smaller x values tend to be associated with smaller y values. Roughly speaking, the general trend seems to be reasonably well represented by the line segment shown in Figure 4-1.   ◊

Of course, it is possible to draw many curves in Figure 4-1, but the straight line is the simplest and most widely used curve for elementary studies of paired data. We can draw many lines in Figure 4-1, but in some sense, the "best" line should be the one that comes closest to each of the points of the scatter diagram. To single out one line as the "best-fitting line," we must find a mathematical criterion for this line and a formula representing the line. This will be done in Section 4.2 by the *method of least squares*.

TABLE 4-1   Results of a Botany Experiment

x = Root Depth (nearest inch)	y = Mean Weight of Watermelon (nearest pound)
26	20
14	10
18	13
10	9
26	19
21	17
7	8
26	15
13	9
19	13
17	12
13	7
16	9
28	17
23	14

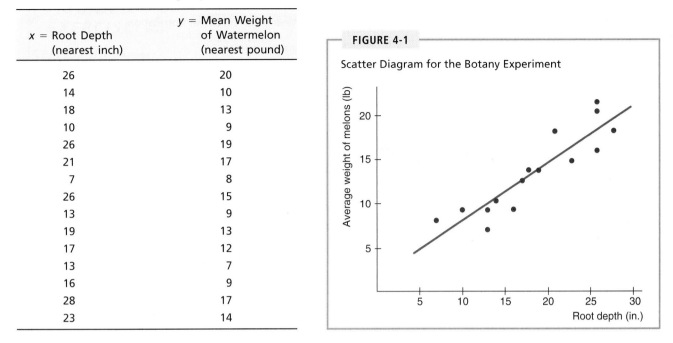

FIGURE 4-1

Scatter Diagram for the Botany Experiment

Introduction to linear correlation

Another problem precedes that of finding the "best-fitting line." That is the problem of determining how well the points of the scatter diagram are suited for fitting *any* line. Certainly, if the points are a very poor fit to *any* line, there is little use in trying to find the "best" line. This problem will be dealt with in Section 4.3 by use of the Pearson product–moment coefficient of correlation.

If the points of a scatter diagram are located so that *no* line is realistically a "good" fit, we then say that the points possess *no linear correlation*. We see some examples of scatter diagrams for which there is no linear correlation in Figure 4-2.

If the points seem close to a straight line, we say the linear correlation is low to moderate, depending on how close the points lie to a line. If all the points do, in fact, lie on a line, then we have *perfect linear correlation*. In Figure 4-3 on the following page, we see some diagrams with perfect linear correlation. In statistical applications, perfect linear correlation almost never occurs.

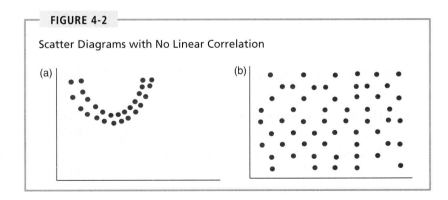

FIGURE 4-2

Scatter Diagrams with No Linear Correlation

(a)

(b)

Looking at the scatter diagram of Figure 4-6, we ask two questions:

1. Can we find a relationship between x and y?

2. If so, how strong is the relationship?

The first step in answering these questions is to try to express the relationship as a mathematical equation. There are many possible equations, but the simplest and most widely used is the linear equation, or the equation of a straight line. Because we will be using this line to predict the y values (temperature) from the x values (chirps per second), we call x the *explanatory variable* and y the *response variable*.

Explanatory variable
Response variable

Our job is to find the "best" linear equation representing the points of the scatter diagram. For our criterion of best-fitting line, we use the *least-squares criterion*, which says that the line we fit to the data points must be such that *the sum of the squares of the vertical distances from the points to the line be made as small as possible*. The least-squares criterion is illustrated in Figure 4-7.

Least-squares criterion

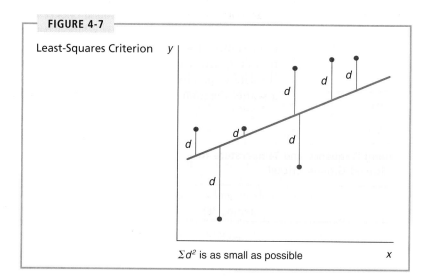

FIGURE 4-7

Least-Squares Criterion

Σd^2 is as small as possible

In Figure 4-7, d represents the difference between the y coordinate of the data point and the corresponding y coordinate on the line. Thus, if the data point lies above the line, d is positive, but if the data point is below the line, d is negative. As a result, the sum of the d values can be small even if the points are widely spread in the scatter diagram. However, the squares d^2 cannot be negative. By minimizing the sum of the squares, we are, in effect, not allowing positive and negative d values to "cancel out" one another in the sum. It is in this way that we can meet the least-squares criterion of minimizing the sum of the squares of the vertical distances between the points and the line over *all* points in the scatter diagram.

Least-squares line

Techniques of calculus can be applied to show that the line that meets the least-squares criterion is as follows:

Least-squares line

$$y = a + bx \tag{1}$$

where $b = \dfrac{SS_{xy}}{SS_x}$ $\quad b$ is the slope $\tag{2}$

$\quad a = \bar{y} - b\bar{x}$ $\quad a$ is the y-intercept $\tag{3}$

and $\bar{y}$ = mean of y values in scatter diagram
$\quad \bar{x}$ = mean of x values in scatter diagram

$$SS_{xy} = \Sigma xy - \frac{(\Sigma x)(\Sigma y)}{n} \tag{4}$$

$$SS_x = \Sigma x^2 - \frac{(\Sigma x)^2}{n} \tag{5}$$

$\quad n$ = number of points in a scatter diagram

In Formulas (4) and (5), the sums are taken over all x or y values in the scatter diagram.

◇ **COMMENT** In other mathematics courses, the slope–intercept form of the equation of a line is usually given as $y = mx + b$, where m refers to the slope of the line and b to the y coordinate of the y intercept. In statistics, it is standard to use the letter b to designate the slope of the least-squares line and the letter a to designate the y coordinate of the intercept. Be attentive! ◇

◇ **COMMENT** The notation SS_x is the same expression used in the computation for the standard deviation of x values (see Section 3.2). Recall that the formula for SS_x is simply a faster way of computing $\Sigma(x - \bar{x})^2$. Likewise, the formula for SS_{xy} is a more efficient way to compute $\Sigma(x - \bar{x})(y - \bar{y})$. ◇

Using the formulas to find the values of b and a

Finding values of b and a on your calculator

Most calculators supporting two-variable statistics provide the values of the slope b and the y intercept a directly. However, computing these values from formulas shows how the raw data are related to the equation of the least-squares line. The following discussion shows you how to use the formulas to find the values of b and a. If you are using your calculator to find the values of b and a directly, then you may omit the discussion regarding the use of the formulas. Go to the margin header "Using values a and b to construct the equation of the least-squares line" on page 137.

● **COMPUTATION NOTES**

1. The formulas used to compute the slope and y intercept of the least-squares line are sensitive to rounding. Answers to the exercises at the end of each section and at the end of the chapter are computer-generated, so you may expect slight differences in your answers, depending on how you round intermediate steps.

2. Do not confuse Σx^2 and $(\Sigma x)^2$. For Σx^2, we *first square* each x value and then find the total sum of these values. For $(\Sigma x)^2$, we *first sum* the x values and then

FIGURE 4-8

Least-Squares Line $y = 24.9 + 3.3x$

Rounded to the nearest whole number, we should predict the temperature to be 88°F. Of course, this is just a prediction, and we would be quite happy if the temperature turned out to be relatively close to our prediction. This brings up the natural question: How *good* are predictions based on the least-squares line? This is a fairly difficult question, and much of the answer requires advanced mathematics; however, a partial answer will be given later, in Section 12.4.

GUIDED EXERCISE 3

Least-squares line

The Quick Sell car dealership has been using 1-minute spot ads on a local TV station. The ads always occur during the evening hours and advertise the different models and price ranges of cars on the lot that week. During a 10-week period, the Quick Sell dealer kept a weekly record of the number x of TV ads versus the number y of cars sold. The results are given in Table 4-5.

Because Quick Sell can afford only 12 ads per week, how many cars can the dealership expect to sell each week?

We'll answer this question in several steps. If you use a calculator to find values of a and b directly, do part (a) and then skip to part (e). To use the formulas to find a and b, complete all parts.

TABLE 4-5

x	y
6	15
20	31
0	10
14	16
25	28
16	20
28	40
18	25
10	12
8	15

Continued

GUIDED EXERCISE 3 continued

(a) Draw a scatter diagram for the data.

⟹ The scatter diagram is shown in Figure 4-9.

(b) Look at Equations (1) to (5) pertaining to the least-squares line. Two of the quantities we need to find b are (Σx) and (Σxy). List the others.

⟹ We also need n, (Σy), (Σx^2), and $(\Sigma x)^2$.

(c) Complete Table 4-6(a).

⟹ The missing table entries are shown in Table 4-6(b).

TABLE 4-6(a)

x	y	x^2	xy
6	15	36	90
20	31	400	620
0	10	0	0
14	16	196	224
25	28	625	700
16	20	256	320
28	40	784	1120
18	25	324	450
10	12	100	120
8	15	64	120
$\Sigma x = 145$	$\Sigma y = 212$	$\Sigma x^2 = 2785$	$\Sigma xy = 3764$

TABLE 4-6(b)

x^2	xy
$(28)^2 = 784$	$28(40) = 1120$
$(18)^2 = 324$	$18(25) = 450$
$(10)^2 = 100$	$10(12) = 120$
$\Sigma x^2 = 2785$	$\Sigma xy = 3764$

FIGURE 4-9 Scatter Diagram and Least-Squares Line for Table 4-5

(d) Use Table 4-6(a) to compute SS_x, SS_{xy}, $\bar{x}$, and $\bar{y}$.

⟹ $$SS_x = \Sigma x^2 - \frac{(\Sigma x)^2}{n} = 2785 - \frac{(145)^2}{10} = 682.5$$

$$SS_{xy} = \Sigma xy - \frac{(\Sigma x)(\Sigma y)}{n} = 3764 - \frac{(145)(212)}{10}$$

$$= 690.0$$

$$\bar{x} = \frac{\Sigma x}{n} = \frac{145}{10} = 14.5, \quad \bar{y} = \frac{\Sigma y}{n} = \frac{212}{10} = 21.2$$

(e) Compute (or find on a calculator) values of a and b in the formula

$$y = a + bx$$

for the least-squares line. What is the equation of the least-squares line?

⟹ $$b = \frac{SS_{xy}}{SS_x} = \frac{690.0}{682.5} \approx 1.01$$

$$a = \bar{y} - b\bar{x} \approx 21.2 - 1.01(14.5) = 6.56$$

The equation for the least-squares line is

$$y = 6.56 + 1.01x$$

Continued

GUIDED EXERCISE 3 continued

(f) Plot the least-squares line on your scatter diagram.

⟹ The least-squares line goes through the point $(\bar{x}, \bar{y})$ = (14.5, 21.2). To get another point on the line, select a value for x and compute the corresponding y value using the equation $y = 6.56 + 1.01x$. For $x = 20$, we get $y = 6.56 + 1.01(20) = 26.8$, so the point (20, 26.8) is also on the line. The least-squares line is shown in Figure 4-9.

(g) Read the y value for $x = 12$ from your graph. Then use the equation of the least-squares line to calculate y when $x = 12$. How many cars can the manager expect to sell if 12 ads per week are aired on TV?

⟹ The graph gives $y \approx 19$. From the equation we get

$y = 6.56 + 1.01x$

$\quad = 6.56 + 1.01(12) \quad$ using 12 in place of x

$\quad = 18.68$

To the nearest whole number, the manager can expect to sell 19 cars when 12 ads are aired on TV each week.

TECH NOTE   When we have more data pairs, it is convenient to use a technology tool such as the TI-83Plus, Excel, or Minitab to find the equation of the least-squares line. The displays show results for data of Guided Exercise 3 regarding car sales and ads.

TI-83Plus Press **STAT**, choose **Calculate**, use option **8:LinReg(a+bx)**. For a graph showing the scatter plot and the least-squares line, press the **STAT PLOT** key, turn on a plot, and highlight the first type. Then press the **Y=** key. To enter the equation of the least-squares line, press **VARS**, select **5:Statistics**, highlight **EQ**, and then **1:RegEQ**. Press **ENTER**. Finally, press **ZOOM** and choose **9:ZoomStat**.

Excel There are several ways to find the equation of the least-squares line in Excel. One way is to make a scatter plot using the menu choices **Chart wizard ➤ Scatter Diagram**. When the diagram is complete, **right click** on one of the points on the diagram, select **trendline**, and, under options, check the box "display equation of line."

TI-83Plus Display

Excel Display

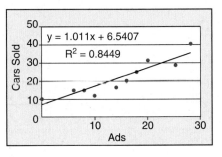

Minitab There are a number of ways to generate the least-squares line. One way is to use the menu selection **Stat ➤ Regression ➤ Fitted Line Plot.** The least-squares equation is shown with the diagram.

Interpolation

We have used the least-squares line to predict y values for x values that were *between* x values observed in the experiment. Predicting y values for x values that are between x values of points in the scatter diagram is called *interpolation*. The least-squares lines can be used for interpolation. Predicting y values for an x value beyond the range of observed x values is a complex problem that is not treated in this book. Prediction beyond the range of observations is called *extrapolation*.

Extrapolation

The *least-squares line*

$$y = a + bx$$

was developed with y as the response variable and x as the explanatory variable. This model can be used only to predict y values from specified x values. If you wish to begin with y values and predict corresponding x values, you must use a different formula. Such a formula would be developed by using a model with x as the response variable and y as the explanatory variable. For our purposes, we'll always arrange to predict y values from given x values.

Effect of extreme data points

The least-squares line can be affected greatly by *extreme* or *influential* data points. This fact is illustrated in Example 2 and in the Data Highlights for this chapter.

E X A M P L E   2

Extreme data points

Consider the two data sets shown below. They are identical except for the y coordinate of the last point.

Set 1	3	5	7	9	15	20	25	27	33	35
	60	55	57	50	47	44	48	40	35	30
Set 2	3	5	7	9	15	20	25	27	33	35
	60	55	57	50	47	44	48	40	35	55

Figure 4-10 shows the scatter plots for these two data sets as well as the equation of the least-squares lines for each set. Compare the slopes of the lines. Notice that the line for data set 2 has a slope closer to zero. This line is not as steep. ◇

FIGURE 4-10

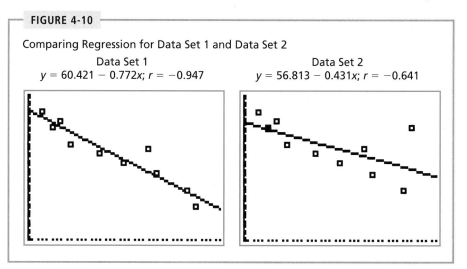

Comparing Regression for Data Set 1 and Data Set 2

Data Set 1
$y = 60.421 - 0.772x; r = -0.947$

Data Set 2
$y = 56.813 - 0.431x; r = -0.641$

Sometimes a scatter diagram clearly indicates the existence of a linear relationship between x and y, but it can happen that the points are widely scattered around the least-squares line. We need a method (besides just looking) for measuring the spread of points about the least-squares line. The next section gives us such a measure. It is the value of r shown in Figure 4-10.

VIEWPOINT   *It's Freezing!*

Can we use average temperatures in January to predict how bad the rest of the winter will be? Can you predict the number of days with freezing temperatures for the entire calendar year by using conditions in January? How good would such a forecast be for predicting growing season or number of frost-free days? Methods of this section can help you answer such questions. For more information, visit the Brase/Brase *Understanding Basic Statistics* site at http://math.college.hmco.com/students and find the link to temperatures.

SECTION 4.2   PROBLEMS

Please do the following for Problems 1–9.
(a) Draw a scatter diagram for the data.
(b) Find $\bar{x}$, $\bar{y}$, and b. Then find the equation of the least-squares line.
(c) Graph the least-squares line on your scatter diagram. Be sure to use the point $(\bar{x}, \bar{y})$ as one of the points on the line.

Answers may vary slightly due to rounding.

1. *Assembly Line: Defects* The number of workers on an assembly line varies due to the level of absenteeism on any given day. In a random sample of production output from several days of work, the following data were obtained, where x = number of workers absent from the assembly line and y = number of defects coming off the line.

x	3	5	0	2	1
y	16	20	9	12	10

   Complete parts (a) through (c).
   (d) On a day when four workers are absent from the assembly line, what would the least-squares line predict for the number of defects coming off the line?

2. *Ranching: Cattle* You are the foreman of the Bar-S cattle ranch in Colorado. A neighboring ranch has calves for sale, and you are going to buy some calves to add to the Bar-S herd. How much should a healthy calf weigh? Let x be the age of the calf (in weeks), and let y be the weight of the calf (in kilograms). The following information is based on data taken from *The Merck Veterinary Manual* (a reference used by many ranchers).

x	1	3	10	16	26	36
y	42	50	75	100	150	200

Complete parts (a) through (c).
(d) The calves you want to buy are 12 weeks old. What does the least-squares line predict for a healthy weight?

3. *Weight of Car: Miles per Gallon* Do heavier cars really use more gasoline? Suppose that a car is chosen at random. Let x be the weight of the car (in hundreds of pounds), and let y be the miles per gallon (mpg). The following information is based on data taken from *Consumer Reports* (vol. 62, no. 4).

x	27	44	32	47	23	40	34	52
y	30	19	24	13	29	17	21	14

Complete parts (a) through (c).
(d) Suppose that a car weighs $x = 38$ (hundred pounds). What does the least-squares line forecast for $y = $ miles per gallon?

4. *Basketball: Fouls* Data for this problem are based on information from *STATS Basketball Scoreboard*. It is thought that basketball teams that make too many fouls in a game tend to lose the game even if they otherwise play well. Let x be the number of fouls more than (i.e., over and above) those of the opposing team. Let y be the percentage of times the team with the larger number of fouls wins the game.

x	0	2	5	6
y	50	45	33	26

Complete parts (a) through (c).
(d) If a team had $x = 4$ fouls over and above those of the opposing team, what does the least-squares equation forecast for y?

5. *College: Income* The following data are based on information from the book *Life in America's Small Cities* (by G. S. Thomas, Prometheus Books). Let x be the percentage of those 25 years or older with 4 or more years of college. Let y be the per capita income in thousands of dollars. Five small cities in South Carolina (Greenwood, Hilton Head Island, Myrtle Beach, Orangeburg, and Sumpter) reported the following information regarding the x and y variables:

x	13.8	21.9	12.5	12.7	11.5
y	9.0	10.8	8.8	6.9	7.2

Complete parts (a) through (c).
(d) In a small city in South Carolina where $x = 20$ percent of the population 25 years or older who have had 4 or more years of college, what would the least-squares equation forecast for $y = $ per capita income (in thousands of dollars) in this community?

Development of Formula for *r*

If there is a *positive* linear relation between variables x and y, then high values of x are paired with high values of y, and low values of x are paired with low values of y. [See Figure 4-11(a).] In the case of *negative* linear correlation, high values of x are paired with low values of y, and low values of x are paired with high values of y. This relation is pictured in Figure 4-11(b). If there is little or no *linear correlation* between x and y, however, then we will find both high and low x values sometimes paired with high y values and sometimes paired with low y values. This relation is shown in Figure 4-11(c).

These observations lead us to the development of the formula for the correlation coefficient r. Taking *high* to mean "above the mean," we can express the relationships pictured in Figure 4-11 by considering the products

$$(x - \bar{x})(y - \bar{y})$$

If both x and y are high, both factors will be positive, and the product will be positive as well. The sign of this product will depend on the relative values of x and y compared with their respective means.

$$(x - \bar{x})(y - \bar{y}) \begin{cases} \text{is positive if } x \text{ and } y \text{ are both "high"} \\ \text{is positive if } x \text{ and } y \text{ are both "low"} \\ \\ \text{is negative if } x \text{ is "low," but } y \text{ is "high"} \\ \text{is negative if } x \text{ is "high," but } y \text{ is "low"} \end{cases}$$

In the case of positive linear correlation, most of the products $(x - \bar{x})(y - \bar{y})$ will be positive and so will the sum over all the data pairs

$$\Sigma(x - \bar{x})(y - \bar{y})$$

For negative linear correlation, the products will tend to be negative, so the sum also will be negative. On the other hand, in the case of little, if any, linear correlation, the sum will tend to be zero.

FIGURE 4-11

Patterns for Linear Correlation

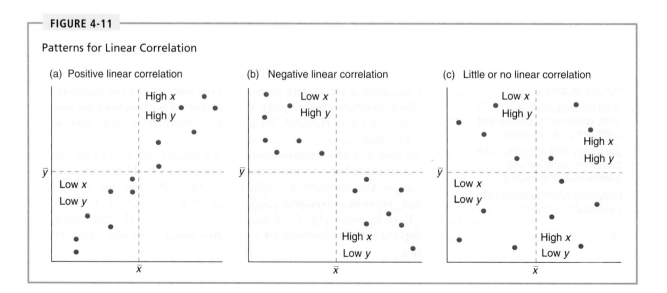

One trouble with the preceding sum is that it will be larger or smaller, ~~~ ing on the units of x and y. Because we want r to be unitless, we standardize both x and y of a data pair by dividing each factor $(x - \bar{x})$ by the sample standard deviation of the x values, s_x. Divide each factor $(y - \bar{y})$ by the standard deviation of the y values, s_y. Finally, we take an average of all the products. For technical reasons, we take the average by dividing by $n - 1$ instead of by n. This process leads us to the desired measurement, r.

$$r = \frac{1}{n-1} \Sigma \frac{(y - \bar{y})}{s_y} \cdot \frac{(x - \bar{x})}{s_x}$$

Meaning of r

Let us delay an example showing how to compute r until we know a little more about the meaning of the correlation coefficient. It can be shown mathematically that r is always a number between $+1$ and -1 ($-1 \le r \le +1$). Table 4-7 gives a quick summary of some basic facts about r. Guided Exercise 4 below shows relationships between scatter diagrams and respective values of r.

TABLE 4-7   Some Facts About the Correlation Coefficient

If r Is	Then	The Scatter Diagram Might Look Something Like
0	There is no linear relation for the points of the scatter diagram. The x and y values have *zero correlation*.	
1 or −1	There is a perfect linear relation between x and y values; all points lie on the least-squares line.	$r = -1$ $r = 1$
Between 0 and 1 ($0 < r < 1$)	The x and y values have a *positive correlation*. By this, we mean that *large x* values are associated with *large y* values, and *small x* values are associated with *small y* values.	As we go from left to right, the least-squares line goes *up*.
Between −1 and 0 ($-1 < r < 0$)	The x and y values have a *negative correlation*. By this, we mean *large x* values are associated with *small y* values, and *small x* values are associated with *large y* values.	As we go from left to right, the least-squares line goes *down*.

GUIDED EXERCISE 4

Scatter diagrams and r

Match the appropriate statement about r to each scatter diagram in Figure 4-12.

(1) $r = 0$.   (2) $r = 1$.   (3) $r = -1$.

(4) r is between 0 and 1.   (5) r is between -1 and 0.

FIGURE 4-12 Scatter Diagrams

(a) $r = 1$ because all the points are on the line and the line goes up from left to right.

(b) r is between -1 and 0 because the points are fairly close to the line, and as we read from left to right, the least-squares line goes down.

(c) $r = 0$ because there is no apparent linear relation among the points.

Computation Formula for r

The defining formula for r is awkward to use because of all the subtractions. As before, we can simplify the formula and produce one that is much easier to use. In fact, it should be no surprise that the sums utilized in computing the slope of the least-squares line are also used in the computational formula for r.

> **Formula to calculate correlation coefficient, r**
>
> $$r = \frac{SS_{xy}}{\sqrt{SS_x SS_y}} \tag{7}$$
>
> where $SS_{xy} = \Sigma xy - \dfrac{(\Sigma x)(\Sigma y)}{n}$ Note: $SS_{xy} = \Sigma(x - \bar{x})(y - \bar{y})$
>
> $SS_x = \Sigma x^2 - \dfrac{(\Sigma x)^2}{n}$ and $SS_y = \Sigma y^2 - \dfrac{(\Sigma y)^2}{n}$
>
> n = number of data pairs in scatter diagram

This formula for r is sensitive to rounding. As in other formulas involving SS_x, SS_{xy}, and SS_y, you want to carry as many digits as is reasonable for your problem until the last step. Again, depending on the rounding process used, answers will vary slightly. The answers for the end-of-section and end-of-chapter exercises were computer-generated, so your answers might differ from them slightly and still be essentially correct.

TABLE 4-8   IQ versus CGA (on a Four-Point Scale) of 12 High School Seniors

IQ, x	117	92	102	115	87	76	107	108	121	91	113	98
CGA, y	3.7	2.6	3.3	2.2	2.4	1.8	2.8	3.2	3.8	3.0	4.0	3.5

EXAMPLE 3

Computing r

Most of us have heard someone say that more intelligent people tend to do better in school. Is this always true? Experienced teachers know that it is only partially true. Students with higher IQs (intelligence quotients) often do better schoolwork, but factors other than IQ can affect academic success. However, let's see if there is a correlation between IQ and cumulative grade averages (CGA). The principal of Delta High School chose 12 students from the senior class at random and compiled the data in Table 4-8.

(a) First, make a scatter diagram, and determine if r is positive, close to 0, or negative.

SOLUTION: The scatter diagram indicates that r is positive. See Figure 4-13.

(b) Compute r. The solution shows the use of the computation formula to compute r. Alternatively, you may read the value of r directly from your calculator or computer display.

SOLUTION: To find r, we must compute Σx, Σy, Σx^2, Σy^2, and Σxy. The values for $(\Sigma x)^2$ and $(\Sigma y)^2$ can be obtained from Σx and Σy. It is easiest to organize our work into a table of five columns (Table 4-9 on the following page). The first two columns are just a repetition of Table 4-8.

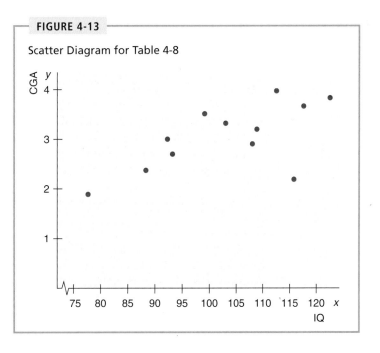

FIGURE 4-13

Scatter Diagram for Table 4-8

TABLE 4-9   Information Necessary to Compute *r*

x(IQ)	y(CGA)	x^2	y^2	xy
117	3.7	13,689	13.7	432.9
92	2.6	8,464	6.8	239.2
102	3.3	10,404	10.9	336.6
115	2.2	13,225	4.8	253.0
87	2.4	7,569	5.8	208.8
76	1.8	5,776	3.2	136.8
107	2.8	11,449	7.8	299.6
108	3.2	11,664	10.2	345.6
121	3.8	14,641	14.4	459.8
91	3.0	8,281	9.0	273.0
113	4.0	12,769	16.0	452.0
98	3.5	9,604	12.3	343.0
$\Sigma x = 1{,}227$	$\Sigma y = 36.3$	$\Sigma x^2 = 127{,}535$	$\Sigma y^2 = 114.9$	$\Sigma xy = 3{,}780.3$
$(\Sigma x)^2 = 1{,}505{,}529$	$(\Sigma y)^2 = 1{,}317.7$			

To use Equation (7) to calculate r, we will first compute SS_{xy}, SS_x, and SS_y.

$$SS_{xy} = \Sigma xy - \frac{(\Sigma x)(\Sigma y)}{n} = 3780.3 - \frac{(1227)(36.3)}{12} \approx 68.63$$

$$SS_x = \Sigma x^2 - \frac{(\Sigma x)^2}{n} = 127{,}535 - \frac{(1227)^2}{12} = 2074.25$$

$$SS_y = \Sigma y^2 - \frac{(\Sigma y)^2}{n} = 114.9 - \frac{(36.3)^2}{12} \approx 5.09$$

Therefore, the correlation coefficient is

$$r = \frac{SS_{xy}}{\sqrt{SS_x SS_y}} = \frac{68.63}{\sqrt{(2074.25)(5.09)}} = 0.6679 \approx 0.67$$

Our correlation coefficient is $r \approx 0.67$. Let's make sure that this answer agrees with what we expect from a quick glance at the scatter diagram (Figure 4-13 on the preceding page). In Figure 4-13, the general trend is upward as we read from left to right, so we would expect a positive value for r. The value 0.67 is in the expected range—that is, it is between 0 and 1.   ◇

It is quite a task to compute r for even 12 data pairs. The use of columns as in Example 3 is extremely helpful. Your value for r should always be between -1 and 1, inclusive. Use a scatter diagram to get a rough idea of the value of r. If your computed value of r is outside the allowable range, or if it disagrees quite a bit with the scatter diagram, recheck your calculations. Be sure you distinguish between expressions such as (Σx^2) and $(\Sigma x)^2$. Negligible rounding errors may occur, depending on how you (or your calculator) round.

GUIDED EXERCISE 5

Computing r

In one of the Boston city parks, there has been a problem with muggings in the summer months. A police cadet took a random sample of 10 days (out of the 90-day summer) and compiled the following data. For each day, x represents the number of police officers on duty in the park and y represents the number of reported muggings on that day.

x	10	15	16	1	4	6	18	12	14	7
y	5	2	1	9	7	8	1	5	3	6

(a) Construct a scatter diagram of x and y values. ⟹ Figure 4-14 shows the scatter diagram.

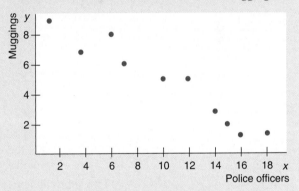

FIGURE 4-14 Scatter Diagram for Number of Police Officers versus Number of Muggings

(b) From the scatter diagram, do you think the computed value of r will be positive, negative, or zero? Explain.

⟹ r will be negative. The general trend is that large x values are associated with small y values and vice versa. From left to right, the least-squares line goes down.

(c) Complete **TABLE 4-10**. ⟹

TABLE 4-11   Completion of Table 4-10.

x	y	x^2	y^2	xy
6	8	36	64	48
18	1	324	1	18
12	5	144	25	60
14	3	196	9	42
		$\Sigma x^2 = 1{,}347$	$\Sigma y^2 = 295$	$\Sigma xy = 343$
		$(\Sigma x)^2 = 10{,}609$	$(\Sigma y)^2 = 2209$	

x	y	x^2	y^2	xy
10	5	100	25	50
15	2	225	4	30
16	1	256	1	16
1	9	1	81	9
4	7	16	49	28
6	8	36	64	48
18	1	324	1	18
12	5	144	25	60
14	3	196	9	42
7	6	49	36	42
$\Sigma x = 103$	$\Sigma y = 47$	$\Sigma x^2 = 1347$	$\Sigma y^2 = 291$	$\Sigma xy = 343$
$(\Sigma x)^2 = 10609$	$(\Sigma y)^2 = 2209$			

Continued

GUIDED EXERCISE 5 continued

(d) Compute SS_{xy}, SS_x, SS_y, and then r.
Alternatively, find the value of r directly by using a calculator or computer software.

$$SS_{xy} = \Sigma xy - \frac{(\Sigma x)(\Sigma y)}{n}$$

$$= 343 - \frac{(103)(47)}{10} = -141.1$$

$$SS_x = \Sigma x^2 - \frac{(\Sigma x)^2}{n}$$

$$= 1347 - \frac{(103)^2}{10} = 286.1$$

$$SS_y = \Sigma y^2 - \frac{(\Sigma y)^2}{n}$$

$$= 295 - \frac{(47)^2}{10} = 74.1$$

$$r = \frac{SS_{xy}}{\sqrt{SS_x SS_y}}$$

$$= \frac{-141.1}{\sqrt{(286.1)(74.1)}} \approx -0.9691 \approx -0.97$$

TECH NOTE Most calculators that support two-variable statistics provide the value of the correlation coefficient r directly. Statistical software provides r, r^2, or both.

TI-83Plus First use **CATALOGUE**, find, **DiagnosticOn**, and press **Enter** twice. Then when you use **STAT, CALC**, option **8:LinReg(a+bx)**, the values of r and r^2 will be given. (Data from Example 3.)

Excel Use the menu selection **Paste Function** (f_x) ➤ **Statistical** ➤ **Correl.**

Minitab Use the menu selection **Stat** ➤ **Basic Statistics** ➤ **Correlation.**

```
LinReg
y=a+bx
a=-.3579
b=.0331
r²=.4415
r=.6645
```

The correlation coefficient can be thought of as another measure of how "good" the least-squares line fits the data points of the scatter diagram. The closer r is to $+1$ or -1, the better the least-squares line "fits" the data. Values of r close to 0 indicate a poor "fit."

Usually, our scatter diagram does not contain *all* possible data points that could be gathered. Most scatter diagrams represent only a *random sample* of data pairs taken from a very large population of possible pairs. Because r is computed by Equation (7) on the basis of a random sample of (x, y) pairs, we expect the values of r to vary from one sample to the next (much as sample means $\bar{x}$ varied from sample to sample). This brings up the question of the *significance* of r. Or put another way, what are the chances that our random sample of data pairs indicates a high correlation when, in fact, the population x and y values are not so strongly correlated? Right now let us just say the significance of r is a separate issue that is treated in Section 12.5.

◊ **COMMENT** As we use computing formulas for the slope of the least-squares line, for r, and for standard deviations s_x and s_y, we see many of the same sums used. There is, in fact, a relationship between the correlation coefficient r and the slope of the least-squares line b. In instances when we know r, s_x, and s_y, we can use the following formula to compute b.

$$b = r\left(\frac{s_y}{s_x}\right) \quad ◊$$

Coefficient of Determination

Coefficient of determination

There is another way to answer the question, How good is the least-squares line as an instrument of regression? The *coefficient of determination* r^2 is the square of the sample correlation coefficient r.

$$\text{Coefficient of determination} = r^2$$

Explained and unexplained variation

The coefficient of determination r^2 is a measure of the proportion of variation in y that is explained by the regression line using x as the predicting variable. If $r = 0.90$, then $r^2 = 0.81$ is the coefficient of determination, and we can say that about 81% of the (variation) behavior of the y variable can be *explained* by the corresponding (variation) behavior of the x variable if we use the equation of the least-squares line. The remaining 19% of the (variation) behavior of the y variable is *unexplained*. It may be due to random chance or to the presence of other variables besides the x that may influence y.

GUIDED EXERCISE 6

Coefficient of determination

(a) In Example 3, we found the correlation coefficient r of the relationship between IQ and cumulative grade averages. In that case, r was 0.67. How would you describe the strength of this relationship?

⟹ The correlation coefficient $r = 0.67$ is moderate but not extremely high. It seems that other factors besides IQ are significant in determining a cumulative grade point average.

(b) Compute the coefficient of determination for the data of Example 3 and comment on the meaning of this number.

⟹ Since $r = 0.67$, then $r^2 = 0.449$ is the value of the coefficient of determination. This says that 44.9% of the variation of y = CGA can be explained by the least-squares line and x = IQ. The remaining $100 - 44.9 = 55.1\%$ of the variation of y is due to random chance or other variables besides x that influence y (possibly, amount of time spent studying!).

Continued

GUIDED EXERCISE 6 continued

(c) In Guided Exercise 5, dealing with the relation between the number of police officers in the park and the number of muggings, we found r to be -0.97. How would you describe the strength of this relationship? Do you think the city is justified in asking for more police officers to be assigned to park duty?

⟹ $r = -0.97$ is a high correlation. The relation between the number of police officers in the park and the number of muggings in the park is a strong and dependable negative correlation. The authors feel that the city would be wise to hire more police officers to patrol the park. But many other aspects of the situation must be considered. Perhaps more crimes would be prevented by putting those officers elsewhere.

(d) Compute the coefficient of determination for the data of Guided Exercise 5 and comment on the meaning of this number.

⟹ Since $r = -0.97$, then $r^2 = 0.941$ is the coefficient of determination. About 94.1% of the variation of y can be explained by the least-squares line and the x variable. The remaining $100 - 94.1 = 5.9\%$ of the variation of y is due to random chance or other variables besides x that influence y.

The correlation coefficient is a mathematical tool for measuring the strength of the linear relationship between two variables. As such, it makes no implication about cause or effect. Just because two variables tend to increase or decrease together does not mean a change in one is *causing* a change in the other. A strong correlation between x and y is sometimes due to other (either known or unknown) confounding variables.

Causation

EXAMPLE 4

Causation

Over a period of years, a certain town observed that the correlation between x, the number of people attending churches, and y, the number of people in the city jail, was $r = 0.90$.

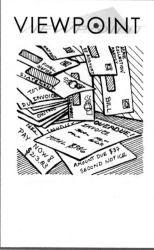

VIEWP●INT   *Low on Credit, High on Cost!!!*

How do you measure automobile insurance risk? One way is to use a little statistics and customer credit ratings. Insurers say statistics show that drivers who have a history of bad credit are more likely to be in serious car accidents. According to a high-level executive at Allstate Insurance Company, financial instability is an extremely powerful predictor of future insurance losses. In short, there seems to be a strong correlation between bad credit ratings and auto insurance claims. Consequently, insurance companies want to charge higher premiums to customers with bad credit ratings. Consumer advocates object strongly because they say bad credit *does not cause* automobile accidents. More than 20 states prohibit or restrict the use of credit ratings to determine auto insurance premiums. Insurance companies respond by saying your best defense is to pay your bills on time!

Does going to church cause people to go to jail? We hope not. During this period, there was a steady increase in population. Therefore, it is not too surprising that both the number of people attending churches and the number of people in jail increased together. The high correlation between x and y is due to the common effect of the increase in the general population. ◊

SECTION 4.3 PROBLEMS

1. Over the past 10 years, there has been a high positive correlation between the number of South Dakota safety inspection stickers issued and the number of South Dakota traffic accidents.
   (a) Do safety inspection stickers cause traffic accidents?
   (b) What third factor might cause traffic accidents and the number of safety stickers to increase together?

2. There is a high positive correlation in the United States between teachers' salaries and annual consumption of liquor.
   (a) Do you think increasing teachers' salaries has caused increased liquor consumption?
   (b) As teachers' salaries have been going up, most other salaries have been going up, too. To some extent, this means an upward trend in buying power for everyone. How might this explain the high correlation between teachers' salaries and liquor consumption?

3. Over the past 30 years in the United States, there has been a strong negative correlation between number of infant deaths at birth and number of people over age 65.
   (a) Is the fact that people are living longer causing a decrease in infant mortalities at birth?
   (b) What third factor might be decreasing infant mortalities and at the same time increasing life span?

4. Over the past few years, there has been a strong positive correlation between the annual consumption of diet soda pop and the number of traffic accidents.
   (a) Do you think that an increasing consumption of diet pop has led to more traffic accidents?
   (b) What third factor or factors might be causing both the annual consumption of diet pop and the number of traffic accidents to increase together?

For each of the Problems 5–10, please do the following:
(a) Draw a scatter diagram for the data.
(b) From the scatter diagram, would you estimate the correlation coefficient r to be closest to 1, 0, or −1?
(c) Compute the correlation coefficient r and the coefficient of determination r^2. What percentage of the variation in y can be *explained* by the corresponding variation in x by using the least-squares line? What percentage of the variation is *unexplained*?
Answers may vary slightly due to rounding.

5. *Economics: Entry-Level Jobs* An economist is studying the job market in Denver area neighborhoods. Let x represent the total number of jobs in a given neighborhood, and let y represent the number of entry-level jobs in the same neighborhood. A sample of 6 Denver neighborhoods gave the following information (units in 100s of jobs).

x	16	33	50	28	50	25
y	2	3	6	5	9	3

Source: Neighborhood Facts, The Piton Foundation. To fiind out more, visit the Brase/Brase*Understanding Basic Statistics* site at http://math.college.hmco.com/students and fiind he link to the Piton Foundation.

Complete parts (a) through (c) for these data, and comment on the meaning of r and r^2 in the context of this problem.

6. *Violent Crimes: Prisons* Does prison really deter violent crime? Let x represent percent change in the rate of violent crime and y represent percent change in the rate of imprisonment in the general U.S. population. For 7 recent years the following data have been obtained (*Source: The Crime Drop in America,* edited by Blumstein and Wallman, Cambridge University Press).

x	6.1	5.7	3.9	5.2	6.2	6.5	11.1
y	−1.4	−4.1	−7.0	−4.0	3.6	−0.1	−4.4

Complete parts (a) through (c) for these data, and comment on the meaning of r and r^2 in the context of this problem.

7. *Auto Accidents: Age* Data for this problem are based on information taken from *The Wall Street Journal.* Let x be the age in years of a licensed automobile driver. Let y be the percentage of all fatal accidents (for a given age) due to speeding. For example, the first data pair indicates that 36% of all fatal accidents of 17-year-olds are due to speeding.

x	17	27	37	47	57	67	77
y	36	25	20	12	10	7	5

Complete parts (a) through (c) for these data, and comment on the meaning of r and r^2 in the context of this problem.

8. *Auto Accidents: Age* Let x be the age of a licensed driver in years. Let y be the percentage of all fatal accidents (for a given age) due to failure to yield the right of way. For example, the first data pair says that 5% of all fatal accidents of 37-year-olds are due to failure to yield the right of way. *The Wall Street Journal* article referenced in Problem 7 reported the following data:

x	37	47	57	67	77	87
y	5	8	10	16	30	43

Complete parts (a) through (c) for these data, and comment on the meaning of r and r^2 in the context of this problem.

9. *Criminology: Skeletons* The state criminology laboratory must sometimes estimate a person's height from partial skeletal remains. How strong a correlation is there between body height and bone size? A random sample of eight adult male x-rays gave the following information, where x = length of femur (thigh bone) and y = body height.

x (in.)	17.5	20	21	19	15.5	18.5	16	18
y (in.)	50	80	78	73	63	71	64	71

Complete parts (a) through (c) for these data, and comment on the meaning of r and r^2 in the context of this problem.

10. *Meteorology: Cyclones* Can a low barometer reading be used to predict maximum wind speed of an approaching tropical cyclone? Data for this problem are based on information taken from *Weatherwise* (vol. 46, no. 1), a publication of the American Meteorological Society. For a random sample of tropical cyclones, let x be the lowest pressure (in millibars) as a cyclone approaches, and let y be the maximum wind speed (in miles per hour) of the cyclone.

x	1004	975	992	935	985	932
y	40	100	65	145	80	150

Complete parts (a) through (c) for these data, and comment on the meaning of r and r^2 in the context of this problem.

11. Examine the computation formula for r, the sample correlation coefficient [Equation (7) of this section].
    (a) In the formula for SS_{xy}, if we exchange the symbols x and y, do we get a different result or do we get the same (i.e., equivalent) result? Explain.
    (b) In the formula for r [see Equation (7) of this section], if we exchange the symbols x and y, do we get a different result or do we get the same (equivalent) result? Explain.
    (c) If we have a set of x and y data values and we exchange each corresponding x and y value to get a new data set, should the sample correlation coefficient be the same for both sets of data? Explain.
    (d) Compute the sample correlation coefficient r for each of the following data sets and show that they are the same.

x	1	3	4
y	2	1	6

x	2	1	6
y	1	3	4

What can you say about the least-squares equation for each data set? Are the equations algebraically equivalent? *Hint:* See Problem 10 of Section 4.2.

SUMMARY

We can often get information about a population of y values by attempting to predict them from known values of another population x. If a random sample of data pairs (x, y) are obtained from an overall population of such data pairs, we may use the equation of the least-squares line

$$y = a + bx$$

to predict y values from given x values. In this equation, a and b are computed from Equations (2) to (5) of Section 4.2. When we use the least-squares line to predict y values from x values, y is the *response variable* and x is the *explanatory variable*.

If data pairs (x, y) are gathered, the extent to which x and y values are related can be measured by the Pearson product–moment correlation coefficient r given in Equation (7) of Section 4.3. The quantity r^2 is called the *coefficient of determination*. We use this value to measure the proportion of variation in y explained by the least-squares line.

IMPORTANT WORDS & SYMBOLS

Section 4.1
Paired data values
Scatter diagram
No linear correlation
Perfect linear correlation

Section 4.2
Explanatory variable
Response variable
Least-squares criterion
Regression

Method of least squares
Interpolation and extrapolation
Least-squares line $y = a + bx$

Section 4.3
Correlation coefficient r (later called sample correlation
    coefficient)
Linear correlation
Zero, positive, and negative correlation
Coefficient of determination
Explained and unexplained variation

VIEWPOINT

Living Arrangements

Male, female, married, single, living alone, living with friends or relatives—all these categories are of interest to the U.S. Census Bureau. In addition to these categories, there are others, such as age, income, and health needs. How strongly correlated are these variables? Can we use one or more of these variables to predict the others? How good is such a prediction? Methods of this chapter can help you answer such questions. For more information regarding such data, visit the Brase/Brase *Understanding Basic Statistics* site at http://math.college.hmco.com/students and find the link to Census Bureau.

CHAPTER REVIEW PROBLEMS

Answers may vary slightly due to rounding.

1. *Desert Ecology: Wildlife* Bighorn sheep are beautiful wild animals found throughout the western United States. Data for this problem are based on information taken from *The Desert Bighorn*, edited by Monson and Sumner (University of Arizona Press). Let x be the age of a bighorn sheep (in years), and let y be the mortality rate (percent that die) in this age group. For example, $x = 1$, $y = 14$ means that 14%

of the bighorn sheep between 1 and 2 years old died. A random sample of Arizona bighorn sheep gave the following information:

x	1	2	3	4	5
y	14	18.9	14.4	19.6	20.0

(a) Draw a scatter diagram.
(b) Find the equation of the least-squares line.
(c) Find r. Find the coefficient of determination r^2. Explain what these measures mean in the context of the problem.

2. *Sociology: Job Changes* A sociologist is interested in the relation between $x =$ number of job changes and $y =$ annual salary (in thousands of dollars) for people living in the Nashville area. A random sample of 10 people employed in Nashville provided the following information:

x (Number of job changes)	4	7	5	6	1	5	9	10	10	3
y (Salary in $1000)	33	37	34	32	32	38	43	37	40	33

(a) Draw a scatter diagram for the data.
(b) Find $\bar{x}$, $\bar{y}$, and b and then the equation of the least-squares line.
(c) Graph the least-squares line on your scatter diagram.
(d) If someone had $x = 2$ job changes, what does the least-squares line predict for y, the annual salary?
(e) Looking at the scatter diagram and least-squares line, do you think the correlation coefficient will be positive, negative, or zero?
(f) Find r. Find the coefficient of determination.

3. *Medical: Fat Babies* Modern medical practice tells us not to encourage babies to become too fat. Is there a positive correlation between the weight x of a 1-year-old baby and the weight y of a mature adult (30 years old)? A random sample of medical files produced the following information for 14 females:

x (lb)	21	25	23	24	20	15	25	21	17	24	26	22	18	19
y (lb)	125	125	120	125	130	120	145	130	130	130	130	140	110	115

(a) Draw a scatter diagram for the data.
(b) Find $\bar{x}$, $\bar{y}$, and b and the equation of the least-squares line.
(c) Graph the least-squares line on your scatter diagram.
(d) If a female baby weighs 20 lb at 1 year, what would you predict she would weigh at 30 years of age?
(e) Looking at the scatter diagram, do you think the correlation coefficient is positive, negative, or zero?
(f) Find r. Find the coefficient of determination. Interpret the meaning of these measurements in the context of the problem.

4. *Sales: Insurance* Dorothy Kelly sells life insurance for the Prudence Insurance Company. She sells insurance by making visits to her clients' homes. Dorothy believes that the number of sales should depend, to some degree, on the number of visits made. For the past several years, she kept careful records of the number of visits (x) she made each week and the number of people (y) who bought

Using TECHNOLOGY

TI-83PLUS • EXCEL • MINITAB • COMPUTERSTAT

Linear Regression

APPLICATION

The data in this section are taken from this reference:

> King, Cuchlaine A. M. *Physical Geography*. Oxford: Basil Blackwell, 1980, 77–86, 196–206. Reprinted with permission of Basil Blackwell Limited, Oxford, England.

Throughout the world, natural ocean beaches are beautiful sights to see. If you have visited natural beaches, you may have noticed that when the gradient or dropoff is steep, the grains of sand tend to be larger. In fact, a manmade beach with the "wrong" size granules of sand tends to be washed away and eventually replaced when the proper size grain is selected by the action of the ocean and the gradient of the bottom. Since manmade beaches are expensive, grain size is an important consideration.

In the data that follow, x = median diameter (in millimeters) of granules of sand, and y = gradient of beach slope in degrees on natural ocean beaches.

x	y
0.17	0.63
0.19	0.70
0.22	0.82
0.235	0.88
0.235	1.15
0.30	1.50
0.35	4.40
0.42	7.30
0.85	11.30

1. Find the sample mean and standard deviation for x and for y.

2. Make a scatter plot. Would you expect a moderately high correlation and good fit for the least-squares line?

3. Find the equation of the least-squares line, and graph the line on the scatter plot.

4. Find the correlation coefficient r and the coefficient of determination r^2.

5. Suppose that you have a truckload of sifted sand in which the median size of granules is 0.38 mm. If you want to put this sand on a beach and you don't want the sand to wash away, then what does the least-squares line predict for the angle of the beach? *Note:* Heavy storms that produce abnormal waves may also wash out the sand. However, in the long run, the size of sand granules that remain on the beach or that are brought back to the beach by long-term wave action are determined to a large extent by the angle at which the beach drops off.

6. Suppose we now have a truckload of sifted sand in which the median size of the granules is 0.45 mm. Repeat Problem 5.

162

Technology Hints (Linear Regression)

TI-83Plus

Be sure to set **DiagnosticOn** (under **Catalogue**).
(a) Scatter diagram: Use **STAT PLOT,** select the first type, use **ZOOM** option **9:ZoomStat.**
(b) Least-squares line and r: Use **STAT, CALC,** option **8:LinReg(a + bx).**
(c) Graph least-squares line and predict: Press **Y=.** Then under **VARS,** select **5:Statistics,** then select **EQ,** and finally select item **1:RegEQ.** Press enter. This sequence of steps will automatically set Y_1 = your regression equation. Press **GRAPH.** To find a predicted value, when the graph is showing, press the **CALC** key and select item **1:Value.** Enter the x value, and the corresponding y value will appear.

Excel

(a) Scatter plot, least-squares line, r^2: Use **Chart wizard.** Select **scatter plot.** Once plot is displayed, *right* click on any data point. Select **trend line.** Under options, check display line and display r^2.
(b) Prediction: Use **Paste Function** (f_x) ➤ **Statistical** ➤ **Forecast.**
(c) Coefficient r: Use **Paste Function** (f_x) ➤ **Statistical** ➤ **Correl.**

Minitab

(a) Scatter plot, least-squares line, r^2: Use menu selection **Stat** ➤ **Regression** ➤ **Fitted line plot.**
(b) Coefficient r: Use menu selection **Stat** ➤ **Basic Statistics** ➤ **Correlation.**

ComputerStat

Scatter plot, least-squares line, r, r^2, prediction: Use the menu item **Linear Regression and Correlation.**

5

Elementary Probability Theory

We see that the theory of probabilities is at bottom only common sense reduced to calculation; it makes us appreciate with exactitude what reasonable minds feel by a sort of instinct, often without being able to account for it.

—Pierre-Simon Laplace

Pierre-Simon Laplace (1749–1827)

This renowned French astronomer and mathematician wrote *The Analytical Theory of Probability.*

This is how the great mathematician Laplace described the theory of mathematical probability. The discovery of the mathematical theory of probability was shared by two Frenchmen: Blaise Pascal and Pierre Fermat. These seventeenth-century scholars were attracted to the subject by the inquiries of the Chevalier de Méré, a gentleman gambler.

Although the first applications of probability were to games of chance and gambling, today the subject seems to pervade almost every aspect of modern life. Everything from the orbits of spacecraft to the social behavior of woodchucks is described in terms of probabilities.

For on-line student resources, visit **math.college.hmco.com/students** and follow the Statistics links to the Brase/Brase, *Understanding Basic Statistics,* 3rd edition Web site.

◊ Why do we need probability? *Hint:* Many important issues in life involve uncertain answers.   (SECTION 5.1)

◊ What are the basic definitions and rules of probability?   (SECTION 5.2)

◊ What are counting techniques, trees, permutations, and combinations? (SECTION 5.3)

FOCUS PROBLEM

How Often Do Lie Detectors Lie?

James Burke is an educator who is known for his interesting science-related radio and television shows aired by the British Broadcasting Corporation. His book *Chances: Risk and Odds in Everyday Life* (Virginia Books, London) contains a great wealth of fascinating information about probabilities. The following quote is from Professor Burke's book:

> *If I take a polygraph test and lie, what is the risk I will be detected?* According to some studies there's about a 72 percent chance you will be caught by the machine.

> *What is the risk that if I take a polygraph test it will incorrectly say that I lied?* At least 1 in 15 will be thus falsely accused.

Both of these statements contain conditional probabilities, which we will study in Section 5.2. Information from that section will allow us to answer the following:

Suppose a person answers 10% of a long battery of questions with lies. Assume that the remaining 90% of the questions are answered truthfully.

1. Estimate the percentage of answers the polygraph will *wrongly* indicate as lies.

2. Estimate the percentage of answers the polygraph will *correctly* indicate as lies.

If the polygraph indicated that 30% of the questions were answered as lies, what would you estimate for the *actual* percentage of questions the person answered as lies?

5.1
What Is Probability?

FOCUS POINTS
✓ Assign probabilities to events.
✓ Explain how the law of large numbers relates to relative frequencies.
✓ Apply basic rules of probability in everyday life.
✓ Explain the relationship between statistics and probability.

Basic concepts

We encounter statements in terms of probability all the time. An excited sports announcer claims that Sheila has a 90% chance of breaking the world record in the upcoming 100-yard dash. Henry figures that if he guesses on a true–false question, the probability of getting it right is 1/2. The Right to Health Lobby claims the probability is 0.40 of getting an erroneous report from a medical laboratory in one low-cost health center. It is consequently lobbying for a federal agency to license and monitor all medical laboratories.

When we use probability in a statement, we're using a *number between 0 and 1* to indicate the likelihood of an event. We'll use the notation $P(A)$ (read, "P of A") to denote the *probability of event A*. The closer to 1 the probability assignment is, the more likely the event is to occur. If the event A is certain to occur, then $P(A)$ should be 1.

It is important to know what probability statements mean and how to compute or assign probabilities to events because probability is the language of inferential statistics. For instance, suppose a college counselor claims that 70% of first-year students receive counseling to help plan their schedules. Because of the high percentage of students needing help, he is requesting that an additional counselor be hired. You want to test the counselor's claim. In doing so, you pick a random sample of 30 first-year students and find that only 3 of them got help from a counselor. Can you challenge the counselor's claim on the basis of this random sample in which only 10% of the students got counselor help with their schedules? To answer this question, we need to find the *probability* of picking a random sample of first-year students in which only 10% got counselor help under the assumption that the counselor's claim is correct. This is the kind of question we will consider in hypothesis testing (Chapter 10).

In the meantime, we need to learn how to find probabilities or assign them to events. There are three major methods. One is *intuition*. The sports announcer probably used Sheila's performances in past track events and his own confidence in her running ability as a basis for his prediction that she has a 90% chance of breaking the world record. In other words, the announcer feels that the probability is 0.90 that Sheila will break the world record.

Relative frequency

The Right to Health Lobby used another method to arrive at its probability statement. It took the *relative frequency* with which erroneous laboratory reports occurred. From a random sample of $n = 100$, it found $f = 40$ erroneous laboratory reports. From this, it computed the relative frequency of erroneous laboratory reports via Formula (1).

Probability formula for relative frequency

$$\text{Probability of an event} = \text{Relative frequency} = \frac{f}{n} \qquad (1)$$

where f is the frequency of an event,
   n is the sample size.

In the case of the laboratory reports, we have

$$\text{Relative frequency} = \frac{f}{n} = \frac{40}{100} = 0.40$$

The relative frequency of erroneous laboratory reports was used as the *probability* of erroneous reports.

  The technique of using the relative frequency of an event as the probability of that event is a common way of assigning probabilities and will be used a great deal in later chapters. The underlying assumption we make is that if events occurred a certain percentage of times in the past, they will occur about the same percentage of times in the future. In fact, this can be strengthened to a very general statement called the *law of large numbers*.

Law of large numbers

Law of large numbers

In the long run, as the sample size increases and increases, the relative frequencies of outcomes get closer and closer to the theoretical (or actual) probability value.

  The law of large numbers is the reason such businesses as health insurance, automobile insurance, and gambling casinos can exist and make a profit. In Central City, Colorado, there are many casinos with many slot machines. The winnings of a gambler on a single play or even a few plays are uncertain (small sample size). This is one of the reasons gambling is exciting. However, on tens of thousands of plays, the theoretical or actual probability of winning favors the casino. That's why the casino and its owners regard gambling as a business. The house is guaranteed a profit in the long run.

  Henry used the third method of assigning probabilities when he determined the probability of correctly guessing the answer to a true–false question. Essentially, he Equally likely outcomes used the probability formula for *equally likely outcomes.*

Probability formula when outcomes are equally likely

$$\text{Probability of an event} = \frac{\text{Number of outcomes favorable to event}}{\text{Total number of outcomes}} \qquad (2)$$

In Henry's case, there are two possible outcomes. A test answer will be either correct or incorrect. Since he is guessing, we assume that the outcomes are equally

Probability Related to Statistics

We conclude this section with a few comments on the nature of statistics versus probability. Although statistics and probability are closely related fields of mathematics, they are nevertheless separate fields. It can be said that probability is the medium through which statistical work is done. In fact, if it were not for probability theory, inferential statistics would not be possible.

Put very briefly, probability is the field of study that makes statements about what will occur when samples are drawn from a *known population*. Statistics is the field of study that describes how samples are to be obtained and how inferences are to be made about *unknown populations*.

A simple but effective illustration of the difference between these two subjects can be made by considering how we treat the following examples:

Example of a probability application

Condition: We *know* the exact makeup of the *entire* population.

Example: Given 3 green marbles, 5 red marbles, and 4 white marbles in a bag, draw 6 marbles at random from the bag. What is the probability none of the marbles is red?

Example of a statistical application

Condition: We have only *samples* from an otherwise *unknown* population.

Example: Draw a random sample of 6 marbles from the (unknown) population of all marbles in a bag and observe the colors. Based on the sample results, make a conjecture about the colors and numbers of marbles in the entire population of all marbles in the bag.

In another sense, probability and statistics are like flip sides of the same coin. On the probability side, you know the overall description of the population. The central problem is to compute the likelihood that a specific outcome will happen. On the statistics side, you know only the results of a sample drawn from the population. The central problem is to describe the sample (descriptive statistic) and to draw conclusions about the population based on the sample results (inferential statistics).

In statistical work, the inferences we draw about an unknown population are not claimed to be absolutely correct. Since the population remains unknown (in a theoretical sense), we must accept a "best guess" for our conclusions and act using the most probable answer rather than absolute certainty.

Probability is the topic of this chapter. However, we will not study probability just for its own sake. Probability is a wonderful field of mathematics, but we will study mainly the ideas from probability that are needed for a proper understanding of statistics.

Even though we have only a brief introduction to probability, we will be able to answer questions raised in situations such as the following:

Salmon moving upstream

- Commercial salmon fishing is very important in Alaska. Since the salmon are sold by weight, it is useful to know the average weight of a freshly caught salmon. How large a sample of freshly caught salmon is needed if we want to be 95% sure the mean weight of the sample is within plus or minus 1 ounce of the mean weight of all catchable salmon in Alaskan waters? For the answer to this type question, see Example 6 in Section 9.4, in which we discuss probability as applied to sample size.

- Many practical situations call on us to choose between two competing statements. *Consumer Reports* did a study on the maintenance records of two competing brands of cellular phones. Each brand claims to have the lowest maintenance costs. Is one brand actually better than the other? If there is an apparent difference, is the difference statistically significant at, say, a 1% level (of risk)? For answers to these kinds of questions, see Sections 11.2 and 11.3, in which we discuss tests for differences.

All the listed examples require some use of probability. This is why we encourage you to study this chapter carefully. Your time in doing so will be well invested.

VIEWP⊙INT   *What Makes a Good Teacher?*

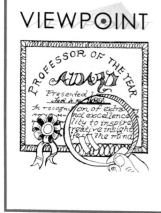

A survey of 735 students at nine colleges in the United States was taken to determine instructor behaviors that help students succeed. Data for this survey can be found by visiting the Brase/Brase *Understanding Basic Statistics* site at http://math.college..hmco.com/students and finding the link to DASL, the Carnegie Mellon University Data and Story Library. Once at the DASL site, select Data Subjects, then Psychology, and then Instructor Behavior. You can estimate the probability of how a student would respond (very positive, neutral, very negative) to different instructor behaviors. For example, more than 90% of the students responded "very positive" to the instructor's use of real-world examples in the classroom.

SECTION 5.1   PROBLEMS

1. *General: Concepts* In your own words, carefully answer the question: What is probability? List three methods of assigning probabilities.

2. *General: Provide Examples* List examples of where probability might be applied in business, medicine, social science, and natural science. Identify some ways that probability will be useful in the study of statistics.

3. *General: Valid Probabilities* Which of the following numbers cannot be the probability of some event?

   (a) 0.71   (b) 4.1   (c) $\frac{1}{8}$ (d) -0.5

   (e) 0.5   (f) 0   (g) 1   (h) 150%

4. *General: Valid Probabilities*
   (a) Explain why −0.41 cannot be the probability of some event.
   (b) Explain why 1.21 cannot be the probability of some event.
   (c) Explain why 120% cannot be the probability of some event.
   (d) Can the number 0.56 be the probability of an event? Explain.

5. *Probability Estimate: Wiggle Your Ears* Can you wiggle your ears? Use the students in your statistics class (or a group of friends) to estimate the percentage of people who can wiggle their ears. How can your result be thought of as an estimate for the probability that a person chosen at random can wiggle his or her ears? *Comment:* National statistics indicate that about 13% of Americans can wiggle their ears (*Source:* Bernice Kanner, *Are You Normal?* St. Martin's Press, New York).

6. *Probability Estimate: Raise One Eyebrow* Can you raise one eyebrow at a time? Use the students in your statistics class (or a group of friends) to estimate the percentage of people who can raise one eyebrow at a time. How can your result be thought of as an estimate for the probability that a person chosen at random can raise one eyebrow at a time? *Comment:* National statistics indicate that about 30% of Americans can raise one eyebrow at a time (*see* source in Problem 5).

7. *Myers–Briggs: Personality Types* Isabel Briggs Myers was a pioneer in the study of personality types. The personality types are broadly defined according to four main preferences. Do married couples choose similar or different personality types in their mates? The following data give an indication (*Source:* I. B. Myers and M. H. McCaulley, *A Guide to the Development and Use of the Myers–Briggs Type Indicators*).

Similarities and Differences in a Random Sample of 375 Married Couples

Number of Similar Preferences	Number of Married Couples
All four	34
Three	131
Two	124
One	71
None	15

Suppose that a married couple is selected at random.
   (a) Use the data to estimate the probability that they will have 0, 1, 2, 3, or 4 personality preferences in common.
   (b) Do the probabilities add up to 1? Why should they? What is the sample space in this problem?

8. *Sociology: Dating Couples* Do couples get engaged or not? If they are engaged, how long did they date before becoming engaged? A poll of 1000 couples conducted by Bruskin and Goldring Research for Korbel Champagne Cellars gave the following information (*USA Today*):

Length of Dating Time Before Engagement

Time	Number of Couples
Never engaged	200
Less than 1 year	240
1 to 2 years	210
More than 2 years	350

(a) Use the data to estimate the probability of each event for a dating couple chosen at random: The couple is not engaged, dated less than 1 year before getting engaged, dated 1 to 2 years before getting engaged, or dated more than 2 years before getting engaged.

(b) Do the probabilities of part (a) add up to 1? Why should they? What is the sample space in this problem?

9. *Psychology: Creativity* When do creative people get their *best* ideas? *USA Today* did a survey of 966 inventors (who hold U.S. patents) and obtained the following information:

Time of Day When Best Ideas Occur

Time	Number of Inventors
6 A.M.–12 noon	290
12 noon–6 P.M.	135
6 P.M.–12 midnight	319
12 midnight–6 A.M.	222

(a) Assuming that the time interval includes the left limit and all the times up to but not including the right limit, estimate the probability that an inventor has a best idea during each time interval: from 6 A.M. to 12 noon, from 12 noon to 6 P.M., from 6 P.M. to 12 midnight, from 12 midnight to 6 A.M.

(b) Do the probabilities of part (a) add up to 1? Why should they? What is the sample space in this problem?

10. *Agriculture: Cotton* A botanist has developed a new hybrid cotton plant that can withstand insects better than other cotton plants. However, there is some concern about the germination of seeds from the new plant. To estimate the probability that a seed from the new plant will germinate, a random sample of 3000 seeds was planted in warm, moist soil. Of these seeds, 2430 germinated.

(a) Use relative frequencies to estimate the probability that a seed will germinate. What is your estimate?

(b) Use relative frequencies to estimate the probability that a seed will *not* germinate. What is your estimate?

(c) Either a seed germinates, or it does not. What is the sample space in this problem? Do the probabilities assigned to the sample space add up to 1? Should they add up to 1? Explain.

(d) Are the outcomes in the sample space of part (c) equally likely?

11. *Business: Customers* John runs a computer software store. Yesterday he counted 127 people who walked by his store, 58 of whom came into the store. Of the 58, only 25 bought something in the store.

(a) Estimate the probability that a person who walks by the store will enter the store.

(b) Estimate the probability that a person who walks into the store will buy something.

(c) Estimate the probability that a person who walks by the store will come in *and* buy something.

(d) Estimate the probability that a person who comes into the store will buy nothing.

5.2
Some Probability Rules—Compound Events

FOCUS POINTS
✓ Compute probabilities of general compound events.
✓ Compute probabilities involving independent events or mutually exclusive events.
✓ Use survey results to compute conditional probabilities.

You roll two dice. What is the probability that you will get a 5 on each die? You draw two cards from a well-shuffled, standard deck without replacing the first card before drawing the second. What is the probability that they will both be aces?

It seems that these two problems are nearly alike. They are alike in the sense that in each case you are to find the probability of two events occurring *together*. In the first problem, you are to find

P(5 on 1st die *and* 5 on 2nd die)

In the second, you want

P(ace on 1st card *and* ace on 2nd card)

Independent events

The two problems differ in one important aspect, however. In the dice problem, the outcome of a 5 on the first die does not have any effect on the probability of getting a 5 on the second die. Because of this, the events are *independent*.

> Two events are **independent** if the occurrence or nonoccurrence of one does *not* change the probability that the other will occur.

In the card problem, the probability of an ace on the first card is 4/52, since there are 52 cards in the deck and 4 of them are aces. If you get an ace on the first card, then the probability of an ace on the second is changed to 3/51, because one ace has already been drawn and only 51 cards remain in the deck. Therefore, the two events in the card-draw problem are *not* independent. They are, in fact, *dependent,* since the outcome of the first draw changes the probability of getting an ace on the second card.

Probability of A and B

Why does the *independence* or *dependence* of two events matter? The type of events determines the way we compute the probability of the two events happening together. If two events A and B are *independent*, then we use Formula (4) to compute the probability of the event A *and* B:

> For independent events,
> $$P(A \ and \ B) = P(A) \cdot P(B) \tag{4}$$

If the events are *dependent,* then we must take into account the changes in the probability of one event caused by the occurrence of the other event. The notation $P(A, \ given \ B)$ denotes the probability that event A will occur, *given* that event B has occurred. This is called a *conditional probability*. We read $P(A, \ given \ B)$ as "probability of A, given B." If A and B are dependent events, then $P(A) \neq P(A, \ given \ B)$ because the occurrence of event B has changed the probability that event A will occur. A standard notation for $P(A, \ given \ B)$ is $P(A \mid B)$. However, we will use the more expanded notation $P(A, \ given \ B)$ to remind you that we assume that event B has already occurred. We use either Formula (5) or Formula (6) to compute the probability of A and B when the events A and B are dependent.

For dependent events,

$$P(A \ and \ B) = P(A) \cdot P(B, \ given \ that \ A \ has \ occurred) \qquad (5)$$

$$P(A \ and \ B) = P(B) \cdot P(A, \ given \ that \ B \ has \ occurred) \qquad (6)$$

We will use either Formula (5) or Formula (6) according to the information available.

Formulas (4), (5), and (6) constitute the *multiplication rules* of probability. They help us compute the probability of events happening together when the sample space is too large for convenient reference or when it is not completely known.

Let's use the multiplication rules to complete the dice and card problems. We'll compare the results with those obtained by using the sample space directly.

EXAMPLE 3

Multiplication rule

Suppose you are going to throw two fair dice. What is the probability of getting a 5 on each die?

SOLUTION USING THE MULTIPLICATION RULE: The two events are independent, so we should use Formula (4). $P(5 \ on \ 1st \ die \ and \ 5 \ on \ 2nd \ die) = P(5 \ on \ 1st) \cdot P(5 \ on \ 2nd)$. To finish the problem, we need only compute the probability of getting a 5 when we throw one die.

There are six faces on a die, and on a fair die each is equally likely to come up when you throw the die. Only one face has five dots, so by Formula (2) for equally likely outcomes,

$$P(5 \ on \ die) = \frac{1}{6}$$

Now we can complete the calculation.

$$P(5 \ on \ 1st \ die \ and \ 5 \ on \ 2nd \ die) = P(5 \ on \ 1st) \cdot P(5 \ on \ 2nd)$$

$$= \frac{1}{6} \cdot \frac{1}{6}$$

$$= \frac{1}{36}$$

SOLUTION USING SAMPLE SPACE: The first task is to write down the sample space. Each die has six equally likely outcomes, and each outcome of the second die can be paired with each of the first. The sample space is shown in Figure 5-1 on the following page. The total number of outcomes is 36, and only one is favorable to a 5 on the first die *and* a 5 on the second. The 36 outcomes are equally likely, so by Formula (2) for equally likely outcomes,

$$P(5 \ on \ 1st \ and \ on \ 2nd) = \frac{1}{36}$$

The two methods yield the same result. The multiplication rule was easier to use because we did not need to look at all 36 outcomes in the sample space for tossing two dice. ◇

FIGURE 5-1

Sample Space for Two Dice

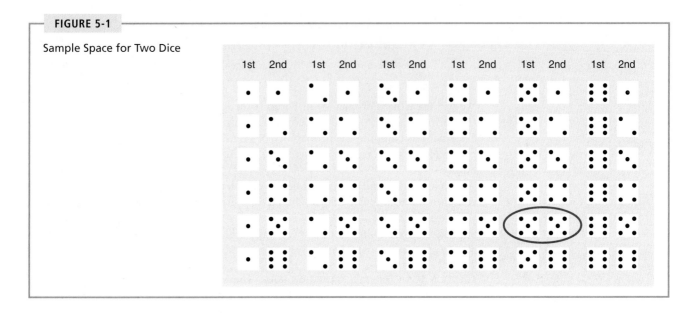

EXAMPLE 4

Dependent events

Compute the probability of drawing two aces from a well-shuffled deck of 52 cards if the first card is not replaced before the second card is drawn.

MULTIPLICATION RULE METHOD: These events are *dependent*. The probability of an ace on the first card is 4/52, but on the second card the probability of an ace is only 3/51 if an ace was drawn for the first card. An ace on the first draw changes the probability for an ace on the second draw. By the multiplication rule for dependent events,

$$P(\text{ace on 1st } and \text{ ace on 2nd}) = P(\text{ace on 1st}) \cdot P(\text{ace on 2nd, } given \text{ ace on 1st})$$

$$= \frac{4}{52} \cdot \frac{3}{51} = \frac{12}{2652} = 0.0045$$

SAMPLE SPACE METHOD: We won't actually look at the sample space because each of the 51 possible outcomes for the second card must be paired with each of the 52 possible outcomes for the first card. This gives us a total of 2652 outcomes in the sample space! We'll just think about the sample space and try to list all the outcomes favorable to the event of aces on both cards. The 12 favorable outcomes are shown in Figure 5-2. By the formula for equally likely outcomes,

$$P(\text{ace on 1st card } and \text{ ace on 2nd card}) = \frac{12}{2652} = 0.0045$$

Again, the two methods agree.   ◇

The multiplication rules apply whenever we wish to determine the probability of two events happening *together*. To indicate together, we use *and* between the events. But before you use a multiplication rule to compute the probability of *A and B*, you must determine if *A* and *B* are independent or dependent events. Let's practice using the multiplication rule.

FIGURE 5-2

Outcomes Favorable to
Drawing Two Aces

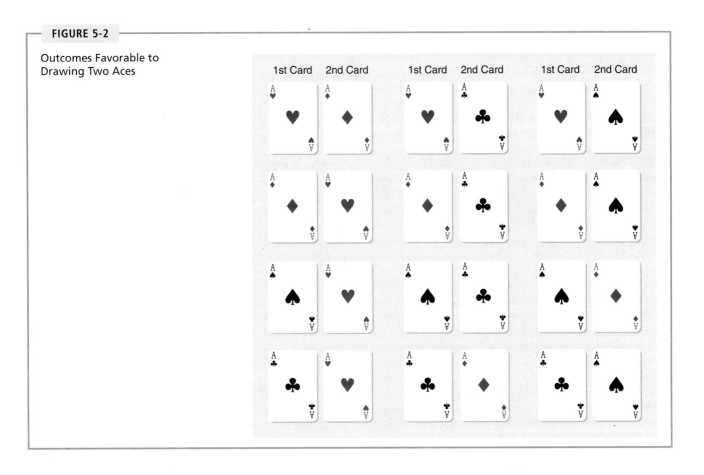

GUIDED EXERCISE 4

Multiplication rule

Andrew is 55, and the probability that he will be alive in 10 years is 0.72. Ellen is 35, and the probability that she will be alive in 10 years is 0.92. Assuming that the life span of one will have no effect on the life span of the other, what is the probability they will both be alive in 10 years?

(a) Are these events dependent or independent?   ⇨   Since the life span of one does not affect the life span of the other, the events are independent.

(b) Use the appropriate multiplication rule to find   ⇨   We use the rule for independent events:

P(Andrew alive in 10 years *and* Ellen alive in 10 years)

$$P(A \text{ and } B) = P(A) \cdot P(B)$$

P(Andrew alive *and* Ellen alive)

$$= P(\text{Andrew alive}) \cdot P(\text{Ellen alive})$$

$$= (0.72)(0.92) = 0.66$$

GUIDED EXERCISE 5

Dependent events

A quality-control procedure for testing Ready-Flash disposable cameras consists of drawing two cameras at random from each lot of 100 without replacing the first camera before drawing the second. If both are defective, the entire lot is rejected. Find the probability that both cameras are defective if the lot contains 10 defective cameras. Since we are drawing the cameras at random, assume that each camera in the lot has an equal chance of being drawn.

(a) What is the probability of getting a defective camera on the first draw?

⟹ The sample space consists of all 100 cameras. Since each is equally likely to be drawn and there are 10 defective ones,

$$P(\text{defective camera}) = \frac{10}{100} = \frac{1}{10}$$

(b) The first camera drawn is not replaced, so there are only 99 cameras for the second draw. What is the probability of getting a defective camera on the second draw if the first camera was defective?

⟹ If the first camera is defective, then there are only 9 defective cameras left among the 99 remaining cameras in the lot.

$P(\text{defective camera on 2nd draw, } given$

$\text{defective camera on 1st}) = \dfrac{9}{99} = \dfrac{1}{11}$

(c) Are the probabilities computed in parts (a) and (b) different? Does drawing a defective camera on the first draw change the probability of getting a defective camera on the second draw? Are the events dependent?

⟹ The answer to all these questions is yes.

(d) Use the formula for dependent events,

$$P(A \text{ and } B) = P(A) \cdot P(B, \text{ given } A \text{ has occurred})$$

to compute P(1st camera defective *and* 2nd camera defective).

⟹ $P(\text{1st defective } and \text{ 2nd defective}) = \dfrac{1}{10} \cdot \dfrac{1}{11}$

$= \dfrac{1}{110}$

$= 0.009$

One of the multiplication rules can be used any time we are trying to find the probability of two events happening *together*. Pictorially, we are looking for the probability of the shaded region in Figure 5-3.

Probability of *A* or *B*

Another way to combine events is to consider the possibility of one event *or* another occurring. For instance, if a sports car saleswoman gets an extra bonus if she sells a convertible or a car with leather upholstery, she is interested in the probability that you will buy a car that is a convertible *or* has leather upholstery. Of course, if you bought a convertible with leather upholstery, that would be fine, too. Pictorially, the shaded portion of Figure 5-4 represents the outcomes satisfying the *or* condition. Notice that the condition *A or B* is satisfied by any one of the following conditions:

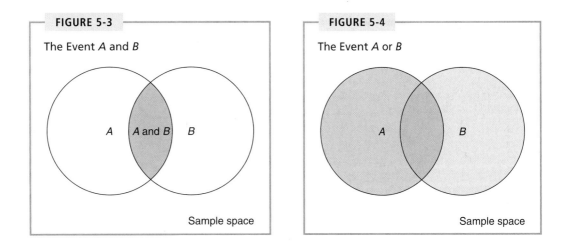

FIGURE 5-3

The Event *A* and *B*

A · A and B · B

Sample space

FIGURE 5-4

The Event *A* or *B*

A · B

Sample space

1. Any outcome in *A* occurs.

2. Any outcome in *B* occurs.

3. Any outcome in both *A* and *B* occurs.

It is important to distinguish between the *or* combinations and the *and* combinations because we apply different rules to compute their probabilities.

GUIDED EXERCISE **6**

Combining events

Indicate how each of the following pairs of events are combined. Use either the *and* combination or the *or* combination.

(a) Satisfying the humanities requirement by taking a course in the history of Japan or by taking a course in classical literature ⟹ Use the *or* combination.

(b) Buying new tires and aligning the tires ⟹ Use the *and* combination.

(c) Getting an A not only in psychology but also in biology ⟹ Use the *and* combination.

(d) Having at least one of these pets: cat, dog, bird, rabbit ⟹ Use the *or* combination.

Once you decide that you are to find the probability of an *or* combination rather than an *and* combination, what formula do you use? Again, it depends on the situation. If you want to compute the probability of drawing either a jack or a king on a single draw from a well-shuffled deck of cards, the formula is simple:

$$P(\text{jack } or \text{ king}) = P(\text{jack}) + P(\text{king}) = \frac{4}{52} + \frac{4}{52} = \frac{8}{52} = \frac{2}{13}$$

since there are 4 jacks and 4 kings in a deck of 52 cards.

If you want to compute the probability of drawing a king or a diamond on a single draw, the formula is a bit more complicated. We have to take the overlap of

FIGURE 5-5

Drawing a King or a
Diamond from a
Standard Deck

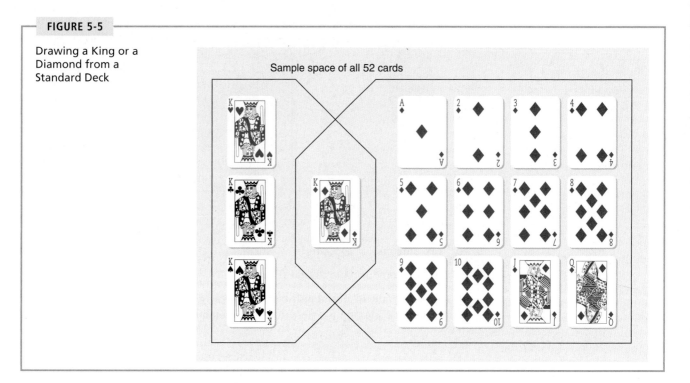

the two events into account so that we do not count the outcomes twice. We can
see the overlap of the two events in Figure 5-5.

$$P(\text{king}) = \frac{4}{52} \quad P(\text{diamond}) = \frac{13}{52} \quad P(\text{king and diamond}) = \frac{1}{52}$$

If we simply add $P(\text{king})$ and $P(\text{diamond})$, we're including $P(\text{king and diamond})$
twice in the sum. To compensate for this double summing, we simply subtract
$P(\text{king and diamond})$ from the sum. Therefore,

$$P(\text{king or diamond}) = P(\text{king}) + P(\text{diamond}) - P(\text{king and diamond})$$

$$= \frac{4}{52} + \frac{13}{52} - \frac{1}{52} = \frac{16}{52} = \frac{4}{13}$$

Mutually exclusive events

We say the events A and B are *mutually exclusive* or *disjoint* if they cannot
occur together. This means that A and B have no outcomes in common or, put
another way, that $P(A \text{ and } B) = 0$. Formula (7) is the *addition rule for mutually
exclusive events A and B.*

For *mutually exclusive* events A and B

$$P(A \text{ or } B) = P(A) + P(B) \tag{7}$$

If the events are not mutually exclusive, we must use the more general Formula
(8), which is the *general addition rule for any events A and B.*

For any events A and B,

$$P(A \text{ or } B) = P(A) + P(B) - P(A \text{ and } B) \tag{8}$$

You may ask: Which formula should we use? The answer is: Use Formula (7) only if you know that A and B are mutually exclusive (i.e., cannot occur together); if you do not know whether A and B are mutually exclusive, then use Formula (8). Formula (8) is valid either way. Notice that when A and B are mutually exclusive, then $P(A \ and \ B) = 0$, so Formula (8) reduces to Formula (7).

GUIDED EXERCISE 7

Mutually exclusive events

The Cost Less Clothing Store carries seconds in slacks. If you buy a pair of slacks in your regular waist size without trying them on, the probability that the waist will be too tight is 0.30 and the probability that it will be too loose is 0.10.

(a) Are the events too tight or too loose mutually exclusive?

⇒ The waist cannot be both too tight and too loose at the same time, so the events are mutually exclusive.

(b) If you choose a pair of slacks at random in your regular waist size, what is the probability that the waist will be too tight or too loose?

⇒ Since the events are mutually exclusive,

P(too tight *or* too loose)

$= P$(too tight) $+ P$(too loose)

$= 0.30 + 0.10$

$= 0.40$

GUIDED EXERCISE 8

General addition rule

Professor Jackson is in charge of a program to prepare people for a high school equivalency exam. Records show that 80% of the students need work in math, 70% need work in English, and 55% need work in both areas.

(a) Are the events need math and need English mutually exclusive?

⇒ These events are not mutually exclusive, since some students need both. In fact,

P(need math *and* need English) $= 0.55$

(b) Use the appropriate formula to compute the probability that a student selected at random needs math *or* needs English.

⇒ Since the events are not mutually exclusive, we use Formula (8):

P(need math *or* need English)

$= P$(need math) $+ P$(need English)

$\quad - P$(need math *and* English)

$= 0.80 + 0.70 - 0.55 = 0.95$

Combination of several events

The addition rule for mutually exclusive events can be extended so that it applies to the situation in which we have more than two events that are each mutually exclusive to all the other events.

EXAMPLE 5

Mutually exclusive events

Laura is playing Monopoly. On her next move she needs to throw a sum bigger than 8 on the two dice in order to land on her own property and pass Go. What is the probability that Laura will roll a sum bigger than 8?

SOLUTION: When two dice are thrown, the largest sum that can come up is 12. Consequently, the only sums larger than 8 are 9, 10, 11, and 12. These outcomes are mutually exclusive, since only one of these sums can possibly occur on one throw of the dice. The probability of throwing more than 8 is the same as

$P(9 \text{ or } 10 \text{ or } 11 \text{ or } 12)$

Since the events are mutually exclusive,

$$P(9 \text{ or } 10 \text{ or } 11 \text{ or } 12) = P(9) + P(10) + P(11) + P(12)$$

$$= \frac{4}{36} + \frac{3}{36} + \frac{2}{36} + \frac{1}{36}$$

$$= \frac{10}{36} = \frac{5}{18}$$

To get the specific values of $P(9)$, $P(10)$, $P(11)$, and $P(12)$, we used the sample space for throwing two dice (see Figure 5-1 on page 178). There are 36 equally likely outcomes—for example, those favorable to 9 are 6, 3; 3, 6; 5, 4; and 4, 5. So $P(9) = 4/36$. The other values can be computed in a similar way.   ◊

The multiplication rule for independent events also extends to more than two independent events. If you toss a fair coin, then roll a fair die, and finally draw a card from a standard deck of bridge cards, the three events are independent. To compute the probability of the outcome heads on the coin *and* five on the die *and* an ace for the card, we use the extended multiplication rule for independent events together with the facts

$$P(\text{head}) = \frac{1}{2} \quad P(5) = \frac{1}{6} \quad P(\text{ace}) = \frac{4}{52} = \frac{1}{13}$$

Then

$$P(\text{head } and \text{ five } and \text{ ace}) = \frac{1}{2} \cdot \frac{1}{6} \cdot \frac{1}{13} = \frac{1}{156}$$

Further Examples

Most of us have been asked to participate in a survey. Schools, retail stores, news media, and government offices all conduct surveys. There are many types of surveys, and it is not our intention to give a general discussion of this topic. Let us study a very popular method called the *simple tally survey*. Such a survey consists of questions for which the responses can be recorded in rows and columns of a table. These questions are appropriate to the information you want and are designed to cover the *entire* population of interest. In addition, the questions should be designed so that we can partition the sample space of responses into distinct (that is, mutually exclusive) sectors.

If the survey includes responses from a reasonably large random sample, then the results should be representative of your population. In this case, we can estimate simple probabilities, conditional probabilities, and the probabilities of some combinations of events directly from the results of the survey.

EXAMPLE 6

Survey

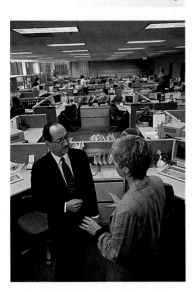

At Hopewell Electronics, all 140 employees were asked about their political affiliation. The employees were grouped by type of work, as executives or production workers. The results with row and column totals are shown in Table 5-2.

Suppose an employee is selected at random from the 140 Hopewell employees. Let us use the following notation to represent different events of choosing: E = executive; PW = production worker; D = Democrat; R = Republican; I = Independent.

(a) Compute $P(D)$ and $P(E)$.

SOLUTION: To find these probabilities, we look at the *entire* sample space.

$$P(D) = \frac{\text{Number of Democrats}}{\text{Number of employees}} = \frac{68}{140} = 0.486$$

$$P(E) = \frac{\text{Number of executives}}{\text{Number of employees}} = \frac{48}{140} = 0.343$$

(b) Compute $P(D, \text{ given } E)$.

SOLUTION: For the conditional probability, we *restrict* our attention to the portion of the sample space satisfying the condition of being an executive.

$$P(D, \text{ given } E) = \frac{\text{Number of executives who are Democrats}}{\text{Number of executives}} = \frac{5}{48} = 0.104$$

(c) Are the events D and E independent?

SOLUTION: One way to determine if the events D and E are independent is to see if $P(D) = P(D, \text{ given } E)$ [or equivalently, if $P(E) = P(E, \text{ given } D)$]. Since $P(D) = 0.486$ and $P(D, \text{ given } E) = 0.104$, we see that $P(D) \neq P(D, \text{ given } E)$. This means that the events D and E are *not* independent. The probability of event D "depends on" whether or not event E has occurred.

TABLE 5-2   Employee Type and Political Affiliation

Employee Type	Political Affiliation			Row Total
	Democrat (*D*)	Republican (*R*)	Independent (*I*)	
Executive (*E*)	5	34	9	48
Production worker (*PW*)	63	21	8	92
Column Total	68	55	17	140 Grand Total

(d) Compute $P(D \text{ and } E)$.

SOLUTION: This probability is not conditional, so we must look at the entire sample space.

$$P(D \text{ and } E) = \frac{\text{Number of executives who are Democrats}}{\text{Total number of employees}} = \frac{5}{140} = 0.036$$

Let's recompute this probability using the rules of probability for dependent events.

$$P(D \text{ and } E) = P(E) \cdot P(D, \text{ given } E) = \frac{48}{140} \cdot \frac{5}{48} = \frac{5}{140} = 0.036$$

The results using the rules are consistent with those using the sample space.

(e) Compute $P(D \text{ or } E)$.

SOLUTION: From part (d) we know that the events Democrat or executive are not mutually exclusive, because $P(D \text{ and } E) \neq 0$. Therefore,

$$P(D \text{ or } E) = P(D) + P(E) - P(D \text{ and } E)$$

$$= \frac{68}{140} + \frac{48}{140} - \frac{5}{140} = \frac{111}{140} = 0.793 \qquad \Diamond$$

GUIDED EXERCISE 9

Survey

Using Table 5-2, let's consider other probabilities regarding the type of employees at Hopewell and their political affiliation. This time let's consider the production worker and the affiliation of Independent. Suppose an employee is selected at random from the group of 140.

(a) Compute $P(I)$ and $P(PW)$.

$$P(I) = \frac{\text{No. of Independents}}{\text{Total no. of employees}}$$

$$= \frac{17}{140} = 0.121$$

$$P(PW) = \frac{\text{No. of production workers}}{\text{Total no. of employees}}$$

$$= \frac{92}{140} = 0.657$$

(b) Compute $P(I, \text{ given } PW)$. This is a conditional probability. Be sure to restrict your attention to production workers since that is the condition given.

$$P(I, \text{ given } PW) = \frac{\text{No. of Independent production workers}}{\text{No. of production workers}}$$

$$= \frac{8}{92} = 0.087$$

Continued

GUIDED EXERCISE 9 continued

(c) Compute $P(I \text{ and } PW)$. In this case, look at the entire sample space and at the number of employees who are both Independent and in production.

$\Rightarrow$ $P(I \text{ and } PW) = \dfrac{\text{No. of Independent production workers}}{\text{Total no. employees}}$

$= \dfrac{8}{140} = 0.057$

(d) Use the multiplication rule for dependent events to calculate $P(I \text{ and } PW)$. Is the result the same as that of part (c)?

$\Rightarrow$ By the multiplication rule,

$P(I \text{ and } PW) = P(PW) \cdot P(I, \text{ given } PW)$

$= \dfrac{92}{140} \cdot \dfrac{8}{92} = \dfrac{8}{140} = 0.057$

The results are the same.

(e) Compute $P(I \text{ or } PW)$. Are the events mutually exclusive?

$\Rightarrow$ Since the events are not mutually exclusive,

$P(I \text{ or } PW) = P(I) + P(PW) - P(I \text{ and } PW)$

$= \dfrac{17}{140} + \dfrac{92}{140} - \dfrac{8}{140} = \dfrac{101}{140} = 0.721$

Basic probability rules

As you apply probability to various settings, keep the following rules in mind.

Summary of basic probability rules for events *A* and *B*

A statistical experiment or statistical observation is any random activity that results in a recordable outcome. The sample space is the set of all possible distinct outcomes. Events are subsets of the sample space.

1. $P(\text{entire sample space}) = 1$

2. For any event A: $0 \le P(A) \le 1$

3. Complement of A: $P(not\ A) = 1 - P(A)$

4. Events A and B are independent events if $P(A) = P(A,\ given\ B)$

5. Multiplication Rules
   If A and B are independent events: $P(A\ and\ B) = P(A) \cdot P(B)$
   If A and B are dependent events: $P(A\ and\ B) = P(A) \cdot P(B,\ given\ A)$
   or equivalently; $P(B,\ given\ A) = P(A\ and\ B)/P(A)$

6. Events A and B are mutually exclusive if $P(A\ and\ B) = 0$

7. Addition Rules
   If A and B are mutually exclusive events: $P(A\ or\ B) = P(A) + P(B)$
   If A and B are not mutually exclusive:

   $P(A\ or\ B) = P(A) + P(B) - P(A\ and\ B)$

In this section, we have studied some important rules that are valid in all probability spaces. The rules and definitions of probability not only are interesting, but they also have extensive *applications* in our everyday lives.

VIEWPOINT   *The Psychology of Odors*

The Smell and Taste Treatment Research Foundation of Chicago collected data on the time required to complete a maze while subjects were smelling different scents. Data for this survey can be found by visiting the Brase/Brase *Understanding Basic Statistics* site at http://math.college.hmco.com/students and finding the link to DASL, the Carnegie Mellon University Data and Story Library. Once at the DASL site, select Data Subjects, then Psychology, and then Scents. You can estimate conditional probabilities regarding response times for smokers, nonsmokers, and types of scents.

SECTION 5.2   PROBLEMS

In Problems 1–12, use the appropriate addition or multiplication rules. When possible, verify results by considering the sample space.

1. *General: Candy Colors* M&M plain candies come in various colors. According to the M&M/Mars Department of Consumer Affairs (link to the Mars company Web site from the Brase/Brase *Understanding Basic Statistics* site at http://math.college.hmco.com/students), the distribution of colors for plain M&M candies is

Color	Purple	Yellow	Red	Orange	Green	Blue	Brown
Percentage	20%	20%	20%	10%	10%	10%	10%

Suppose you have a large bag of plain M&M candies and you take one candy at random. Find
   (a) *P*(green *or* blue candy). Are these outcomes mutually exclusive? Why?
   (b) *P*(yellow candy *or* red candy). Are these outcomes mutually exclusive? Why?
   (c) *P*(*not* purple candy).

2. *General: Candy Colors* According to the Department of Consumer Affairs of M&M/Mars, the color distribution of almond M&M candies is

Color	Purple	Yellow	Red	Green	Blue	Brown
Percentage	20%	20%	20%	20%	10%	10%

Suppose you have a large bag of almond M&M candies and you take one candy at random. Compute the probabilities in parts (a) through (c) of Problem 1 for almond M&M candies. Compare the results with those of plain M&M candies. Do you expect any differences? Why or why not?

3. *General: Candy Colors* Dulce de Leche~Caramel M&M candies have another color distribution, utilizing only six colors. The color distribution for these candies is uniform. There are only six colors: brown, red, yellow, orange, green, and blue. Each color comprises about 16.6% of the mix. Compute the probabilities in parts (a) through (c) of Problem 1 for these candies. Compare the results with those of plain M&M candies. Do you expect any differences? Why or why not?

4. *Environmental: Land Formations* Arches National Park is located in southern Utah. The park is famous for its beautiful desert landscape and its many natural sandstone arches. Park Ranger Edward McCarrick started an inventory (not yet complete) of natural arches within the park that have an opening of at least 3 feet. The following table is based on information taken from the book *Canyon Country Arches and Bridges*, by F. A. Barnes. The height of the arch opening is rounded to the nearest foot.

Height of arch, feet	3–9	10–29	30–49	50–74	75 and higher
Number of arches in park	111	96	30	33	18

For an arch chosen at random in Arches National Park, use the preceding information to estimate the probability that the height of the arch opening is
(a) 3 to 9 feet tall   (d) 10 to 74 feet tall
(b) 30 feet or taller   (e) 75 feet or taller
(c) 3 to 49 feet tall

5. *General: Roll Two Dice* You roll two fair dice, a green one and a red one.
   (a) Are the outcomes on the dice independent?
   (b) Find $P(5$ on green die *and* 3 on red die$)$.
   (c) Find $P(3$ on green die *and* 5 on red die$)$.
   (d) Find $P((5$ on green die *and* 3 on red die$)$ *or* $(3$ on green die *and* 5 on red die$))$.

6. *General: Roll Two Dice* You roll two fair dice, a green one and a red one.
   (a) Are the outcomes on the dice independent?
   (b) Find $P(1$ on green die *and* 2 on red die$)$.
   (c) Find $P(2$ on green die *and* 1 on red die$)$.
   (d) Find $P((1$ on green die *and* 2 on red die$)$ *or* $(2$ on green die *and* 1 on red die$))$.

7. *General: Roll Two Dice* You roll two fair dice, a green one and a red one.
   (a) What is the probability of getting a sum of 6?
   (b) What is the probability of getting a sum of 4?
   (c) What is the probability of getting a sum of 6 *or* 4? Are these outcomes mutually exclusive?

8. *General: Roll Two Dice* You roll two fair dice, a green one and a red one.
   (a) What is the probability of getting a sum of 7?
   (b) What is the probability of getting a sum of 11?
   (c) What is the probability of getting a sum of 7 *or* 11? Are these outcomes mutually exclusive?

9. *General: Deck of Cards* You draw two cards from a standard deck of 52 cards without replacing the first one before drawing the second.
   (a) Are the outcomes on the two cards independent? Why?
   (b) Find $P($ace on 1st card *and* king on 2nd$)$.
   (c) Find $P($king on 1st card *and* ace on 2nd$)$.
   (d) Find the probability of drawing an ace *and* a king in either order.

10. *General: Deck of Cards* You draw two cards from a standard deck of 52 cards without replacing the first one before drawing the second.
    (a) Are the outcomes on the two cards independent? Why?
    (b) Find $P(3$ on 1st card *and* 10 on 2nd$)$.
    (c) Find $P(10$ on 1st card *and* 3 on 2nd$)$.
    (d) Find the probability of drawing a 10 *and* a 3 in either order.

11. *General: Deck of Cards* You draw two cards from a standard deck of 52 cards, but before you draw the second card, you put the first one back and reshuffle the deck.
    (a) Are the outcomes on the two cards independent? Why?
    (b) Find P(ace on 1st card *and* king on 2nd).
    (c) Find P(king on 1st card *and* ace on 2nd).
    (d) Find the probability of drawing an ace *and* a king in either order.

12. *General: Deck of Cards* You draw two cards from a standard deck of 52 cards, but before you draw the second card, you put the first one back and reshuffle the deck.
    (a) Are the outcomes on the two cards independent? Why?
    (b) Find P(3 on 1st card *and* 10 on 2nd).
    (c) Find P(10 on 1st card *and* 3 on 2nd).
    (d) Find the probability of drawing a 10 *and* a 3 in either order.

13. *Marketing: Toys* USA Today gave the information shown in the table about ages of children receiving toys. The percentages represent all toys sold.

    What is the probability that a toy is purchased for someone
    (a) 6 years or older?
    (b) 12 years or younger?
    (c) between 6 and 12 years old?
    (d) between 3 and 9 years old?

Age (years)	Percentage of Toys
2 and under	15%
3–5	22%
6–9	27%
10–12	14%
13 and over	22%

    A child between 10 and 12 years old looks at this probability distribution and asks, "Why are people more likely to buy toys for kids older than I am (13 and over) than for kids in my age group (10–12)?" How would you respond?

14. *Health Care: Flu* Based on data from the *Statistical Abstract of the United States*, 112th Edition, only about 14% of senior citizens (65 years or older) get the flu each year. However, about 24% of the people under 65 years old get the flu each year. In the general population, there are 12.5% senior citizens (65 years or older).
    (a) What is the probability that a person selected at random from the general population is a senior citizen who will get the flu this year?
    (b) What is the probability that a person selected at random from the general population is a person under age 65 who will get the flu this year?
    (c) Answer parts (a) and (b) for a community that has 95% senior citizens.
    (d) Answer parts (a) and (b) for a community that has 50% senior citizens.

15. *Psychology: Lie Detector Test* In this problem, you are asked to solve part of the Focus Problem at the beginning of this chapter. In his book *Chances: Risk and Odds in Everyday Life,* James Burke says that there is a 72% chance a polygraph test (lie detector test) will catch a person who is in fact lying. Furthermore, there is approximately a 7% chance that the polygraph will falsely accuse someone of lying.
    (a) Suppose that a person answers 90% of a long battery of questions truthfully. What percentage of the story will the polygraph *wrongly* indicate is a lie?
    (b) Suppose that a person answers 10% of a long battery of questions with lies. What percentage of the story will the polygraph *correctly* indicate is a lie?
    (c) Repeat parts (a) and (b) if 50% of the questions are answered truthfully and 50% are answered with lies.
    (d) Repeat parts (a) and (b) if 15% of the questions are answered truthfully and the rest are answered with lies.

16. *Psychology: Lie Detector Test* This problem continues the Focus Problem. The solution involves applying several basic probability rules and a little algebra to solve an equation.

(a) If the polygraph (of Problem 15) said 30% of the questions were answered with lies, what would you estimate for the actual percentage of lies in the story? *Hint:* Let B = event detector indicates a lie. We are given $P(B) = 0.30$. Let A = event person is lying, so *not* A = event person is not lying. Then

$$P(B) = P(A \text{ and } B) + P(not \ A \text{ and } B)$$
$$P(B) = P(A)P(B, \text{ given } A) + P(not \ A)P(B, \text{ given not } A)$$

Replacing $P(not \ A)$ by $1 - P(A)$ gives

$$P(B) = P(A)P(B, \text{ given } A) + [1 - P(A)]P(B, \text{ given not } A)$$

Substitute known values for $P(B)$, $P(B, \text{ given } A)$, and $P(B, \text{ given not } A)$ into the last equation and solve for $P(A)$.

(b) If the polygraph (of Problem 15) said 70% of questions were answered with lies, what would you estimate for the actual percentage of lies in the story?

17. *Survey: Sales Approach* In a sales effectiveness seminar, a group of sales representatives tried two approaches to selling a customer a new automobile: the aggressive approach and the passive approach. From 1160 customers, the following record was kept:

	Sale	No Sale	Row Total
Aggressive	270	310	580
Passive	416	164	580
Column Total	686	474	1160

Suppose that a customer is selected at random from the 1160 participating customers. Let us use the following notation for events: A = aggressive approach, Pa = passive approach, S = sale, N = no sale. So $P(A)$ is the probability that an aggressive approach was used, and so on.

(a) Compute $P(S)$, $P(S, \text{ given } A)$, and $P(S, \text{ given } Pa)$.
(b) Are the events S = sale and Pa = passive approach independent? Explain.
(c) Compute $P(A \text{ and } S)$ and $P(Pa \text{ and } S)$.
(d) Compute $P(N)$ and $P(N, \text{ given } A)$.
(e) Are the events N = no sale and A = aggressive approach independent? Explain.
(f) Compute $P(A \text{ or } S)$.

18. *Survey: Medical Tests* Diagnostic tests of medical conditions have several results. The test result can be positive or negative, whether or not a patient has the condition (+ indicates a patient has the condition). Consider a random sample of 200 patients, some of whom have a medical condition and some of whom do not. Results of a new diagnostic test for the condition are shown.

	Condition Present	Condition Absent	Row Total
Test Result +	110	20	130
Test Result −	20	50	70
Column Total	130	70	200

Assume the sample is representative of the entire population. For a person selected at random, compute the following probabilities:

(a) $P(+,$ *given* condition present); this is known as the *sensitivity* of a test.
(b) $P(-,$ *given* condition present); this is known as the false-negative rate.
(c) $P(-,$ *given* condition absent); this is known as the *specificity* of a test.
(d) $P(+,$ *given* condition absent); this is known as the false-positive rate.
(e) $P($condition present *and* $+)$; this is the predictive value of the test.
(f) $P($condition present *and* $-)$.

19. *Survey: Lung/Heart* In an article entitled "Diagnostic accuracy of fever as a measure of postoperative pulmonary complications" (*Heart Lung* 10, No. 1:61), J. Roberts and colleagues discuss using a fever of 38°C or higher as a diagnostic indicator of postoperative atelectasis (collapse of the lung) as evidenced by x-ray observation. For fever $\geq$ 38°C as the diagnostic test, the results for postoperative patients are

	Condition Present	Condition Absent	Row Total
Test Result +	72	37	109
Test Result −	82	79	161
Column Total	154	116	270

Complete parts (a) through (f) from Problem 18.

20. *Survey: Customer Loyalty* Are customers more loyal in the East or in the West? The following table is based on information from *Trends in the United States*, published by the Food Marketing Institute, Washington, D.C. The columns represent length of customer loyalty (years) at a primary supermarket. The rows represent regions of the United States.

	Less Than 1 Year	1–2 Years	3–4 Years	5–9 Years	10–14 Years	15 or More Years	Row Total
East	32	54	59	112	77	118	452
Midwest	31	68	68	120	63	173	523
South	53	92	93	158	106	158	660
West	41	56	67	78	45	86	373
Column Total	157	270	287	468	291	535	2008

What is the probability that a customer chosen at random
(a) has been loyal 10 to 14 years?
(b) has been loyal 10 to 14 years, given that he or she is from the East?
(c) has been loyal *at least* 10 years?
(d) has been loyal *at least* 10 years, given that he or she is from the West?
(e) is from the West, given that he or she has been loyal less than 1 year?
(f) is from the South, given that he or she has been loyal less than 1 year?
(g) has been loyal *1 or more years,* given that he or she is from the East?
(h) has been loyal *1 or more years,* given that he or she is from the West?
(i) Are the events from the East and loyal 15 or more years independent? Explain.

5.3
Trees and Counting Techniques

Tree diagrams

When outcomes are equally likely, we compute the probability of an event by using the formula

$$P(A) = \frac{\text{Number of outcomes favorable to the event } A}{\text{Number of outcomes in the sample space}}$$

The probability formula requires that we be able to determine the number of outcomes in the sample space. In the problems we have done in previous sections, this task has not been difficult because the number of outcomes was small or the sample space consisted of fairly straightforward events. The tools we present in this section will help you count the number of possible outcomes in larger sample spaces or those formed by more complicated events.

A *tree diagram* helps us display the outcomes of an experiment consisting of a series of activities. The total number of outcomes corresponds to the total number of final branches in the tree. Perhaps the best way to learn to make a tree diagram is to see one. In the next example, we will see a tree diagram and analyze its parts.

EXAMPLE 7

Tree diagram

Jacqueline is in the nursing program and is required to take a course in psychology and one in anatomy and physiology (A and P) next semester. She also wants to take Spanish II. If there are four sections of psychology, two of A and P, and three of Spanish, how many different class schedules can Jacqueline choose from? (Assume that the times of the sections do not conflict with each other.) Figure 5-6 shows a tree diagram for Jacqueline's possible schedules.

FIGURE 5-6

Tree Diagram for Selecting Class Schedules

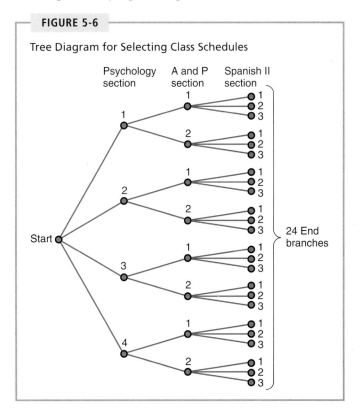

TABLE 5-3 Schedules Utilizing Section 1 of Psychology

Psychology Section	A and P Section	Spanish II Section
1	1	1
1	1	2
1	1	3
1	2	1
1	2	2
1	2	3

TABLE 5-4 Schedules Utilizing Section 2 of Psychology

Psychology Section	A and P Section	Spanish II Section
2	1	1
2	1	2
2	1	3
2	2	1
2	2	2
2	2	3

SOLUTION: Let's study the tree diagram and see how it shows Jacqueline's schedule choices. There are four branches from Start. These branches indicate the four possible choices for psychology sections. No matter which section of psychology Jacqueline chooses, she can choose from the two available A and P sections. Therefore, we have two branches leading from *each* psychology branch. Finally, after the psychology and A and P sections are selected, there are three choices for Spanish II. That is why there are three branches from *each* A and P section.

The tree ends with a total of 24 branches. This number of end branches tells us the number of possible schedules. The outcomes themselves can be listed from the tree by following each series of branches from Start to End. For instance, the top branch from Start generates the schedules shown in Table 5-3.

Following the second branch from Start, we see all the possible schedules utilizing Section 2 of psychology (see Table 5-4). The other 12 schedules can be listed in a similar manner. ◇

We draw a tree diagram in stages, indicating the possible outcomes for the first event, second event, and so forth. Guided Exercise 10 will lead you through the process.

GUIDED EXERCISE 10

Tree diagram

Louis plays three tennis matches. Use a tree diagram to list the possible win and loss sequences Louis can experience for the set of three matches.

(a) On the first match Louis can win or lose. From Start, indicate these two branches.

FIGURE 5-7 W = Win, L = Lose

Continued

GUIDED EXERCISE 10 continued

(b) Regardless of whether Louis wins or loses the first match, he plays the second and can again win or lose. Attach branches representing these two outcomes to *each* of the first match results.

➡ **FIGURE 5-8**

(c) Louis may win or lose the third match. Attach branches representing these two outcomes to *each* of the second match results.

➡ **FIGURE 5-9**

(d) How many possible win–lose sequences are there for the three matches?

➡ Since there are eight branches at the end, there are eight sequences.

(e) Complete this list of win–lose sequences.

1st	2nd	3rd
W	W	W
W	W	L
W	L	W
W	L	L
___	___	___
___	___	___
___	___	___
___	___	___

➡ The last four sequences all involve a loss on Match 1.

1st	2nd	3rd
L	W	W
L	W	L
L	L	W
L	L	L

Tree diagrams help us display the outcomes of an experiment involving several stages. If we label each branch of the tree with an appropriate probability, we can use the tree diagram to help us compute probabilities of an outcome displayed on the tree. One of the easiest ways to illustrate this feature of tree diagrams is to use an experiment of drawing balls out of an urn. We do this in the next example.

EXAMPLE 8

Probability

Suppose there are five balls in an urn. They are identical except in color. Three of the balls are red and two are blue. You are instructed to draw out one ball, note its color, and set it aside. Then you are to draw out another ball and note its color. What are the outcomes of the experiment? What is the probability of each outcome?

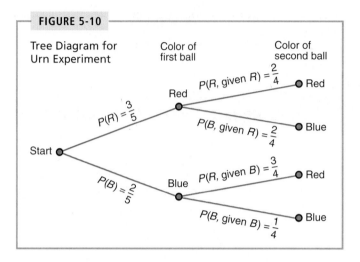

3 red
2 Blue

FIGURE 5-10

Tree Diagram for Urn Experiment

SOLUTION: The tree diagram in Figure 5-10 will help us answer these questions. Notice that since you did not replace the first ball before drawing the second one, the two stages of the experiment are dependent. The probability associated with the color of the second ball depends on the color of the first ball. For instance, on the top branches, the color of the first ball drawn is red, so we compute the probabilities of the colors on the second ball accordingly. The tree diagram helps us organize the probabilities.

From the diagram, we see that there are four possible outcomes to the experiment. They are

> RR = red on 1st *and* red on 2nd
> RB = red on 1st *and* blue on 2nd
> BR = blue on 1st *and* red on 2nd
> BB = blue on 1st *and* blue on 2nd

To compute the probability of each outcome, we will use the multiplication rule for dependent events. As we follow the branches for each outcome, we will find the necessary probabilities.

$$P(R \text{ on 1st } and \text{ R on 2nd}) = P(R) \cdot P(R, \textit{ given } R)$$
$$= \frac{3}{5} \cdot \frac{2}{4} = \frac{3}{10}$$

$$P(R \text{ on 1st } and \text{ B on 2nd}) = P(R) \cdot P(B, \textit{ given } R)$$
$$= \frac{3}{5} \cdot \frac{2}{4} = \frac{3}{10}$$

$$P(B \text{ on 1st } and \text{ R on 2nd}) = P(B) \cdot P(R, \textit{ given } B)$$
$$= \frac{2}{5} \cdot \frac{3}{4} = \frac{3}{10}$$

$$P(B \text{ on 1st } and \text{ B on 2nd}) = P(B) \cdot P(B, \textit{ given } B)$$
$$= \frac{2}{5} \cdot \frac{1}{4} = \frac{1}{10}$$

Notice that the probabilities of the outcomes in the sample space add to 1, as they should.   ◊

Probability

Repeat the urn experiment with the five balls, three of which are red and two of which are blue. This time replace the first ball before drawing the second.

(a) Draw a tree diagram for the outcomes of this experiment. Show the probabilities of each stage on the appropriate branch. (*Hint:* Are the stages dependent or independent?)

FIGURE 5-11 Tree Diagram for Urn Experiment (with replacement)

(b) List the four possible outcomes of the experiment (i.e., list the sample space).

Red on 1st *and* red on 2nd
Red on 1st *and* blue on 2nd
Blue on 1st *and* red on 2nd
Blue on 1st *and* blue on 2nd

(c) Use the multiplication rule for independent events and the probabilities shown on your tree to compute the probability of each outcome.

$$P(\text{1st } R \text{ and 2nd } R) = P(R) \cdot P(R)$$
$$= \frac{3}{5} \cdot \frac{3}{5} = \frac{9}{25}$$

$$P(\text{1st } R \text{ and 2nd } B) = P(R) \cdot P(B)$$
$$= \frac{3}{5} \cdot \frac{2}{5} = \frac{6}{25}$$

$$P(\text{1st } B \text{ and 2nd } R) = P(B) \cdot P(R)$$
$$= \frac{2}{5} \cdot \frac{3}{5} = \frac{6}{25}$$

$$P(\text{1st } B \text{ and 2nd } B) = P(B) \cdot P(B)$$
$$= \frac{2}{5} \cdot \frac{2}{5} = \frac{4}{25}$$

Continued

GUIDED EXERCISE 11 continued

(d) Do the probabilities of the outcomes in the sample space add up to 1?

⟹ Yes, as they should.

(e) Compare the tree diagram of this exercise with that of the previous example, in which the first ball was not replaced before the second was drawn. Are the outcomes the same? Are the probabilities of the corresponding outcomes the same?

⟹ The outcomes are the same: *RR, RB, BR,* and *BB.* However, because one experiment did not permit replacement of the first ball before the second was drawn and the other experiment required replacement, the corresponding probabilities of the second stages of the experiment are different. The corresponding probabilities of the final outcomes are also different.

Multiplication rule of counting

When an outcome is composed of a series of events, tree diagrams tell us how many possible outcomes there are. They also help us list the individual outcomes and organize the probabilities associated with each stage of the outcomes. However, if we are interested only in the number of different outcomes created by a series of events, the multiplication rule will give us the total number of outcomes more directly. We state the multiplication rule for an outcome composed of a series of two events.

Multiplication rule of counting

If there are n possible outcomes for event E_1 and m possible outcomes for event E_2, then there are a total of $n \times m$ or nm possible outcomes for the series of events E_1 followed by E_2.

The rule extends to outcomes created by a series of three, four, or more events. We simply multiply the number of outcomes possible for each step in the series of events to get the total number of outcomes for the series.

EXAMPLE 9
Multiplication rule

The Night Hawk is the new car model produced by Limited Motors, Inc. It comes with a choice of two body styles, three interior package options, and four different colors, as well as the choice of automatic or standard transmission. Select-an-Auto Car Dealership wants to carry one of each of the different types of Night Hawks. How many cars are required?

SOLUTION: There are four items to select. We take the product of the number of choices for each item.

$$\left(\begin{array}{c}\text{No. of body}\\\text{styles}\end{array}\right)\left(\begin{array}{c}\text{No. of}\\\text{interiors}\end{array}\right)\left(\begin{array}{c}\text{No. of}\\\text{colors}\end{array}\right)\left(\begin{array}{c}\text{No. of transmission}\\\text{types}\end{array}\right) = (2)(3)(4)(2) = 48$$

Select-an-Auto must stock 48 cars to have one of each possible type.   ◊

GUIDED EXERCISE 12

Multiplication rule of counting

The Old Sage Inn offers a special dinner menu each night. There are two appetizers to choose from, three main courses, and four desserts. A customer can select one item from each category. How many different meals can be ordered from the special dinner menu?

(a) Each special dinner consists of three items. List the item and the number of choices per item.

⟹ Appetizer—2; main course—3; dessert—4

(b) Find the number of different dinners composed of the three items.

⟹ (2)(3)(4) = 24   There are 24 different dinners.

Sometimes when we consider n items, we need to know the number of different *ordered arrangements* of the n items that are possible. The multiplication rules can help us find the number of possible ordered arrangements. Let's consider the classic example of determining the number of different ways in which eight people can be seated at a dinner table. For the first chair at the head of the table, there are eight choices. For the second chair, there are seven choices, since one person is already seated. For the third chair, there are six choices, since two people are already seated. By the time we get to the last chair, there is only one person left for that seat. We can view each arrangement as an outcome of a series of eight events. Event 1 is *fill the first chair,* event 2 is *fill the second chair,* and so forth. The multiplication rule will tell us the number of different outcomes.

Choices for	1st	2nd	3rd	4th	5th	6th	7th	8th	Chair position
	↓	↓	↓	↓	↓	↓	↓	↓	
	(8)	(7)	(6)	(5)	(4)	(3)	(2)	(1)	= 40,320

In all, there are 40,320 different seating arrangements for eight people. It is no wonder that it takes a little time to seat guests at a dinner table!

Factorial notation

The multiplication pattern shown above is not unusual. In fact, it is an example of the multiplication indicated by the factorial notation 8!.

! is read "factorial"

8! is read "8 factorial"

$8! = 8 \cdot 7 \cdot 6 \cdot 5 \cdot 4 \cdot 3 \cdot 2 \cdot 1$

In general, $n!$ indicates the product of n with each of the positive counting numbers less than n. By *special definition* $0! = 1$.

Factorial notation

For a counting number n,

$n! = n(n - 1)(n - 2) \cdots 1$

$0! = 1$

$1! = 1$

GUIDED EXERCISE 13

Factorial

(a) Evaluate 3!.

⟹ $3! = 3 \cdot 2 \cdot 1 = 6$

(b) How many different ways can three objects be arranged in order? How many choices do you have for the first position? for the second position? for the third position?

⟹ We have three choices for the first position, two for the second position, and one for the third position. By the multiplication rule, we have

$(3)(2)(1) = 3! = 6$ arrangements

(c) Verify step (b) with a three-stage tree diagram.

⟹ **FIGURE 5-12** Three Choices

Permutations

We have considered the number of ordered arrangements of n objects taken as an entire group. Specifically, we considered a dinner party for eight and found the number of ordered seating arrangements for all eight people. However, suppose you have an open house and have only five chairs. How many ways can five of the eight people seat themselves in the chairs? The formula we use to compute this number is called the *permutation formula*. As we see in the next example, the *permutations rule* is really another version of the multiplication rule.

Counting rule for permutations

The number of ways to *arrange in order* n distinct objects, taking them r at a time, is

$$P_{n,r} = \frac{n!}{(n-r)!} \tag{9}$$

where n and r are whole numbers and $n \geq r$. Another commonly used notation for permutations is nPr.

EXAMPLE 10

Permutations rule

Let's compute the number of ordered seating arrangements we have for eight people in five chairs.

SOLUTION: In this case, we are considering a total of $n = 8$ different people, and we wish to arrange $r = 5$ of these people. Substituting into Formula (9), we have

$$P_{n,r} = \frac{n!}{(n-r)!}$$

$$P_{8,5} = \frac{8!}{(8-5)!} = \frac{8!}{3!} = \frac{40,320}{6} = 6720$$

Using the multiplication rule, we get the same results

Chair	1		2		3		4		5	
Choices for	8	×	7	×	6	×	5	×	4	= 6720

The permutations rule has the advantage of using factorials. Most scientific calculators have a factorial key ! as well as a permutations key nPr (see the Tech Note). ◊

TECH NOTE Most scientific calculators have a factorial key, often designated x! or n!. Many of these same calculators have the permutation function built in, often labeled nPr. They also have the combination function that is discussed after the next exercise. The combination function is often labeled nCr.

TI-83Plus The factorial, permutation, and combination functions are all under **MATH**, then **PRB**.

Excel Use the **Paste Function** [fₓ] then **all. Fact** gives factorials, **Permut** gives permutations, **Combin** gives combinations.

GUIDED EXERCISE 14

Permutation

The board of directors of Belford Community Hospital has 12 members. Three officers—president, vice president, and treasurer—must be elected from the members. How many different possible slates of officers are there? We will view a slate of officers as a list of three people with one person for president listed first, one person for vice president listed second, and one person for treasurer listed third. For instance, if Mr. Acosta, Ms. Hill, and Mr. Smith wish to be on a slate together, there are several different slates possible, depending on which one will run for president, which for vice president, and which for treasurer. Not only are we asking for the number of different groups of three names for a slate, we are also concerned about order, since it makes a difference which name is listed in which position.

(a) What is the size of the group from which the slates of officers will be selected? This is the value of *n*.

⟹ $n = 12$

Continued

GUIDED EXERCISE 14 continued

(b) How many people will be selected for each slate of officers? This is the value of r.

$\Rightarrow$ $r = 3$

(c) Each slate of officers is composed of three candidates. Different slates occur as we arrange the three candidates in the positions of president, vice president, and treasurer. For this reason, we need to consider the number of *permutations* of 12 items arranged in groups of 3. Compute $P_{n,r}$.

$\Rightarrow$ $P_{n,r} = \dfrac{n!}{(n-r)!}$

$P_{12,3} = \dfrac{12!}{(12-3)!} = \dfrac{12!}{9!} = \dfrac{479,001,600}{362,880} = 1320$

There are 1320 different possible slates of officers. An alternative is to use your calculator to compute $P_{12,3}$ directly.

Combinations

In each of our previous counting formulas, we have taken the *order* of the objects or people into account. But what if order is not important? For instance, suppose we need to choose 3 members from the 12-member board of directors of Belford Community Hospital to go to a convention. We are interested in *different groupings* of 12 people so that each group contains 3 people. The order is of no concern, since all 3 will go to the convention. In other words, we need to consider the number of different *combinations* of 12 people taken 3 at a time. Our next formula will help us compute this number of different combinations.

Counting rule for combinations

The number of *combinations* of n objects taken r at a time is

$$C_{n,r} = \frac{n!}{r!(n-r)!} \tag{10}$$

where n and r are whole numbers and $n \geq r$. Other commonly used notations for combinations include nCr and $\binom{n}{r}$.

Notice the difference between the concepts of permutations and combinations. When we consider permutations, we are considering groupings *and order*. When we consider combinations, we are considering only the number of different groupings. For combinations, order within the groupings is not considered. As a result, the number of combinations of n objects taken r at a time is generally smaller than the number of permutations of the same n objects taken r at a time. In fact, the combinations formula is simply the permutations formula with the number of permutations of each distinct group divided out. In the formula for combinations, notice the factor of $r!$ in the denominator.

Now let's look at an example in which we use the *combinations rule* to compute the number of *combinations* of 12 people taken 3 at a time.

EXAMPLE 11

Combinations

Three members from the group of 12 on the board of directors at Belford Community Hospital will be selected to go to a convention with all expenses paid. How many different groups of 3 are there?

SOLUTION: In this case, we are interested in *combinations* rather than permutations of 12 people taken 3 at a time. Using Formula (10), we get

$$C_{n,r} = \frac{n!}{r!(n-r)!} \quad \text{or}$$

$$C_{12,3} = \frac{12!}{3!(12-3)!} = \frac{12!}{3!9!} = \frac{479,001,600}{(6)(362,880)} = 220$$

There are 220 different possible groups of 3 to go to the convention.

Another way to get the solution is to use your calculator to evaluate $C_{12,3}$ directly. Since order is not considered, this number is much smaller than the number of different slates of 3 officers we computed in Guided Exercise 14.   ◇

We have different formulas for permutations and combinations of n objects taken r at a time. How do you decide which one to use? Always ask yourself if order in the groups of r objects is relevant. If it is, use $P_{n,r}$. If order is not relevant, use $C_{n,r}$.

We have introduced you to three counting formulas: the multiplication rule, the permutations rule, and the combinations rule. Other rules apply when the objects are not distinct. Many counting problems are easy to state and fairly difficult to solve. Some have you combine several counting rules. However, the problems for this section are all straightforward. Some ask you to use your counting abilities to compute probabilities.

GUIDED EXERCISE 15

Combinations

In your political science class, you are given a list of 10 books. You are to select 4 to read during the semester. How many different *combinations* of 4 books are available from the list of 10?

(a) Is the order in which you read the books relevant to the task of selecting the books?

⟹ No.

(b) Do we use the number of permutations or combinations of 10 books taken 4 at a time?

⟹ Since consideration of order in which the books are selected is not relevant, we compute the number of *combinations* of 10 books taken 4 at a time.

(c) How many books are available from which to select? How many must you read? What is the value of n? of r?

⟹ There are 10 books among which you must select 4 to read. $n = 10$ and $r = 4$.

(d) Compute $C_{10,4}$ to determine the number of different groups of 4 books from the list of 10.

⟹ $$C_{n,r} = \frac{n!}{r!(n-r)!}$$

$$C_{10,4} = \frac{10!}{4!(10-4)!} = \frac{10!}{4!6!} = \frac{3,628,800}{(24)(720)} = 210$$

There are 210 different groups of 4 books to select from the list of 10. An alternate method of solution is to use the nCr key on your calculator.

VIEWPOINT

Powerball

Powerball is a multistate lottery game that consists of drawing five distinct whole numbers from numbers 1 through 53. Then one more number from numbers 1 through 42 is selected as the Powerball number (this number could be one of the original five). Powerball numbers are drawn every Wednesday and Saturday. If you match all six numbers, you win the jackpot, which is worth at least 10 million dollars. Use methods of this section to show that there are 120,526,770 possible Powerball plays. This means that the odds of winning the jackpot are 1 in 120,526,770. For more information about the game of Powerball and the probability of winning different prizes, visit the Brase/Brase *Understanding Basic Statistics* site at http://math.college.hmco.com/students and find the link to the Multi-State Lottery Association. Then select Powerball.

SECTION 5.3 PROBLEMS

1. *Tree Diagram*
   (a) Draw a tree diagram to display all the possible head–tail sequences that can occur when you flip a coin three times.
   (b) How many sequences contain exactly two heads?
   (c) *Probability extension:* Assuming the sequences are all equally likely, what is the probability that you will get exactly two heads when you toss a coin three times?

2. *Tree Diagram*
   (a) Draw a tree diagram to display all the possible outcomes that can occur when you flip a coin and then toss a die.
   (b) How many outcomes contain a head and a number greater than four?
   (c) *Probability extension:* Assuming the outcomes displayed in the tree diagram are all equally likely, what is the probability that you will get a head *and* a number greater than four when you flip a coin and toss a die?

3. *Tree Diagram* There are six balls in an urn. They are identical except for color. Two are red, three are blue, and one is yellow. You are to draw a ball from the urn, note its color, and set it aside. Then you are to draw another ball from the urn and note its color.
   (a) Make a tree diagram to show all possible outcomes of the experiment. Label the probability associated with each stage of the experiment on the appropriate branch.
   (b) *Probability extension:* Compute the probability for each outcome of the experiment.

4. *Tree Diagram*
   (a) Make a tree diagram to show all the possible sequences of answers for three multiple-choice questions, each with four possible responses.
   (b) *Probability extension:* Assuming that you are guessing the answers so that all outcomes listed in the tree are equally likely, what is the probability that you will guess the one sequence that contains all three correct answers?

5. *Multiplication Rule* Four wires (red, green, blue, and yellow) need to be attached to a circuit board. A robotic device will attach the wires. The wires can be attached

in any order, and the production manager wishes to determine which order would be fastest for the robot to use. Use the multiplication rule of counting to determine the number of all the possible sequences of assembly that must be tested. (*Hint:* There are four choices for the first wire, three for the second, two for the third, and only one for the fourth.)

6. *Multiplication Rule* A sales representative must visit four cities: Omaha, Dallas, Wichita, and Oklahoma City. There are direct air connections between each of the cities. Use the multiplication rule of counting to determine the number of different choices the sales representative has for the order in which to visit the cities. How is this problem similar to Problem 5?

7. *Multiplication Rule* You have two decks of cards (52 cards per deck), and you draw one card from each deck.
    (a) Use the multiplication rule of counting to determine the number of pairs of cards possible.
    (b) There are four kings in each deck. How many pairs of kings are possible?
    (c) *Probability extension:* Assuming all pairs are equally likely to be drawn, what is the probability of drawing two kings?

8. *Multiplication Rule* You toss a pair of dice.
    (a) Use the multiplication rule of counting to determine the number of possible pairs of outcomes. (Recall that there are six possible outcomes for each die.)
    (b) There are three even numbers on each die. How many outcomes are possible with even numbers appearing on each die?
    (c) *Probability extension:* What is the probability that both dice will show an even number?

9. Compute $P_{5,2}$.

10. Compute $P_{8,3}$.

11. Compute $P_{7,7}$.

12. Compute $P_{9,9}$.

13. Compute $C_{5,2}$.

14. Compute $C_{8,3}$.

15. Compute $C_{7,7}$.

16. Compute $C_{8,8}$.

17. *Permutation: Hiring* There are three nursing positions to be filled at Lilly Hospital. Position one is the day nursing supervisor; position two is the night nursing supervisor; and position three is the nursing coordinator position. There are 15 candidates qualified for all three of the positions. Use the permutations rule to determine the number of different ways the positions can be filled by these applicants.

18. *Counting: Sales* During the Computer Daze special promotion, a customer purchasing a computer and printer is given a choice of three free software packages. There are 10 different software packages from which to select. How many different groups of software packages can be selected?

19. *Counting: Hiring* There are 15 qualified applicants for five trainee positions in a fast-food management program. How many different groups of trainees can be selected? (*Hint:* Is order important? If not, use the formula for combinations.)

20. *Counting: Grading* One professor grades homework by randomly choosing 5 out of 12 homework problems to grade.
    (a) How many different groups of 5 problems are there from the 12 problems?
    (b) *Probability extension:* Jerry did only 5 problems of one assignment. What is the probability that the problems he did comprised the group that was selected to be graded?

(c) Silvia did 7 problems. How many different groups of 5 did she complete? What is the probability that one of the groups of 5 she completed comprised the group selected to be graded?

21. *Counting: Hiring* The qualified applicant pool for six management trainee positions consists of seven women and five men.
    (a) How many different groups of applicants can be selected for the positions?
    (b) How many different groups of trainees would consist entirely of women?
    (c) *Probability extension:* If the applicants are equally qualified and the trainee positions are selected by drawing the names at random so that all groups of six are equally likely, what is the probability that the trainee class will consist entirely of women?

22. *Counting: Lottery* In the Colorado State Lotto game, there are 42 numbers. Players choose any 6. Then the state selects 6 of the numbers at random. The winning tickets (for the grand prize) are those on which the player's 6 numbers match the state's 6 numbers.
    (a) From 42 numbers, how many groups of 6 are possible?
    (b) *Probability extension:* If you buy one lottery ticket, what is the probability of winning the grand prize?
    (c) *Probability extension:* If you buy 10 lottery tickets, what is the probability of winning the grand prize?

SUMMARY

In this chapter, we first examined the question: What is probability? We found that probabilities can be assigned to events by intuition, by the method of relative frequency, or by the method of equally likely outcomes. Next, we studied some probability rules. The most important rules are the multiplication rules for independent and dependent events and the addition rules for mutually exclusive and general events. We also looked at some counting techniques useful in computing probabilities. These techniques included tree diagrams, the multiplication rule for counting, combinations, and permutations.

IMPORTANT WORDS & SYMBOLS

Section 5.1
Probability of an event A, $P(A)$
Relative frequency
Law of large numbers
Equally likely outcomes
Sample space
Statistical experiment
Complement of A

Section 5.2
Independent events
Dependent events
A, *given* B
Conditional probability

Multiplication rules of probability (for independent and dependent events)
A *and* B
Mutually exclusive events
A *or* B
Addition rules (for mutually exclusive and general events)

Section 5.3
Tree diagram
Multiplication rule of counting
Permutations rule
Combinations rule

VIEWPOINT   *Deathday and Birthday*

Can people really postpone death? If so, how much can the timing of death be influenced by psychological, social, or other influential factors? One special event is a birthday. Will famous people try to postpone their death until an important birthday? Both Thomas Jefferson and John Adams died on July 4, 1826, when the United States was celebrating its 50th birthday. Is this only a strange coincidence, or is there an unexpected connection between birthdays and deathdays? The probability associated with a death rate decline of famous people just before important birthdays has been studied by Professor D. P. Phillips, of the State University of New York, and is presented in the book *Statistics, A Guide to the Unknown*, edited by J. M. Tanur.

CHAPTER REVIEW PROBLEMS

1. *Salary Raise: Women* Does it pay to ask for a raise? A national survey of heads of households showed the percentage of those who asked for a raise and the percentage who got one (*USA Today*). According to the survey, of the women interviewed, 24% had asked for a raise, and of those women who had asked for a raise, 45% received the raise. If a woman is selected at random from the survey population of women, find the following probabilities: *P*(woman asked for a raise); *P*(woman received raise, *given* she asked for one); *P*(woman asked for raise *and* received raise).

2. *Salary Raise: Men* According to the same survey quoted in Problem 1, of the men interviewed, 20% had asked for a raise and 59% of the men who had asked for a raise *received* the raise. If a man is selected at random from the survey population of men, find the following probabilities: *P*(man asked for a raise); *P*(man received raise, *given* he asked for one); *P*(man asked for raise *and* received raise).

3. *General: Deck of Cards* Two cards are drawn at random from a standard deck. (A standard deck has 52 cards: 13 hearts, 13 diamonds, 13 clubs, and 13 spades.)
   (a) Are the outcomes of the two cards independent? Why?
   (b) If the first card is replaced before the second is drawn, what is the probability that both cards will be hearts?
   (c) If the first card is not replaced before the second is drawn, what is the probability that both cards will be hearts?

4. *General: Die and Coin* Suppose you throw a fair die and flip a fair coin. Let's represent the outcomes of 3 on the die and heads on the coin by 3H.
   (a) One outcome is 3H. What are the other outcomes? What is the sample space?
   (b) Are all outcomes in the sample space equally likely? Explain.
   (c) What is the probability of getting heads and a number less than 3?

5. *General: Thumbtack*
   (a) Describe how you could use a relative frequency to estimate the probability that a thumbtack will land with its flat side down.
   (b) What is the sample space of outcomes for the thumbtack?
   (c) How would you make a probability assignment to this sample space if, when you drop 500 tacks, 340 land flat side down?

TABLE 5-5   Time in Which Oil Was Washed Off

Reaction	Within 5 Minutes	After 5 Minutes	Row Total
None	420	50	470
Mild	60	330	390
Strong	20	120	140
Column Total	500	500	1000

6. *Survey: Reaction to Poison Ivy* Allergic reactions to poison ivy can be miserable. Plant oils cause the reaction. Researchers at Allergy Institute did a study to see the effects of washing the oil off within 5 minutes of exposure. A random sample of 1000 people with known allergies to poison ivy participated in the study. Oil from the poison ivy plant was rubbed on a patch of skin. For 500 of the subjects, it was washed off *within* 5 minutes. For the other 500 subjects, the oil was washed off *after* 5 minutes. The results are summarized in Table 5-5.

   Let's use the following notation for the various events: W = washing oil off within 5 minutes, A = washing oil off after 5 minutes, N = no reaction, M = mild reaction, S = strong reaction. Find the following probabilities for a person selected at random from this sample of 1000 subjects.
   (a) $P(N)$, $P(M)$, $P(S)$
   (b) $P(N, given\ W)$, $P(S, given\ W)$
   (c) $P(N, given\ A)$, $P(S, given\ A)$
   (d) $P(N\ and\ W)$, $P(M\ and\ W)$
   (e) $P(N\ or\ M)$ Are the events N = no reaction and M = mild reaction mutually exclusive? Explain.
   (f) Are the events N = no reaction and W = washing oil off within 5 minutes independent? Explain.

7. *General: Two Dice* In a game of craps you roll two fair dice. Whether you win or lose depends on the sum of the numbers occurring on the tops of the dice. Let x be the random variable that is the sum of the numbers on the tops of the dice.
   (a) What values can x take on?
   (b) What is the probability distribution of these x values (that is, what is the probability that $x = 2$, 3, etc.)?

8. *Academic: Passing French* Class records at Rockwood College indicate a student selected at random has probability 0.77 of passing French 101. For the student who passes French 101, the probability is 0.90 that he or she will pass French 102. What is the probability that a student selected at random will pass both French 101 and 102?

9. *Combination: City Council* There is money to send two of eight city council members to a conference in Honolulu. All want to go, so they decide to choose the members to go to the conference by a random process. How many different combinations of two council members can be selected from the eight who want to go to the conference?

10. Compute
    (a) $P_{7,2}$ (b) $C_{7,2}$ (c) $P_{3,3}$ (d) $C_{4,4}$

11. *Inspection: Food Processing* Freeze Dry Food, Inc., packages all its foods in clear plastic that is sealed. The quality control for the packaging process checks for three items: (1) that the weight shown is correct, (2) that the label is correct, and (3) that the package is properly sealed. These three processes can be done in any order. A computer-operated device directs the packages to the three inspection stations

according to backlog in that area. If there is a larger backlog in one area, products are sent to one of the other two areas first. In how many different ways can a package be cycled through the three inspection stations?

12. *Scheduling: College Courses* A student must satisfy the literature, social science, and philosophy requirements this semester. There are four literature courses to select from, three social science courses, and two philosophy courses. Make a tree diagram showing all the possible sequences of literature, social science, and philosophy courses.

13. *General: Multiplication Rule* There are five multiple-choice questions on an exam, each with four possible answers. Use the multiplication rule of counting to determine the number of possible answer sequences for the five questions. Only one of the sets can contain all five correct answers. If you are guessing, so that you are as likely to choose one sequence of answers as another, what is the probability of getting all five answers correct?

14. *General: Multiplication Rule* A coin is tossed six times. Use the multiplication rule of counting to determine the number of possible head–tail sequences that can occur.

DATA HIGHLIGHTS: GROUP PROJECTS

Break into small groups and discuss the following topics. Organize a brief outline in which you summarize the main points of your group discussion.

1. Look at Figure 5-13, "Peeking at workers' e-mail." What group of people was surveyed? Estimate the probability that an executive selected at random from the survey population works at a company that uses e-mail. Estimate the probability that an executive peeks at employee e-mail, given that the company uses e-mail. Compute the probability that an executive from the survey population works at a company that uses e-mail *and* peeks at the employees' e-mail.

FIGURE 5-13

USA SNAPSHOTS®

A look at statistics that shape the nation

Peeking at workers' e-mail

Nearly 80% of executives say their company uses e-mail; 36% say they peek at employees' mail. Percentage who say an employer has a right to read anything in the company's electronic communication system:

Yes 75%
No 23%
Uncertain/refused 2%

Source: Society for Human Resource Management survey of human resource executives

By Anne R. Carey and Marcy E. Mullins, USA TODAY

Copyright © 1996 USA TODAY. Reprinted with permission.

FIGURE 5-14

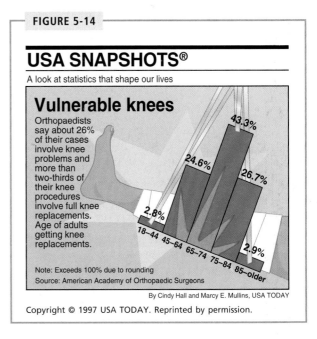

USA SNAPSHOTS®

A look at statistics that shape our lives

Vulnerable knees

Orthopaedists say about 26% of their cases involve knee problems and more than two-thirds of their knee procedures involve full knee replacements. Age of adults getting knee replacements.

2.8% 18–44
24.6% 45–64
43.3% 65–74
26.7% 75–84
2.9% 85–older

Note: Exceeds 100% due to rounding
Source: American Academy of Orthopaedic Surgeons

By Cindy Hall and Marcy E. Mullins, USA TODAY

Copyright © 1997 USA TODAY. Reprinted by permission.

2. Consider the information given in Figure 5-14, "Vulnerable knees," on the preceding page. What is the probability an orthopaedic case selected at random involves knee problems? Of those cases, estimate the probability that the case requires full knee replacement. Compute the probability that an orthopaedic case selected at random involves a knee problem *and* requires a full knee replacement. Next, look at the probability distribution for ages of patients requiring full knee replacement. Medicare insurance coverage begins when a person reaches age 65. What is the probability that the age of a person receiving a knee replacement is 65 or older?

LINKING CONCEPTS: WRITING PROJECTS

Discuss each of the following topics in class or review the topics on your own. Then write a brief but complete essay in which you summarize the main points. Please include formulas as appropriate.

1. Discuss the following concepts and give examples from everyday life where you encounter each concept. *Hint:* For instance, consider the "experiment" of arriving for class. Some possible outcomes are not arriving (that is, missing class), arriving on time, and arriving late.

   (a) Sample space.

   (b) Probability assignment to a sample space. In your discussion, be sure to include answers to the following questions.

      (i) Is there more than one valid way to assign probabilities to a sample space? Explain and give an example.

      (ii) How can probabilities be estimated by relative frequencies? How could probabilities be computed if events are equally likely?

2. Discuss the concepts of mutually exclusive events and independent events. List several examples of each type of event from everyday life.

   (a) If *A* and *B* are mutually exclusive events, does it follow that *A* and *B* *cannot* be independent events? Give an example to demonstrate your answer. *Hint:* Discuss an election where only one person can win the election. Let *A* be the event that party A's candidate wins, let *B* be the event that party B's candidate wins. Does the outcome of one event determine the outcome of the other event? Are *A* and *B* mutually exclusive events?

   (b) Discuss conditions under which $P(A \ and \ B) = P(A) \cdot P(B)$ is true. Under what conditions is this not true?

   (c) Discuss conditions under which $P(A \ or \ B) = P(A) + P(B)$ is true. Under what conditions is this not true?

Using TECHNOLOGY

TI-83PLUS • *MINITAB* • *EXCEL* • *COMPUTERSTAT*

Demonstration of the Law of Large Numbers

Computers can be used to simulate experiments. In packages such as Minitab and Excel, programs using random-number generators can be designed (see the *Technology Guides*) to simulate activities such as tossing a die. In ComputerStat, such a program exists (menu selection: ➤ **Descriptive Statistics** ➤ **Simulate the Experiment of Tossing One Die**). The following printouts show the simulations for tossing a die 6, 500, 50,000, 500,000, and 1,000,000 times. Notice how the relative frequencies of the outcomes approach the theoretical probabilities of 1/6 or 0.16667 for each outcome. Do you expect the same results every time the simulation is done? Why or why not?

Results of tossing one die 6 times

Outcome	Number of Occurrences	Relative Frequency
⚀	0	.00000
⚁	1	.16667
⚂	2	.33333
⚃	0	.00000
⚄	1	.16667
⚅	2	.33333

Results of tossing one die 500 times

Outcome	Number of Occurrences	Relative Frequency
⚀	87	.17400
⚁	83	.16600
⚂	91	.18200
⚃	69	.13800
⚄	87	.17400
⚅	83	.16600

Results of tossing one die 50,000 times

Outcome	Number of Occurrences	Relative Frequency
⚀	8528	.17056
⚁	8354	.16708
⚂	8246	.16492
⚃	8414	.16828
⚄	8178	.16356
⚅	8280	.16560

Results of tossing one die 500,000 times

Outcome	Number of Occurrences	Relative Frequency
⚀	83644	.16729
⚁	83368	.16674
⚂	83398	.16680
⚃	83095	.16619
⚄	83268	.16654
⚅	83227	.16645

Results of tossing one die 1,000,000 times

Outcome	Number of Occurrences	Relative Frequency
⚀	166643	.16664
⚁	166168	.16617
⚂	167391	.16739
⚃	165790	.16579
⚄	167243	.16724
⚅	166765	.16677

Suppose you have classes with six different professors.

1. What is the probability that all six are extroverts?

2. What is the probability that none of your professors is an extrovert?

3. What is the probability that at least two of your professors are extroverts?

4. In a group of six professors selected at random, what is the *expected number* of extroverts? What is the *standard deviation* of the distribution?

◇ **COMMENT** Both extroverted and introverted professors can be excellent teachers.   ◇

6.1   Introduction to Random Variables and Probability Distributions

FOCUS POINTS
✓ Distinguish between discrete and continuous random variables.
✓ Graph discrete probability distributions.
✓ Compute μ and σ for a discrete probability distribution.

For our purposes, we will say that a *statistical experiment* or *observation* is any process by which measurements are obtained. Examples are

1. Counting the number of eggs in a robin's nest

2. Measuring daily rainfall in inches

3. Counting the number of defective light bulbs in a case of bulbs

4. Measuring the weight in kilograms of a polar bear cub

Let x represent a quantitative variable that is measured or observed in an experiment. We are interested in the numerical values that x can take on. So x = number of eggs in a robin's nest and x = weight in kilograms of a polar bear cub would be examples of such quantitative variables. Furthermore, we say that the quantitative variable x is a *random variable* because the value that x takes on in a given experiment is a chance or random outcome. We will study two types of random variables: *discrete random variables* and *continuous random variables*.

Discrete random variable

> **Definition**
>
> When the observations of a quantitative random variable can take on only a finite number of values or a countable number of values, we say that the variable is a *discrete random variable*.

We know what a *finite* number of values is, but what is a *countable* number of values? As an example, let the random variable x be the number of wells an oil prospector drills until the first productive well is found. Then x could be any of the values 1, 2, 3. . . . In theory, we have an infinite number of possibilities for the values of x. The set of values of x corresponds to the set of counting numbers. Therefore, this type of infinity is called *countable,* and we say x has a countable number of values. This is an intuitive approach to the concept of a countable set. The reader interested in a more rigorous discussion is referred to the advanced text *Introduction to Mathematical Statistics* by Hogg and Craig (Macmillan Publishing).

In most of the cases we will consider, a *discrete random variable* will be the result of a count (the terms *countable* and *discrete*, however, have different mathematical meanings). For instance, the number of students in a certain section of a statistics course this term is a discrete random variable. The value must be a counting number such as 25, or 57, or 135, and so forth. The values 25.34 or $25\frac{1}{2}$ are not possible. The cost of tuition to the nearest dollar or nearest cent is another example of a discrete random variable. In this case, we are counting dollars or cents.

Continuous random variable

> **Definition**
>
> When the observations of a quantitative random variable can take on any of the countless number of values in a line interval, we say that the variable is a *continuous random variable*.

For our purposes, we will see most *continuous random variables* occurring as the result of a measurement. For example, the air pressure in an automobile tire represents a continuous random variable. The air pressure could in theory take on any value from 0 lb/in^2 (psi) to the bursting pressure of the tire. Values such as 20.126 psi, 20.12678 psi, and so forth are possible. Another example is the height of students in your statistics class. The height could in theory take on any value from a low of, say, 3 feet to a high of, say, 7.25 feet.

The distinction between discrete and continuous random variables is important because of the different mathematical techniques associated with the two kinds of random variables. Although we will not discuss these techniques at great length in this book, the distinction is very important in the study of advanced mathematical statistics.

In general, measurements of quantities such as length, weight, volume, temperature, or time yield continuous random variables. If the temperature changes from 12°C to 13°C, for example, it must take on all the temperature values between 12 and 13. Temperatures cannot just jump from one reading to the next. Discrete random variables often come from counts, such as the number of passing scores on an exam or the number of weeds in a garden.

GUIDED EXERCISE 1

Discrete or continuous random variables

Which of the following random variables are discrete and which are continuous?

(a) *Measure* the time it takes a student selected at random to register for the fall term.

⟹ Time can take on any value, so this is a continuous random variable.

(b) *Count* the number of bad checks drawn on Upright Bank on a day selected at random.

⟹ The number of bad checks can be only a whole number such as 0, 1, 2, 3, etc. This is a discrete variable.

Continued

GUIDED EXERCISE 1 continued

(c) *Measure* the amount of gasoline needed to drive your car 200 miles.

➡ We are measuring volume, which can assume any value, so this is a continuous random variable.

(d) Pick a random sample of 50 registered voters in a district and find the number who voted in the last county election.

➡ This is a count, so the variable is discrete.

Probability distribution

A random variable has a probability distribution whether it is discrete or continuous. The *probability distribution* is simply an assignment of probabilities to the specific values of the random variable or to a range of values of the random variable.

1. The probability distribution of a *discrete* random variable has a probability assigned to *each* value of the random variable.

2. The sum of these probabilities must be 1.

Let's look at a discrete probability distribution and its graph.

EXAMPLE 1

Discrete probability distribution

Dr. Fidgit developed a test to measure boredom tolerance. He administered it to a group of 20,000 adults between the ages of 25 and 35. The possible scores were 0, 1, 2, 3, 4, 5, and 6, with 6 indicating the highest tolerance for boredom. The test results for this group are shown in Table 6-1.

(a) If a subject is chosen at random from this group, the probability that he or she will have a score of 3 is 6000/20,000, or 0.30. In a similar way, we can use the relative frequency to compute the probabilities for the other scores (Table 6-2). These probability assignments make up the probability distribution. Notice that the scores are mutually exclusive: No one subject has two scores. The sum of the probabilities of all the scores is 1.

(b) The graph of this distribution is simply a relative-frequency histogram (see Figure 6-1) in which the height of the bar over a score represents the probability

TABLE 6-1 Boredom Tolerance Test Scores for 20,000 Subjects

Score	Number of Subjects
0	1400
1	2600
2	3600
3	6000
4	4400
5	1600
6	400

TABLE 6-2 Probability Distribution of Scores on Boredom Tolerance Test

Score x	Probability $P(x)$
0	0.07
1	0.13
2	0.18
3	0.30
4	0.22
5	0.08
6	0.02
	$\Sigma P(x) = 1$

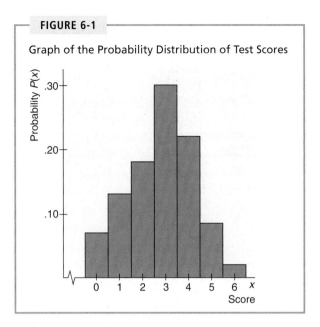

FIGURE 6-1

Graph of the Probability Distribution of Test Scores

of that score. Since each bar is one unit wide, the area of the bar over a score equals the height and thus represents the probability of that score. Since the sum of the probabilities is 1, the area under the graph is also 1.

(c) The Topnotch Clothing Company needs to hire someone with a score on the boredom tolerance test of 5 or 6 to operate the fabric press machine. Since the scores 5 and 6 are mutually exclusive, the probability that someone in the group who took the boredom tolerance test made either a 5 or a 6 is the sum

$$P(5 \ or \ 6) = P(5) + P(6)$$
$$= 0.08 + 0.02 = 0.10$$

Notice that to find $P(5 \ or \ 6)$, we could have simply added the *areas* of the bars over 5 and over 6. One out of 10 of the group who took the boredom tolerance test would qualify for the position at Topnotch Clothing.   ◊

GUIDED EXERCISE 2

Discrete probability distribution

One of the elementary tools of cryptanalysis (the science of code breaking) is to use relative frequencies of occurrence of different letters in the alphabet to break standard English alphabet codes. Large samples of plain text such as newspaper stories generally yield about the same relative frequencies for letters. A sample 1000 letters long yielded the information in Table 6-3.

Continued

GUIDED EXERCISE 2 continued

(a) Use the relative frequencies to compute the omitted probabilities in Table 6-3.

⟹ Table 6-4 shows the completion of Table 6-3.

TABLE 6-3   Frequency of Letters in a 1000-Letter Sample

Letter	Freq.	Prob.	Letter	Freq.	Prob.
A	73	——	N	78	0.078
B	9	0.009	O	74	——
C	30	0.030	P	27	0.027
D	44	0.044	Q	3	0.003
E	130	——	R	77	0.077
F	28	0.028	S	63	0.063
G	16	0.016	T	93	0.093
H	35	0.035	U	27	——
I	74	——	V	13	0.013
J	2	0.002	W	16	0.016
K	3	0.003	X	5	0.005
L	35	0.035	Y	19	0.019
M	25	0.025	Z	1	0.001

Source: From Elementary Cryptanalysis. A Mathematical Approach, by Abraham Sinkov. Copyright © 1968 by Yale University. Reprinted by permission of Random House, Inc.

TABLE 6-4   Completion of Table 6-3

Letter	Relative Frequency	Probability
A	$\dfrac{73}{1,000}$	0.073
E	$\dfrac{130}{1,000}$	0.130
I	$\dfrac{74}{1,000}$	0.074
O	$\dfrac{74}{1,000}$	0.074
U	$\dfrac{27}{1,000}$	0.027

(b) Do the probabilities of all the individual letters add up to 1?

⟹ Yes.

(c) If a letter is selected at random from a newspaper story, what is the probability that the letter will be a vowel?

⟹ If a letter is selected at random,

$$P(a, e, i, o, \text{ or } u) = P(a) + P(e) + P(i) + P(o) + P(u)$$

$$= 0.073 + 0.130 + 0.074 + 0.074 + 0.027$$

$$= 0.378$$

Mean and standard deviation of a discrete probability distribution

A probability distribution can be thought of as a relative-frequency distribution based on a very large *n*. As such, it has a mean and standard deviation. If we are referring to the probability distribution of a *population,* then we use the Greek letters μ for the mean and σ for the standard deviation. When we see the Greek letters used, we know the information given is from the *entire population* rather than just a sample. If we have a sample probability distribution, we use $\bar{x}$ (*x* bar) and *s*, respectively, for the mean and standard deviation. For a given population, μ and σ are fixed numbers and are sometimes called the *parameters* of the population.

Expected value

The mean of a probability distribution is often called the *expected value* of the distribution. This terminology reflects the idea that the mean represents a "central point" or "cluster point" for the entire distribution. Of course, the mean or expected value is an average value, and as such, it *need not be a point of the sample space*.

Definition

The *mean* and the *standard deviation of a discrete population probability distribution* are found by using these formulas:

$\mu = \Sigma x P(x)$ is the expected value of x

$\sigma = \sqrt{\Sigma(x - \mu)^2 P(x)}$ is the standard deviation of x

where x is the value of a random variable,
   $P(x)$ is the probability of that variable,
   the sum Σ is taken for all the values of the random variable.

Note: μ is the *population mean* and σ is the underlying *population standard deviation* because the sum Σ is taken over *all* values of the random variable (i.e., the entire sample space).

EXAMPLE 2

Expected value, standard deviation

Are we influenced to buy a product by an ad we saw on TV? National Infomercial Marketing Association determined the number of times *buyers* of a product watched a TV infomercial *before* purchasing the product. The results are shown here:

Number of Times Buyers Saw Infomercial	1	2	3	4	5*
Percentage of Buyers	27%	31%	18%	9%	15%

*This category was 5 or more, but will be treated as 5 in this example.

We can treat the information shown as an estimate of the probability distribution because the events are mutually exclusive and the sum of the percentages is 100%. Compute the mean and standard deviation of the distribution.

SOLUTION: We put the data in the first two columns of a computation table and then fill in the other entries (see Table 6-5).

TABLE 6-5   Number of Times Buyers View Infomercial Before Making Purchase

x (number of viewings)	$P(x)$	$xP(x)$	$x - \mu$	$(x - \mu)^2$	$(x - \mu)^2 P(x)$
1	0.27	0.27	−1.54	2.372	0.640
2	0.31	0.62	−0.54	0.292	0.091
3	0.18	0.54	0.46	0.212	0.038
4	0.09	0.36	1.46	2.132	0.192
5	0.15	0.75	2.46	6.052	0.908
	$\mu = \Sigma x P(x) = 2.54$				$\Sigma(x - \mu)^2 P(x) = 1.869$

The average number of times a buyer views the infomercial before purchase is

$$\mu = \Sigma x P(x) = 2.54 \text{ (sum of column 3)}$$

To find the standard deviation, we take the square root of the sum of column 6:

$$\sigma = \sqrt{\Sigma (x - \mu)^2 P(x)} \approx \sqrt{1.869} \approx 1.37 \qquad \diamond$$

● **CALCULATOR NOTE**   Some calculators, including the TI-83, accept fractional frequencies. If yours does, you can get μ and σ directly by using techniques for grouped data and the calculator's STAT mode.

GUIDED EXERCISE 3

Expected value

At a carnival, you pay $2.00 to play a coin-flipping game with three fair coins. On each coin one side has the number 0 and the other side has the number 1. You flip the three coins at one time and you win $1.00 for every 1 that appears on top. Are your expected earnings equal to the cost to play? We'll answer this question in several steps.

(a) In this game, the random variable of interest counts the number of 1s that show. What is the sample space for the values of this random variable?

⟹   The sample space is {0, 1, 2, 3}, since any of these numbers of 1s can appear.

(b) There are eight equally likely outcomes for throwing three coins. They are 000, 001, 010, 011, 100, 101, _____, and _____.

⟹   110 and 111.

(c) Complete Table 6-6.

TABLE 6-6

Number of 1s, x	Frequency	P(x)	xP(x)
0	1	0.125	0
1	3	0.375	_____
2	3	_____	_____
3	_____	_____	_____

⟹

TABLE 6-7   Completion of Table 6-6

x	Frequency	P(x)	xP(x)
0	1	0.125	0
1	3	0.375	0.375
2	3	0.375	0.750
3	1	0.125	0.375

(d) The expected value is the sum

$$\mu = \Sigma x P(x)$$

Sum the appropriate column of Table 6-6 to find this value. Are your expected earnings less than, equal to, or more than the cost of the game?

⟹   The expected value can be found by summing the last column of Table 6-6. The expected value is $1.50. It cost $2.00 to play the game; the expected value is less than the cost. The carnival is making money. In the long run, the carnival can expect to make an average of about 50 cents per player.

We have seen probability distributions of discrete variables and the formulas to compute the mean and standard deviation of discrete population probability distributions. Probability distributions of continuous random variables are similar except that the probability assignments are made to *intervals of values* rather than to specific values of the random variable. We will see an important example of a discrete probability distribution, the binomial distribution, in the next section and one of a continuous probability distribution in Chapter 7 when we study the normal distribution.

VIEWPOINT

The Rosetta Project

Around 196 B.C. Egyptian priests inscribed a decree on a granite slab affirming the rule of 13-year-old Ptolemy V. The proclamation was in Egyptian hieroglyphics with another translation in a form of ancient Greek. By 1799, the meaning of Egyptian hieroglyphics had been lost for many centuries. However, Napoleon's troops discovered the granite slab (Rosetta Stone). Linguists used the Rosetta Stone and their knowledge of ancient Greek to unlock the meaning of the Egyptian hieroglyphics.

Linguistic experts say that because of industrialization and globalization, by the year 2100 as many as 90% of the world's languages may be extinct. To help preserve some of these languages for future generations, 1000 translations of the first three chapters of Genesis have been inscribed in tiny text onto 3-inch nickel disks and encased in hardened glass balls that are expected to last at least 1000 years. Why Genesis? Because it is the most translated text in the world. The Rosetta Project is sending the disks to libraries and universities all over the world. It is very difficult to send information into the future. However, if in the year 2500 linguists are using the "Rosetta Disks" to unlock the meaning of a lost language, you may be sure they will use statistical methods of cryptanalysis (*see* Guided Exercise 2). To find out more about the Rosetta Project, visit the Brase/Brase *Understanding Basic Statistics* site at http://math.college.hmco.com/students and find the link to the Rosetta Project site.

SECTION 6.1   PROBLEMS

1. *Random Variables: Classification* Which of the following are continuous variables, and which are discrete?
   (a) Number of traffic fatalities per year in the state of Florida
   (b) Distance a golf ball travels after being hit with a driver
   (c) Time required to drive from home to college on any given day
   (d) Number of ships in Pearl Harbor on any given day
   (e) Your weight before breakfast each morning

2. *Random Variables: Classification* Which of the following are continuous variables, and which are discrete?
   (a) Speed of an airplane
   (b) Age of a college professor chosen at random

(c) Number of books in the college bookstore
(d) Weight of a football player chosen at random
(e) Number of lightning strikes in Rocky Mountain National Park on a given day

3. *Probability Distribution* Consider each distribution. Determine if it is a valid probability distribution or not, and explain your answer.

(a)

x	0	1	2
P(x)	0.25	0.60	0.15

Yes =1

(b)

x	0	1	2
P(x)	0.25	0.60	0.20

No

4. *Marketing: Age* What is the age distribution of promotion-sensitive shoppers? A *supermarket super shopper* is defined as a shopper for whom at least 70% of the items purchased were on sale or purchased with a coupon. The following table is based on information taken from *Trends in the United States* (Food Marketing Institute, Washington, D.C.).

Age range, years	18–28	29–39	40–50	51–61	62 and over
Midpoint x	23	34	45	56	67
Percent of super shoppers	7%	44%	24%	14%	11%

For the 62 and over group, use the midpoint 67 years.
(a) Using the age midpoints x and the percentage of super shoppers, do we have a valid probability distribution? Explain.
(b) Use a histogram to graph the probability distribution of part (a).
(c) Compute the expected age μ of a super shopper.
(d) Compute the standard deviation σ for ages of super shoppers.

5. *Marketing: Income* What is the income distribution of super shoppers (*see* Problem 4)? In the following table, income units are in thousands of dollars, and each interval goes up to but does not include the given high value. The midpoints are given to the nearest thousand dollars.

Income range	5–15	15–25	25–35	35–45	45–55	55 or more
Midpoint x	10	20	30	40	50	60
Percent of super shoppers	21%	14%	22%	15%	20%	8%

(a) Using the income midpoints x and the percent of super shoppers, do we have a valid probability distribution? Explain.
(b) Use a histogram to graph the probability distribution of part (a).
(c) Compute the expected income μ of a super shopper.
(d) Compute the standard deviation σ for the income of super shoppers.

6. *History: Florence Nightingale* What was the age distribution of nurses in Great Britain at the time of Florence Nightingale? Thanks to Florence Nightingale and the British census of 1851, we have the following information (based on data from the classic text *Notes on Nursing*, by Florence Nightingale). *Note:* In 1851, there were 25,466 nurses in Great Britain. Furthermore, Nightingale made a strict distinction between nurses and domestic servants.

Age range (yr)	20–29	30–39	40–49	50–59	60–69	70–79	80+
Midpoint x	24.5	34.5	44.5	54.5	64.5	74.5	84.5
Percent of nurses	5.7%	9.7%	19.5%	29.2%	25.0%	9.1%	1.8%

(a) Using the age midpoints x and the percent of nurses, do we have a valid probability distribution? Explain.

(b) Use a histogram to graph the probability distribution of part (a).

(c) Find the probability that a British nurse selected at random in 1851 would be 60 years of age or older.

(d) Compute the expected age μ of a British nurse contemporary to Florence Nightingale.

(e) Compute the standard deviation σ for ages of nurses shown in the distribution.

7. *Fishing: Trout* The following data are based on information taken from *Daily Creel Summary*, published by the Paiute Indian Nation, Pyramid Lake, Nevada. Movie stars and U.S. presidents have fished Pyramid Lake. It is one of the best places in the lower 48 states to catch trophy cutthroat trout. In this table, x = number of fish caught in a 6-hour period. The percentage data are the percentages of fishermen who caught x fish in a 6-hour period while fishing from shore.

x	0	1	2	3	4 or more
%	44%	36%	15%	4%	1%

Pyramid Lake, Nevada

(a) Convert the percentages to probabilities and make a histogram of the probability distribution.

(b) Find the probability that a fisherman selected at random fishing from shore catches one or more fish in a 6-hour period.

(c) Find the probability that a fisherman selected at random fishing from shore catches two or more fish in a 6-hour period.

(d) Compute μ, the expected value of the number of fish caught per fisherman in a 6-hour period (round 4 or more to 4).

(e) Compute σ, the standard deviation of the number of fish caught per fisherman in a 6-hour period (round 4 or more to 4).

8. *Criminal Justice: Parole* USA Today reported that approximately 25% of all state prison inmates released on parole become repeat offenders while on parole. Suppose the parole board is examining five prisoners up for parole. Let x = number of prisoners out of five on parole who become repeat offenders. The methods of Section 6.2 can be used to compute the probability assignments for the x distribution.

x	0	1	2	3	4	5
$P(x)$	0.237	0.396	0.264	0.088	0.015	0.001

(a) Find the probability that one or more of the five parolees will be repeat offenders. How does this number relate to the probability that none of the parolees will be repeat offenders?

(b) Find the probability that two or more of the five parolees will be repeat offenders.

(c) Find the probability that four or more of the five parolees will be repeat offenders.

(d) Compute μ, the expected number of repeat offenders out of five.

(e) Compute σ, the standard deviation of the number of repeat offenders out of five.

9. *Fund Raiser: Hiking Club* The college hiking club is having a fund raiser to buy new equipment for fall and winter outings. The club is selling Chinese fortune cookies at a price of $1 per cookie. Each cookie contains a piece of paper with a different number written on it. A random drawing will determine which number is the winner of a dinner for two at a local Chinese restaurant. The dinner is valued at $35. Since the fortune cookies were donated to the club, we can ignore the cost of the cookies. The club sold 719 cookies before the drawing.
   (a) Lisa bought 15 cookies. What is the probability she will win the dinner for two? What is the probability she will not win?
   (b) Lisa's expected earnings can be found by multiplying the value of the dinner by the probability that she will win. What are Lisa's expected earnings? How much did she effectively contribute to the hiking club?

10. *Spring Break: Caribbean Cruise* The college student senate is sponsoring a spring break Caribbean cruise raffle. The proceeds are to be donated to the Samaritan Center for the Homeless. A local travel agency donated the cruise, valued at $2000. The students sold 2852 raffle tickets at $5 per ticket.
    (a) Kevin bought six tickets. What is the probability that Kevin will win the spring break cruise to the Caribbean? What is the probability that Kevin will not win the cruise?
    (b) Expected earnings can be found by multiplying the value of the cruise by the probability that Kevin will win. What are Kevin's expected earnings? Is this more or less than the amount Kevin paid for the six tickets? How much did Kevin effectively contribute to the Samaritan Center for the Homeless?

11. *Expected Value: Life Insurance* Jim is a 60-year-old Anglo male in reasonably good health. He wants to take out a $50,000 term (that is, straight death benefit) life insurance policy until he is 65. The policy will expire on his 65th birthday. The probability of death in a given year is provided by the Vital Statistics Section of the *Statistical Abstract of the United States* (116th Edition).

x = age	60	61	62	63	64
P(death at this age)	0.01191	0.01292	0.01396	0.01503	0.01613

Jim is applying to Big Rock Insurance Company for his term insurance policy.
    (a) What is the probability that Jim will die in his 60th year? Using this probability and the $50,000 death benefit, what is the expected loss to Big Rock Insurance?
    (b) Repeat part (a) for years 61, 62, 63, and 64. What would be the total expected loss to Big Rock Insurance over the years 60 through 64?
    (c) If Big Rock Insurance wants to make a profit of $700 above the expected total loss paid out for Jim's death, how much should it charge for the policy?
    (d) If Big Rock Insurance Company charges $5000 for the policy, how much profit does the company expect to make?

12. *Expected Value: Life Insurance* Sara is a 60-year-old Anglo female in reasonably good health. She wants to take out a $50,000 term (that is, straight death benefit) life insurance policy until she is 65. The policy will expire on her 65th birthday. The probability of death in a given year is provided by the Vital Statistics Section of the *Statistical Abstract of the United States* (116th Edition).

x = age	60	61	62	63	64
P(death at this age)	0.00756	0.00825	0.00896	0.00965	0.01035

Sara is applying to Big Rock Insurance Company for her term insurance policy.

(a) What is the probability that Sara will die in her 60th year? Using this probability and the $50,000 death benefit, what is the expected loss to Big Rock Insurance?

(b) Repeat part (a) for years 61, 62, 63, and 64. What would be the total expected loss to Big Rock Insurance over the years 60 through 64?

(c) If Big Rock Insurance wants to make a profit of $700 above the expected total loss paid out for Sara's death, how much should it charge for the policy?

(d) If Big Rock Insurance Company charges $5000 for the policy, how much profit does the company expect to make?

6.2
Binomial Probabilities

FOCUS POINTS

✓ List the defining features of a binomial experiment.

✓ Compute binomial probabilities using the formula $P(r) = C_{n,r}\, p^r q^{n-r}$

✓ Use the binomial table to find $P(r)$.

✓ Use the binomial probability distribution to solve real-world applications.

Binomial Experiment

On a TV quiz show each contestant has a try at the wheel of fortune. The wheel of fortune is a roulette wheel with 36 slots, one of which is gold. If the ball lands in the gold slot, the contestant wins $50,000. No other slot pays. What is the probability that the quiz show will have to pay the fortune to three contestants out of 100?

In this problem, the contestant and the quiz show sponsors are concerned about only two outcomes from the wheel of fortune: The ball lands on the gold, or the ball does not land on the gold. This problem is typical of an entire class of problems that are characterized by the feature that there are exactly two possible outcomes (for each trial) of interest. These problems are called *binomial experiments*, or *Bernoulli experiments*, after the Swiss mathematician Jacob Bernoulli, who studied them extensively in the late 1600s.

Features of a binomial experiment

Features of a binomial experiment

1. There are a fixed number of trials. We denote this number by the letter n.

2. The n trials are independent and repeated under identical conditions.

3. Each trial has only two outcomes: success, denoted by S, and failure, denoted by F.

4. For each individual trial, the probability of success is the same. We denote the probability of success by p and that of failure by q. Since each trial results in either success or failure, $p + q = 1$ and $q = 1 - p$.

5. The central problem of a binomial experiment is to find the probability of r successes out of n trials.

EXAMPLE 3

Binomial experiment

Let's see how the wheel of fortune problem meets the criteria of a binomial experiment. We'll take the criteria one at a time.

1. Each of the 100 contestants has a trial at the wheel, so there are $n = 100$ trials in this problem.

2. Assuming that the wheel is fair, the *trials are independent,* since the result of one spin of the wheel has no effect on the results of other spins.

3. We are interested in only two outcomes on each spin of the wheel: The ball either lands on the gold, or it does not. Let's call landing on the gold *success* (*S*) and not landing on the gold *failure* (*F*). In general, the assignment of the terms *success* and *failure* to outcomes does not imply good or bad results. These terms are assigned simply for the user's convenience.

4. On each trial the probability *p* of success (landing on the gold) is 1/36, since there are 36 slots and only one of them is gold. Consequently, the probability of failure is

$$q = 1 - p = 1 - \frac{1}{36} = \frac{35}{36}$$

on each trial.

5. We want to know the probability of 3 successes out of 100 trials, so $r = 3$ in this example. It turns out that the probability the quiz show will have to pay the fortune to 3 contestants out of 100 is about 0.23. Later in this section we'll see how this probability was computed.   ◊

Anytime we make selections from a population *without replacement, we do not have independent trials.* However, replacement is often not practical. If the number of trials is quite small with respect to the population, we *almost* have independent trials, and we can say the situation is *closely approximated* by a binomial experiment. For instance, suppose we select 20 tuition bills at random from a collection of 10,000 bills issued at one college and observe if the bill is in error or not. If 600 of the 10,000 bills are in error, then the probability that the first one selected is in error is 600/10,000, or 0.0600. If the first is in error, then the probability that the second is in error is 599/9999, or 0.0599. Even if the first 19 bills selected are in error, the probability that the 20th is also in error is 581/9981, or 0.0582. All these probabilities round to 0.06, and we can say that the independence condition is approximately satisfied.

GUIDED EXERCISE 4

Binomial experiment

Let's analyze the following binomial experiment to determine *p*, *q*, *n*, and *r*:

According to the *Textbook of Medical Physiology*, 5th Edition, by Arthur Guyton, 9% of the population has blood type B. Suppose we choose 18 people at random from the population and test the blood type of each. What is the probability that 3 of these people have blood type B? (*Note:* Independence is approximated because 18 people is an extremely small sample with respect to the entire population.)

(a) In this experiment, we are observing whether or not a person has type B blood. We will say we have a success if the person has type B blood. What is failure?

➡ Failure occurs if a person does not have type B blood.

Continued

GUIDED EXERCISE 4 continued

(b) The probability of success is 0.09, since 9% of the population has type B blood. What is the probability of failure, q?

⟹ The probability of failure is
$$q = 1 - p$$
$$= 1 - 0.09 = 0.91$$

(c) In this experiment, there are $n = $ _____ trials.

⟹ In this experiment, $n = 18$.

(d) We wish to compute the probability of 3 successes out of 18 trials. In this case, $r = $ _____.

⟹ In this case, $r = 3$.

Next, we will see how to compute the probability of r successes out of n trials when we have a binomial experiment.

Computing Probabilities for a Binomial Experiment

The central problem of a binomial experiment is to find the probability of r successes out of n trials. Now we'll see how to find these probabilities.

A model with three trials

Suppose you are taking a timed final exam. You have three multiple-choice questions left to do. Each question has four suggested answers, and only one of the answers is correct. You have only 5 seconds left to do these three questions, so you decide to mark answers on the answer sheet without even reading the questions. Assuming that your answers are randomly selected, what is the probability that you get zero, one, two, or all three questions correct?

This is a binomial experiment. Each question can be thought of as a trial, so there are $n = 3$ trials. The possible outcomes on each trial are success S, indicating a correct response, or failure F, meaning a wrong answer. The trials are independent—the outcome of any one trial does not affect the outcome of the others.

What is the probability of success on any question? Since you are guessing and there are four answers from which to select, the probability of a correct answer is 0.25. The probability q of a wrong answer is then 0.75. In short, we have a binomial experiment with $n = 3$, $p = 0.25$, and $q = 0.75$.

Now what are the possible outcomes in terms of success or failure for these three trials? Let's use the notation *SSF* to mean success on the first question, success on the second, and failure on the third. There are eight possible combinations of *S*'s and *F*'s. They are

SSS   SSF   SFS   FSS   SFF   FSF   FFS   FFF

To compute the probability of each outcome, we can use the multiplication law because the trials are independent. For instance, the probability of success on the first two questions and failure on the last is

$$P(SSF) = P(S) \cdot P(S) \cdot P(F) = p \cdot p \cdot q = p^2q = (0.25)^2(0.75) = 0.047$$

In a similar fashion, we can compute the probability of each of the eight outcomes. These are shown in Table 6-8 on the following page, along with the number of successes r associated with each trial.

TABLE 6-8 Outcomes for a Binomial Experiment with $n = 3$ Trials

Outcome	Probability of Outcome				r (number of successes)
SSS	$P(SSS) = P(S)P(S)P(S)$	$= p^3$	$= (0.25)^3$	≈ 0.016	3
SSF	$P(SSF) = P(S)P(S)P(F)$	$= p^2q$	$= (0.25)^2(0.75)$	≈ 0.047	2
SFS	$P(SFS) = P(S)P(F)P(S)$	$= p^2q$	$= (0.25)^2(0.75)$	≈ 0.047	2
FSS	$P(FSS) = P(F)P(S)P(S)$	$= p^2q$	$= (0.25)^2(0.75)$	≈ 0.047	2
SFF	$P(SFF) = P(S)P(F)P(F)$	$= pq^2$	$= (0.25)(0.75)^2$	≈ 0.141	1
FSF	$P(FSF) = P(F)P(S)P(F)$	$= pq^2$	$= (0.25)(0.75)^2$	≈ 0.141	1
FFS	$P(FFS) = P(F)P(F)P(S)$	$= pq^2$	$= (0.25)(0.75)^2$	≈ 0.141	1
FFF	$P(FFF) = P(F)P(F)P(F)$	$= q^3$	$= (0.75)^3$	≈ 0.422	0

Now we can compute the probability of r successes out of three trials for $r = 0, 1, 2,$ or 3. Let's compute $P(1)$. The notation $P(1)$ stands for the probability of one success. For three trials, there are three different outcomes that show exactly one success. They are the outcomes *SFF, FSF,* and *FFS.* Since the outcomes are mutually exclusive, we can add the probabilities. So

$$P(1) = P(SFF \ or \ FSF \ or \ FFS) = P(SFF) + P(FSF) + P(FFS)$$
$$= pq^2 + pq^2 + pq^2$$
$$= 3pq^2$$
$$= 3(0.25)(0.75)^2$$
$$= 0.422$$

In the same way, we can find $P(0)$, $P(2)$, and $P(3)$. These values are shown in Table 6-9.

We have done quite a bit of work to determine your chances of $r = 0, 1, 2,$ or 3 successes on three multiple-choice questions if you are just guessing. Now we see that there is only a small chance (about 0.016) that you will get them all correct.

The model we constructed in Table 6-9 to compute the probability of r successes out of three trials can be used for any binomial experiment with $n = 3$ trials. Simply change the values of p and q to fit the experiment. In Guided Exercise 5 we use this model again.

TABLE 6-9 $P(r)$ for $n = 3$ Trials, $p = 0.25$

r (number of successes)	P(r) (probability of r successes in 3 trials)		P(r) for $p = 0.25$
0	$P(0) = P(FFF)$	$= q^3$	0.422
1	$P(1) = P(SFF) + P(FSF) + P(FFS)$	$= 3pq^2$	0.422
2	$P(2) = P(SSF) + P(SFS) + P(FSS)$	$= 3p^2q$	0.141
3	$P(3) = P(SSS)$	$= p^3$	0.016

GUIDED EXERCISE 5

Compute P(r)

Maria is doing a study on the issue of the quarter system versus the semester system. To obtain faculty input, she mails out questionnaires to the faculty. The probability that a faculty member returns the completed questionnaire is 0.65. Three faculty members chosen at random from the foreign language department are sent questionnaires. Compute the probability that *exactly two* completed questionnaires are returned and the probability that *all three* are returned. We'll do these computations in steps.

(a) In this problem, what are the values of n, p, q?

⟹ $n = 3$, $p = 0.65$, $q = 1 - 0.65 = 0.35$

(b) The probability that exactly two questionnaires will be returned is $P(\underline{\hspace{1cm}})$. In this case, $r = \underline{\hspace{1cm}}$. By Table 6-9,

$P(2) = 3p^2q$ for $n = 3$ trials

Use this formula to compute $P(2)$.

⟹ We want $P(2)$, so $r = 2$:

$P(2) = 3(0.65)^2(0.35) = 0.444$

(c) Use the appropriate formula from Table 6-9 to compute the probability that all three questionnaires will be returned.

⟹ $P(3) = p^3 = (0.65)^3 = 0.275$

Table 6-9 can be used as a model for computing the probability of r successes out of only *three* trials. How can we compute the probability of 7 successes out of 10 trials? We can develop a table for $n = 10$, but this would be a tremendous task because there are 1024 possible combinations of successes and failures on 10 trials. Fortunately, mathematicians have given us a direct formula to compute the probability of r successes for any number of trials.

General formula for binomial probability distribution

Formula for the binomial probability distribution

$$P(r) = C_{n,r}p^rq^{n-r}$$

where $C_{n,r} = \dfrac{n!}{r!(n-r)!}$ is the *binomial coefficient*. Values of $C_{n,r}$ for various values of n and $r = 0, 1, 2, \ldots, n$ can be found in Table 2 of the Appendix.

Those of you who studied the section on counting techniques (Section 5.3) will recognize the symbol $C_{n,r}$ as the symbol used for the number of combinations of n objects taken r at a time.

Table for $C_{n,r}$

Table 2 of the Appendix gives the values of the binomial coefficient $C_{n,r}$ for selected values of n and r. However, you can compute $C_{n,r}$ directly from a formula or on your calculator. (See Section 5.3 for the formula for $C_{n,r}$ and examples showing how to use the formula.)

In the meantime, let's look more carefully at the formula itself. There are two main parts. The expression $p^r q^{n-r}$ is the probability of getting one outcome with r successes and $n - r$ failures. The binomial coefficient $C_{n,r}$ counts the number of outcomes that have r successes and $n - r$ failures. For instance, in the case of $n = 3$ trials, we saw in Table 6-8 that the probability of getting an outcome with one success and two failures was pq^2. This is the value of $p^r q^{n-r}$ when $r = 1$ and $n = 3$. We also observed that there were three outcomes with one success and two failures, so $C_{3,1}$ is 3.

Now let's take a look at an application of the binomial distribution formula in Example 4.

EXAMPLE 4

Compute P(r)

Privacy is a concern for many users of the Internet. One survey showed that 59% of Internet users are somewhat concerned about the confidentiality of their e-mail. Based on this information, what is the probability that for a random sample of 10 Internet users, 6 are concerned about the privacy of their e-mail?

SOLUTION:

(a) This is a binomial experiment with 10 trials. If we assign success to an Internet user being concerned about the privacy of e-mail, the probability of success is 59%. We are interested in the probability of 6 successes. We have

$$n = 10 \qquad p = 0.59 \qquad q = 0.41 \qquad r = 6$$

By the formula,

$$P(6) = C_{10,6}(0.59)^6(0.41)^{10-6}$$

$\qquad = 210(0.59)^6(0.41)^4$ Use Table 2 of the Appendix or a calculator.

$\qquad = 210(0.0422)(0.0283)$ Use a calculator.

$\qquad = 0.25$

There is a 25% chance that *exactly* 6 of the 10 Internet users are concerned about the privacy of e-mail.

(b) Many calculators have a built-in combinations function. On the TI-83Plus you can find the combinations function under the math menu. It is designated nCr. Figure 6-2 displays the process for computing $P(6)$ directly on the TI-83Plus. ◊

In many cases, we will be interested in the probability of a range of successes rather than in the probability of an exact number of successes. For instance, we might wish to compute the probability that *at least* 6 of the 10 Internet users have

FIGURE 6-2

T1-83Plus Display

```
10 nCr 6*.59^6*.
41^(10-6)
         .250303245
```

concerns about the privacy of e-mail. In such a case, we need to use the addition rule for mutually exclusive events.

$$P(at\ least\ 6\ \text{successes}) = P(r \geq 6)$$
$$= P(r = 6\ or\ 7\ or\ 8\ or\ 9\ or\ 10)$$
$$= P(6) + P(7) + P(8) + P(9) + P(10)$$

Table for P(r)

It would be quite a task to compute all the required probabilities by using the formula. Table 3 of the Appendix gives values for $P(r)$ for selected p and values of n through 20. To use the table, find the section labeled with your value of n. Then find the entry in the column headed by your value of p and the row labeled by the r value of interest.

Table 3 of the Appendix has only a limited selection of values for p. In fact, for Example 4 the probability of success, $p = 0.59$, is not available in the table. However, for other problems in this text, you will find the specified value of p. Example 5 demonstrates the use of Table 3 (Appendix) to find binomial probabilities.

EXAMPLE 5

Using the binomial distribution table to find P(r)

A biologist is studying a new hybrid tomato. It is known that the seeds of this hybrid tomato have probability 0.70 of germinating. The biologist plants 10 seeds.

(a) What is the probability that *exactly* 8 seeds will germinate?

SOLUTION: This is a binomial experiment with $n = 10$ trials. Each seed planted represents an independent trial. We'll say germination is success, so the probability for success on each trial is 0.70.

$$n = 10 \qquad p = 0.70 \qquad q = 0.30 \qquad r = 8$$

We wish to find $P(8)$, the probability of exactly eight successes.

In Table 3 of the Appendix, find the section with $n = 10$. Then find the entry in the column headed by $p = 0.70$ and the row headed by the r value 8. This entry is 0.233.

$$P(8) = 0.233$$

(b) What is the probability that *at least* 8 seeds will germinate?

SOLUTION: In this case, we are interested in the probability of 8 or more seeds germinating. This means we are to compute $P(r \geq 8)$. Since the events are mutually exclusive, we can use the addition rule

$$P(r \geq 8) = P(r = 8 \quad or \quad r = 9 \quad or \quad r = 10) = P(8) + P(9) + P(10)$$

We already know the value of $P(8)$. We need to find $P(9)$ and $P(10)$.

Use the same part of the table but find the entries in the row headed by the r value 9 and then the r value 10. Be sure to use the column headed by the value of p, 0.70.

$$P(9) = 0.121 \quad \text{and} \quad P(10) = 0.028$$

Now we have all the parts necessary to compute $P(r \geq 8)$.

$$P(r \geq 8) = P(8) + P(9) + P(10)$$
$$= 0.233 + 0.121 + 0.028 = 0.382 \qquad \diamond$$

In Guided Exercise 6 you'll practice using the formula for $P(r)$ in one part and then use Table 3 (Appendix) for $P(r)$ values in the second part.

GUIDED EXERCISE 6

Find P(r)

A rarely performed and somewhat risky eye operation is known to be successful in restoring the eyesight of 30% of the patients who undergo the operation. A team of surgeons has developed a new technique for this operation that has been successful for four of six operations. Does it seem likely that the new technique is much better than the old? We'll use the binomial probability distribution to answer this question. We'll compute the probability of at least four successes in six trials for the old technique.

(a) Each operation is a binomial trial. In this case,
$n = $ _____, $p = $ _____, $q = $ _____, $r = $ _____.

➡ $n = 6, p = 0.30, q = 1 - 0.30 = 0.70, r = 4$

(b) Use your values of n, p, and q, as well as Table 2 of the Appendix (or your calculator) to compute $P(4)$ from the formula

$$P(r) = C_{n,r}p^r q^{n-r}$$

➡ $P(4) = C_{6,4}(0.30)^4(0.70)^2$
$= 15(0.0081)(0.490)$
$= 0.060$

(c) Compute the probability of *at least* four successes out of the six trials.

$$P(r \geq 4) = P(r = 4 \text{ or } r = 5 \text{ or } r = 6)$$
$$= P(4) + P(5) + P(6)$$

Use Table 3 of the Appendix to find values of $P(4)$, $P(5)$, and $P(6)$. Then use these values to compute $P(r \geq 4)$.

➡ To find $P(4)$, $P(5)$, and $P(6)$ in Table 3, we look in the section labeled $n = 6$. Then we find the column headed by $p = 0.30$. To find $P(4)$, we use the row labeled $r = 4$. For the values of $P(5)$ and $P(6)$, use the same column but change the row headers to $r = 5$ and $r = 6$, respectively.

$$P(r \geq 4) = P(4) + P(5) + P(6)$$
$$= 0.060 + 0.010 + 0.001 = 0.071$$

(d) Under the older operation technique, the probability that at least four patients out of six regain their eyesight is _____. Does it seem that the new technique is better than the old? Would you encourage the surgeon team to do more work on the new technique?

➡ It seems the new technique is better than the old, since, by pure chance, the probability of four or more successes out of six trials is only 0.071 for the old technique. This means one of the following two things may be happening:

(i) The new method is no better than the old method, and our surgeons have encountered a rare event (probability 0.071), or

(ii) The new method is in fact better. We think it is worth encouraging the surgeons to do more work on the new technique.

Common expressions and corresponding inequalities

Many times we are asked to compute the probability of a range of successes. For instance, in a binomial experiment with n trials, we may be asked to compute the probability of four or more successes. Table 6-10 shows how common English expressions such as "four or more successes" translate to inequalities involving r.

FIGURE 6-3

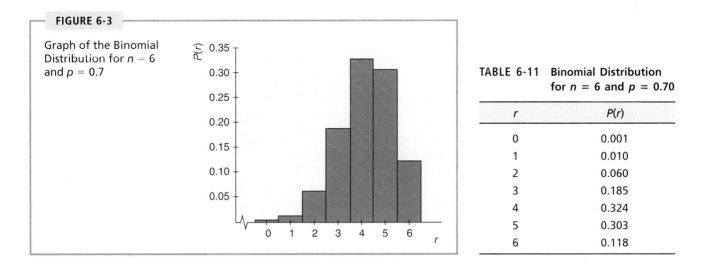

Graph of the Binomial Distribution for $n = 6$ and $p = 0.7$

TABLE 6-11 Binomial Distribution for $n = 6$ and $p = 0.70$

r	$P(r)$
0	0.001
1	0.010
2	0.060
3	0.185
4	0.324
5	0.303
6	0.118

Guided Exercise 7 illustrates another binomial distribution with $n = 6$ trials. The graph will be different from that of Figure 6-3 because the probability of success p is different.

GUIDED EXERCISE 7

Graph of a binomial distribution

Jim enjoys playing basketball. He figures that he makes about 50% of the field goals he attempts during a game. Make a histogram showing the probability that Jim will make 0, 1, 2, 3, 4, 5, or 6 shots out of six attempted field goals.

(a) This is a binomial experiment with $n =$ _____ trials. In this situation, we'll say success occurs when Jim makes an attempted field goal. What is the value of p?

⟹ In this example, $n = 6$ and $p = 0.5$.

(b) Use Table 3 of the Appendix to complete Table 6-12 of $P(r)$ values for $n = 6$ and $p = 0.5$.

TABLE 6-12

r	$P(r)$
0	0.016
1	0.094
2	0.234
3	_____
4	_____
5	_____
6	_____

⟹ TABLE 6-13 Completion of Table 6-12

r	$P(r)$
.	.
.	.
.	.
3	0.312
4	0.234
5	0.094
6	0.016

Continued

240   Chapter 6   The Binomial Probability Distribution and Related Topics

GUIDED EXERCISE 7 continued

(c) Use the values of $P(r)$ given in Table 6-13 to complete the histogram in Figure 6-4.

FIGURE 6-4 Beginning of Graph of Binomial Distribution for $n = 6$ and $p = 0.5$

FIGURE 6-5 Completion of Figure 6-4

(d) The area of the bar over $r = 2$ is 0.234. What is the area of the bar over $r = 4$? How does the probability that Jim makes exactly two field goals out of six compare with the probability that he makes exactly four field goals out of six?

The area of the bar over $r = 4$ is also 0.234. Jim is as likely to make two out of six field goals attempted as he is to make four out of six.

(e) Is the graph of the binomial distribution for $n = 6$ and $p = 0.5$ symmetrical?

Yes.

In Example 6 and Guided Exercise 7, we see the graphs of two binomial distributions associated with $n = 6$ trials. The two graphs are different because the probability of success p is different in the two cases. In Example 6, $p = 0.7$ and the graph is skewed to the left—that is, the left tail is longer. In Guided Exercise 7, p is equal to 0.5 and the graph is symmetrical—that is, if we fold it in half, the two halves coincide exactly. Whenever p *equals 0.5, the graph of the binomial distribution will be symmetrical no matter how many trials we have.* In Chapter 7, we will see that if the number of trials n is quite large, the binomial distribution is almost symmetrical even when p is not close to 0.5.

Mean and standard deviation of binomial probability distributions

Two other features that help describe the graph of any distribution are the balance point of the distribution and the spread of the distribution about that balance point. The *balance point* is the mean μ of the distribution, and the *measure of spread* that is most commonly used is the standard deviation σ. The mean μ is the *expected value* of the number of successes.

For the binomial distribution, we can use two special formulas to compute the mean μ and the standard deviation σ. These are easier to use than the general formulas in Section 6.1 for μ and σ of any discrete probability distribution.

For the binomial distribution

$\mu = np$ is the expected number of successes r

$\sigma = \sqrt{npq}$ is the standard deviation for the number of successes r

where n is the number of trials,

p is the probability of success,

q is the probability of failure ($q = 1 - p$).

EXAMPLE 7

Compute μ and σ

Let's compute the mean and standard deviation for the distribution of Example 6 that describes the probabilities of lone diners leaving tips at the Green Spot Restaurant.

SOLUTION: In that example,

$$n = 6 \qquad p = 0.7 \qquad q = 0.3$$

For the binomial distribution,

$$\mu = np = 6(0.7) = 4.2$$

The balance point of the distribution is at $\mu = 4.2$. The standard deviation is given by

$$\sigma = \sqrt{npq} = \sqrt{6(0.7)(0.3)} = \sqrt{1.26} \approx 1.12 \qquad \Diamond$$

The mean μ is not only the balance point of the distribution; it is also the *expected value* of r. Specifically, in Example 6, the waiter can expect 4.2 lone diners out of 6 to leave a tip. (The waiter would probably round the expected value to 4 tippers out of 6.)

GUIDED EXERCISE **8**

Expected value and standard deviation

When Jim (of Guided Exercise 7) shoots field goals in basketball games, the probability that he makes a shot is only 0.5.

(a) The mean of the binomial distribution is the expected value of r successes out of n trials. Out of six throws, what is the expected number of goals Jim will make?

The expected value is the mean μ:

$$\mu = np = 6(0.5) = 3$$

Jim can expect to make three goals out of six tries.

(b) For six trials, what is the standard deviation of the binomial distribution of the number of successful field goals Jim makes?

$$\sigma = \sqrt{npq} = \sqrt{6(0.5)(0.5)} = \sqrt{1.5} \approx 1.22$$

VIEWP⊙INT   *Kodiak Island, Alaska*

Kodiak Island is famous for its giant brown bears. The sea surrounding the island is also famous for its king crab. The state of Alaska, Department of Fish and Game, has collected a huge amount of data regarding ocean latitude, longitude, and size of king crab. Of special interest to commercial fishing skippers is the size of crab. Those too small must be returned to the sea. To find locations and sizes of king crab catches near Kodiak Island, visit the Brase/Brase *Understanding Basic Statistics* site at http://math.college.hmco.com/students and find the link to the StatLib site hosted by the Department of Statistics at Carnegie Mellon University. Once at StatLib, go to crab data. From this information, it is possible to use methods of this chapter and Chapter 9 to estimate the proportion of legal crab in a sea skipper's catch.

SECTION 6.3   PROBLEMS

1. *Binomial Distribution: Histograms* Consider a binomial distribution with $n = 5$ trials. Use the probabilities given in Table 3 of the Appendix to make histograms showing the probability of $r = 0, 1, 2, 3, 4, 5$ successes for each of the following. Comment on the skewness of each distribution.
   (a) The probability of success is $p = 0.50$. *symmetry*
   (b) The probability of success is $p = 0.25$. *Right*
   (c) The probability of success is $p = 0.75$. *left*   *Mirror Image*
   (d) What is the relationship between the distributions shown in parts (b) and (c)?
   (e) If the probability of success is $p = 0.73$, do you expect the distribution to be skewed to the right or to the left? Why?

2. *Binomial Distributions: Histograms* Figure 6-6 shows histograms of several binomial distributions with $n = 6$ trials. Match the given probability of success with the best graph.
   (a) $p = 0.30$ goes with graph _____.
   (b) $p = 0.50$ goes with graph _____.
   (c) $p = 0.65$ goes with graph _____.
   (d) $p = 0.90$ goes with graph _____.
   (e) In general, when the probability of success p is close to 0.5, would you say that the graph is more symmetrical or more skewed? In general, when the probability of success p is close to 1, would you say that the graph is skewed to the right or to the left? What about when p is close to 0?

3. *Marketing: Photography* Does the *kid factor* make a difference? If you are talking photography, the answer may be yes! The following table is based on information from *American Demographics* (Vol. 19, No. 7).

Age of children in household, years	Under 2	None under 21
Percent of U.S. households that buy film	80%	50%

Let us say you are a market research person who interviews a random sample of 10 households.
   (a) Suppose that the 10 households you interview are chosen to have children under the age of 2 years. Let r represent the number of such households that buy film.

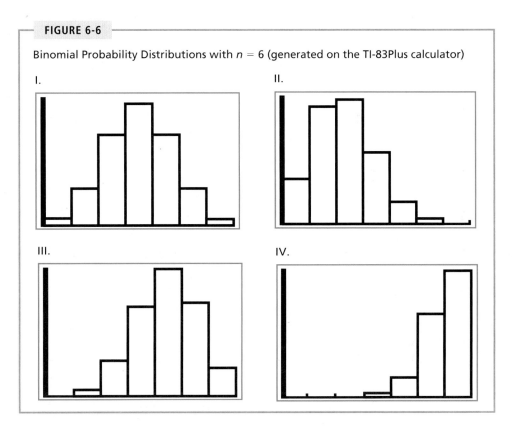

FIGURE 6-6

Binomial Probability Distributions with $n = 6$ (generated on the TI-83Plus calculator)

I.

II.

III.

IV.

Make a histogram showing the probability distribution of r for $r = 0$ through $r = 10$. Find the mean and standard deviation of this probability distribution.

(b) Suppose that the 10 households are chosen to have no children under 21 years old. Let r represent the number of such households that buy film. Make a histogram showing the probability distribution of r for $r = 0$ through $r = 10$. Find the mean and standard deviation of this probability distribution.

(c) Compare the distributions shown in parts (a) and (b). You are designing TV ads to sell film. Could you justify featuring ads of parents taking pictures of toddlers? Explain your answer.

4. *Quality Control: Syringes* The quality-control inspector of a production plant will reject a batch of syringes if two or more defectives are found in a random sample of eight syringes taken from the batch. Suppose the batch contains 1% defective syringes.

(a) Make a histogram showing the probability of $r = 0, 1, 2, 3, 4, 5, 6, 7, 8$ defective syringes in a random sample of eight syringes.

(b) Find μ. What is the expected number of defective syringes the inspector will find?

(c) What is the probability that the batch will be accepted?

(d) Find σ.

5. *Hype: Improved Products* The *Wall Street Journal* reported that approximately 25% of the people who are told a product is *improved* will believe that it is in fact improved. The remaining 75% believe that this is just hype (the same old thing with no real improvement). Suppose a marketing study consisted of a random sample of eight people who are given a sales talk about a new, *improved* product.

(a) Make a histogram showing the probability that $r = 0$ to 8 people believe the product is in fact improved.

(b) Compute the mean and standard deviation of this probability distribution.

6. *Insurance: Auto* The Mountain States Office of State Farm Insurance Company reports that approximately 85% of all automobile damage liability claims were made by people under 25 years of age. A random sample of five automobile insurance liability claims is under study.
   (a) Make a histogram showing the probability that $r = 0$ to 5 claims are made by people under 25 years of age.
   (b) Find the mean and standard deviation of this probability distribution. For samples of size 5, what is the expected number of claims made by people under 25 years of age?

7. *Criminal Justice: Convictions* Innocent until proven guilty? In Japanese criminal trials, about 95% of the defendants are found guilty. In the United States, about 60% of the defendants are found guilty in criminal trials (*Source: The Book of Risks,* by Larry Laudan, John Wiley and Sons). Suppose you are a news reporter following seven criminal trials.
   (a) If the trials were in Japan, what is the probability that all the defendants would be found guilty? What is this probability if the trials were in the United States?
   (b) Of the seven trials, what is the expected number of guilty verdicts in Japan? What is the expected number in the United States? What is the standard deviation in each case?

8. *Focus Problem: Personality Types* Solve the Focus Problem at the beginning of this chapter.

9. *Lifestyle: ATM Users* Do you talk to machines? A recent survey for NCR Corporation showed that 47% of automated teller machine (ATM) users in the USA tell the ATM to hurry up.
   (a) For a random sample of 11 people, the distribution of those telling the machine to hurry up is shown in the following Excel display. Is the probability of success (saying "hurry up" to the ATM) close to 50%? Is the histogram more skewed or more symmetrical? Compute the expected number out of 11 who tell the ATM to hurry up. Is the expected value μ close to the center of the graph?

r	$P(r)$
0	0.0009
1	0.0090
2	0.0401
3	0.1067
4	0.1892
5	0.2348
6	0.2083
7	0.1319
8	0.0585
9	0.0173
10	0.0031
11	0.0002

USA: $n = 11$, $p = 0.47$

   (b) In Great Britain, the ATM users are not so vocal or impatient. The probability is only 36% that a British ATM user tells the machine to hurry up. For a random sample of 11 British ATM users, the following display shows the probability distribution of telling the machine to hurry up. Is the probability of success less than 50%? Is the histogram skewed? In which direction? Compute the expected number out of 11 who tell the ATM to hurry up. Is the expected value on the left or right side of the graph?

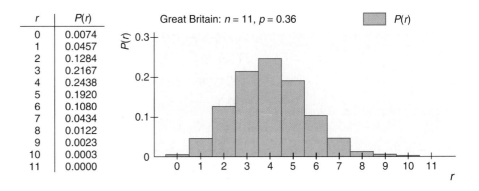

r	P(r)
0	0.0074
1	0.0457
2	0.1284
3	0.2167
4	0.2438
5	0.1920
6	0.1080
7	0.0434
8	0.0122
9	0.0023
10	0.0003
11	0.0000

(c) What if we change success to "does not tell the ATM to hurry up." In Great Britain, what is the probability that an ATM user does not tell the machine to hurry up? The next display shows this distribution. Is this distribution skewed? In what direction? Compute the expected value. Does it lie on the left side or right side of the graph?

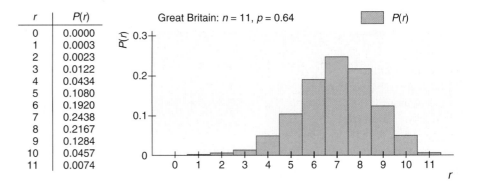

r	P(r)
0	0.0000
1	0.0003
2	0.0023
3	0.0122
4	0.0434
5	0.1080
6	0.1920
7	0.2438
8	0.2167
9	0.1284
10	0.0457
11	0.0074

(d) Compare the P(r) entries in the table in part (b) with those of the entries of the table of part (c). How do the entries in part (b) for r = 0, 1, 2, 3 compare to those in part (c) for r = 11, 10, 9, and 8, respectively?

(e) Compute the standard deviations of each the distributions in parts (b) and (c). Are they the same? Why would you expect this?

SUMMARY

The concepts of discrete and continuous random variables were introduced in this chapter. We looked at the general probability distribution for a discrete random variable and saw how to compute the expected value μ and the standard deviation σ of a discrete probability distribution.

Then we turned our attention to a special discrete probability distribution, the binomial probability distribution. This distribution is used to determine the probability of outcomes from a binomial experiment.

A binomial experiment must meet the following criteria:

1. There are a fixed number of trials denoted by n.

2. The trials are independent and repeated under identical conditions.

3. Each trial has only two outcomes: success S and failure F.

4. For each trial the probability p of success remains the same. The probability of failure is $q = 1 - p$.

5. The central problem is to find the probability of r successes out of n trials.

   The formula for the binomial probability distribution is

   $$P(r) = C_{n,r}p^r q^{n-r}$$

where $C_{n,r}$ is the binomial coefficient as found in Table 2 of the Appendix. For certain values of p and n, $P(r)$ can be found directly in Table 3 of the Appendix.

   The mean or expected value and standard deviation of the binomial distribution are given by the formulas

   $$\mu = np$$
   $$\sigma = \sqrt{npq} \qquad \text{where } q = 1 - p$$

IMPORTANT WORDS & SYMBOLS

Section 6.1
Random variable
  Discrete
  Continuous
Population parameters
Mean μ of a probability distribution
Standard deviation σ of a probability distribution
Expected value μ

Section 6.2
Binomial experiment
Independent trials

Successes and failures in a binomial experiment
Probability of success $P(S) = p$
Probability of failure $P(F) = q = 1 - p$
Binomial coefficient $C_{n,r}$
Binomial probability distribution $P(r) = C_{n,r}p^r q^{n-r}$

Section 6.3
Mean for the binomial distribution $\mu = np$
Standard deviation for binomial distribution
  $\sigma = \sqrt{npq}$

VIEWPOINT   *What's Your Type?*

Are students and professors *really* compatible? One way of answering this question is to look at Myers-Briggs Type Indicators for personality preferences. What is the probability that your professor is introverted and judgmental? What is the probability that you are extroverted and perceptive? Are most of the leaders in student government extroverted and judgmental? Is it true that members of Phi Beta Kappa have personality types more like the professors'? We will consider questions such as these in more detail in Chapter 9 (Estimation) and Chapter 10 (Hypothesis Testing), where we will continue our work with binomial probabilities. In the meantime, you can find many answers regarding careers, probability, and personality types in *Applications of the Myers-Briggs Type Indicator in Higher Education*, edited by J. Provost and S. Anchors.

CHAPTER REVIEW PROBLEMS

1. *Probability Distribution: Auto Leases* The Consumer Banker Association released a report showing the length of automobile leases for new automobiles. The results are

Lease Length in Months	Percent of Leases
13–24	12.7%
25–36	37.1%
37–48	28.5%
49–60	21.5%
More than 60	0.2%

   (a) Use the midpoint of each class, and call the midpoint of the last class 66.5 months for purposes of computing the expected lease term. Also find the standard deviation of the distribution.
   (b) Sketch a graph of the probability for the duration of new auto leases.

2. *Ecology: Predator and Prey* Isle Royale, an island in Lake Superior, has provided an important study site of wolves and prey. In the National Park Service Scientific Monograph Series 11, *Wolf Ecology and Prey Relationships on Isle Royale*, Peterson gives results of many wolf–moose studies. Of special interest is the study of the number of moose killed by wolves. In the period from 1958 to 1974, there were 296 moose deaths identified as wolf kills. The age distribution of the kills is

Age of Moose in Years	Number Killed by Wolves
Calf (0.5 yr)	112
1–5	53
6–10	73
11–15	56
16–20	2

   (a) For each age group, compute the probability that a moose in that age group is killed by a wolf.
   (b) Consider all ages in a class equal to the class midpoint. Find the expected age of a moose killed by a wolf and the standard deviation of the ages.

3. *Insurance: Auto* State Farm Insurance studies show that in Colorado, 55% of the auto insurance claims submitted for property damage were submitted by males under 25 years of age. Suppose 10 property damage claims involving automobiles are selected at random.
   (a) Let r be the number of claims that are made by males under age 25. Make a histogram for the r-distribution probabilities.
   (b) What is the probability that six or more claims are made by males under age 25?
   (c) What is the expected number of claims made by males under age 25? What is the standard deviation of the r-probability distribution?

FIGURE 6-8

USA SNAPSHOTS®

A look at statistics that shape the nation

U.S. future on final frontier

Percentages of adults who agree with these statements about the USA's space program:

Youth should want to be astronauts	77%
Humanity should explore planets	64%
Space exploration impacts daily life	57%
Given a chance I'd travel in space	44%
Space will be colonized in my lifetime	18%

Source: Opinion Research for Space Day Partners

By Cindy Hall and Julie Stacey, USA TODAY

(b) Suppose that the jackpot increased to $25 million (and that there were to be only one winner). Compute your expected winnings if you buy one ticket. Does the probability of winning the jackpot change because the jackpot is higher?

(c) Pretend that you are going to buy 10 Powerball tickets when the jackpot is $10 million. Use the random-number table to select your numbers. Check the Multi-State Lottery Association Web site (or any other Powerball site) for the most recent drawing results to see if you would have won a prize.

(d) The probability of winning any prize is about 0.0277. Suppose that you decide to buy five tickets. Use the binomial distribution to compute the probability of winning (any prize) at least once. *Note:* You will need to use the binomial formula. Carry at least three digits after the decimal.

2. Would you like to travel in space, if given a chance? According to Opinion Research for Space Day Partners, you are not alone. Forty-four percent of adults surveyed agreed that they would travel in space if given a chance. Look at Figure 6-8, and use the information presented to answer the following questions.

(a) According to Figure 6-8, the probability that an adult selected at random agrees with the statement that humanity should explore planets is 64%. Round this probability to 65%, and use this estimate with the binomial distribution table to determine the probability that of 10 adults selected at random, at least half agree that humanity should explore planets.

(b) Does space exploration have an impact on daily life? Find the probability that of 10 adults selected at random, at least 9 agree that space exploration does have an impact on daily life. *Hint:* Use the formula for the binomial distribution.

(c) In a room of 35 adults, what is the expected number who would travel in space, given a chance? What is the standard deviation?

LINKING CONCEPTS: WRITING PROJECTS

Discuss each of the following topics in class or review the topics on your own. Then write a brief but complete essay in which you summarize the main points. Please include formulas and graphs as appropriate.

1. Discuss what we mean by a binomial experiment. As you can see, a binomial process or binomial experiment involves a lot of assumptions! For example, all the trials are supposed to be independent and repeated under identical conditions. Is this always true? Can we always be completely certain the probability of success does not change from one trial to the next? In the real world, there is almost nothing we can be absolutely sure about, so the *theoretical* assumptions of the binomial probability distribution often will not be completely satisfied. Does that mean we cannot use the binomial distribution to solve practical problems? Looking at this chapter, the answer that appears is that we can indeed use the binomial distribution even if not all the assumptions are *exactly* met. We find in practice that the conclusions are sufficiently accurate for our intended application. List three applications of the binomial distribution for which you think, although some of the assumptions are not exactly met, there is adequate reason to apply the binomial distribution anyhow.

2. Why do we need to learn the formula for the binomial probability distribution? Using the formula repeatedly can be very tedious. To cut down on tedious calculations, most people will use a binomial table such as found in the Appendix of this book.
   (a) However, there are many applications where a table in the back of *any* book is not adequate. For instance, compute

   $$P(r = 3) \quad \text{where } n = 5 \text{ and } p = 0.735$$

   Do you find the result in the table? Do the calculation by using the formula. List some other situations in which a table might not be adequate to solve a particular binomial distribution problem.
   (b) The formula itself also has limitations. For instance, consider the difficulty of computing

   $$P(r \geq 285) \quad \text{where } n = 500 \text{ and } p = 0.6$$

   What are some of the difficulties you run into? Consider the calculation of $P(r = 285)$. You will be raising 0.6 and 0.4 to very high powers; this will give you very, very small numbers. Then you need to compute $C_{500,285}$, which is a very, very large number. When combining extremely large and extremely small numbers in the same calculation, most accuracy is lost unless you carry a huge number of significant digits. If this isn't tedious enough, then consider the steps that you need to compute

   $$P(r \geq 285) = P(r = 285) + P(r = 286) + \cdots + P(r = 500)$$

   Does it seem clear that we need a better way to compute $P(r \geq 285)$? In Chapter 7, you will find a much better way to compute binomial probabilities when the number of trials is large.

3. In Chapter 3, we learned about means and standard deviations. In Section 6.1, we learned that probability distributions also can have a mean and standard deviation. Discuss what is meant by the expected value and standard deviation of a binomial distribution. How does this relate back to the material we learned in Chapter 3 and Section 6.1?

4. In Chapter 2, we looked at the shapes of distributions. Review the concepts of skewness and symmetry; then categorize the following distributions as to skewness or symmetry:

(a) A binomial distribution with $n = 11$ trials and $p = 0.50$
(b) A binomial distribution with $n = 11$ trials and $p = 0.10$
(c) A binomial distribution with $n = 11$ trials and $p = 0.90$

In general, does it seem true that binomial probability distributions in which the probability of success is close to 0 are skewed right, whereas those with probability of success close to 1 are skewed left?

Using TECHNOLOGY

Binomial Distributions

Although tables of binomial probabilities can be found in most libraries, such tables are often inadequate. Either the value of p (the probability of success on a trial) you are looking for is not in the table, or the value of n (the number of trials) you are looking for is too large for the table. In Chapter 7, we will study the normal approximation to the binomial. This approximation is a great help in many practical applications. Even so, we sometimes use the formula for the binomial probability distribution on a computer or graphing calculator to compute the probability we want.

APPLICATIONS

The following percentages were obtained over many years of observation by the U.S. Weather Bureau. All data listed are for the month of December.

Location	Long-Term Mean % of Clear Days in Dec.
Juneau, Alaska	18%
Seattle, Washington	24%
Hilo, Hawaii	36%
Honolulu, Hawaii	60%
Las Vegas, Nevada	75%
Phoenix, Arizona	77%

Adapted from *Local Climatological Data*, U.S. Weather Bureau publication, "Normals, Means, and Extremes" Table.

In the locations listed, the month of December is a relatively stable month with respect to weather. Since weather patterns from one day to the next are more or less the same, it is reasonable to use a binomial probability model.

1. Let r be the number of clear days in December. Since December has 31 days, $0 \leq r \leq 31$. Using appropriate computer software or calculators available to you, find the probability $P(r)$ for each of the listed locations when $r = 0, 1, 2, \ldots, 31$.

2. For each location, what is the expected value of the probability distribution? What is the standard deviation?

You may find that using cumulative probabilities and appropriate subtraction of probabilities will make the solution of Applications 3 to 7 easier than adding probabilities.

3. Estimate the probability that Juneau will have at most 7 clear days in December.

4. Estimate the probability that Seattle will have from 5 to 10 (including 5 and 10) clear days in December.

5. Estimate the probability that Hilo will have at least 12 clear days in December.

6. Estimate the probability that Phoenix will have 20 or more clear days in December.

7. Estimate the probability that Las Vegas will have from 20 to 25 (including 20 and 25) clear days in December.

Technology Hints

The Tech Note in Section 6.2 gives specific instructions for binomial distribution functions on the TI-83Plus, Excel, and Minitab. For ComputerStat, select Binomial Coefficients and Probability Distributions under the Probability Distribution menu.

7 *Normal Distributions*

One cannot escape the feeling that these mathematical formulas have an independent existence and an intelligence of their own, that they are wiser than we are, wiser even than their discoverers, that we get more out of them than was originally put into them.

—Heinrich Hertz

How can it be that mathematics, a product of human thought independent of experience, is so admirably adapted to the objects of reality?

—Albert Einstein

Heinrich Rudolf Hertz (1857–1894)

This German physicist was largely responsible for doing pioneering work in electromagnetic theory.

 For on-line student resources, visit **math.college.hmco.com/students** and follow the Statistics links to the Brase/Brase, *Understanding Basic Statistics,* 3rd edition Web site.

Heinrich Hertz was a pioneer in the study of radio waves. His work and the later work of Maxwell and Marconi led the way to modern radio, television, and radar. Albert Einstein is world renowned for his great discoveries in relativity and quantum mechanics. Everyone who has worked in both mathematics and real-world applications cannot help but marvel at how the "pure thought" of the mathematical sciences can predict and explain events in other realms. In this chapter, we will study the most important type of probability distribution in all of mathematical statistics: the normal distribution. Why is the normal distribution so important? Two of the reasons are that it applies to a wide variety of situations, and other distributions tend to become normal under certain conditions.

◇ What are some characteristics of a normal distribution? What does the empirical rule tell us about the data spread around the mean? How can this information be used in quality control?   (SECTION 7.1)

◇ Can we compare apples and oranges, or maybe elephants and butterflies? In most cases, the answer is no, . . . unless we first *standardize* our measurements. What are a *standard normal distribution* and a *standard z score*?   (SECTION 7.2)

◇ How do we convert *any* normal distribution to a *standard normal* distribution? How do we find probabilities of "standardized events"?   (SECTION 7.3)

◇ The binomial and normal are two of the most important probability distributions in statistics. Under certain limiting conditions, the binomial can be thought to evolve (or envelope) into the normal distribution. How can we apply this in the real world?   (SECTION 7.4)

FOCUS PROBLEM

Large Auditorium Shows: How Many Will Attend?

For many years, Denver, as well as most other cities, has hosted large exhibition shows in big auditoriums. These shows include house and gardening shows, fishing and hunting shows, car shows, boat shows, Native American powwows, and so on. Information provided by Denver exposition sponsors indicates that most shows have an average attendance of about 8000 people per day with an estimated standard deviation of about 500 people. Suppose that the daily attendance figures follow a normal distribution.

1. What is the probability that the daily attendance will be fewer than 7200 people?

2. What is the probability that the daily attendance will be more than 8900 people?

3. What is the probability that the daily attendance will be between 7200 and 8900 people?

Most exhibition shows open in the morning and close in the late evening. A study of Saturday arrival times showed that the average arrival time was 3 hours and 48 minutes after the doors open, and the standard deviation was estimated at about 52 minutes. Suppose that the arrival times follow a normal distribution.

4. At what time after the doors open will 90% of the people who are coming to the Saturday show have arrived?

5. At what time after the doors open will only 15% of the people who are coming to the Saturday show have arrived?

6. Do you think the probability distribution of arrival times for Friday might be different from the distribution of arrival times for Saturday? Explain.

7.1
Graphs of Normal Probability Distributions

FOCUS POINTS

✓ Graph a normal curve and summarize its important properties.

✓ Apply the empirical rule to solve real-world problems.

✓ Use control limits to construct control charts. Examine the chart for three possible out-of-control signals.

One of the most important examples of a continuous probability distribution is the *normal distribution*. This distribution was studied by the French mathematician Abraham de Moivre (1667–1754) and later by the German mathematician Carl Friedrich Gauss (1777–1855), whose work is so important that the normal distribution is sometimes called *gaussian*. The work of these mathematicians provided a foundation on which much of the theory of statistical inference is based.

Applications of a normal probability distribution are so numerous that some mathematicians refer to it as "a veritable Boy Scout knife of statistics." However, before we can apply it, we must examine some of the properties of a normal distribution.

A rather complicated formula, presented later in this section, defines a normal distribution in terms of μ and σ, the mean and standard deviation of the population distribution. It is only through this formula that we can verify if a distribution is normal. However, we can look at the graph of a normal distribution and get a good pictorial idea of some of the essential features of any normal distribution.

Normal curve

The graph of a normal distribution is called a *normal curve*. It possesses a shape very much like the cross section of a pile of dry sand. Because of its shape, blacksmiths would sometimes use a pile of dry sand in the construction of a mold for a bell. Thus the normal curve is also called a *bell-shaped curve* (see Figure 7-1).

FIGURE 7-1

A Normal Curve

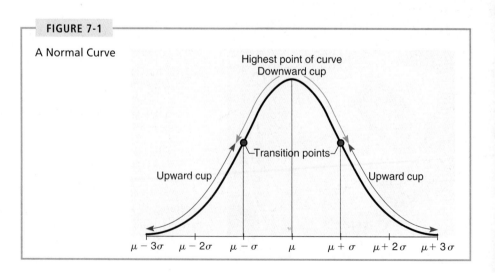

We see that a general normal curve is smooth and symmetrical about the vertical line over the mean μ. Notice that the highest point of the curve occurs over μ. If the distribution were graphed on a piece of sheet metal, cut out, and placed on a knife edge, the balance point would be at μ. We also see that the curve tends to level out and approach the horizontal (x axis) like a glider making a landing. However, in mathematical theory, such a glider would never quite finish its landing because a normal curve never touches the horizontal axis.

The parameter σ controls the spread of the curve. The curve is quite close to the horizontal axis at $\mu + 3\sigma$ and $\mu - 3\sigma$. Thus, if the standard deviation σ is large, the curve will be more spread out; if it is small, the curve will be more peaked. Figure 7-1 shows the normal curve cupped downward for an interval on either side of the mean μ. Then it begins to cup upward as we go to the lower part of the bell. The exact places where the *transition* between the upward and downward cupping occur are above the points $\mu + \sigma$ and $\mu - \sigma$. In the terminology of calculus, transition points such as these are called *inflection points*.

Important properties of a normal curve

1. The curve is bell-shaped with the highest point over the mean μ.

2. It is symmetrical about a vertical line through μ.

3. The curve approaches the horizontal axis but never touches or crosses it.

4. The transition (inflection) points between cupping upward and downward occur above $\mu + \sigma$ and $\mu - \sigma$.

GUIDED EXERCISE **1**

Curves that are not normal

Each of the curves in Figure 7-2 fails to be a normal curve. Give reasons why these curves are not normal curves.

FIGURE 7-2

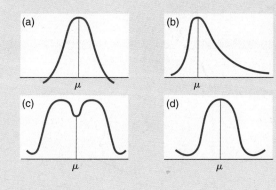

(a) A normal curve gets closer and closer to the horizontal axis, but it never touches it or crosses it.

(b) A normal curve must be symmetrical. This curve is not.

(c) A normal curve is bell-shaped with one peak. Because this curve has two peaks, it is not normal.

(d) The tails of a normal curve must get closer and closer to the x axis. In this curve the tails are going away from the x axis.

GUIDED EXERCISE 2

Identify μ and σ on a normal curve

The points *A*, *B*, and *C* are indicated on the normal curve in Figure 7-3. One of these points is μ, one is $\mu + \sigma$, and one is $\mu - 2\sigma$.

FIGURE 7-3 A Normal Curve

(a) Which point corresponds to the mean? What is the value of μ?

⟹ The mean μ is under the peak of the normal curve. The point *B* corresponds to the mean, so $\mu = 10$.

(b) Which point corresponds to $\mu + \sigma$? Use the values of $\mu + \sigma$ and μ to compute σ.

⟹ The point *C* where the curve changes from cupped down to cupped up is one standard deviation σ from the mean. The point *C* is $\mu + \sigma$. Since $\mu + \sigma = 12$ and $\mu = 10$, $\sigma = 2$.

(c) Which point corresponds to $\mu - 2\sigma$?

⟹ Since $\mu = 10$ and $\sigma = 2$, we see that

$$\mu - 2\sigma = 10 - 2(2) = 6$$

Point *A* corresponds to $\mu - 2\sigma$.

The parameters that control the shape of a normal curve are the mean μ and the standard deviation σ. When both μ and σ are specified, a specific normal curve is determined. In brief, μ locates the balance point, and σ determines the extent of the spread.

GUIDED EXERCISE 3

Identify μ and σ on a normal curve

Look at the normal curves in Figure 7-4.

Continued

GUIDED EXERCISE 3 continued

FIGURE 7-4

(a) Do these distributions have the same mean? If so, what is it?

⇒ The means are the same, since both graphs have the high point over 6. $\mu = 6$.

(b) One of the curves corresponds to a normal distribution with $\sigma = 3$ and the other to one with $\sigma = 1$. Which curve has which σ?

⇒ Curve A has $\sigma = 1$, and curve B has $\sigma = 3$. (Since curve B is more spread out, it has the larger σ value.)

◊ **COMMENT** The normal distribution curve is always above the horizontal axis. The area beneath the curve and above the axis is exactly one. As such, the normal distribution curve is an example of a *density curve*. The formula used to generate the shape of the normal distribution curve is called the *normal density function*. If x is a normal random variable with mean μ and standard deviation σ, the formula for the normal density function is

$$f(x) = \frac{e^{-\frac{1}{2}\left(\frac{x-\mu}{\sigma}\right)^2}}{\sigma\sqrt{2\pi}}$$

In this text, we will not use this formula explicitly. However, we will use tables of areas based on the normal density function. ◊

Empirical rule

The total area under any normal curve studied in this book will *always* be 1. The graph of the normal distribution is important because the portion of the *area* under the curve above a given interval represents the *probability* that a measurement will lie in that interval.

In Section 3.2, we studied Chebyshev's theorem. This theorem gives us information about the *smallest* proportion of data that lies within 2, 3, or k standard deviations of the mean. This result applies to *any* distribution. However, for normal distributions, we can get a much more precise result, which is given by the *empirical rule*.

Empirical rule

For a distribution that is symmetrical and bell-shaped (in particular, for a normal distribution):

Approximately 68% of the data values will lie within one standard deviation on each side of the mean.

Approximately 95% of the data values will lie within two standard deviations on each side of the mean.

Approximately 99.7% (or almost all) of the data values will be within three standard deviations on each side of the mean.

The preceding statement is called the *empirical rule* because, for symmetrical, bell-shaped distributions, the given percentages are observed in practice. Furthermore, for the normal distribution, the empirical rule is a direct consequence of the very nature of the distribution (see Figure 7-5). Notice that the empirical rule is a stronger statement than Chebyshev's theorem in that it gives *definite percentages*, not just lower limits. Of course, the empirical rule applies only to normal or symmetrical, bell-shaped distributions, whereas Chebyshev's theorem applies to all distributions.

EXAMPLE 1

Empirical rule

The playing life of a Sunshine radio is normally distributed with mean $\mu = 600$ hours and standard deviation $\sigma = 100$ hours. What is the probability that a radio selected at random will last from 600 to 700 hours?

SOLUTION: The probability that the playing time will be between 600 and 700 hours is equal to the percentage of the total area under the curve that is shaded in Figure 7-6. Since $\mu = 600$ and $\mu + \sigma = 600 + 100 = 700$, we see that the shaded area is simply the area between μ and $\mu + \sigma$. The area from μ to $\mu + \sigma$ is 34% of the total area. This tells us that the probability a Sunshine radio will last between 600 and 700 playing hours is about 0.34. ◇

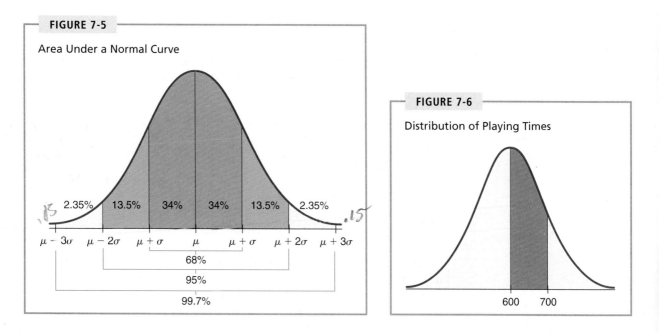

FIGURE 7-5

Area Under a Normal Curve

2.35%   13.5%   34%   34%   13.5%   2.35%

$\mu - 3\sigma$ $\mu - 2\sigma$ $\mu + \sigma$ μ $\mu + \sigma$ $\mu + 2\sigma$ $\mu + 3\sigma$

68%

95%

99.7%

FIGURE 7-6

Distribution of Playing Times

600   700

GUIDED EXERCISE 4

Empirical rule

The yearly wheat yield per acre on a particular farm is normally distributed with mean $\mu = 35$ bushels and standard deviation $\sigma = 8$ bushels.

(a) Shade the area under the curve in Figure 7-7 that represents the probability that an acre will yield between 19 and 35 bushels.

 See Figure 7-8.

(b) Is the area the same as the area between $\mu - 2\sigma$ and μ?

Yes, since $\mu = 35$ and $\mu - 2\sigma = 35 - 2(8) = 19$.

FIGURE 7-7

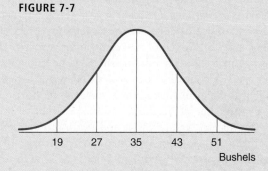

FIGURE 7-8 Completion of Figure 7-7

Bushels

(c) Use Figure 7-5 to find the percentage of area over the interval between 19 and 35.

The area between the values $\mu - 2\sigma$ and μ is 47.5% of the total area.

(d) What is the probability that the yield will be between 19 and 35 bushels per acre?

It is 47.5% of the total area, which is 1. Therefore, the probability is 0.475 that the yield will be between 19 and 35 bushels.

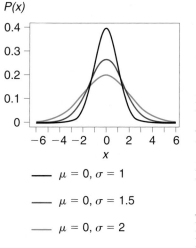

$\mu = 0, \sigma = 1$

$\mu = 0, \sigma = 1.5$

$\mu = 0, \sigma = 2$

TECH NOTE We can graph normal distributions on the TI-83Plus, Excel, and Minitab. In each technology, set the range of x values between -3.5σ and 3.5σ. Then use the built-in normal density functions to generate the corresponding y values.

TI-83Plus Press the **Y =** key. Then, under **DISTR**, select **1:normalpdf(x,μ,σ)** and fill in desired μ and σ values. Press the **WINDOW** key. Set **Xmin** to $\mu - 3\sigma$ and **Xmax** to $\mu + 3\sigma$. Finally, press the **ZOOM** key and select option **0:ZoomFit.**

Excel In one column, enter x values from -3.5σ to 3.5σ in increments of 0.2σ. In the next column, enter y values by using the menu choices **Paste Function** [f_x] ➤ **Statistical** ➤ **NORMDIST(x, μ, σ, false).** Next, use the chart wizard, and select **XY(scatter).** Choose the first picture with the dots connected and fill in the dialogue boxes.

Minitab In one column, enter x values from -3.5σ to 3.5σ in increments of 0.2σ. In the next column, enter y values by using the menu choices **Calc** ➤ **Probability Distribution** ➤ **Normal.** Fill in the dialogue box. Next, use menu choices **Graph** ➤ **Plot.** Fill in the dialogue box. Under Display, select connect.

Control Charts

If we are examining data over a period of equally spaced time intervals or in some sequential order, then *control charts* are especially useful. Business managers and people in charge of production processes are aware that there exists an inherent amount of variability in any sequential set of data. For example, the sugar content of bottled drinks taken sequentially off a production line, the extent of clerical errors in a bank from day to day, advertising expenses from month to month, or even the number of new customers from year to year are examples of sequential data. There is a certain amount of variability in each.

A random variable x is said to be in *statistical control* if it can be described by the *same* probability distribution when it is observed at successive points in time. Control charts combine graphic and numerical descriptions of data with probability distributions.

Control charts were invented in the 1920s by Walter Shewhart at Bell Telephone Laboratories. Since a control chart is a *warning device*, it is not absolutely necessary that our assumptions and probability calculations be precisely correct. For example, the x distributions need not follow a normal distribution exactly. Any mound-shaped and more or less symmetrical distribution will be good enough.

How do we make a control chart? A control chart for a variable x is a plot of the observed x values (on the vertical scale) in time sequence order (the horizontal scale represents time). There is a center line at height μ and dashed control limits at $\mu \pm 2\sigma$ and at $\mu \pm 3\sigma$. How do we pick values for μ and σ? In most practical cases, values for μ (population mean) and σ (population standard deviation) are computed from past data for which the process we are studying was known to be *in control*. Methods for choosing the sample size to fit given error tolerances can be found in Chapter 9.

Sometimes values for μ and σ are chosen as *target values*. That is, μ and σ values are chosen as set goals or targets that reflect the production level or service level at which a company hopes to perform. To be realistic, such target assignments for μ and σ should be reasonably close to actual data taken when the process was operating at a satisfactory production level. In Example 2, we will make a control chart; then we discuss ways to analyze it to see if a process or service is "in control."

EXAMPLE 2 *Control chart*	Susan Tamara is director of personnel at the Antlers Lodge in Denali National Park, Alaska. Every summer Ms. Tamara hires many part-time employees from all over the United States. Most are college students seeking summer employment. Perhaps the biggest activity for Ms. Tamara's staff is that of "making up" the rooms each day. The lodge has 385 rooms, and from long experience Ms. Tamara has determined that by 3:30 P.M. each day the average number of rooms not made up is $\mu = 19.3$ with standard deviation $\sigma = 4.7$. Although the lodge has a policy that guest rooms be made up by 3:30 P.M., Ms. Tamara knows that there will always be a few rooms not made up by this time because there is a high personnel turnover and corresponding reassignment of staff to new jobs for which they are in a training period. Every 15 days Ms. Tamara has a general staff meeting. Before the meeting, she examines several control charts for the restaurant, the gift shop, and, of course, housekeeping. For the past 15 days, the housekeeping unit has reported the following number of rooms not ready by 3:30 P.M. (Table 7-1). Make a control chart for these data.

TABLE 7-1   Number of Rooms not Made Up by 3:30 P.M.

Day	1	2	3	4	5	6	7	8
x = number of rooms	11	20	25	23	16	19	8	25

Day	9	10	11	12	13	14	15
x = number of rooms	17	20	23	29	18	14	10

Mt. McKinley, Denali National Park

Before we make a control chart, we need to know a few things about the distribution of rooms not made up by 3:30 P.M. Ms. Tamara is aware from long experience that the distribution is symmetrical and bell-shaped. It is approximately normal, with mean $\mu = 19.3$ and standard deviation $\sigma = 4.7$. In addition, this distribution of x values is acceptable to the top administration of the Antlers Lodge.

SOLUTION: A control chart for a variable x is a plot of the observed x values (vertical scale) in time sequence order (the horizontal scale represents time). Place horizontal lines at

The mean $\mu = 19.3$

The control limits $\mu \pm 2\sigma = 19.3 \pm 2(4.7)$ or 9.90 and 28.70

The control limits $\mu \pm 3\sigma = 19.3 \pm 3(4.7)$ or 5.20 and 33.40

Then plot the data from Table 7-1. (See Figure 7-9.)   ◇

Once we have made a control chart, the main question is the following: As time goes on, is the x variable continuing in this same distribution, or is the distribution of x values changing? If the x distribution is continuing more or less the same, we say it is in *statistical control*. If it is not, we say it is *out of control*.

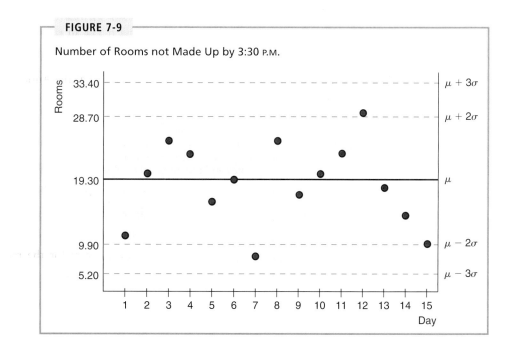

FIGURE 7-9

Number of Rooms not Made Up by 3:30 P.M.

FIGURE 7-11

Number of Rooms not Made
Up by 3:30 P.M.

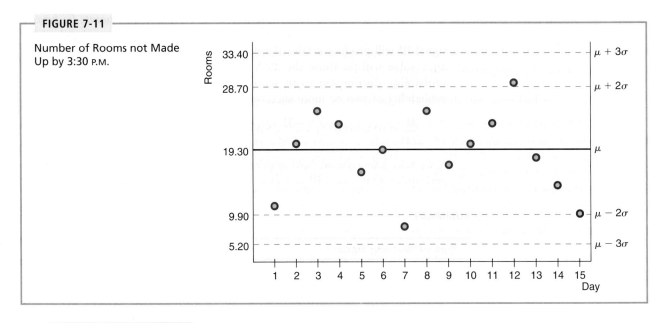

GUIDED EXERCISE 5

Control chart

Over the next 15-day period, let's suppose that housekeeping again reported the number of rooms not made up by 3:30 P.M. to Ms. Tamara of the Antlers Lodge. The data in Table 7-2 show the results.

TABLE 7-2   Next 15-Day Report of Rooms not Made Up by 3:30 P.M.

Day	1	2	3	4	5	6	7	8
x = number of rooms	25	8	23	15	26	24	31	21
Day	9	10	11	12	13	14	15	
x = number of rooms	27	20	25	21	27	11	16	

(a) We assume that we are still working with the symmetrical, bell-shaped distribution of x values, with mean $\mu = 19.3$ and $\sigma = 4.7$. Compute the "control limits" of $\mu \pm 2\sigma$ and $\mu \pm 3\sigma$. Draw a control chart showing the solid line at the mean and the dashed lines at the control limits. Plot the data for the 15-day period.

FIGURE 7-12 Next 15-Day Report of Rooms not Made Up by 3:30 P.M.

Continued

GUIDED EXERCISE 5 continued

(b) Interpret the control chart of part (a).

⟹ Days 5 to 13 are above $\mu = 19.3$. We have nine consecutive days on one side of the mean. This is a warning signal! It would appear that the mean μ is slowly drifting up beyond the target value of 19.3. The chart indicates that housekeeping is "out of control." Ms. Tamara should take corrective measures at her next staff meeting.

(c) Over another 15-day period Ms. Tamara obtained the data shown in Table 7-3 for housekeeping. Make a control chart using target values. $\mu = 19.3$ and $\sigma = 4.7$.

⟹ **FIGURE 7-13** 3rd Housekeeping Data Report

TABLE 7-3   3rd Housekeeping Data Report

Day	1	2	3	4	5	6	7	8
Number of rooms	29	14	18	21	11	20	35	24
Day	9	10	11	12	13	14	15	
Number of rooms	19	12	19	6	8	11	20	

(d) Interpret the control chart of part (c).

⟹ On day 7, we have a data value beyond $\mu + 3\sigma$ (i.e., above 33.40). On days 11, 12, and 13, we have two of three data values beyond $\mu - 2\sigma$ (i.e., below 9.90). The occurrences on both these periods are out-of-control warning signals. Ms. Tamara might ask her staff about both these periods. There may be a lesson to be learned about day 7 when housekeeping apparently had a lot of trouble. Also, days 11, 12, and 13 were very good days. Perhaps a lesson could be learned about why things went so well.

VIEWPOINT

In Control? Out of Control?

If you care about quality, you also must care about control! Dr. Walter Shewhart invented control charts when he was working for Bell Laboratories. The great contribution of control charts is to separate variation into two sources: (1) random or chance causes (in control) and (2) special or assignable causes (out of control). A process is said to be in *statistical control* when it is no longer afflicted with special or assignable causes. The performance of a process that is in statistical control is predictable. Predictability and quality control tend to be closely associated.

(*Source:* Adapted from the classic text *Statistical Methods from the Viewpoint of Quality Control,* by W. A. Shewhart, with Foreword by W. E. Deming, Dover Publications.)

SECTION 7.1   PROBLEMS

1. *General: Curves* Which, if any, of the curves in Figure 7-14 look(s) like a normal curve? If a curve is not a normal curve, tell why.

2. *General: Normal Curves* Look at the normal curve in Figure 7-15, and find μ, $\mu + \sigma$, and σ.

3. *General: Normal Curves* Look at the two normal curves in Figures 7-16 and 7-17. Which has the larger standard deviation? What is the mean of the curve in Figure 7-16? What is the mean of the curve in Figure 7-17?

4. *General: Normal Curves* Sketch a normal curve
   (a) with mean 15 and standard deviation 2.
   (b) with mean 15 and standard deviation 3.
   (c) with mean 12 and standard deviation 2.
   (d) with mean 12 and standard deviation 3.
   (e) Consider two normal curves. If the first one has a larger mean than the second one, must it have a larger standard deviation than the second one as well? Explain your answer.

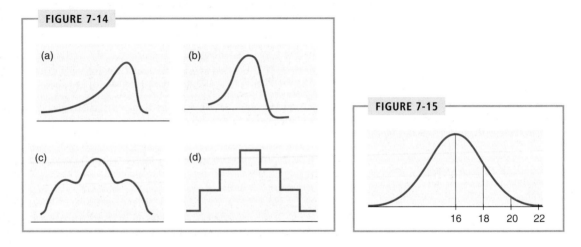

FIGURE 7-14

(a) (b) (c) (d)

FIGURE 7-15

16   18   20   22

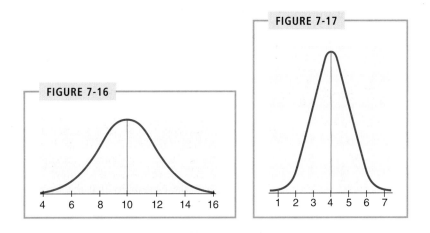

FIGURE 7-16

FIGURE 7-17

5. *General: Normal Curves* By the empirical rule, what percentage of the area under the normal curve lies
   (a) to the left of μ?   50%
   (b) between $\mu - \sigma$ and $\mu + \sigma$?   68%
   (c) between $\mu - 3\sigma$ and $\mu + 3\sigma$?   99.7

6. *General: Normal Curves* By the empirical rule, what percentage of the area under a normal curve lies
   (a) to the right of μ?
   (b) between $\mu - 2\sigma$ and $\mu + 2\sigma$?
   (c) to the right of $\mu + 3\sigma$?

7. *Distribution: Heights of Coeds* Assuming that the heights of college women are normally distributed, with mean 65 in. and standard deviation 2.5 in. (based on information from *Statistical Abstract of the United States*, 112th Edition), answer the following questions. (*Hint:* Use Problems 5 and 6 and Figure 7-5.)
   (a) What percentage of women are taller than 65 in.?
   (b) What percentage of women are shorter than 65 in.?
   (c) What percentage of women are between 62.5 in. and 67.5 in.?
   (d) What percentage of women are between 60 in. and 70 in.?

8. *Distribution: Rhode Island Red Chickens* The incubation time for Rhode Island Red chicks is normally distributed with mean of 21 days and standard deviation of approximately 1 day (based on information from *World Book Encyclopedia*). Look at Figure 7-5 and answer the following questions. If 1000 eggs are being incubated, how many chicks do we expect will hatch
   (a) in 19 to 23 days?
   (b) in 20 to 22 days?
   (c) in 21 days or fewer?
   (d) in 18 to 24 days? (Assume all eggs eventually hatch.)
   (*Note:* In this problem, let us agree to think of a single day or a succession of days as a continuous interval of time.)

9. *Archaeology: Tree Rings* At Burnt Mesa Pueblo, archaeological studies have used the method of tree-ring dating in an effort to determine when prehistoric people lived in the pueblo. Wood from several excavations gave a mean of (year) 1243 with a standard deviation of 36 years (*Bandelier Archaeological Excavation Project: Summer 1989 Excavations at Burnt Mesa Pueblo*, edited by Kohler, Washington State University Department of Anthropology). The distribution of dates was more or less mound-shaped and symmetrical about the mean.

Use the empirical rule to

(a) estimate a range of years centered about the mean in which about 68% of the data (tree-ring dates) will be found.

(b) estimate a range of years centered about the mean in which about 95% of the data (tree-ring dates) will be found.

(c) estimate a range of years centered about the mean in which almost all the data (tree-ring dates) will be found.

10. *Vending Machine: Soft Drinks* A vending machine automatically pours soft drinks into cups. The amount of soft drink dispensed into a cup is normally distributed with mean of 7.6 oz and standard deviation of 0.4 oz. Examine Figure 7-5 and use the empirical rule to answer the following questions.

(a) Estimate the probability that the machine will overflow an 8-oz cup.

(b) Estimate the probability that the machine will not overflow an 8-oz cup.

(c) The machine has just been loaded with 850 cups. How many of these do you expect will overflow when served?

11. *Pain Management: Laser Therapy* "Effect of Helium-Neon Laser Auriculotherapy on Experimental Pain Threshold" is the title of an article in the journal *Physical Therapy* (Vol. 70, No. 1, pp. 24–30). In this article, laser therapy was discussed as a useful alternative to drugs in pain management of chronically ill patients. To measure pain threshold, a machine was used that delivered low-voltage direct current to different parts of the body (wrist, neck, and back). The machine measured current in milliamperes (mA). The pretreatment experimental group in the study had an average threshold of pain (pain was first detectable) at $\mu = 3.15$ mA with standard deviation $\sigma = 1.45$ mA. Assume that the distribution of threshold pain so measured in milliamperes is symmetrical and more or less mound-shaped. Use the empirical rule to

(a) estimate a range of milliamperes centered about the mean in which about 68% of the experimental group will have a threshold of pain.

(b) estimate a range of milliamperes centered about the mean in which about 95% of the experimental group will have a threshold of pain.

12. *Control Charts: Yellowstone National Park* Yellowstone Park Medical Services (YPMS) provides emergency health care for park visitors. Such health care includes treatment for everything from indigestion and sunburn to more serious injuries. A recent issue of *Yellowstone Today* (National Park Service Publication) indicated that the average number of visitors treated each day by YPMS was 21.7. The estimated standard deviation was 4.2 (summer data). The distribution of numbers treated is approximately mound-shaped and symmetrical.

(a) For a 10-day summer period, the following data show the number of visitors treated each day by YPMS:

Day	1	2	3	4	5	6	7	8	9	10
Number treated	25	19	17	15	20	24	30	19	16	23

Make a control chart for the daily number of visitors treated by YPMS, and plot the data on the control chart. Do the data indicate that the number of visitors treated by YPMS is "in control"? Explain your answer.

(b) For another 10-day summer period, the following data were obtained:

Day	1	2	3	4	5	6	7	8	9	10
Number treated	20	15	12	21	24	28	32	36	35	37

Make a control chart, and plot the data on the chart. Do the data indicate that the number of visitors treated by YPMS is "in control" or "out of control"? Explain your answer. Identify all out-of-control signals by type (I, II, or III). If you were the park superintendent, do you think YPMS might need some (temporary) extra help? Explain.

13. *Control Charts: Bank Loans* Tri-County Bank is a small independent bank in central Wyoming. This is a rural bank that makes loans on items as small as horses and pickup trucks to items as large as ranch land. Total monthly loan requests are used by bank officials as an indicator of economic business conditions in this rural community. The mean monthly loan request for the past several years has been 615.1 (in thousands of dollars) with a standard deviation of 11.2 (in thousands of dollars). The distribution of loan requests is approximately mound-shaped and symmetrical.

(a) For 12 months, the following monthly loan requests (in thousands of dollars) were made to Tri-County Bank:

Month	1	2	3	4	5	6
Loan request	619.3	625.1	610.2	614.2	630.4	615.9

Month	7	8	9	10	11	12
Loan request	617.2	610.1	592.7	596.4	585.1	588.2

Make a control chart for the total monthly loan requests, and plot the preceding data on the control chart. From the control chart, would you say the local business economy is heating up or cooling down? Explain your answer by referring to any trend you may see on the control chart. Identify all out-of-control signals by type (I, II, or III).

(b) For another 12-month period, the following monthly loan requests (in thousands of dollars) were made to Tri-County Bank:

Month	1	2	3	4	5	6
Loan request	608.3	610.4	615.1	617.2	619.3	622.1

Month	7	8	9	10	11	12
Loan request	625.7	633.1	635.4	625.0	628.2	619.8

Make a control chart for the total monthly loan requests, and plot the preceding data on the control chart. From the control chart, would you say the local business economy is heating up, cooling down, or about normal? Explain your answer by referring to the control chart. Identify all out-of-control signals by type (I, II, or III).

14. *Control Charts: Model Rooms* The manager of Sun Motel has 316 rooms in Palo Alto, California. From observations over a period of time, she knows that on an average night 268 rooms will be rented. The long-term standard deviation is 12 rooms. This distribution is approximately mound-shaped and symmetrical. Minitab has a number of control chart options. Among them is the one called I-chart, which gives a control chart for individual measurements. The figures below show such

The resulting standard distribution will always have mean $\mu = 0$ and standard deviation $\sigma = 1$.

Areas Under the Standard Normal Curve

We have seen how to convert *any* normal distribution to the *standard* normal distribution. We can change any x value to a z value and back again. But what is the advantage of all this work? The advantage is that there are extensive tables that show the *area under the standard normal curve* for almost any interval along the z axis. The areas are important because they are equal to the *probability* that the measurement of an item selected at random falls in this interval. Thus, the *standard* normal distribution can be a tremendously helpful tool.

For instance, Sunshine Stereo guarantees their cassette decks for a period of 2 years. The company statistician has computed that the cassette deck life is normally distributed with a mean of 2.3 years and a standard deviation 0.4 year. What is the probability that a cassette deck will stop working during the guarantee period?

To answer questions of this type, we convert the given normal distribution to the standard normal distribution. Then we use a table to find the area over the interval in question and, hence, the probability an item selected at random will fall into that interval. Before we can carry out this plan, though, we must practice using Table 4 of the Appendix to find areas under the standard normal curve.

Using a Standard Normal Distribution Table

Using a table to find areas and probabilities associated with the standard normal distribution is a fairly straightforward activity. However, it is important to first observe the range of z values for which areas are given. This range is usually depicted in a picture that accompanies the table. Two commonly found styles of standard normal distribution tables are shown in Figure 7-21.

Middle-of-curve style table

Each style of table has its advantages. The middle-of-curve style is a compact table that fits on one page since it shows only z values on the right side of the distribution. Using this type of table relies heavily on the inherent symmetry of the normal distribution and the fact that the area under the right half of the curve is 0.5. To find such a table along with brief instructions for use, visit the Brase/Brase

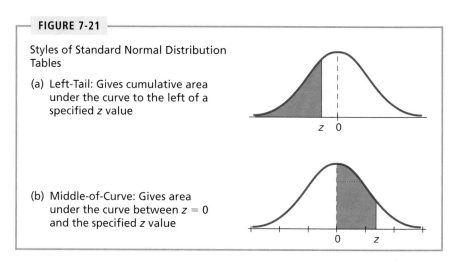

FIGURE 7-21

Styles of Standard Normal Distribution Tables

(a) Left-Tail: Gives cumulative area under the curve to the left of a specified z value

(b) Middle-of-Curve: Gives area under the curve between $z = 0$ and the specified z value

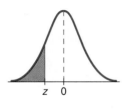

TABLE 7-5   Excerpt from Table 4 of the Appendix Showing Negative z Values

z	.00	.01	...	.07	.08	.09
−3.4	.0003	.0003	...	.0003	.0003	.0002
:						
−1.1	.1357	.1335	...	.1210	.1190	.1170
−1.0	.1587	.1562	...	.1423	.1401	.1379
−0.9	.1841	.1814	...	.1660	.1635	.1611
:						
−0.0	.5000	.4960	...	.4721	.4681	.4641

Understanding Basic Statistics site at http://math.college.hmco.com/students and use the link to the tables for this text.

Left-tail style table

In this text, *we will use the left-tail style table.* This style table gives cumulative areas to the left of a specified z. Determining other areas under the curve utilizes the fact that the area under the entire curve is 1. Taking advantage of the symmetry of the normal distribution is also useful. The procedures you learn for using the left-tail style normal distribution table apply directly to cumulative normal distribution areas found on calculators and in computer software packages such as Excel and Minitab.

EXAMPLE 6

Standard normal distribution table

Use Table 4 of the Appendix to find the described areas under the standard normal curve.

(a) Find the area under the standard normal curve to the left of $z = -1.00$.

SOLUTION: First, shade the area to be found on the standard normal distribution curve, as shown in Figure 7-22. Notice that the z value we are using is negative. This means we will look at the portion of Table 4 in the Appendix for which the z values are negative. In the upper-left corner of the table we see the letter z. The column under z gives us the units value and tenths for z. The other column headings indicate the hundredths value of z. Table entries give areas under the standard normal curve to the left of the listed z values. To find the area to the left of $z = -1.00$, we use the row headed by -1.0 and then move to the column headed by the hundredths position 0.00. This entry is shaded in Table 7-5. We see the area is 0.1587.

(b) Find the area to the left of $z = 1.18$ illustrated in Figure 7-23.

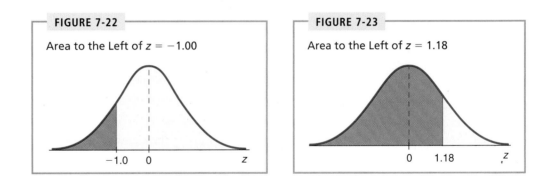

FIGURE 7-22

Area to the Left of $z = -1.00$

FIGURE 7-23

Area to the Left of $z = 1.18$

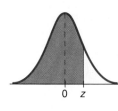

TABLE 7-6   Excerpt from Table 4 of the Appendix Showing Positive z Values

z	.00	.01	.02	...	.08	.09
0.0	.5000	.5040	.5080	...	.5319	.5359
:						
0.9	.8159	.8186	.8212	...	.8365	.8359
1.0	.8413	.8438	.8461	...	.8599	.8621
1.1	.8643	.8665	.8686	...	.8810	.8830
:						
3.4	.9997	.9997	.9997	...	.9997	.9998

SOLUTION: In this case, we are looking for an area to the left of a positive z value, so we look in the portion of Table 4 that shows positive z values. Again we first sketch the area to be found on a standard normal curve as shown in Figure 7-23 on the previous page. Look in the row headed by 1.1 and move to the column headed by .08. The desired area is shaded (see Table 7-6). We see that the area to the left of 1.18 is 0.8810.   ◇

GUIDED EXERCISE 8

Using the standard normal distribution table

Table 4, Areas of a Standard Normal Distribution, is located in the Appendix as well as in the endpapers of the text. Spend a little time studying the table, and then answer these questions.

(a) As z values increase, do the areas to the left of z increase?   ⟹   Yes, as z values increase, we move to the right on the normal curve, and the areas increase.

(b) If a z value is negative, is the area to the left of z less than 0.5000?   ⟹   Yes. Remember that a negative z value is on the left side of the standard normal distribution. The entire left half of the normal distribution has area 0.5, so any area to the left of $z = 0$ will be less than 0.5.

(c) If a z value is positive, is the area to the left of z greater than 0.5000?   ⟹   Yes. Positive z values are on the right side of the standard normal distribution, and any area to the left of a positive z value includes the entire left half of the normal distribution.

Using Table 4 to find other areas   Table 4 gives areas under the standard normal distribution that are to the *left of a z value*. How do we find other areas under the standard normal curve?

Using left-tail style standard normal distribution table

1. For areas to the left of a specified z value, use the table entry directly.

2. For areas to the right of a specified z value, look up the table entry for z and subtract the area from 1.
   Note: Another way to find the same area is to use the symmetry of the normal curve and look up the table entry for $-z$.

3. For areas between two z values, z_1 and z_2 (where $z_2 > z_1$), *subtract* the table area for z_1 from the table area for z_2.

Figure 7-24 illustrates the procedure for using Table 4, Areas of a Standard Normal Distribution, to find any specified area under the standard normal distribution. Again, it is useful to sketch the area in question before you use Table 4.

◊ **COMMENT:** Notice that the z values shown in Appendix Table 4 are formatted to the hundredths position. It is convenient to *round or format z values to the hundredths position* before using the table. The areas are all given to 4 places after the decimal, so give your answers to 4 places after the decimal.   ◊

FIGURE 7-24

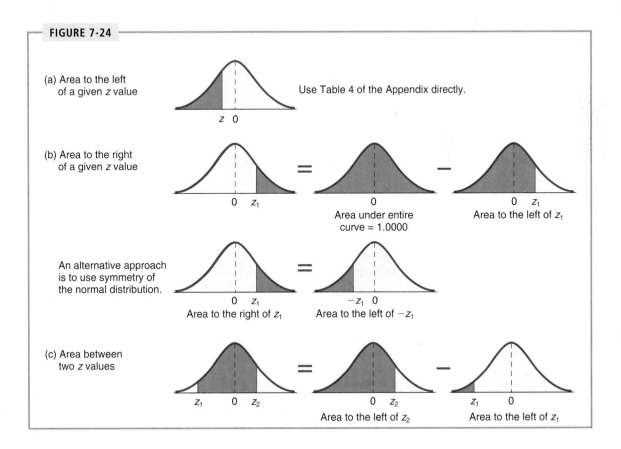

(a) Area to the left of a given z value

Use Table 4 of the Appendix directly.

(b) Area to the right of a given z value

Area under entire curve = 1.0000

Area to the left of z_1

An alternative approach is to use symmetry of the normal distribution.

Area to the right of z_1

Area to the left of $-z_1$

(c) Area between two z values

Area to the left of z_2

Area to the left of z_1

◊ **COMMENT:** The smallest z value shown in Table 4 is -3.49, while the largest value is 3.49. These values are, respectively, far to the left and far to the right on the standard normal distribution, with very little area beyond either value. We will follow the common convention of treating any area to the left of a z value smaller than -3.49 as 0.000. Similarly, we will consider any area to the right of a z value greater than 3.49 as 0.000. We understand that there is some area in these extreme tails. However, these areas are each less than 0.0002. Now let's get real about this! Some very specialized applications, beyond the scope of this book, do need to measure areas and corresponding probabilities in these extreme tails. But in most practical applications, *we follow the convention of treating the areas in the extreme tails as zero.* ◊

EXAMPLE 7

Using table to find areas

Use Table 4 of the Appendix to find the specified areas.

(a) Find the area between $z = 1.00$ and $z = 2.70$.

SOLUTION: First, sketch a diagram showing the area (see Figure 7-25). Because we are finding the area between two z values, we subtract corresponding table entries.

$$\text{(Area between 1.00 and 2.70)} = \text{(Area left of 2.70)} - \text{(Area left of 1.00)}$$
$$= 0.9965 - 0.8413$$
$$= 0.1552$$

(b) Find the area to the right of $z = 0.94$.

SOLUTION: First, sketch the area to be found (see Figure 7-26).

$$\text{(Area to right of 0.94)} = \text{(Area under entire curve)} - \text{(Area to left of 0.94)}$$
$$= 1.000 - 0.8264$$
$$= 0.1736$$

Alternatively,

$$\text{(Area to the right of 0.94)} = \text{(Area to the left of } -0.94)$$
$$= 0.1736$$ ◊

Probabilities associated with the standard normal distribution

We have practiced the skill of finding areas under the standard normal curve for various intervals along the z axis. This skill is important since *the probability*

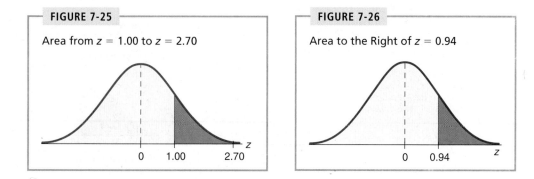

FIGURE 7-25

Area from $z = 1.00$ to $z = 2.70$

0   1.00   2.70 z

FIGURE 7-26

Area to the Right of $z = 0.94$

0   0.94 z

that z lies in an interval *is given by the area* under the standard normal curve above that interval.

Because the normal distribution is continuous, there is no area under the curve exactly over a specific z. Therefore, probabilities such as $P(z \geq z_1)$ are the same as $P(z > z_1)$. When dealing with probabilities or areas under a normal curve that are specified with inequalities, *strict inequality* symbols can be used *interchangeably* with *inequality-or-equal* symbols.

GUIDED EXERCISE 9

Probabilities associated with the standard normal distribution

Let z be a random variable with a standard normal distribution.

(a) $P(z \geq 1.15)$ refers to the probability that z values lie to the right of 1.15. Shade the corresponding area under the standard normal curve and find $P(z \geq 1.15)$.

▷ **FIGURE 7-27** Area to Be Found

$$P(z \geq 1.15) = 1.000 - P(z \leq 1.15) = 1.000 - 0.8749 = 0.1251$$

Alternatively,

$$P(z \geq 1.15) = P(z \leq -1.15) = 0.1251$$

(b) Find $P(-1.78 \leq z \leq 0.35)$. First, sketch the area under the standard normal curve corresponding to the area.

▷ **FIGURE 7-28** Area to Be Found

$$P(-1.78 \leq z \leq 0.35) = P(z \leq 0.35) - P(z \leq -1.78)$$
$$= 0.6368 - 0.0375 = 0.5993$$

🔵 **TECH NOTE** The TI-83Plus, Excel, and Minitab all provide cumulative areas under any normal distribution, including the standard normal. The Tech Note of Section 7.3 shows examples.

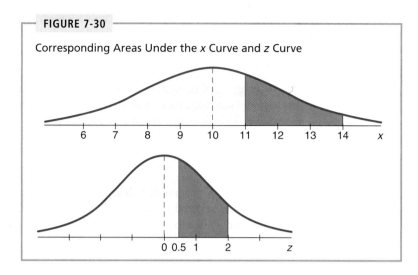

FIGURE 7-30

Corresponding Areas Under the *x* Curve and *z* Curve

The probability is 0.2857 that an *x* value selected at random from a normal distribution with mean 10 and standard deviation 2 lies between 11 and 14.   ◊

GUIDED EXERCISE **10**

Normal distribution probability

In Section 7.2, we talked about Sunshine Stereo cassette decks. The cassette deck life was normally distributed with a mean of 2.3 years and a standard deviation of 0.4 year. We wanted to know the probability that a cassette deck will break down during the guarantee period of 2 years.

(a) Let *x* represent the life of a cassette deck. The statement that the cassette deck breaks during the 2-year guarantee period means the life is less than 2 years, or $x \leq 2$. Convert this to a statement about *z*.

$\Rightarrow$ $z = \dfrac{x - \mu}{\sigma} = \dfrac{2 - 2.3}{0.4} = -0.75$

So $x \leq 2$ means $z \leq -0.75$.

(b) Indicate the area to be found in Figure 7-31. Does this area correspond to the probability that $z \leq -0.75$?

$\Rightarrow$ See Figure 7-32.
Yes, the shaded area does correspond to the probability that $z \leq -0.75$.

FIGURE 7-31

FIGURE 7-32 $z \leq -0.75$

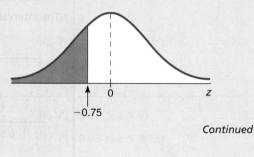

Continued

GUIDED EXERCISE 10 continued

(c) Use Table 4 of the Appendix to find $P(z \leq -0.75)$.

⟹ 0.2266

(d) What is the probability that the cassette deck will break before the end of the guarantee period? [*Hint:* $P(x \leq 2) = P(z \leq -0.75)$.]

⟹ The probability is

$$P(x \leq 2) = P(z \leq -0.75)$$
$$\approx 0.23$$

This means the company will repair or replace about 23% of the cassette decks.

TECH NOTE The TI-83Plus, Excel, and Minitab all provide areas under any normal distribution. Excel and Minitab give the left-tail area to the left of a specified x value. The TI-83Plus has you specify an interval from a lower bound to an upper bound and provides the area under the normal curve for that interval. For example, to solve Guided Exercise 10 regarding the probability a cassette deck will break during the guarantee period, we find $P(x \leq 2)$ for a normal distribution with $\mu = 2.3$ and $\sigma = 0.4$.

TI-83Plus Press **DISTR** Key, select **2:normalcdf (lower bound, upper bound, μ, σ)** and press enter. Type in the specified values. For a left-tail area, use a lower bound setting at about 4 standard deviations below the mean. Likewise, for a right-tail area use an upper bound about 4 standard deviations above the mean. For our example, use a lower bound of $\mu - 4\sigma = 2.3 - 4(0.4) = 0.7$.

```
normalcdf(.7,2,2.3,
.4)
          .226595934
```

Excel Press the **Paste Function** (f_x) ➤ **Statistical** ➤ **NORMDIST.** Fill in the dialogue box, using True for cumulative.

=	=NORMDIST(2,2.3,0.4, TRUE)		
	C	D	E
0.226627			

Minitab Use the menu selection **Calc** ➤ **Probability Distribution** ➤ **Normal.** Fill in the dialogue box, marking cumulative.

```
Cumulative Distribution Function
Normal with mean = 2.3 and
standard deviation = 0.4
      x      P(X <= x)
     2.0       0.2266
```

Finding *z* or *x*, given a probability

Inverse Normal Distribution

Sometimes we need to find z or x values that correspond to a given area under the normal curve. This situation arises when we want to specify a guarantee period so that a given percentage of the total products produced by a company lasts at least as long as the duration of the guarantee period. In such cases, we use the standard normal distribution table "in reverse." When we look up an area and find the corresponding z value, we are using the *inverse normal probability distribution*.

EXAMPLE 9
Find x, given probability

Magic Video Games, Inc., sells an expensive video computer games package. Because the package is so expensive, the company wants to advertise an impressive guarantee for the life expectancy of its computer control system. The guarantee policy will refund the full purchase price if the computer fails during the guarantee period. The research department has done tests that show that the mean life for the computer is 30 months, with standard deviation of 4 months. The computer life is normally distributed. How long can the guarantee period be if management does not want to refund the purchase price on more than 7% of the Magic Video packages?

SOLUTION: Let us look at the distribution of lifetimes for the computer control system, and shade the portion of the distribution in which the computer lasts fewer months than the guarantee period. (See Figure 7-33.)

If a computer system lasts fewer months than the guarantee period, a full-price refund will have to be made. The lifetimes requiring a refund are in the shaded region in Figure 7-33. This region represents 7% of the total area under the curve.

We can use Table 4 of the Appendix to find the z value so that 7% of the total area under the *standard* normal curve lies to the left of the z value. Then we convert the z value to its corresponding x value to find the guarantee period.

We want to find the z value with 7% of the area under the standard normal curve to the left of z. Since we are given the area in a left tail, we can use Table 4 of the Appendix directly to find z. The area value is 0.0700. However, this area is not in our table, so we use the closest area, which is 0.0694, and the corresponding z value of $z = -1.48$ (see Table 7-7).

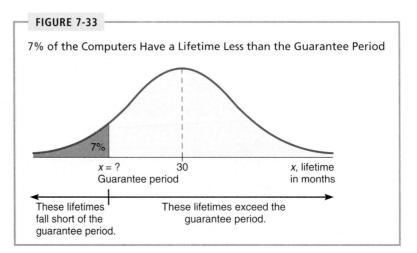

FIGURE 7-33

7% of the Computers Have a Lifetime Less than the Guarantee Period

TABLE 7-7   Excerpt from Table 4 of the Appendix

z	.00	...	.07		.08	.09
:						
−1.4	.0808		.0708		.0694	.0681
				↑		
				0.0700		

To translate this value back to an x value (in months), we use the formula

$$x = z\sigma + \mu$$
$$\qquad = -1.48(4) + 30 \qquad \text{(Use } \sigma = 4 \text{ months and } \mu = 30 \text{ months.)}$$
$$\qquad = 24.08 \text{ months}$$

The company can guarantee the Magic Video Games package for $x = 24$ months. For this guarantee period, they expect to refund the purchase price of no more than 7% of the video games packages.   ◇

Example 9 had us find a z value corresponding to a given area to the left of z. What if the specified area is to the right of z or between $-z$ and z? Figure 7-34 shows us how to proceed.

◇ **COMMENT** When we use Table 4 in the Appendix to find a z value corresponding to a given area, we usually use the nearest area value rather than interpolation

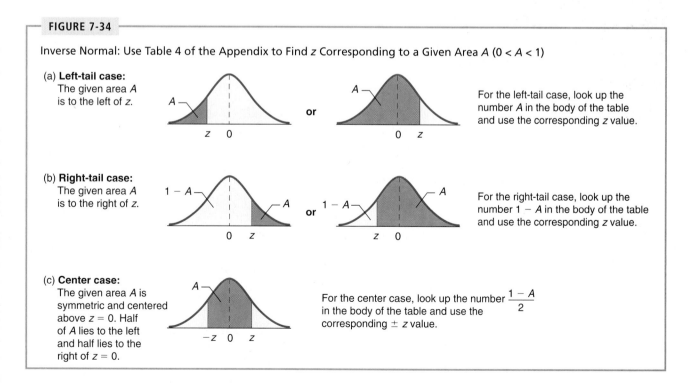

FIGURE 7-34

Inverse Normal: Use Table 4 of the Appendix to Find z Corresponding to a Given Area A (0 < A < 1)

(a) **Left-tail case:**
The given area A is to the left of z.

For the left-tail case, look up the number A in the body of the table and use the corresponding z value.

(b) **Right-tail case:**
The given area A is to the right of z.

For the right-tail case, look up the number 1 − A in the body of the table and use the corresponding z value.

(c) **Center case:**
The given area A is symmetric and centered above z = 0. Half of A lies to the left and half lies to the right of z = 0.

For the center case, look up the number $\frac{1-A}{2}$ in the body of the table and use the corresponding ± z value.

between values. However, when the area value given is exactly halfway between two area values of the table, we use the z value halfway between z values of the corresponding table areas. Example 10 demonstrates this procedure. However, this interpolation convention is not always used, especially if the area is changing slowly as it does in the tail ends of the distribution. *When the z value corresponding to an area is smaller than −2, the standard convention is to use the z value corresponding to the smaller area. Likewise, when the z value is larger than 2, the standard convention is to use the z values corresponding to the larger area.* We will see an example of this special case in Example 1 of Chapter 9. ◇

EXAMPLE 10

Find z

Find the z value so that 90% of the area under the standard normal curve lies between $-z$ and z.

SOLUTION: Sketch a picture showing the described area (see Figure 7-35).

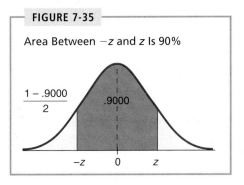

FIGURE 7-35

Area Between $-z$ and z Is 90%

We find the corresponding area in the left tail.

$$\text{(Area left of } -z) = \frac{1 - 0.9000}{2}$$

$$= 0.0500$$

Looking in Table 7-8, we see that 0.0500 lies exactly between areas 0.0495 and 0.0505. The halfway value between $z = -1.65$ and $z = -1.64$ is $z = -1.645$. Therefore, we conclude that 90% of the area under the standard normal curve lies between the z values -1.645 and 1.645. ◇

TABLE 7-8 Excerpt from Table 4 of the Appendix

z	...	.04		.05
:				
−1.6		.0505	↑	.0495
			0.0500	

GUIDED EXERCISE 11

Find z

Find the z value so that 3% of the area under the standard normal curve lies to the right of z.

(a) Draw a sketch of the standard normal distribution showing the described area.

FIGURE 7-36 3% of Total Area Lies to the Right of z

(b) Find the area to the left of z.

Area to the left of $z = 1 - 0.0300 = 0.9700$.

(c) Look up the area in Table 7-9 and find the corresponding z value.

The closest area is 0.9699. This area is to the left of $z = 1.88$.

TABLE 7-9   Excerpt from Table 4 of the Appendix

z	.00	.01	.02	.03	.04	.05	.06	.07	.08	.09
1.8	.9641	.9649	.9656	.9664	.9671	.9678	.9686	.9693	.9699	.9706
1.9	.9713	.9719	.9726	.9732	.9738	.9744	.9750	.9756	.9761	.9767

(d) Suppose the time to complete a test is normally distributed with $\mu = 40$ minutes and $\sigma = 5$ minutes. After how many minutes can we expect all but about 3% of the tests to be completed?

We are looking for an x value so that 3% of the normal distribution lies to the right of x. In part (c), we found that 3% of the standard normal curve lies to the right of $z = 1.88$. We convert $z = 1.88$ to an x value.

$$x = z\sigma + \mu$$
$$= 1.88(5) + 40 = 49.4 \text{ minutes}$$

All but about 3% of the tests will be complete after 50 minutes.

(e) Use Table 7-10 to find a z value so that 3% of the area under the standard normal curve lies to the left of z.

The closest area is 0.0301. This is the area to the left of $z = -1.88$.

TABLE 7-10   Excerpt from Table 4 of the Appendix

z	.00	.01	.02	.03	.04	.05	.06	.07	.08	.09
−1.9	.0287	.0281	.0274	.0268	.0262	.0256	.0250	.0244	.0239	.0233
−1.8	.0359	.0351	.0344	.0336	.0329	.0322	.0314	.0307	.0301	.0294

(f) Compare the z value of part (c) with the z value of part (e). Is there any relationship between the z values?

One z value is the negative of the other. This result is expected because of the symmetry of the normal distribution.

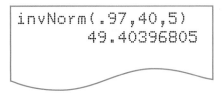

TECH NOTE   When we are given a z value and we find an area to the left of z, we are using a normal distribution function. When we are given an area to the left of z and we find the corresponding z, we are using an inverse normal distribution function. The TI-83Plus, Excel, and Minitab all have inverse normal distribution functions for any normal distribution. For instance, to find an x value from a normal distribution with mean 40 and standard deviation 5 such that 97% of the area lies to the left of x, use the described instructions.

TI-83Plus Press the **DISTR** key and select **3:invNorm(area,μ,σ)**.

```
invNorm(.97,40,5)
        49.40396805
```

Excel Press the **Paste Function** (f_x) ➤ **Statistical** ➤ **NORMINV**. Fill in the dialogue box.

=	=NORMINV(0.97,40,5)	
C	D	
49.40395		

Minitab Use the menu selection **Calc ➤ Probability Distribution ➤ Normal**. Fill in the dialogue box, marking Inverse Cumulative.

```
Inverse Cumulative Distribution Function
Normal with mean = 40.000 and
  standard deviation = 5.00000
P(X <= x)          x
  0.9700      49.4040
```

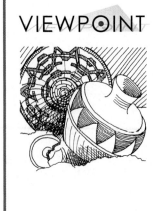

VIEWPOINT   *Want to Be an Archaeologist?*

Each year about 4500 students work with professional archaeologists in scientific research at the Crow Canyon Archaeological Center, Cortez, Colorado. In fact, Crow Canyon was included in *The Princeton Review Guide to America's Top 100 Internships*. The nonprofit, multidisciplinary program at Crow Canyon enables students and laypeople with little or no background to get started in archaeological research. The only requirement is that you be interested in Native American culture and history. By the way, a knowledge of introductory statistics could come in handy in this internship. For more information about the program, visit the Brase/Brase *Understanding Basic Statistics* site at http://math.college.hmco.com/students and find the link to Crow Canyon.

SECTION 7.3   PROBLEMS

In Problems 1–10, assume that x has a normal distribution, with the specified mean and standard deviation. Find the indicated probabilities.

1. $P(3 \leq x \leq 6)$; $\mu = 4$; $\sigma = 2$

2. $P(10 \leq x \leq 26)$; $\mu = 15$; $\sigma = 4$

3. $P(50 \leq x \leq 70)$; $\mu = 40$; $\sigma = 15$

4. $P(7 \leq x \leq 9)$; $\mu = 5$; $\sigma = 1.2$

5. $P(8 \leq x \leq 12)$; $\mu = 15$; $\sigma = 3.2$

6. $P(40 \leq x \leq 47)$; $\mu = 50$; $\sigma = 15$

7. $P(x \geq 30)$; $\mu = 20$; $\sigma = 3.4$

8. $P(x \geq 120)$; $\mu = 100$; $\sigma = 15$

9. $P(x \geq 90)$; $\mu = 100$; $\sigma = 15$

10. $P(x \geq 2)$; $\mu = 3$; $\sigma = 0.25$

In Problems 11–20, find the z value described and sketch the area described.

11. Find z so that 6% of the standard normal curve lies to the left of z.

12. Find z so that 5.2% of the standard normal curve lies to the left of z.

13. Find z so that 55% of the standard normal curve lies to the left of z.

14. Find z so that 97.5% of the standard normal curve lies to the left of z.

15. Find z so that 8% of the standard normal curve lies to the right of z.

16. Find z so that 5% of the standard normal curve lies to the right of z.

17. Find z so that 82% of the standard normal curve lies to the right of z.

18. Find z so that 95% of the standard normal curve lies to the right of z.

19. Find the z value so that 98% of the standard normal curve lies between $-z$ and z.

20. Find the z value so that 95% of the standard normal curve lies between $-z$ and z.

21. *Medical: Blood Glucose* The level of blood glucose and diabetes are closely related. Let x be a random variable measured in milligrams of glucose per deciliter (1/10th of a liter) of blood. After a 12-hour fast, the random variable x will have a distribution that is approximately normal with mean $\mu = 85$ and standard deviation $\sigma = 25$ (*Diagnostic Tests with Nursing Implications*, edited by S. Loeb, Springhouse Press). *Note:* After 50 years of age, both the mean and standard deviation tend to increase. What is the probability that for an adult (under 50 years old) after a 12-hour fast
    (a) x is more than 60?
    (b) x is less than 110?
    (c) x is between 60 and 110?
    (d) x is greater than 140 (borderline diabetes starts at 140)?

22. *Budget: Maintenance* The weekly amount of money spent on cleaning, maintenance, and repairs at a large restaurant was observed over a long period of time to be approximately normally distributed with mean $\mu = \$615$ and standard deviation $\sigma = \$42$.
    (a) If \$646 is budgeted for next week, what is the probability the actual costs will exceed the budgeted amount?
    (b) How much should be budgeted for weekly repairs, cleaning, and maintenance so that the probability the budgeted amount will be exceeded in a given week is only 0.10?

23. *Education: SAT and ACT Scores* For a given population of high school seniors, the Scholastic Aptitude Test (SAT) in mathematics has a mean score of 500 with a standard deviation of 100. Another widely used test is American College Testing (ACT) exam. The mathematics portion of the ACT has a mean of 18 and a standard deviation of 6. (Visit the Brase/Brase *Understanding Basic Statistics* site at http://math.college.hmco.com/students and find the link to the College Board.) Both SAT and ACT scores are normally distributed. What is the probability that a randomly selected high school senior's score on the mathematics part of the SAT will be
    (a) more than 675?   (b) less than 450?   (c) between 450 and 675?

    What is the probability that a randomly selected high school senior's score on the mathematics part of the ACT will be
    (d) more than 28?   (e) more than 12?   (f) between 12 and 28?

24. *Education: SAT and ACT Scores* Please refer to SAT and ACT information from Problem 23.
    (a) Suppose that an engineering school honors program will accept only high school seniors with a mathematics SAT or ACT score in the top 10%. What is the minimum SAT score in mathematics for this program? What is the minimum ACT score in mathematics for this program?
    (b) Suppose that an engineering school will accept only high school seniors with a mathematics SAT or ACT score in the top 20%. What is the minimum SAT score in mathematics for this program? What is the minimum ACT score in mathematics for this program?
    (c) Suppose that an engineering school will accept only high school seniors with a mathematics SAT or ACT score in the top 60%. What is the minimum SAT score in mathematics for this program? What is the minimum ACT score in mathematics for this program?

25. *Guarantee: Batteries* Quick Start Company makes 12-volt car batteries. After many years of product testing, the company knows that the average life of a Quick Start battery is normally distributed, with a mean of 45 months and a standard deviation of 8 months.
    (a) If Quick Start guarantees a full refund on any battery that fails within the 36-month period after purchase, what percentage of its batteries will the company expect to replace?
    (b) If Quick Start does not want to make refunds for more than 10% of its batteries under the full-refund guarantee policy, for how long should the company guarantee the batteries (to the nearest month)?

26. *Guarantee: Watches* Accrotime is a manufacturer of quartz crystal watches. Accrotime researchers have shown that the watches have an average life of 28 months before certain electronic components deteriorate, causing the watch to become unreliable. The standard deviation of watch lifetimes is 5 months, and the distribution of lifetimes is normal.
    (a) If Accrotime guarantees a full refund on any defective watch for 2 years after purchase, what percentage of total production will the company expect to replace?
    (b) If Accrotime does not want to make refunds on more than 12% of the watches it makes, how long should the guarantee period be (to the nearest month)?

27. *Consumer: TV Replacement Consumer Reports* gave information about the age at which various household products are replaced. For example, color TVs are replaced

at an average age of $\mu = 8$ years after purchase, and the (95% of data) range was from 5 to 11 years. Thus, the range was $11 - 5 = 6$ years. Let x be the age (years) at which a color TV is replaced. Assume that x has a distribution that is approximately normal.

(a) Read the empirical rule (Section 7.1); then explain why approximately four standard deviations will cover a 95% range of data values centered at the mean. Since the range is 6 years, explain why 1.5 years would be a good approximation for the standard deviation of x values.

(b) What is the probability that someone will keep a color TV more than 5 years before replacement?

(c) What is the probability that someone will keep a color TV fewer than 10 years before replacement?

(d) Assume that the average life of a color TV is 8 years with a standard deviation of 1.5 years before it breaks. Suppose that a company guarantees color TVs and will replace a TV that breaks while under guarantee with a new one. However, the company does not want to replace more than 10% of the TVs under guarantee. For how long should the guarantee be made (round to the nearest tenth of a year)?

28. *Veterinary Science: Horses* The resting heart rate for an adult horse should average about $\mu = 46$ beats per minute with (95% of data) range from 22 to 70 beats per minute, based on information from *The Merck Veterinary Manual* (a classic reference used in most veterinary colleges). Let x be a random variable that represents the resting heart rate for an adult horse. Assume that x has a distribution that is approximately normal.

(a) Estimate the standard deviation of the x distribution. *Hint: See* Problem 27.

(b) What is the probability that the heart rate is less than 25 beats per minute?

(c) What is the probability that the heart rate is greater than 60 beats per minute?

(d) What is the probability that the heart rate is between 25 and 60 beats per minute?

(e) A horse whose resting heart rate is in the upper 10% of the probability distribution of heart rates may have a secondary infection or illness that needs to be treated. What is the heart rate corresponding to the upper 10% cutoff point of the probability distribution?

29. *Insurance: Satellites* A relay microchip in a telecommunications satellite has a life expectancy that follows a normal distribution with a mean of 90 months and a standard deviation of 3.7 months. When this computer-relay microchip malfunctions, the entire satellite is useless. A large London insurance company is going to insure the satellite for 50 million dollars. Assume that the only part of the satellite in question is the microchip. All other components will work indefinitely.

(a) For how many months should the satellite be insured to be 99% confident that it will last beyond the insurance date?

(b) If the satellite is insured for 84 months, what is the probability that it will malfunction before the insurance coverage ends?

(c) If the satellite is insured for 84 months, what is the expected loss to the insurance company?

(d) If the insurance company charges $3 million for 84 months of insurance, how much profit does the company expect to make?

30. *Focus Problem: Show Attendance* Solve the Focus Problem stated at the beginning of this chapter.

7.4
Normal Approximation to the Binomial Distribution

FOCUS POINTS

✓ State the assumptions needed for the normal approximation to the binomial.

✓ Compute μ and σ for the normal approximation.

✓ Use the continuity correction to convert a range of r values to a corresponding range of normal x values.

✓ Convert the x values to a range of standardized z scores and find desired probabilities.

Criteria $np > 5$ and $nq > 5$

The probability that a new vaccine will protect adults from cholera is known to be 0.85. It is administered to 300 adults who must enter an area where the disease is prevalent. What is the probability that more than 280 of these adults will be protected from cholera by the vaccine?

This question falls into the category of a binomial experiment with number of trials n equal to 300, the probability of success p equal to 0.85, and the number of successes r greater than 280. It is possible to use the formula for the binomial distribution to compute the probability that r is greater than 280. However, this approach would involve a number of tedious and long calculations. There is an easier way to do this problem, for under the conditions stated below, the normal distribution can be used to approximate the binomial distribution.

Again, let p be the probability of success and let q be the probability of failure in a single binomial trial. Let n be the number of trials in the binomial experiment. If n, p, and q are such that *both* $np > 5$ and $nq > 5$, then the normal probability distribution with $\mu = np$ and $\sigma = \sqrt{npq}$ will be a good *normal approximation to the binomial distribution*. As n gets larger, the approximation becomes better.

◇ **SUMMARY** Consider the binomial distribution with

133 n = number of trials

95 r = number of successes

.66 p = probability of success

.34 q = probability of failure = $1 - p$

If $np > 5$ and $nq > 5$, then r has a binomial distribution that is approximated by a *normal* distribution with

$$\mu = np$$

and

$$\sigma = \sqrt{npq} \quad \diamond$$

EXAMPLE 11

Binomial distribution graphs

Graph the binomial distributions where $p = 0.25$, $q = 0.75$, and the number of trials is first $n = 3$, then $n = 10$, $n = 25$, and finally $n = 50$.

SOLUTION: The authors used the computer program ComputerStat to obtain the binomial distributions for the given values of p, q, and n. The results have been organized and graphed in Figures 7-37, 7-38, 7-39, and 7-40.

When $n = 3$, the outline of the histogram does not even begin to take the shape of a normal curve. But when $n = 10$, 25, or 50, it does begin to take a normal shape, indicated by the red curve.

From a theoretical point of view, the histograms in Figures 7-38, 7-39, and 7-40 would have bars for all values of r from $r = 0$ to $r = n$. However, in the construction of these histograms, the bars of height less than 0.001 unit have been omitted—that is, in this example, probabilities less than 0.001 have been rounded to 0. ◇

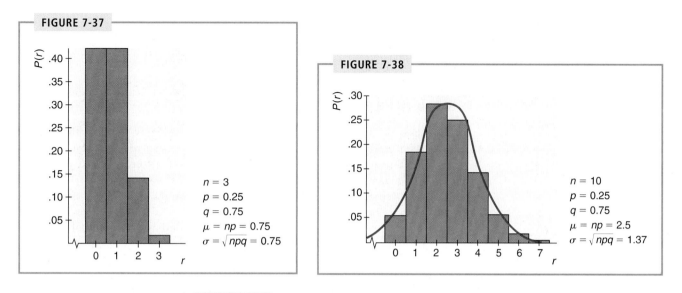

FIGURE 7-37

$n = 3$
$p = 0.25$
$q = 0.75$
$\mu = np = 0.75$
$\sigma = \sqrt{npq} = 0.75$

FIGURE 7-38

$n = 10$
$p = 0.25$
$q = 0.75$
$\mu = np = 2.5$
$\sigma = \sqrt{npq} = 1.37$

FIGURE 7-39

$n = 25$
$p = 0.25$
$q = 0.75$
$\mu = np = 6.25$
$\sigma = \sqrt{npq} = 2.17$

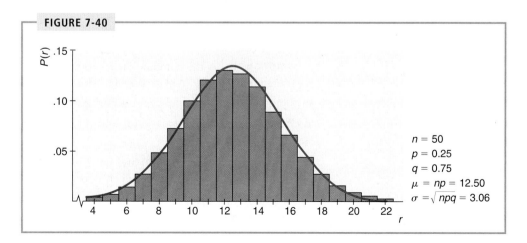

FIGURE 7-40

$n = 50$
$p = 0.25$
$q = 0.75$
$\mu = np = 12.50$
$\sigma = \sqrt{npq} = 3.06$

EXAMPLE 12

Normal approximation

The owner of a new apartment building must install 25 water heaters. From past experience in other apartment buildings, she knows that Quick Hot is a good brand. It is guaranteed for 5 years only, but from her past experience, she knows that the probability it will last 10 years is 0.25.

(a) What is the probability that 8 or more of the 25 water heaters will last at least 10 years?

SOLUTION: In this example, $n = 25$ and $p = 0.25$, so Figure 7-39 (on the preceding page) represents the probability distribution we will use. Let r be the binomial random variable corresponding to the number of successes out of $n = 25$ trials. We want to find $P(r \geq 8)$ by using the normal approximation. This probability is represented graphically (Figure 7-39) by the area of the bar over 8 and all bars to the right of the bar over 8.

Let x be a normal random variable corresponding to a normal distribution with $\mu = np = 25(0.25) = 6.25$ and $\sigma = \sqrt{npq} = \sqrt{25(0.25)0.75} \approx 2.17$. This normal curve is represented by the red line in Figure 7-39 on the preceding page. The area under the normal curve from $x = 7.5$ to the right is approximately the same as the area of the bars from the bar over $r = 8$ to the right. It is important to notice that we start with $x = 7.5$ because the bar over $r = 8$ really starts at $x = 7.5$.

The area of the bars and the area under the corresponding red (normal) curve are approximately equal, so we conclude that $P(r \geq 8)$ is approximately equal to $P(x \geq 7.5)$.

When we convert $x = 7.5$ to standard units, we get

$$z = \frac{x - \mu}{\sigma}$$

$$= \frac{7.5 - 6.25}{2.17} \quad \text{(use } \mu = 6.25 \text{ and } \sigma = 2.17\text{)}$$

$$= 0.58$$

The probability we want is

$$P(x \geq 7.5) = P(z \geq 0.58) = 1 - P(z \leq 0.58)$$
$$= 1 - 0.7190$$
$$= 0.2810$$

(b) How does this result compare with the result we can obtain by using the formula for the binomial probability distribution with $n = 25$ and $p = 0.25$?

SOLUTION: Using the binomial distribution function on the TI-83Plus, the authors computed that $P(r \geq 8) \approx 0.2735$. This means that the probability is approximately 0.27 that 8 or more water heaters will last at least 10 years.

(c) How do the results of parts (a) and (b) compare?

SOLUTION: The error of approximation is the difference between the approximate normal value (0.2810) and the binomial value (0.2735). The error is only $0.2810 - 0.2735 = 0.0075$, which is negligible for most practical purposes.

FIGURE 7-41

$P(6 \leq r \leq 10)$ Is Approximately Equal to $P(5.5 \leq x \leq 10.5)$

We knew in advance that the normal approximation to the binomial probability would be good, since $np = 25\ (0.25) = 6.25$ and $nq = 25(0.75) = 18.75$ are both greater than 5. These are the conditions that assure us that the normal approximation will be sufficiently close to the binomial probability for most practical purposes.   ◊

Continuity correction: converting *r* values to *x* values

Remember that when using the normal distribution to approximate the binomial, we are computing the areas under bars. The bar over *r* goes from $r - 0.5$ to $r + 0.5$. If *r* is a *left* endpoint of an interval, we *subtract* 0.5 to get the corresponding normal variable *x*. If *r* is a *right* endpoint of an interval, we *add* 0.5 to get the corresponding variable *x*. For instance, $P(6 \leq r \leq 10)$, where *r* is a binomial variable, is approximated by $P(5.5 \leq x \leq 10.5)$, where *x* is the corresponding normal variable (see Figure 7-41). Adding or subtracting 0.5 to or from *r* to obtain a corresponding range of normal values *x* is called the *continuity correction*. It is needed because we are approximating a discrete probability distribution by a continuous probability distribution.

◊ **COMMENT** The binomial distribution is for *discrete* random variables. However, the normal distribution is for a *continuous* random variable. In this case, adding or subtracting 0.5 to or from (as appropriate) *r* will improve the approximation of the normal to the binomial distribution, especially when *np* or *nq* is small.   ◊

GUIDED EXERCISE 12

Continuity correction

From many years of observation, a biologist knows the probability is only 0.65 that any given Arctic tern will survive the migration from its summer nesting area to its winter feeding grounds. A random sample of 500 Arctic terns were banded at their summer nesting area. Use the normal approximation to the binomial and the following steps to find the probability that between 310 and 340 of the banded Arctic terns will survive the migration. Let r be the number of surviving terns.

Arctic tern

(a) To approximate $P(310 \leq r \leq 340)$, we use the normal curve with $\mu =$ _____ and $\sigma =$ _____.

⟹ We use the normal curve with

$$\mu = np = 500(0.65) = 325$$
$$\sigma = \sqrt{npq} = \sqrt{500(0.65)(0.35)} \approx 10.67$$

(b) $P(310 \leq r \leq 340)$ is approximately equal to $P(\underline{\hspace{1cm}} \leq x \leq \underline{\hspace{1cm}})$, where x is a variable from the normal distribution described in part (a).

⟹ Since 310 is the left endpoint, we subtract 0.5, and since 340 is the right endpoint, we add 0.5. Consequently,

$$P(310 \leq r \leq 340) \approx P(309.5 \leq x \leq 340.5)$$

(c) Convert the condition $309.5 \leq x \leq 340.5$ to a condition in standard units.

⟹ Since $\mu = 325$ and $\sigma = 10.67$, the condition $309.5 \leq x \leq 340.5$ becomes

$$\frac{309.5 - 325}{10.67} \leq z \leq \frac{340.5 - 325}{10.67}$$

or

$$-1.45 \leq z \leq 1.45$$

(d) $P(310 \leq r \leq 340) = P(309.5 \leq x \leq 340.5)$
$$= P(-1.45 \leq z \leq 1.45)$$
$$= \underline{\hspace{1cm}}$$

⟹ $P(-1.45 \leq z \leq 1.45) = P(z \leq 1.45) - P(z \leq -1.45)$
$$= 0.9265 - 0.0735$$
$$= 0.8530$$

(e) Will the normal distribution make a good approximation to the binomial for this problem? Explain your answer.

⟹ Since

$$np = 500(0.65) = 325$$

and

$$nq = 500(0.35) = 175$$

are both greater than 5, the normal distribution will be a good approximation to the binomial.

VIEWPOINT   *Sunspots, Tree Rings, and Statistics*

Ancient Chinese astronomers recorded extreme sunspot activity with a peak around 1200 A.D. Mesa Verde tree rings in the period between 1276 and 1299 were unusually narrow, indicating a drought and/or a severe cold spell in the region at that time. A cooling trend could have narrowed the window of frost-free days below the approximately 80 days needed for cultivation of aboriginal corn and beans. Is this the reason the ancient Anasazi dwellings in Mesa Verde were abandoned? Is there a connection with the extreme sunspot activity? Much research and statistical work continues to be done on this topic.

Source: Prehistoric Astronomy in the Southwest, by J. McKim Malville and C. Putnam, Department of Astronomy, University of Colorado.

SECTION 7.4   PROBLEMS

Note: When we say *between a* and *b*, we mean every value from *a* to *b*, *including a and b*. Due to rounding, your answers might vary slightly from answers given in the text.

1. *Health: Lead Contamination* More than a decade ago high levels of lead in the blood put 88% of children at risk. A concerted effort was made to remove lead from the environment. Now, according to the *Third National Health and Nutrition Examination Survey (NHANES III)* conducted by the Centers for Disease Control, only 9% of children in the United States are at risk of high blood-lead levels.
   (a) In a random sample of 200 children taken more than a decade ago, what is the probability that 50 or more had high blood-lead levels?
   (b) In a random sample of 200 children taken now, what is the probability that 50 or more have high blood-lead levels?

2. *Insurance: Claims* Do you try to *pad* an insurance claim to cover your deductible? About 40% of all U.S. adults will try to pad their insurance claims! (*Source: Are You Normal?*, by Bernice Kanner, St. Martin's Press.) Suppose that you are the director of an insurance adjustment office. Your office has just received 128 insurance claims to be processed in the next few days. What is the probability that
   (a) half or more of the claims have been padded?
   (b) fewer than 45 of the claims have been padded?
   (c) from 40 to 64 of the claims have been padded?
   (d) more than 80 of the claims are *not* padded?

3. *Law Enforcement: Arrests* In Colorado Springs, a local newspaper ran a full page of nearly 100 mug shots of people the police wanted to arrest for serious crimes. Within 1 week, the police received enough information to locate and arrest about 17% of these "wanted" people (reported in *Rocky Mountain News*). If next month the newspaper runs a full page of 125 mug shots of fugitives, what is the probability that the police will receive enough information to locate and arrest (within 1 week)
   (a) at least 15 fugitives?
   (b) 28 or more fugitives?

(c) between 15 and 28 fugitives?

(d) In the solution to this problem, what was n? p? q? Does it appear that both np and nq are larger than 5? Why is this an important consideration?

4. *Novels: Romance* USA Today reported that 11% of all books sold are of the romance genre. If a local bookstore sells 316 books on a given day, what is the probability that

(a) fewer than 40 are romances?

(b) at least 25 are romances?

(c) between 25 and 40 are romances?

(d) In the solution to this problem, what was n? p? q? Does it appear that both np and nq are larger than 5? Why is this an important consideration?

5. *Longevity: 90th Birthday* It is estimated that 3.5% of the general population will live past their 90th birthday (*Statistical Abstract of the United States*, 112th Edition). In a graduating class of 753 high school seniors, what is the probability that

(a) 15 or more will live beyond their 90th birthday?

(b) 30 or more will live beyond their 90th birthday?

(c) between 25 and 35 will live beyond their 90th birthday?

(d) more than 40 will live beyond their 90th birthday?

6. *Fishing: Billfish* Ocean fishing for billfish is very popular in the Cozumel Region of Mexico. In *World Record Game Fishes* (published by the International Game Fish Association), it was stated that in the Cozumel Region about 44% of strikes (while trolling) resulted in a catch. Suppose that on a given day a fleet of fishing boats got a total of 24 strikes. What is the probability that the number of fish caught was

(a) 12 or fewer?

(b) 5 or more?

(c) between 5 and 12?

(d) In the solution to this problem, what was n? p? q? Does it appear that both np and nq are larger than 5? Why is this an important consideration?

7. *Grocery Stores: New Products* The *Denver Post* stated that 80% of all new products introduced in grocery stores fail (are taken off the market) within 2 years. If a grocery store chain introduces 66 new products, what is the probability that within 2 years

(a) 47 or more fail?

(b) 58 or fewer fail?

(c) 15 or more succeed?

(d) fewer than 10 succeed?

8. *Crime: Murder* What are the chances that a person who is murdered actually knew the murderer? The answer to this question explains why a lot of police detective work begins with relatives and friends of the victim! About 64% of the people who are murdered actually knew the person who did the murder (*Chances: Risk and Odds in Everyday Life*, by James Burke). Suppose that a detective file in New Orleans has 63 current unsolved murders. What is the probability that

(a) at least 35 of the victims knew their murderer?

(b) at most 48 of the victims knew their murderer?

(c) fewer than 30 victims did *not* know their murderer?

(d) more than 20 victims did *not* know their murderer?

9. *Private Investigation: Finding People* Old Friends Information Service is a California company that finds addresses for people who have lost track of each other. Old Friends claims to be 70% successful in reuniting people (*Wall Street Journal*). In December, Old Friends had 430 requests for addresses of lost acquaintances. What is the probability that the number of addresses found was

(a) more than 280?

(b) at least 320?

(c) between 280 and 320?

(d) In the solution to this problem, what was n? p? q? Does it appear that both np and nq are larger than 5? Why is this an important consideration?

10. *Archaeology: Pottery* Santa Fe black on white is a style of pottery that occurs in about 61% of the pot shards found in the Bandelier National Monument area (*Bandelier Archaeological Excavation Project: Summer 1990 Excavations at Burnt Mesa Pueblo,* edited by Kohler, Washington State University). At one excavation site 8641 pot shards have been found that have not yet been cleaned and identified. What is the probability that

(a) fewer than 5200 are Santa Fe black on white?

(b) more than 5400 are Santa Fe black on white?

(c) between 5200 and 5400 are Santa Fe black on white?

(d) In the solution to this problem, what was n? p? q? Does it appear that both np and nq are larger than 5? Why is this an important consideration?

11. *Lawyers: Bar Exam* Over the years, it has been observed that of all the lawyers who take the state bar exam, only 57% pass (information from the National Conference on Bar Examiners, referenced in *The Book of Odds,* by Shook and Shook, Signet). Suppose that this year 850 lawyers are going to take the Ohio bar exam. What is the probability that

(a) 540 or more pass?

(b) 500 or fewer pass?

(c) between 485 and 525 pass?

12. *Coupons: Redemption* More than 200 billion grocery coupons are distributed each year for discounts exceeding $84 billion. However, according to a report in *USA Today,* only 3.2% of the coupons are redeemed. If a company distributes 5000 coupons, what is the probability that

(a) more than 100 are redeemed?

(b) fewer than 200 are redeemed?

(c) between 100 and 200 are redeemed?

SUMMARY

In this chapter, we have examined graphs of normal distributions, control charts, standard units, z scores, and areas under the standard normal curve.

It is important to remember that a normal distribution with mean μ and standard deviation σ can be transformed into the standard normal distribution with mean 0 and standard deviation 1 by the formula

$$z = \frac{x - \mu}{\sigma}$$

Values of z and Table 4 of the Appendix can be used to obtain the area under the standard normal curve to the left of z. Areas under the normal curve represent probabilities that a z value will fall into the interval over which the area lies.

Given a probability associated with an interval on the standard normal curve, we can use Table 4 of the Appendix to obtain the associated z values. When x follows a normal distribution, we can find the x values associated with a given probability by (1) finding the corresponding z values, and (2) converting to x values by using the formula

$$x = z\sigma + \mu$$

In the last section, we saw that we can use the normal distribution to approximate the binomial distribution if $np > 5$ and $nq > 5$, where n is the number of trials, p is the probability of success on a single trial, and $q = 1 - p$.

IMPORTANT WORDS & SYMBOLS

Section 7.1
Normal distributions
Normal curves
Upward cup and downward cup on normal curves
Symmetry of normal curves
Empirical rule
Control chart
Out-of-control signals

Section 7.2
z value or z score

Standard units
Standard normal distribution ($\mu = 0$ and $\sigma = 1$)
Raw score, x
Area under the standard normal curve

Section 7.3
Areas under any normal curve

Section 7.4
Normal approximation to the binomial distribution
Continuity correction

VIEWPOINT   *Nenana Ice Classic*

The Nenana Ice Classic is a betting pool offering a large cash prize to the lucky winner who can guess the time, to the nearest minute, of the ice breakup on the Tanana River at the town of Nenana, Alaska. Official breakup time is defined when the surging river dislodges a tripod on the ice. This breaks an attached line and stops a clock set to Yukon Standard Time. The event is so popular that the first state legislature of Alaska (1959) made the Nenana Ice Classic an official statewide lottery. Since 1918, the earliest breakup was April 20, 1940, at 3:27 P.M., and the latest recorded breakup was May 20, 1964, at 11:41 A.M. Want to make a statistical guess predicting when the ice will break up? Breakup times from 1918 to 1996 are recorded in *The Alaska Almanac*, published by Alaska Northwest Books, Anchorage.

CHAPTER REVIEW PROBLEMS

1. Given that z is the standard normal variable (with mean 0 and standard deviation 1), find
   (a) $P(0 \leq z \leq 1.75)$ (b) $P(-1.29 \leq z \leq 0)$
   (c) $P(1.03 \leq z \leq 1.21)$ (d) $P(z \geq 2.31)$
   (e) $P(z \leq -1.96)$ (f) $P(z \leq 1.00)$

2. Given that z is the standard normal variable (with mean 0 and standard deviation 1), find

(a) $P(0 \le z \le 0.75)$ (b) $P(-1.50 \le z \le 0)$
(c) $P(-2.67 \le z \le -1.74)$ (d) $P(z \ge 1.56)$
(e) $P(z \le -0.97)$ (f) $P(z \le 2.01)$

3. Given that x is a normal variable with mean $\mu = 47$ and standard deviation $\sigma = 6.2$, find
   (a) $P(x \le 60)$ (b) $P(x \ge 50)$ (c) $P(50 \le x \le 60)$

4. Given that x is a normal variable with mean $\mu = 110$ and standard deviation $\sigma = 12$, find
   (a) $P(x \le 120)$ (b) $P(x \ge 80)$ (c) $P(108 \le x \le 117)$

5. Find z so that 5% of the area under the standard normal curve lies to the right of z.

6. Find z so that 1% of the area under the standard normal curve lies to the left of z.

7. Find z so that 95% of the area under the standard normal curve lies between $-z$ and z.

8. Find z so that 99% of the area under the standard normal curve lies between $-z$ and z.

9. *Nursing: Exams* On a practical nursing licensing exam, the mean score is 79 and the standard deviation is 9 points.
   (a) What is the standardized score of a student with a raw score of 87?
   (b) What is the standardized score of a student with a raw score of 79?
   (c) Assuming the scores follow a normal distribution, what is the probability that a score selected at random is above 85?

10. *Aptitude Tests: Mechanical* On an auto mechanic aptitude test, the mean score is 270 points and the standard deviation is 35 points.
    (a) If a student has a standardized score of 1.9, how many points is that?
    (b) If a student has a standardized score of -0.25, how many points is that?
    (c) Assuming the scores follow a normal distribution, what is the probability that a student selected at random will get between 200 and 340 points?

11. *Recycling: Aluminum Cans* One environmental group did a study of recycling habits in a California community. It found that 70% of the aluminum cans sold in the area were recycled.
    (a) If 400 cans are sold today, what is the probability that 300 or more will be recycled?
    (b) Of the 400 cans sold, what is the probability that between 260 and 300 will be recycled?

12. *Guarantee: Disc Players* Future Electronics makes compact disc players. Their research department found that the life of the laser beam device is normally distributed with mean 5000 hours and standard deviation 450 hours.
    (a) Find the probability that the laser beam device will wear out in 5000 hours or sooner.
    (b) Future Electronics wants to place a guarantee on the players so that no more than 5% fail during the guarantee period. Because the laser pickup is the part most likely to wear out first, the guarantee period will be based on the life of the laser beam device. How many playing hours should the guarantee cover? (Round to the next playing hour.)

13. *Guarantee: Package Delivery* Express Courier Service has found that the delivery time for packages is normally distributed with mean 14 hours and standard deviation 2 hours.

(a) For a package selected at random, what is the probability that it will be delivered in 18 hours or less?

(b) What should be the guaranteed delivery time on all packages in order to be 95% sure that the package will be delivered before this time? (*Hint:* Note that 5% of the packages will be delivered at a time beyond the guarantee time period.)

14. *Control Chart: Landing Gear* Hydraulic pressure in the main cylinder of the landing gear of a commercial jet is very important for a safe landing. If the pressure is not high enough, the landing gear may not lower properly. If it is too high, the connectors in the hydraulic line may spring a leak.

   In-flight landing tests show the actual pressure in the main cylinders is a variable with mean 819 pounds per square inch and standard deviation 23 pounds per square inch. Assume that these values for the mean and standard deviation are considered safe values by engineers.

   (a) For nine consecutive test landings, the pressure in the main cylinder is recorded as follows:

Landing number	1	2	3	4	5	6	7	8	9
Pressure	870	855	830	815	847	836	825	810	792

   Make a control chart for the pressure in the main cylinder of the hydraulic landing gear, and plot the data on the control chart. Looking at the control chart, would you say the pressure is "in control" or "out of control"? Explain your answer. Identify any out-of-control signals by type (I, II, or III).

   (b) For 10 consecutive test landings, the pressure was recorded on another plane as follows:

Landing number	1	2	3	4	5	6	7	8	9	10
Pressure	865	850	841	820	815	789	801	765	730	725

   Make a control chart and plot the data on the chart. Would you say the pressure is "in control" or not? Explain your answer. Identify any out-of-control signals by type (I, II, or III).

15. *Electronic Scanners: Errors* Instead of hearing the jingle of prices being rung up manually on cash registers, we now hear the beep of prices being scanned by electronic scanners. How accurate is price scanning? There are errors, and according to a *Denver Post* article, when the error occurs in the store's favor, it is larger than when it occurs in the customer's favor. An investigation of large discount stores by the Colorado state inspectors showed that the average error in the store's favor was $2.66. Assume that the distribution of scanner errors is more or less mound-shaped. If the standard deviation of scanner errors (in the store's favor) is $0.85, use the empirical rule to

   (a) estimate a range of scanner errors centered about the mean in which 68% of the errors will lie.

   (b) estimate a range of scanner errors centered about the mean in which 95% of the errors will lie.

   (c) estimate a range of scanner errors centered about the mean in which almost all the errors will lie.

16. *Customer Complaints: Time* The Customer Service Center in a large New York department store has determined that the amount of time spent with a customer about a complaint is normally distributed with a mean of 9.3 minutes and a standard

deviation of 2.5 minutes. What is the probability that for a randomly chosen customer with a complaint the amount of time spent resolving the complaint will be
(a) shorter than 10 minutes?
(b) more than 5 minutes?
(c) between 8 and 15 minutes?

17. *Medical: Flight for Life* The Flight for Life emergency helicopter service is available for medical emergencies occurring from 15 to 90 miles from the hospital. Emergencies that occur closer to the hospital can be handled effectively by ambulance service. A long-term study of the service shows that the response time from receipt of the dispatch call to arrival at the scene of the emergency is normally distributed with a mean of 42 minutes and a standard deviation of 8 minutes. For a randomly received call, what is the probability that the response time will be
(a) between 30 and 45 minutes?
(b) shorter than 30 minutes?
(c) more than 60 minutes?

18. *Unlisted Phones: Sacramento* How easy is it to get into contact with a person in Sacramento, California? If you don't know the telephone number, it could be difficult. The data from Survey Sampling of Fairfield, Connecticut, reported in *American Demographics,* show that 68% of the telephone-owning households in Sacramento have unlisted numbers. For a random sample of 150 Sacramento households with telephones, what is the probability that
(a) 100 or more have unlisted numbers?
(b) fewer than 100 have unlisted numbers?
(c) between 50 and 65 (including 50 and 65) households have *listed* numbers?

19. *Medical: Blood Type* Blood type AB is found in only 3% of the population (*Textbook of Medical Physiology,* by A. Guyton, M.D.). If 250 people are chosen at random, what is the probability that
(a) 5 or more will have this blood type?
(b) between 5 and 10 will have this blood type?

DATA HIGHLIGHTS: GROUP PROJECTS

Break into small groups and discuss the following topics. Organize a brief outline in which you summarize the main points of your group discussion.

1. Examine Figure 7-42 on the next page. Government documents and the Census Bureau show that the age at first marriage for U.S. citizens is approximately normally distributed for both men and women. For men, the average age is about 27 years, and for women, the average is about 24 years. For both sexes, the standard deviation is about 2.5 years.
   (a) If the distribution is symmetrical and mound-shaped (such as the normal distribution), why would you expect the median, mean, and mode to be equal?
   (b) Consider the age at first marriage in 1995. What is the probability that a man selected at random was over age 30 at the time of his first marriage? What is the probability that he was under age 20? What is the probability that he was between 20 and 30?
   (c) Consider the age at first marriage in 1995. What is the probability that a woman selected at random was over age 28 at the time of her first marriage? What is the probability that she was under age 18? What is the probability that she was between 18 and 28?

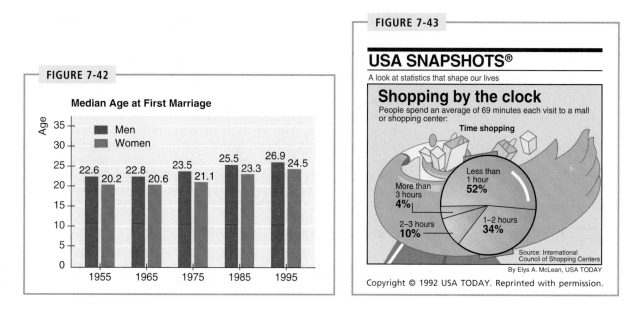

FIGURE 7-42

Median Age at First Marriage

FIGURE 7-43

USA SNAPSHOTS®

A look at statistics that shape our lives

Shopping by the clock

People spend an average of 69 minutes each visit to a mall or shopping center:

Time shopping

Less than 1 hour **52%**

More than 3 hours **4%**

2–3 hours **10%**

1–2 hours **34%**

Source: International Council of Shopping Centers

By Elys A. McLean, USA TODAY

Copyright © 1992 USA TODAY. Reprinted with permission.

(d) At what age were only 10% of eligible men (who had never been married before) left? At what age were only 5% left?

(e) At what age were only 10% of eligible women (who had never been married before) left? At what age were only 5% left?

(f) The Census Bureau tracks data about marriages and age at first marriage. The *Statistical Abstract of the United States* is published each year and contains tables giving this information. For more information, visit the Brase/Brase *Understanding Basic Statistics* site at http://math.college.hmco.com/students and find the link to the Census Bureau. Look for marriage under the subject index, and follow the links to the table. Using either the Abstract or the Census Bureau Web site, find the most recent information about median age at first marriage. Although the median age changes from year to year, the standard deviation usually does not change much. Under the assumption that the age at first marriage follows a distribution that is approximately normal, the mean age at first marriage equals the median age. Using standard deviation of about 2.5 years, repeat parts (b) through (e) for the most recent year for which median age at first marriage is available.

2. Examine Figure 7-43, "Shopping by the clock" (*USA Today*).

(a) Notice that 52% of the people who go to a shopping center spend less than 1 hour. However, the figure also states that people spend an average of 69 minutes on each visit to a shopping center. How could *both* these claims be correct? Write a brief, complete essay in which you discuss mean, median, and mode in the context of symmetrical distribution. Also discuss mean, median, and mode for distributions that are skewed left, skewed right, or general distributions. Then answer the question: How could 52% of the people spend less than 1 hour in the shopping center while the average (we don't know which average was used) time spent was 69 minutes?

(b) Ala Moana Shopping Center in Honolulu is sometimes advertised as the largest shopping center for 2500 miles and the best shopping center in the middle of the Pacific Ocean. There are many interesting Hawaiian, Asian, and other ethnic shops in Ala Moana, so this center is a favorite of tourists. Suppose that a tour group of 75 people has just arrived at the shopping center. The International Council of Shopping Centers indicates that 86% (52% plus 34%) of the people will spend less than 2 hours in a shopping center. Let us assume this applies to our tour group.

 (i) For the tour group of 75 people, what is the expected number who finish shopping on or before 2 hours? What is the standard deviation?

 (ii) For the tour group of 75 people, what is the probability that 55 or more will finish shopping in 2 hours or less?

 (iii) For the tour group, what is the probability that 70 or more will finish shopping in 2 hours or less?

 (iv) What is the probability that between 50 and 70 (including 50 and 70) people in the tour group will finish shopping in 2 hours or less?

 (v) If you are a tour director, how could you use this information to plan an appropriate length of time for the stop at Ala Moana shopping center? You do not want to frustrate your group by allowing too little time for shopping, but you also do not want to spend too much time at the center.

LINKING CONCEPTS: WRITING PROJECTS

Discuss each of the following topics in class or review the topics on your own. Then write a brief but complete essay in which you summarize the main points. Please include formulas and graphs as appropriate.

1. If you look up the word *normal* in a dictionary, you will find it is synonymous with the words *standard* or *usual*. Consider the very wide and general applications of the normal probability distribution. Comment on why good synonyms for *normal probability distribution* might be the *standard probability distribution* or the *usual probability distribution*. List at least three random variables from everyday life for which you think the normal probability distribution could be applicable.

2. Why are standard z values so important? Is it true that z values have no units of measurement? Why would this be desirable for comparing data sets with *different* units of measurement? How can we assess differences in quality or performance by simply comparing z values under a standard normal curve? Examine the formula to compute standard z values. Notice it involves *both* the mean and standard deviation. Recall that in Chapter 3 we commented that the mean of a data collection was not entirely adequate to describe the data; you need the standard deviation as well. Discuss this topic again in the light of what you now know about normal distributions and standard z values.

3. If you look up the word *empirical* in a dictionary, you find it means relying on experiment and observation rather than on theory. Discuss the empirical rule in this context. The empirical rule certainly applies to the normal distribution, but does it also apply to a wide variety of other distributions that are not *exactly* (theoretically) normal? Discuss the terms *mound-shaped* and *symmetrical*. Draw several sketches of distributions that are mound-shaped *and* symmetrical. Draw sketches of distributions that are not mound-shaped or not symmetrical. To which distributions will the empirical rule apply?

4. Most companies that manufacture a product have a division of quality control or quality assurance. The purpose of the quality-control division is to make reasonably certain that the products manufactured are up to company standards. Write a brief essay in which you describe how the statistics you have learned so far could be applied to an industrial application (such as control charts and the Antlers Lodge example).

8
Introduction to Sampling Distributions

No one wants to learn by mistakes, but we cannot learn enough from success to go beyond the state of the art. . . . Such is the nature not only of science and engineering, but of all human endeavors.

—Henry Petroski

Experience isn't what happens to you. It's what you make out of what happens to you.

—Aldous Huxley

**Aldous Huxley
(1894–1963)**

This British novelist and critic wrote about the theme of the human being confronted by the modern world.

8.1 Sampling Distributions

8.2 The Central Limit Theorem

8.3 Class Project Illustrating the Central Limit Theorem

For on-line student resources, visit **math.college.hmco.com/students** and follow the Statistics links to the Brase/Brase, *Understanding Basic Statistics,* 3rd edition Web site.

Henry Petroski is a professor of engineering at Duke University, and Aldous Huxley was a well-known modern writer. Both quotes imply that experience, mistakes, information, and life itself are closely related. Life and uncertainty appear to be inseparable. Only those who are no longer living can escape chance happenings. Mistakes are bound to occur. However, not all mistakes are bad. The discovery of penicillin was a "mistake" when mold (penicillin) was accidentally introduced into a bacterial culture by Alexander Fleming in 1928.

Most of the really important decisions in life will involve incomplete information. In one lifetime, we simply cannot experience *everything*. Nor should we even want to! This is one reason why *experience by way of sampling* is so important. Statistics can help you have the experiences and yet maintain some control over mistakes. Remember, it is what you make out of experience (sample data) that is of real value.

In this chapter, we will study how information from samples relates to information about populations. We cannot be certain that the information from a sample reflects corresponding information about the entire population, but we can describe likely differences. Study this chapter and the following material carefully. We believe that your effort will be rewarded by helping you appreciate the joy and wonder of living in an uncertain universe.

PREVIEW QUESTIONS

◇ As humans, our experiences are finite and limited. Consequently, most of the important decisions in our lives are based on sample (incomplete) information. What do we mean by a probability sampling distribution? How will this help us? (SECTION 8.1)

◇ There is an old saying: All roads lead to Rome. In statistics, we could recast this saying: All probability distributions average out to be normal distributions (as the sample size increases). How can we take advantage of this in our study of sampling distributions? (SECTION 8.2)

◇ How can random digits 0 through 9 help us visualize the central limit theorem in action? (SECTION 8.3)

FOCUS PROBLEM

Impulse Buying

The Food Marketing Institute, Progressive Grocer, New Products News, and Point of Purchaser Advertising Institute are organizations that analyze supermarket sales. One of the interesting discoveries was that the average amount of impulse buying in a grocery store was very time-dependent. As reported in the *Denver Post*, "when you dilly dally in a store for 10 unplanned minutes, you can kiss nearly $20 goodbye." For this reason, it is in the best interest of the supermarket to keep you in the store longer. In the *Post* article, it was pointed out that long checkout lines (near end-aisle displays), "samplefest" events of tasting free samples, video kiosks, magazine and book sections, and so on help keep customers in the store longer. On average, a single customer who strays from his or her grocery list can plan on impulse spending of $20 for every 10 minutes spent wandering about in the supermarket.

Let x represent the dollar amount spent on supermarket impulse buying in a 10-minute (unplanned) shopping interval. Based on the *Post* article, the mean of the x distribution is about $20 and the (estimated) standard deviation is about $7.

315

(a) Consider a random sample of $n = 100$ customers, each of whom has 10 minutes of unplanned shopping time in a supermarket. From the central limit theorem, what can you say about the probability distribution of $\bar{x}$, the *average* amount spent by these customers due to impulse buying? Is the $\bar{x}$ distribution approximately normal? What are the mean and standard deviation of the $\bar{x}$ distribution? Is it necessary to make any assumption about the x distribution? Explain.

(b) What is the probability that $\bar{x}$ is between $18 and $22?

(c) Let us assume that x has a distribution that is approximately normal. What is the probability that x is between $18 and $22?

(d) In part (b), we used $\bar{x}$, the *average* amount spent, computed for 100 customers. In part (c), we used x, the amount spent by only *one* individual customer. The answers for parts (b) and (c) are very different. Why would this happen? In this example, $\bar{x}$ is a much more predictable or reliable statistic than x. Consider that almost all marketing strategies and sales pitches are designed for the *average* customer and *not* the *individual* customer. How does the central limit theorem tell us that the average customer is much more predictable than the individual customer?

8.1
Sampling Distributions

FOCUS POINTS

✓ Review such commonly used terms as random sample, relative frequency, parameter, statistic, and sampling distribution.
✓ From raw data, construct a relative frequency distribution for $\bar{x}$ values and compare the result to a theoretical sampling distribution.

Parameter

Statistic

Let us begin with some common statistical terms. Most of these have been discussed before, but this is a good time to review them.

From a statistical point of view, a *population* can be thought of as a set of measurements (or counts), either existing or conceptual. We discussed populations at some length in Chapter 1. A *sample* is a subset of measurements from the population. For our purposes, the most important samples are *random samples*, which were discussed in Section 1.2.

A *parameter* is a numerical descriptive measure of a *population*. Examples of *population parameters* are the population mean μ, population variance σ^2, population standard deviation σ, and the population proportion of successes p in a binomial distribution.

A *statistic* is a numerical descriptive measure of a *sample* (usually a random sample). Examples of statistics are the sample mean $\bar{x}$, sample variance s^2, sample standard deviation s, and the sample estimate $\hat{p} = r/n$ (read $\hat{p}$ as "p hat") for the proportion of successes r in a binomial distribution with n trials. Notice that for a given population a specified parameter is a *fixed* quantity, while the statistic might vary depending on which sample has been selected.

Some commonly used statistics and corresponding parameters

Measure	Statistic	Parameter
Mean	$\bar{x}$ (x bar)	μ (mu)
Variance	s^2	σ^2 (sigma squared)
Standard deviation	s	σ (sigma)
Proportion	$\hat{p}$ (p hat)	p

Often we do not have access to all the measurements of an entire population because of time, money, effort constraints, etc. So we must use measurements from a sample instead. In such cases, we will use a statistic (such as $\bar{x}$, s, or $\hat{p}$) to make *inferences* about corresponding population parameters (e.g., μ, σ, or p). The principal types of inferences we will make are

1. To *estimate* the value of a population parameter.

2. To formulate a *decision* about the value of a population parameter.

To evaluate the reliability of our inferences, we will need to know the probability distribution for the statistic we are using. Such a probability distribution is called a *sampling distribution*. Perhaps an example will help clarify this discussion.

Sampling distribution

EXAMPLE 1

Sampling distribution for $\bar{x}$

Pinedale, Wisconsin, is a rural community with a children's fishing pond. Posted rules say that all fish under 6 inches must be returned to the pond, only children under 12 years old may fish, and a limit of five fish may be kept per day. Susan is a college student who was hired by the community last summer to make sure the rules were obeyed and to see that the children were safe from accidents. The pond contains only rainbow trout and has been well stocked for many years. Each child has no difficulty catching his or her limit of five trout.

As a project for her biometrics class, Susan kept a record of the lengths (to the nearest inch) of all trout caught last summer. Hundreds of children visited the pond and caught their limit of five trout, so Susan has a lot of data. To make Table 8-1 on the following page, Susan selected 100 children at random and listed the length of each of the five trout caught by a child in the sample. Then, for each child, she listed the mean length of the five trout that child caught.

Now let us turn our attention to the following question: What is the average (mean) length of a trout taken from the Pinedale children's pond last summer?

SOLUTION: We can get an idea of the average length by looking at the far-right column of Table 8-1. But just looking at 100 of the $\bar{x}$ values doesn't tell us much. Let's organize our $\bar{x}$ values into a frequency table. We used a class width of 0.38 to make Table 8-2.

Note: Techniques of Section 2.2 dictate a class width of 0.4. However, this choice results in the tenth class being beyond the data. Consequently, we shortened the class width slightly and also started the first class with a value slightly lower than the smallest data value.

The far-right column of Table 8-2 contains relative frequencies $f/100$. Recall that the relative frequencies may be thought of as probabilities, so we effectively have a probability distribution. Because $\bar{x}$ represents the mean length of a trout (based on samples of five trout caught by each child), we estimate the probability of $\bar{x}$ falling into each class by using the relative frequencies. Figure 8-1 is a relative-frequency or probability distribution of the $\bar{x}$ values.

The bars of Figure 8-1 represent our estimated probabilities of $\bar{x}$ values based on the data of Table 8-1. The bell-shaped curve represents the theoretical probability distribution that would be obtained if the number of children (i.e., number of $\bar{x}$ values) were much larger.

Figure 8-1 represents a *probability sampling distribution* for the sample mean $\bar{x}$ of trout lengths based on random samples of size 5. We see that the distribution is mound-shaped and even somewhat bell-shaped. Irregularities are due to the small number of samples used (only 100 sample means) and the rather small sample size

TABLE 8-1 Length Measurements of Trout Caught by a Random Sample of 100 Children at the Pinedale Children's Pond

Sample	Length (to nearest inch)					$\bar{x}$ = Sample Mean	Sample	Length (to nearest inch)					$\bar{x}$ = Sample Mean
1	11	10	10	12	11	10.8	51	9	10	12	10	9	10.0
2	11	11	9	9	9	9.8	52	7	11	10	11	10	9.8
3	12	9	10	11	10	10.4	53	9	11	9	11	12	10.4
4	11	10	13	11	8	10.6	54	12	9	8	10	11	10.0
5	10	10	13	11	12	11.2	55	8	11	10	9	10	9.6
6	12	7	10	9	11	9.8	56	10	10	9	9	13	10.2
7	7	10	13	10	10	10.0	57	9	8	10	10	12	9.8
8	10	9	9	9	10	9.4	58	10	11	9	8	9	9.4
9	10	10	11	12	8	10.2	59	10	8	9	10	12	9.8
10	10	11	10	7	9	9.4	60	11	9	9	11	11	10.2
11	12	11	11	11	13	11.6	61	11	10	11	10	11	10.6
12	10	11	10	12	13	11.2	62	12	10	10	9	11	10.4
13	11	10	10	9	11	10.2	63	10	10	9	11	7	9.4
14	10	10	13	8	11	10.4	64	11	11	12	10	11	11.0
15	9	11	9	10	10	9.8	65	10	10	11	10	9	10.0
16	13	9	11	12	10	11.0	66	8	9	10	11	11	9.8
17	8	9	7	10	11	9.0	67	9	11	11	9	8	9.6
18	12	12	8	12	12	11.2	68	10	9	10	9	11	9.8
19	10	8	9	10	10	9.4	69	9	9	11	11	11	10.2
20	10	11	10	10	10	10.2	70	13	11	11	9	11	11.0
21	11	10	11	9	12	10.6	71	12	10	8	8	9	9.4
22	9	12	9	10	9	9.8	72	13	7	12	9	10	10.2
23	8	11	10	11	10	10.0	73	9	10	9	8	9	9.0
24	9	12	10	9	11	10.2	74	11	11	10	9	10	10.2
25	9	9	8	9	10	9.0	75	9	11	14	9	11	10.8
26	11	11	12	11	11	11.2	76	14	10	11	12	12	11.8
27	10	10	10	11	13	10.8	77	8	12	10	10	9	9.8
28	8	7	9	10	8	8.4	78	8	10	13	9	8	9.6
29	11	11	8	10	11	10.2	79	11	11	11	13	10	11.2
30	8	11	11	9	12	10.2	80	12	10	11	12	9	10.8
31	11	9	12	10	10	10.4	81	10	9	10	10	13	10.4
32	10	11	10	11	12	10.8	82	11	10	9	9	12	10.2
33	12	11	8	8	11	10.0	83	11	11	10	10	10	10.4
34	8	10	10	9	10	9.4	84	11	10	11	9	9	10.0
35	10	10	10	10	11	10.2	85	10	11	10	9	7	9.4
36	10	8	10	11	13	10.4	86	7	11	10	9	11	9.6
37	11	10	11	11	10	10.6	87	10	11	10	10	10	10.2
38	7	13	9	12	11	10.4	88	9	8	11	10	12	10.0
39	11	11	8	11	11	10.4	89	14	9	12	10	9	10.8
40	11	10	11	12	9	10.6	90	9	12	9	10	10	10.0
41	11	10	9	11	12	10.6	91	10	10	8	6	11	9.0
42	11	13	10	12	9	11.0	92	8	9	11	9	10	9.4
43	10	9	11	10	11	10.2	93	8	10	9	9	11	9.4
44	10	9	11	10	9	9.8	94	12	11	12	13	10	11.6
45	12	11	9	11	12	11.0	95	11	11	9	9	9	9.8
46	13	9	11	8	8	9.8	96	8	12	8	11	10	9.8
47	10	11	11	11	10	10.6	97	13	11	11	12	8	11.0
48	9	9	10	11	11	10.0	98	10	11	8	10	11	10.0
49	10	9	9	10	10	9.6	99	13	10	7	11	9	10.0
50	10	10	6	9	10	9.0	100	9	9	10	12	12	10.4

TABLE 8-2   Frequency Table for 100 Values of $\bar{x}$

Class	Class Limits Lower	Class Limits Upper	f = Frequency	$f/100$ = Relative Frequency
1	8.39	8.76	1	0.01
2	8.77	9.14	5	0.05
3	9.15	9.52	10	0.10
4	9.53	9.90	19	0.19
5	9.91	10.28	27	0.27
6	10.29	10.66	18	0.18
7	10.67	11.04	12	0.12
8	11.05	11.42	5	0.05
9	11.43	11.80	3	0.03

(5 trout per child). These irregularities would become less obvious and even disappear if the sample of children became much larger, if we used a larger number of classes in Figure 8-1, and if the number of trout used in each sample became larger. In fact, the curve would eventually become a perfect bell-shaped curve. We will discuss this property at some length in the next section, which introduces the *central limit theorem*.   ◇

There are other sampling distributions besides the $\bar{x}$ distribution. In the chapters ahead, we will see that other statistics have different sampling distributions. However, the $\bar{x}$ sampling distribution is very important. It will serve us well in our inferential work in Chapters 9, 10, and 11 on estimation and testing.

Let us summarize the information about sampling distributions in the following exercise.

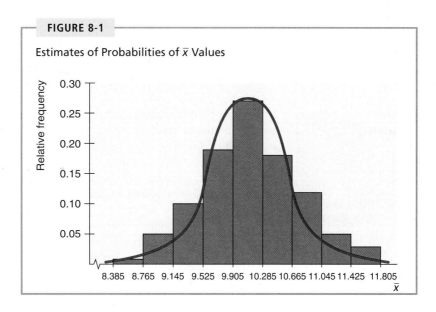

FIGURE 8-1

Estimates of Probabilities of $\bar{x}$ Values

GUIDED EXERCISE 1

Terminology

(a) What is a population parameter? Give an example.

➡ A population parameter is a numerical descriptive measure of a population. Examples are μ, σ, and p. (There are many others.)

(b) What is a sample statistic? Give an example.

➡ A sample statistic or a statistic is a numerical descriptive measure of a sample. Examples are $\bar{x}$, s, and $\hat{p}$.

(c) What is a sampling distribution?

➡ A sampling distribution is a probability distribution for the sample statistic we are using.

(d) In Table 8-1, what makes up the members of the sample? What is the sample statistic corresponding to each sample? What is the sampling distribution? To which population parameter does this sampling distribution correspond?

➡ There are 100 samples, each of which has five trout lengths. The first sample of five trout has lengths 11, 10, 10, 12, and 11. The sample statistic is the sample mean $\bar{x} = 10.8$. The sampling distribution is shown in Figure 8-1. This sampling distribution relates to the population mean μ of all lengths of trout taken from the Pinedale children's pond (i.e., trout over 6 inches long).

(e) Where will sampling distributions be used in our study of statistics?

➡ Sampling distributions will be used for statistical inference. (Chapter 9 will concentrate on the method of inference called *estimation*. Chapter 10 will concentrate on a method of inference called *testing*.)

VIEWP◉INT   *"Chance Favors the Prepared Mind"*
—Louis Pasteur

It also has been said that a discovery is nothing more than an accident that meets a prepared mind. Sampling can be one of the best forms of preparation. In fact, sampling may be the primary way we humans venture into the unknown. Probability sampling distributions can provide new information for the sociologist, scientist, or economist. In addition, ordinary human sampling of life can help writers and artists develop preferences, style, and insight. Ansel Adams became famous for photographing lyrical, unforgettable landscapes such as "Moonrise, Hernandez, New Mexico." Adams claimed that he was a strong believer in the quote by Pasteur. In fact, he claims that the Hernandez photograph was just such a favored chance happening that his prepared mind readily grasped. During his lifetime, Adams made over $25 million from sales and royalties on the Hernandez photograph.

SECTION 8.1   PROBLEMS

This is a good time to review several important concepts, some of which we have studied earlier. Please write out a careful but brief answer to each of the following questions.

1. What is a population? Give three examples. *[handwritten: Equal chance of being selected]*

2. What is a random sample from a population? (*Hint:* See Section 1.2.)

3. What is a population parameter? Give three examples. *[handwritten: Numerical descriptor of a population]*

4. What is a sample statistic? Give three examples. *[handwritten: numerical measure]*

5. What is the meaning of the term *statistical inference*? What types of inferences will we make about population parameters? *[handwritten: estimation & testing]*

6. What is a sampling distribution? *[handwritten: prob. distribution for population]*

7. How do frequency tables, relative frequencies, and histograms using relative frequencies help us understand sampling distributions? *[handwritten: organize data]*

8. How can relative frequencies be used to help us estimate probabilities occurring in sampling distributions?

9. Give an example of a specific sampling distribution we studied in this section. Outline other possible examples of sampling distributions from areas such as business administration, economics, finance, psychology, political science, sociology, biology, medical science, sports, engineering, chemistry, linguistics, and so on.

8.2
The Central Limit Theorem

The $\bar{x}$ Distribution, Given x Is Normal

FOCUS POINTS

✓ For a normal distribution, use μ and σ to construct the theoretical sampling distribution for the statistic $\bar{x}$.

✓ For large samples, use sample estimates to construct a good approximate sampling distribution for the statistic $\bar{x}$.

✓ Learn the statement and underlying meaning of the central limit theorem well enough to explain it to a friend who is intelligent, but (unfortunately) doesn't know much about statistics.

In Section 8.1, we began a study of the distribution of $\bar{x}$ values, where $\bar{x}$ was the (sample) mean length of five trout caught by children at the Pinedale children's fishing pond. Let's consider this example again in the light of a very important theorem of mathematical statistics.

THEOREM 8.1 Let x be a random variable with a *normal distribution* whose mean is μ and standard deviation is σ. Let $\bar{x}$ be the sample mean corresponding to random samples of size n taken from the x distribution. Then the following are true:

(a) The $\bar{x}$ distribution is a *normal distribution*.

(b) The mean of the $\bar{x}$ distribution is μ.

(c) The standard deviation of the $\bar{x}$ distribution is $\sigma/\sqrt{n}$.

We conclude from Theorem 8.1 that when x has a normal distribution, the $\bar{x}$ distribution will be normal *for any sample size n*. Furthermore, we can convert the $\bar{x}$ distribution to the standard normal z distribution using the following formulas.

$$\mu_{\bar{x}} = \mu$$

$$\sigma_{\bar{x}} = \frac{\sigma}{\sqrt{n}}$$

$$z = \frac{\bar{x} - \mu_{\bar{x}}}{\sigma_{\bar{x}}} = \frac{\bar{x} - \mu}{\sigma/\sqrt{n}}$$

where n is the sample size,
   μ is the mean of the x distribution, and
   σ is the standard deviation of the x distribution.

Theorem 8.1 is a wonderful theorem! It says that the $\bar{x}$ distribution will be normal provided the x distribution is normal. The sample size n could be 2, 3, 4, or any (fixed) sample size we wish. Furthermore, the mean of the $\bar{x}$ distribution is μ (same as for the x distribution), but the standard deviation is $\sigma/\sqrt{n}$ (which is, of course, smaller than σ). The next example illustrates Theorem 8.1.

EXAMPLE 2

*Probability regarding x;
regarding x̄*

Suppose that a team of biologists has been studying the Pinedale children's fishing pond. Let x represent the length of a single trout taken at random from the pond. This group of biologists has determined that x has a normal distribution with mean $\mu = 10.2$ inches and standard deviation $\sigma = 1.4$ inches.

(a) What is the probability that a *single trout* taken at random from the pond is between 8 and 12 inches long?

   SOLUTION: We use the methods of Chapter 7 with $\mu = 10.2$ and $\sigma = 1.4$ to get

   $$z = \frac{x - \mu}{\sigma} = \frac{x - 10.2}{1.4}$$

   Therefore,

   $$P(8 < x < 12) = P\left(\frac{8 - 10.2}{1.4} < z < \frac{12 - 10.2}{1.4}\right)$$
   $$= P(-1.57 < z < 1.29)$$
   $$= 0.9015 - 0.0582$$
   $$= 0.8433$$

   Therefore, the probability is about 0.8433 that a *single* trout taken at random is between 8 and 12 inches long.

(b) What is the probability that the *mean length* $\bar{x}$ of five trout taken at random is between 8 and 12 inches?

   SOLUTION: If we let $\mu_{\bar{x}}$ represent the mean of the distribution, then Theorem 8.1 part (b) tells us that

   $$\mu_{\bar{x}} = \mu = 10.2$$

   If $\sigma_{\bar{x}}$ represents the standard deviation of the $\bar{x}$ distribution, then Theorem 8.1 part (c) tells us that

   $$\sigma_{\bar{x}} = \sigma/\sqrt{n} = 1.4/\sqrt{5} \approx 0.63$$

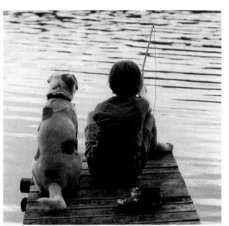

To create a standard z variable from $\bar{x}$, we subtract $\mu_{\bar{x}}$ and divide by $\sigma_{\bar{x}}$:

$$z = \frac{\bar{x} - \mu_{\bar{x}}}{\sigma_{\bar{x}}} = \frac{\bar{x} - \mu}{\sigma/\sqrt{n}} = \frac{\bar{x} - 10.2}{0.63}$$

To standardize the interval $8 < \bar{x} < 12$, we use 8 and then 12 in place of $\bar{x}$ in the preceding formula for z.

$$8 < \bar{x} < 12$$
$$\frac{8 - 10.2}{0.63} < z < \frac{12 - 10.2}{0.63}$$
$$-3.49 < z < 2.86$$

Theorem 8.1 part (a) tells us that $\bar{x}$ has a normal distribution. Therefore,

$$P(8 < \bar{x} < 12) = P(-3.49 < z < 2.86) = 0.9979 - 0.0002 = 0.9977$$

The probability is about 0.9977 that the mean length based on a sample size 5 is between 8 and 12 inches.

(c) Looking at the results of parts (a) and (b), we see that the probabilities (0.8433 and 0.9977) are quite different. Why is this the case?

SOLUTION: According to Theorem 8.1, both x and $\bar{x}$ have a normal distribution, and both have the same mean of 10.2 inches. The difference is in the standard deviation for x and $\bar{x}$. The standard deviation of the x distribution is $\sigma = 1.4$. The standard deviation of the $\bar{x}$ distribution is

$$\sigma_{\bar{x}} = \sigma/\sqrt{n} = 1.4/\sqrt{5} \approx 0.63$$

The standard deviation of $\bar{x}$ is less than half the standard deviation of x. Figure 8-2 shows the distribution of x and $\bar{x}$.

Looking at Figure 8-2(a) and (b), we see that both curves use the same scale on the horizontal axis. The means are the same, and the shaded area is above the interval from 8 to 12 on each graph. It becomes clear that the smaller standard deviation of the $\bar{x}$ distribution has the effect of gathering together much more of the total probability into the region over its mean. Therefore, the region from 8 to 12 has a much higher probability for the $\bar{x}$ distribution. ◇

FIGURE 8-2

General Shapes of the x and $\bar{x}$ Distributions

(a) The x distribution with $\mu = 10.2$ and $\sigma = 1.4$

(b) The $\bar{x}$ distribution with $\mu_{\bar{x}} = 10.2$ and $\sigma_{\bar{x}} = 0.63$ for samples of size $n = 5$

Standard error of the mean

Theorem 8.1 describes the distribution of a particular statistic: namely, the distribution of sample means $\bar{x}$. The standard deviation of a statistic is referred to as the *standard error* of that statistic. For the $\bar{x}$ sampling distribution, the standard error of $\bar{x}$ is $\sigma_{\bar{x}}$ or $\sigma/\sqrt{n}$. In other words, the *standard error of the mean* is $\sigma/\sqrt{n}$. The expression *standard error* appears commonly on software printouts and refers to the standard deviation of the sampling distribution being used. (In Minitab, the expression SE MEAN refers to the standard error of the mean.)

Statistical software

The $\bar{x}$ Distribution, Given x Follows Any Distribution

Theorem 8.1 gives complete information about the $\bar{x}$ distribution, provided the original x distribution is known to be normal. What happens if we don't have information about the shape of the original x distribution? The *central limit theorem* tells us what to expect.

Central limit theorem

◇ **THEOREM 8.2 The Central Limit Theorem** If x possesses *any* distribution with mean μ and standard deviation σ, then the sample mean $\bar{x}$ based on a random sample of size n will have a distribution that approaches the distribution of a normal random variable with mean μ and standard deviation $\sigma/\sqrt{n}$ as n increases without limit.   ◇

The central limit theorem is indeed surprising! It says that x can have *any* distribution whatever, but as the sample size gets larger and larger, the distribution of $\bar{x}$ will approach a *normal* distribution. From this relation, we begin to appreciate the scope and significance of the normal distribution.

In the central limit theorem, the degree to which the distribution of $\bar{x}$ values fits a normal distribution depends on both the selected value of n and the original distribution of x values. A natural question is: How large should the sample size be if we want to apply the central limit theorem? After a great deal of theoretical as well as empirical study, statisticians agree that if n is 30 or larger, the $\bar{x}$ distribution will appear to be normal and the central limit theorem will apply. However, this rule should not be applied blindly. If the x distribution is definitely not symmetrical about its mean, then the $\bar{x}$ distribution also will display a lack of symmetry. In such a case, a sample size larger than 30 may be required to get a reasonable approximation to the normal.

Large sample convention: $n \geq 30$

In practice, it is a good idea, when possible, to make a histogram of sample x values. If the histogram is approximately mound-shaped, and if it is more or less symmetrical, then we may be assured that, for all practical purposes, the $\bar{x}$ distribution will be well approximated by a normal distribution and the central limit theorem will apply when the sample size is 30 or larger. The main thing to remember is that in almost all practical applications, a sample size of 30 or more is adequate for the central limit theorem to hold. However, in a few rare applications, you may need a sample size larger than 30 to get reliable results.

Let's summarize this information for convenient reference:

For almost all x distributions, if we use a random sample of size 30 or larger, the $\bar{x}$ distribution will be approximately normal, and the larger the sample size becomes, the closer the $\bar{x}$ distribution gets to the normal. Furthermore, we may convert the $\bar{x}$ distribution to a standard normal distribution using the following formulas.

$$\mu_{\bar{x}} = \mu$$

$$\sigma_{\bar{x}} = \frac{\sigma}{\sqrt{n}}$$

$$z = \frac{\bar{x} - \mu_{\bar{x}}}{\sigma_{\bar{x}}} = \frac{\bar{x} - \mu}{\sigma/\sqrt{n}}$$

where n is the sample size ($n \geq 30$),
μ is the mean of the x distribution, and
σ is the standard deviation of the x distribution.

GUIDED EXERCISE 2

Central limit theorem

(a) Suppose x has a *normal* distribution with mean $\mu = 18$ and standard deviation $\sigma = 3$. If we draw random samples of size 5 from the x distribution and $\bar{x}$ represents the sample mean, what can you say about the $\bar{x}$ distribution? How could you standardize the $\bar{x}$ distribution?

⟹ Since the x distribution is given to be *normal*, the $\bar{x}$ distribution also will be normal even though the sample size is much less than 30. The mean is $\mu_{\bar{x}} = \mu = 18$. The standard deviation is

$$\sigma_{\bar{x}} = \sigma/\sqrt{n} = 3/\sqrt{5} \approx 1.3$$

We could standardize $\bar{x}$ as follows:

$$z = \frac{\bar{x} - \mu}{\sigma/\sqrt{n}} = \frac{\bar{x} - 18}{1.3}$$

(b) Suppose we know that the x distribution has mean $\mu = 75$ and standard deviation $\sigma = 12$, but we have no information as to whether or not the x distribution is normal. If we draw samples of size 30 from the x distribution and $\bar{x}$ represents the sample mean, what can you say about the $\bar{x}$ distribution? How could you standardize the $\bar{x}$ distribution?

⟹ Since the sample size is large enough, the $\bar{x}$ distribution will be approximately a normal distribution. The mean of the $\bar{x}$ distribution is

$$\mu_{\bar{x}} = \mu = 75$$

The standard deviation of the distribution is

$$\sigma_{\bar{x}} = \sigma/\sqrt{n} = 12/\sqrt{30} \approx 2.2$$

We could standardize $\bar{x}$ as follows:

$$z = \frac{\bar{x} - \mu}{\sigma/\sqrt{n}} = \frac{\bar{x} - 75}{2.2}$$

(c) Suppose you did not know that x had a normal distribution. Would you be justified in saying that the $\bar{x}$ distribution is approximately normal if the sample size was $n = 8$?

⟹ No, the sample size should be 30 or larger if we don't know that x has a normal distribution.

Now let's look at an example that demonstrates the use of the central limit theorem in a decision-making process.

EXAMPLE 3

Central limit theorem

A certain strain of bacteria occurs in all raw milk. Let x be the bacteria count per milliliter of milk. The health department has found that if the milk is not contaminated, then x has a distribution that is more or less mound-shaped and symmetrical. The mean of the x distribution is $\mu = 2500$, and the standard deviation is $\sigma = 300$. In a large commercial dairy, the health inspector takes 42 random samples of the milk produced each day. At the end of the day, the bacteria count in each of the 42 samples is averaged to obtain the sample mean bacteria count $\bar{x}$.

(a) Assuming that the milk is not contaminated, what is the distribution of $\bar{x}$?

> **SOLUTION:** The sample size is $n = 42$. Since this value exceeds 30, the central limit theorem applies, and we know that $\bar{x}$ will be approximately normal with mean and standard deviation
>
> $$\mu_{\bar{x}} = \mu = 2500$$
> $$\sigma_{\bar{x}} = \sigma/\sqrt{n} = 300/\sqrt{42} \approx 46.3$$

(b) Assuming that the milk is not contaminated, what is the probability that the average bacteria count $\bar{x}$ for one day is between 2350 and 2650 bacteria per milliliter?

> **SOLUTION:** We convert the interval
>
> $$2350 \le \bar{x} \le 2650$$
>
> to a corresponding interval on the standard z *axis*.
>
> $$z = \frac{\bar{x} - \mu}{\sigma/\sqrt{n}} = \frac{\bar{x} - 2500}{46.3}$$

> $\bar{x} = 2350$ converts to $z = \dfrac{2350 - 2500}{46.3} = -3.24$
>
> $\bar{x} = 2650$ converts to $z = \dfrac{2650 - 2500}{46.3} = 3.24$
>
> Therefore,
>
> $$P(2350 \le \bar{x} \le 2650) = P(-3.24 \le z \le 3.24)$$
> $$= 0.9994 - 0.0006$$
> $$= 0.9988$$
>
> The probability is 0.9988 that $\bar{x}$ is between 2350 and 2650.

(c) At the end of each day, the inspector must decide to accept or reject the accumulated milk that has been held in cold storage awaiting shipment. Suppose that the 42 samples taken by the inspector have a mean bacteria count $\bar{x}$ that is *not* between 2350 and 2650. If you were the inspector, what would be your comment on this situation?

> **SOLUTION:** The probability that $\bar{x}$ is between 2350 and 2650 is very high. If the inspector finds that the average bacteria count for the 42 samples is not between 2350 and 2650, then it is reasonable to conclude that there is something wrong with the milk. If $\bar{x}$ is less than 2350, you might suspect someone added chemicals to the milk to artificially reduce the bacteria count. If $\bar{x}$ is above 2650, you might suspect some other kind of biologic contamination.   ◇

GUIDED EXERCISE 3

Probability regarding x̄

In mountain country, major highways sometimes use tunnels instead of long, winding roads over high passes. However, too many vehicles in a tunnel at the same time can cause a hazardous situation. Traffic engineers are studying a long tunnel in Colorado. If x represents the time for a vehicle to go through the tunnel, it is known that the x distribution has mean $\mu = 12.1$ minutes and standard deviation $\sigma = 3.8$ minutes under ordinary traffic conditions. From a histogram of x values, it was found that the x distribution is mound-shaped with some symmetry about the mean.

Engineers have calculated that, *on average*, vehicles should spend from 11 to 13 minutes in the tunnel. If the time is less than 11 minutes, traffic is moving too fast for safe travel in the tunnel. If the time is more than 13 minutes, there is a problem of bad air (too much carbon monoxide and other pollutants).

Under ordinary conditions, there are about 50 vehicles in the tunnel at one time. What is the probability that the mean time for 50 vehicles in the tunnel will be from 11 to 13 minutes?

We will answer this question in steps.

(a) Let $\bar{x}$ represent the sample mean based on samples of size 50. Describe the $\bar{x}$ distribution.

From the central limit theorem we expect the $\bar{x}$ distribution to be approximately normal with mean and standard deviation

$$\mu_{\bar{x}} = \mu = 12.1 \qquad \sigma_{\bar{x}} = \frac{\sigma}{\sqrt{n}} = \frac{3.8}{\sqrt{50}} \approx 0.54$$

(b) Find $P(11 < \bar{x} < 13)$.

We convert the interval

$$11 < \bar{x} < 13$$

to a standard z interval and use the standard normal probability table to find our answer. Since

$$z = \frac{\bar{x} - \mu}{\sigma/\sqrt{n}} = \frac{\bar{x} - 12.1}{0.54}$$

$\bar{x} = 11$ converts to $z = \dfrac{11 - 12.1}{0.54} = -2.04$

and $\bar{x} = 13$ converts to $z = \dfrac{13 - 12.1}{0.54} = 1.67$

Therefore,

$$P(11 < \bar{x} < 13) = P(-2.04 < z < 1.67)$$
$$= 0.9525 - 0.0207$$
$$= 0.9318$$

(c) Comment on your answer for part (b).

It would seem that about 93% of the time there should be no safety hazard for average traffic flow.

by G. W. Mierau and J. L. Schmidt, Mesa Verde Museum Association). Suppose that a doe that weighs less than 54 kg is considered undernourished.

(a) What is the probability that a single doe captured at random in December (weighed and released) is undernourished?

(b) If the park has about 2200 does, what number do you expect to be undernourished in December?

(c) To estimate the health of the December doe population, park rangers use the rule that the average weight of $n = 50$ does should be more than 60 kg. If the average weight is less than 60 kg, it is thought that the entire population of does might be undernourished. What is the probability the average weight $\bar{x}$ for a random sample of 50 does is less than 60 kg (assume a healthy population)?

(d) Compute the probability that $\bar{x} < 64.2$ kg for 50 does (assume a healthy population). Suppose park rangers captured, weighed, and released 50 does in December, and the average weight was $\bar{x} = 64.2$ kg. Do you think the doe population is undernourished or not? Explain.

10. *General: Graphical Interpretation*

(a) Figure 8-3 (Minitab-generated) shows two sampling distributions for $\bar{x}$ taken from the same normal population of x values. One distribution came from samples of size 49 and the other from samples of size 100. Which curve goes with sample size 49? Which goes with sample size 100?

(b) Consider the area under each graph to the right of 51.5. For which sample size is the area to the right of 51.5 greater?

(c) Consider the statement "The smaller the sample size, the more likely it is that the sample mean will be greater than a specified value to the right of the mean μ of the population from which the sample is taken." Do you think this statement is true based on your answer to part (b)? Explain.

11. *Finance: Templeton Funds* Templeton World is a mutual fund that invests in both U.S. and foreign markets. Let x be a random variable that represents the monthly percentage return for the Templeton World fund. Based on information from

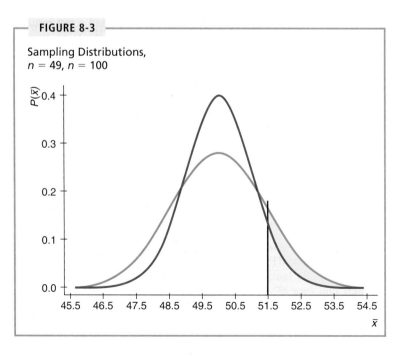

FIGURE 8-3

Sampling Distributions,
$n = 49$, $n = 100$

the *Morningstar Guide to Mutual Funds* (available in most libraries), x has mean $\mu = 1.6\%$ and standard deviation $\sigma = 0.9\%$.

(a) Templeton World fund has over 250 stocks that combine together to give the overall monthly percentage return x. We can consider the monthly return of the stocks in the fund to be a sample from the population of monthly returns of all world stocks. Then we see that the overall monthly return x for Templeton World fund is itself an average return computed using all 250 stocks in the fund. Why would this indicate that x has an approximately normal distribution? Explain. *Hint: See* discussion after Theorem 8.2.

(b) After 6 months, what is the probability that the *average* monthly percentage return $\bar{x}$ will be between 1 and 2%? *Hint: See* Theorem 8.1, and assume that x has a normal distribution as based on part (a).

(c) After 2 years, what is the probability that $\bar{x}$ will be between 1 and 2%?

(d) Compare your answers for parts (b) and (c). Did the probability increase as n (number of months) increased? Why would this happen?

(e) If after 2 years the average monthly percentage return $\bar{x}$ was less than 1%, would that tend to shake your confidence in the statement that $\mu = 1.6\%$? Might you suspect that μ has slipped below 1.6%? Explain.

12. *Focus Problem: Impulse Buying* Solve the Focus Problem at the beginning of this chapter.

13. *General: Distribution of Sample Means*
(a) If we have a distribution of x values that is more or less mound-shaped and somewhat symmetrical, what is the size of sample needed to claim that the distribution of sample means $\bar{x}$ from random samples of that size is approximately normal?
(b) If the original distribution of x values is known to be normal, do we need to make any restriction about sample size in order to claim that the distribution of sample means $\bar{x}$ taken from random samples of a given size is normal?

8.3
Class Project Illustrating the Central Limit Theorem

FOCUS POINTS

✓ Examine different samples of the same size from a uniform distribution.

✓ Observe that the sample statistic $\bar{x}$ varies from sample to sample.

✓ Use the empirical rule to compare the sample distribution of $\bar{x}$ values to the normal distribution.

As we have seen in the preceding two sections, the value of a sample statistic such as $\bar{x}$ can vary from sample to sample. The central limit theorem describes the distribution of the sample statistic $\bar{x}$ when samples are sufficiently large (usually of size 30 or more).

In this section, we explore a sampling distribution of $\bar{x}$ values computed for samples drawn from the uniform distribution of digits 0 through 9. The in-class project follows.

STEP 1: Generate Random Samples of Specified Size n from Digits 0 Through 9

Have each student use the random-number table (Table 1 in the Appendix) or a random-number generator on a calculator or computer to select a random sample of $n = 5$ digits from the list 0, 1, 2, 3, 4, 5, 6, 7, 8, 9. In our samples, we allow repetition of digits. Therefore, 3, 6, 1, 6, and 8 would be a random sample of size $n = 5$ digits. Be sure each student uses the same sample size. It is best to have at least 50 samples, so ask each student to create an appropriate number of samples so that there are at least 50 samples in all. It is important that students using a random-number table select the starting place in the table at random.

STEP 2: Compute the Sample Mean $\bar{x}$ of the Digits in Each Sample

Have each student compute $\bar{x}$ for each of his or her samples of size n.

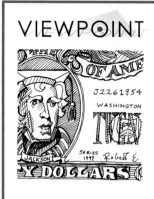

VIEWPOINT   *Why Wait? Apply Now for a College Loan!*

The cost of education is high. The cost of not having an education is higher! What can you expect? What about tuition and student fees? What about room and board? What is the total cost for 1 year at college? Perhaps some averages based on random samples of colleges could be useful. For more information, visit the Brase/Brase *Understanding Basic Statistics* site at http://math.college.hmco .com/students and find the link to the U.S. News site. Then select education. Search for the geographic regions of the colleges of interest.

CHAPTER REVIEW PROBLEMS

1. *General Discussion* Let x be a random variable representing the amount of sleep each adult in New York City got last night. Consider a sampling distribution of sample means $\bar{x}$.
   (a) As the sample size becomes increasingly large, what distribution does the $\bar{x}$ distribution approach?
   (b) As the sample size becomes increasingly large, what value will the mean $\mu_{\bar{x}}$ of the $\bar{x}$ distribution approach?
   (c) What value will the standard deviation $\sigma_{\bar{x}}$ of the sampling distribution approach?
   (d) How do the two $\bar{x}$ distributions for sample size $n = 50$ and $n = 100$ compare?

2. *Normal Distributions: General Discussion* If x has a normal distribution with mean $\mu = 15$ and standard deviation $\sigma = 3$, describe the distribution of $\bar{x}$ values for sample size n, where $n = 4$, $n = 16$, $n = 100$. How do the $\bar{x}$ distributions compare for the various sample sizes?

3. *Job Interview: Length* The personnel office at a large electronics firm regularly schedules job interviews and maintains records of the interviews. From the past records, they have found that the length of a first interview is normally distributed with mean $\mu = 35$ minutes and standard deviation $\sigma = 7$ minutes.
   (a) What is the probability that a first interview will last 40 minutes or longer?
   (b) Nine first interviews are usually scheduled per day. What is the probability that the average length of time for the nine interviews will be 40 minutes or longer?

4. *Drugs: Effects* A new muscle relaxant is available. Researchers of the firm developing the relaxant have done studies that indicate that the time lapse between administration of the drug and beginning effects of the drug is normally distributed with mean $\mu = 38$ minutes and standard deviation $\sigma = 5$ minutes.
   (a) The drug is administered to one patient selected at random. What is the probability that the time it takes to go into effect is 35 minutes or less?
   (b) The drug is administered to a random sample of 10 patients. What is the probability that the average time before it is effective for all 10 patients is 35 minutes or less?
   (c) Comment on the differences of the results in parts (a) and (b).

5. *Psychology: IQ Scores* Assume that IQ scores are normally distributed with standard deviation of 15 points and mean of 100 points. If 100 people are chosen at random, what is the probability that the sample mean of IQ scores will not differ from the population mean by more than 2 points?

6. *Hatchery Fish: Length* A large tank of fish from a hatchery is being delivered to a lake. The hatchery claims that the mean length of fish in the tank is 15 inches, and the standard deviation is 2 inches. A random sample of 36 fish is taken from the tank. Let $\bar{x}$ be the mean sample length of these fish. What is the probability that $\bar{x}$ is within 0.5 inch of the claimed population mean?

7. *Light Bulbs: Hours of Life* A company that makes light bulbs claims that its bulbs have an average life of 750 hours with standard deviation of 20 hours. A random sample of 64 light bulbs is taken. Let $\bar{x}$ be the mean life of this sample.
   (a) What is the probability that $\bar{x} \geq 750$?
   (b) What is the probability that $745 \leq \bar{x} \leq 755$?

8. *Meteorology: Miami and Fairbanks* Let x be a random variable that represents daily high temperatures (degrees Fahrenheit) in January. The following information is based on a report from the U.S. Department of Commerce Environmental Data Services. For Miami, Florida, the mean of the x distribution is $\mu = 76$, and the standard deviation is approximately $\sigma = 1.9$. For Fairbanks, Alaska, the mean of the x distribution is $\mu = 0$ with approximate standard deviation $\sigma = 5.3$. Assume that x has a normal distribution.
   (a) For one day chosen at random in January, what is the probability that the high temperature in Miami will be less than 77°F? What is the probability the high temperature in Fairbanks will be less than 3°F?
   (b) If we choose $n = 7$ days in January, what can we say about the probability distribution of $\bar{x}$, the average high temperature? What is the probability that $\bar{x}$ is less than 77°F for Miami? less than 3°F for Fairbanks?
   (c) Suppose that we cannot assume that x has a normal distribution, but we can say that the distribution is approximately symmetrical and mound-shaped. In this case, what can we say about the $\bar{x}$ probability distribution? If we use all 31 days in January, what is the probability $\bar{x} < 77$°F in Miami? less than 3°F in Fairbanks?

DATA HIGHLIGHTS: GROUP PROJECTS

Break into small groups and discuss the following topics. Organize a brief outline in which you summarize the main points of your group discussion.

Iris setosa is a beautiful wildflower that is found in such diverse places as Alaska, the Gulf of St. Lawrence, much of North America, and even in English meadows and parks. R. A. Fisher, with his colleague Dr. Edgar Anderson, studied these flowers extensively. Dr. Anderson described how he collected information on irises:

> I have studied such irises as I could get to see, in as great detail as possible, measuring iris standard after iris standard and iris fall after iris fall, sitting squat-legged with record book and ruler in mountain meadows, in cypress swamps, on lake beaches, and in English parks. [Anderson, E., "The Irises of the Gaspé Peninsula," *Bulletin, American Iris Society,* 59:2–5, 1935.]

The data in Table 8-4 were collected by Dr. Anderson and were published by his friend and colleague R. A. Fisher in a paper entitled "The Use of Multiple Measurements in Taxonomic Problems" (*Annals of Eugenics,* part II, 179–188, 1936). To find these

TABLE 8-4   Petal Length in Centimeters for
             Iris setosa

1.4	1.4	1.3	1.5	1.4
1.7	1.4	1.5	1.4	1.5
1.5	1.6	1.4	1.1	1.2
1.5	1.3	1.4	1.7	1.5
1.7	1.5	1	1.7	1.9
1.6	1.6	1.5	1.4	1.6
1.6	1.5	1.5	1.4	1.5
1.2	1.3	1.4	1.3	1.5
1.3	1.3	1.3	1.6	1.9
1.4	1.6	1.4	1.5	1.4

Wild iris

data, visit the Brase/Brase *Understanding Basic Statistics* site at http://math.college.hmco .com/students and find the link to DASL, the Carnegie Mellon University Data and Story Library. From the DASL site, look under famous data sets.

Let x be a random variable representing petal length. Using the TI-83 calculator, it was found that the sample mean is $\bar{x} = 1.46$ cm and the sample standard deviation is $s = 0.17$ cm. Figure 8-4 shows a histogram for the given data generated on a TI-83 calculator.

(a) Examine the histogram for petal lengths. Would you say that the distribution is approximately mound-shaped and symmetrical? Our sample has only 50 irises; if many thousands of irises had been used, do you think that the distribution would look even more like a normal curve? Let x be the petal length of *Iris setosa*. Research has shown that x has an approximately normal distribution with mean $\mu = 1.5$ cm and standard deviation $\sigma = 0.2$ cm.

(b) Use the empirical rule with $\mu = 1.5$ and $\sigma = 0.2$ to get an interval into which approximately 68% of the petal lengths will fall. Repeat this for 95% and 99.7%. Examine the raw data and compute the percentage of the raw data that actually falls into each of these intervals (the 68% interval, the 95% interval, and the 99.7% interval). Compare your computed percentages with those given by the empirical rule.

(c) Compute the probability that a petal length is between 1.3 and 1.6 cm. Compute the probability that the petal length is greater than 1.6 cm.

FIGURE 8-4

Petal Length (cm) *Iris setosa* (TI-83)

(d) Suppose that a random sample of 30 irises is obtained. Compute the probability that the average petal length for this sample is between 1.3 and 1.6 cm. Compute the probability that the average petal length is greater than 1.6 cm.

(e) Compare your answers for parts (c) and (d). Do you notice any differences? Why would these differences occur?

LINKING CONCEPTS: WRITING PROJECTS

Discuss each of the following topics in class, or review the topics on your own. Then write a brief but complete essay in which you summarize the main points. Please include formulas and graphs as appropriate.

1. Most people would agree that increased information should give better predictions. Discuss how sampling distributions actually enable better predictions by providing more information. Examine Theorem 8.1 again. Suppose that x is a random variable with a *normal* distribution. Then $\bar{x}$, the sample mean based on random samples of size n, also will have a normal distribution for *any* value of $n = 1, 2, 3, \ldots$.

   What happens to the standard deviation of the $\bar{x}$ distribution as n (the sample size) increases? Consider the following table for different values of n.

n	1	2	3	4	10	50	100
$\sigma/\sqrt{n}$	1σ	0.71σ	0.58σ	0.50σ	0.32σ	0.14σ	0.10σ

   In this case, "increased information" means a larger sample size n. Give a brief explanation why a large *standard deviation* will usually result in poor statistical predictions, whereas a small standard deviation usually results in much better predictions. Since the standard deviation of the sampling distribution $\bar{x}$ is $\sigma/\sqrt{n}$, we can decrease the standard deviation by increasing n. In fact, if we look at the preceding table, we see that if we use a sample of only size $n = 4$, we cut the standard deviation of $\bar{x}$ by 50% of the standard deviation σ of x. If we were to use a sample of size $n = 100$, we would cut the standard deviation of $\bar{x}$ to 10% of the standard deviation σ of x.

   Give the preceding discussion some thought and explain why you should get much better predictions for μ by using $\bar{x}$ from a sample of size n rather than by just using x. Write a brief essay in which you explain why sampling distributions are an important tool in statistics.

2. In a way, the central limit theorem can be thought of as a kind of "grand central station." It is a connecting hub or center for a great deal of statistical work. We will use it extensively in Chapters 9, 10, and 11. Put in a very elementary way, it says that as the sample size n increases, the mean $\bar{x}$ will always approach a normal distribution, no matter where the original x variable came from. For most people, it is the complete generality of the central limit theorem that is so awe inspiring: It applies to practically everything. List and discuss at least three variables from everyday life where you expect the variable x itself does *not* follow a normal or bell-shaped distribution. Then discuss what would happen to the sampling distribution $\bar{x}$ as the sample size increased. Sketch diagrams of the $\bar{x}$ distributions as the sample size n increases.

TI-83PLUS • **EXCEL** • **MINITAB** • **COMPUTERSTAT**

As we have seen in this chapter, the value of a sample statistic such as $\bar{x}$ varies from one sample to another. The central limit theorem describes the distribution of the sample statistic $\bar{x}$ when samples are sufficiently large.

We can use technology tools to generate samples of the same size from the same population. Then we can look at the statistic $\bar{x}$ for each sample, and the resulting $\bar{x}$ distribution.

ComputerStat Computer Demonstration

The software package ComputerStat has the program Central Limit Theorem Demonstration. This program carries out the suggested project for samples from the uniform distribution. The user may select the sample size n to be any integer between 2 and 50. Figure 8-5 shows some displays from this program.

Technology Hints

The TI-83Plus, Excel, and Minitab all support the process of drawing random samples from a variety of distributions. Macros can be written in Excel and Minitab to repeat the six steps of the class project described in Section 8.3.

TI-83Plus

You can generate random samples from uniform, normal, and binomial distributions. Press **MATH**, and select **PRB**. Selection **5:randInt(lower, upper, sample size m)** generates m random integers from the specified interval. Selection **6:randNorm(μ, σ, sample size m)** generates m random numbers from a normal distribution with mean μ and standard deviation σ. Selection **7:randBin(number of trials n, p, sample size m)** generates m random values (number of successes out of n

FIGURE 8-5 ComputerStat Central Limit Theorem Demonstration

(a) $n = 10$ (b) $n = 35$

trials) for a binomial distribution with probability of success p on each trial. You can put these values in lists by using **Edit** under **Stat.** Highlight the list header, press Enter, and then select one of the options discussed.

Excel

Use the menu selection **Tools ➤ Data Analysis ➤ Random Number Generator.** The dialogue box provides choices for the population distribution, including uniform, binomial, and normal distributions. Fill in the required parameters and designate the location for the output.

Random Number Generation	? X
Number of Variables:	1	OK
Number of Random Numbers:	30	Cancel
Distribution:	Normal ▼	Help
Parameters	
Mean =	0
Standard Deviation =	1
Random Seed:	
Output options	
● Output Range:	$A$1:$A$30
○ New Worksheet Ply:	
○ New Workbook	

Minitab

Use the menu selections **Calc ➤ Random Data.** Then select the population distribution. The choices include integer, uniform, binomial, and normal distributions. Fill in the dialogue box, where the number of rows indicates the number of data in the sample.

Integer Distribution	X
	Generate 30 rows of data
	Store in column[s]:
	C1
	Minimum value: 0
	Maximum value: 9
Select	
Help	OK	Cancel

9.1
Estimating μ with Large Samples

As we observed in Chapter 8, we often do not have access to all measurements of an entire population because of time, money, difficulty in finding population members, etc. Instead, we rely on information from a sample. We use the sample statistic to estimate the corresponding population parameter.

An estimate of a population parameter given by a single number is called a *point estimate* of that parameter. It should be no great surprise that we use $\bar{x}$ (the sample mean) as a point estimate for μ (the population mean) and s (the sample standard deviation) as a point estimate for σ (the population standard deviation). In this section, we will discuss estimates of μ and σ from *large samples* ($n \geq 30$).

Statistical theory and empirical results show that if a distribution is approximately mound-shaped and symmetrical, then when the sample size is 30 or larger, we are safe, for most practical purposes, if we estimate σ by s. The error resulting from taking the population standard deviation σ to be equal to the sample standard deviation s is negligible. However, when the sample size is less than 30, we will use special small sample methods, which we will study in Section 9.2.

For large samples of size $n \geq 30$,

$$\sigma \approx s$$

is a good estimate, for most practical purposes.

Basic terminology

Using $\bar{x}$ to estimate μ is not quite so simple, even when we have a large sample size. The *error of estimate* is the magnitude of the difference between the point estimate and the true parameter value. Using $\bar{x}$ as a *point estimate* for μ, the error of estimate is the magnitude of $\bar{x} - \mu$. If we use absolute-value notation, we can indicate the error of estimate for μ by the notation $|\bar{x} - \mu|$.

We cannot say exactly how close $\bar{x}$ is to μ when μ is unknown. Therefore, the exact error of estimate is unknown when the population parameter is unknown. Of course, μ is usually not known or there would be no need to estimate it. In this section, we will use the language of probability to give us an idea of the size of the error of estimate when we use $\bar{x}$ as a point estimate of μ.

First, we need to learn about *confidence levels*. The reliability of an estimate will be measured by the confidence level.

Finding the critical value

Suppose we want a confidence level of c (see Figure 9-1). Theoretically, you can choose c to be any value between 0 and 1, but usually c is equal to a number such as 0.90, 0.95, or 0.99. In each case, the value z_c is the number such that the area under the standard normal curve falling between $-z_c$ and z_c is equal to c. The value z_c is called the *critical value* for a confidence level of c.

The area under the normal curve from $-z_c$ to z_c is the probability that the standardized normal variable z lies in that interval. This means that

$$P(-z_c < z < z_c) = c$$

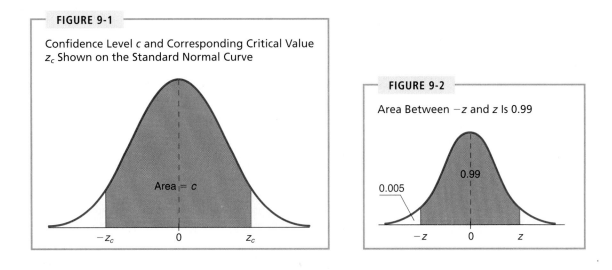

FIGURE 9-1

Confidence Level c and Corresponding Critical Value z_c Shown on the Standard Normal Curve

FIGURE 9-2

Area Between $-z$ and z Is 0.99

EXAMPLE 1

Find a critical value

Let us use Table 4 in the Appendix to find a number $z_{0.99}$ so that 99% of the area under the standard normal curve lies between $-z_{0.99}$ and $z_{0.99}$. That is, we will find $z_{0.99}$ so that

$$P(-z_{0.99} < z < z_{0.99}) = 0.99$$

SOLUTION: In Section 7.3, we saw how to find the z value when we were given an area between $-z$ and z. The first thing we did was to find the corresponding area to the left of $-z$. If A is the area between $-z$ and z, then $(1 - A)/2$ is the area to the left of z. In our case, the area between $-z$ and z is 0.99. The corresponding area in the left tail is $(1 - 0.99)/2 = 0.005$ (see Figure 9-2).

Next, we use Table 4 of the Appendix to find the z value corresponding to a left-tail area of 0.0050.

TABLE 9-1   Excerpt from Table 4 of the Appendix

z	.00	...	.07	.08	.09
-3.4	.0003		.0003	.0003	.0002
$\vdots$					
-2.5	.0062		.0051	.0049	.0048
				↑	
				.0050	

From Table 9-1, we see that the desired area, 0.0050, is exactly halfway between areas corresponding to $z = -2.58$ and $z = -2.57$. Because the two area values are so close together, we use the more extreme z value -2.58 rather than interpolate. For $z_{0.99} = 2.58$, we have

$$P(-2.58 < z < 2.58) \approx 0.99$$

◇

The results of Example 1 will be used a great deal in our later work. For convenience, Table 9-2 gives some levels of confidence and corresponding critical values z_c. The same information is provided in Table 4(b) of the Appendix.

Error of estimate

An estimate is not very valuable unless we have some kind of measure of how "good" it is. The language of probability can give us an idea of the size of the error of estimate caused by using the sample mean $\bar{x}$ as an estimate for the population mean.

Remember that $\bar{x}$ is a random variable. Each time we draw a sample of size n from a population, we can get a different value for $\bar{x}$. According to the central limit theorem, if the sample size is large, then $\bar{x}$ has a distribution that is approximately normal with mean $\mu_{\bar{x}} = \mu$, the population mean we are trying to estimate. The standard deviation is $\sigma_{\bar{x}} = \sigma/\sqrt{n}$.

This information, together with our work on confidence levels, leads us to the probability statement

$$P\left(-z_c \frac{\sigma}{\sqrt{n}} < \bar{x} - \mu < z_c \frac{\sigma}{\sqrt{n}}\right) = c \tag{1}$$

◇ **COMMENT** To derive Equation (1), we start with the probability statement $P(-z_c < z < z_c) = c$. Since $n \geq 30$, we can use the central limit theorem and replace z by $(\bar{x} - \mu)/(\sigma/\sqrt{n})$. Finally, we multiply all parts of the inequality by $(\sigma/\sqrt{n})$ to obtain Equation (1). ◇

Equation (1) uses the language of probability to give us an idea of the size of the error of estimate for the corresponding confidence level c. In words, Equation (1) says that the probability is c that our point estimate $\bar{x}$ is within a distance $\pm z_c(\sigma/\sqrt{n})$ of the population mean μ. This relationship is shown in Figure 9-3.

The *error of estimate* (or absolute error) using $\bar{x}$ as a point estimate for μ is $|\bar{x} - \mu|$. In most practical problems, μ is unknown, so the error of estimate is also

TABLE 9-2   Some Levels of Confidence and Their Corresponding Critical Values

Level of Confidence c	Critical Value z_c
0.70, or 70%	1.04
0.75, or 75%	1.15
0.80, or 80%	1.28
0.85, or 85%	1.44
0.90, or 90%	1.645
0.95, or 95%	1.96
0.98, or 98%	2.33
0.99, or 99%	2.58

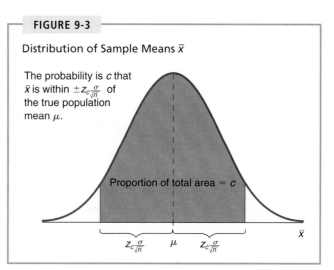

FIGURE 9-3

Distribution of Sample Means $\bar{x}$

The probability is c that $\bar{x}$ is within $\pm z_c \frac{\sigma}{\sqrt{n}}$ of the true population mean μ.

Proportion of total area $= c$

$z_c \frac{\sigma}{\sqrt{n}}$ μ $z_c \frac{\sigma}{\sqrt{n}}$ $\bar{x}$

unknown. However, Equation (1) allows us to compute an *error tolerance E*, which serves as a bound on the error of estimate. Using a $c\%$ level of confidence, we can say that the point estimate $\bar{x}$ differs from the population mean μ by a *maximal error tolerance* of

$$E = z_c \frac{\sigma}{\sqrt{n}}$$

Using the symbol E in place of $z_c\,(\sigma/\sqrt{n})$, we can rewrite Equation (1) as

$$P(-E < \bar{x} - \mu < E) = c \tag{2}$$

Applying a little algebra to Equation (2) produces

$$P(\bar{x} - E < \mu < \bar{x} + E) = c \tag{3}$$

Confidence intervals (large sample)

Equation (3) says that there is a chance of c that the population mean μ lies in the interval from $\bar{x} - E$ to $\bar{x} + E$. We say the interval from $\bar{x} - E$ to $\bar{x} + E$ is a c *confidence interval.* We may get a different confidence interval for each different sample that is taken. Some intervals will contain the population mean μ and others will not. However, in the long run, the proportion of confidence intervals that contains μ is c.

---

c Confidence interval for μ (large samples, $n \geq 30$)

$$\bar{x} - E < \mu < \bar{x} + E \tag{4}$$

where $\bar{x}$ = sample mean

$E = z_c \dfrac{\sigma}{\sqrt{n}}$ if the population standard deviation σ is known

$E \approx z_c \dfrac{s}{\sqrt{n}}$ if σ is not known and we use the sample standard deviation s as an approximation for σ

c = confidence level $(0 < c < 1)$

z_c = critical value (see Table 9-2 or Table 4(b) of the Appendix for commonly used values)

n = sample size $(n \geq 30)$

◇ **IMPORTANT COMMENT** What if you could assume

　　(i) you were drawing your samples from a *normal* population *and*

　　(ii) the population standard deviation σ was known to you?

Under these assumptions, a c confidence interval for the population mean would be $\bar{x} - E < \mu < \bar{x} + E$, where

$$E = z_c \frac{\sigma}{\sqrt{n}}$$

and the condition that we must have a large sample $(n \geq 30)$ could be dropped.　◇

In most practical situations, we don't know the population standard deviation σ. In these situations, we use Equation (4) with large samples ($n \geq 30$) to find a c confidence interval, or we use small samples ($n < 30$) and the methods of the next section to find a c confidence interval.

EXAMPLE 2

Confidence interval for μ

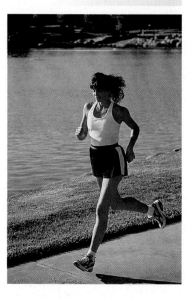

Julia enjoys jogging. She has been jogging over a period of several years, during which time her physical condition has remained constantly good. Usually, she jogs 2 miles per day. During the past year Julia has sometimes recorded her times required to run 2 miles. She has a sample of 90 of these times. For these 90 times the mean was $\bar{x} = 15.60$ minutes and the standard deviation was $s = 1.80$ minutes. Let μ be the mean jogging time for the entire distribution of Julia's 2-mile running times (taken over the past year). Find a 0.95 confidence interval for μ.

SOLUTION: The interval from $\bar{x} - E$ to $\bar{x} + E$ will be a 95% confidence interval for μ. In this case, $c = 0.95$, so $z_c = 1.96$ (see Table 9-2). The sample size $n = 90$ is large enough that we may approximate σ as $s = 1.80$ minutes. Therefore,

$$E \approx z_c \frac{s}{\sqrt{n}}$$

$$E = 1.96\left(\frac{1.80}{\sqrt{90}}\right)$$

$$E = 0.37$$

Using Equation (4), the given value of $\bar{x}$, and our computed value for E, we get the 95% confidence interval for μ.

$$\bar{x} - E < \mu < \bar{x} + E$$
$$15.60 - 0.37 < \mu < 15.60 + 0.37$$
$$15.23 < \mu < 15.97$$

We conclude with 95% confidence that the population mean μ of jogging times for Julia is between 15.23 and 15.97 minutes.   ◊

A few comments are in order about the general meaning of the term *confidence interval*. It is important to realize that the endpoints $\bar{x} \pm E$ are really statistical *variables*. Equation (3) says that we have a chance c of obtaining a sample such that the interval, once it is computed, will contain the parameter μ. Of course, after the confidence interval is numerically fixed, it either does or does not contain μ. So the probability is 1 or 0 that the interval, when it is fixed, will contain μ. A nontrivial probability statement can be made only about variables, not constants. Therefore, Equation (3) really says that if we repeat the experiment many times and get lots of confidence intervals (for the same sample size), then the proportion of all intervals that will turn out to contain the mean μ is c.

In Figure 9-4, the horizontal lines represent 0.90 confidence intervals for various samples of the same size from a distribution. Some of these intervals contain μ, and others do not. Since the intervals are 0.90 confidence intervals, about 90% of all such intervals should contain μ. For each sample, the interval goes from $\bar{x} - E$ to $\bar{x} + E$.

◊ **COMMENT** Please see Using Technology at the end of this chapter for a computer demonstration of this discussion about confidence intervals.   ◊

FIGURE 9-4

0.90 Confidence Intervals for Samples of the Same Size

For each sample the interval goes from
$\bar{x} - E$ to $\bar{x} + E$

μ

GUIDED
EXERCISE 1

Confidence interval for μ

Walter usually meets Julia at the track. He prefers to jog 3 miles. While Julia kept her record, he also kept one for his time required to jog 3 miles. For his 90 times, the mean was $\bar{x} = 22.50$ minutes and the standard deviation was $s = 2.40$ minutes. Let μ be the mean jogging time for the entire distribution of Walter's 3-mile running times over the past several years. Find a 0.99 confidence interval for μ.

(a) What is the value of $z_{0.99}$? (See Table 9-2.)

$\Longrightarrow$ $z_{0.99} = 2.58$

(b) Since the sample size is large, what can we use for σ?

$\Longrightarrow$ $\sigma \approx s = 2.40$

(c) What is the value of E?

$\Longrightarrow$ $E = z_c \dfrac{\sigma}{\sqrt{n}}$

$$\approx 2.58 \left(\frac{2.40}{\sqrt{90}} \right)$$

$$\approx 0.65$$

(d) What are the endpoints for a 0.99 confidence interval for μ?

$\Longrightarrow$ The endpoints are given by

$$\bar{x} - E \approx 22.50 - 0.65$$
$$= 21.85$$
$$\bar{x} + E \approx 22.50 + 0.65$$
$$= 23.15$$

**GUIDED
EXERCISE 2**

Sample size

We have said that a sample of size 30 or larger is a large sample. In this section, we indicated two important reasons why our methods require large samples.

What are these reasons? ⟹ *Reason 1:* Our methods require $\bar{x}$ to have approximately a normal distribution. We know from the central limit theorem that this will be the case for large samples.

Reason 2: Unless we somehow know σ, our methods require us to approximate σ with the sample standard deviation s. This approximation will be good only if the sample size is large.

When we use samples to estimate the mean of a population, we generate a small error. However, samples are useful even when it is possible to survey the entire population because the use of a sample may yield savings of time or effort in collecting data.

TECH NOTE The TI-83Plus, Excel, and Minitab all support confidence intervals for μ from large samples. The level of support varies according to the technology. When a confidence interval is given, the standard mathematical notation (lower value, upper value) is used. For instance, the notation (15.23, 15.97) means the interval from 15.23 to 15.97.

TI-83Plus This calculator gives the most extensive support. The user can opt to enter raw data or just summary statistics. In each case, the value of σ (or the sample estimate) must be specified. Press the **STAT** key and select **TESTS**, use **7:ZInterval**. The T1-83Plus output shows the results for Example 2.

```
ZInterval
 Inpt:Data Stats
 σ:1.8
 x̄:15.6
 n:90
 C-Level:95
 Calculate
```

```
ZInterval
 (15.228, 15.972)
 x̄=15.6
 n=90
```

Excel Excel gives only the value of the maximal error of estimate E. Use the menu choice **Paste Function** (f_x) ➤ **Statistics** ➤ **Confidence**(alpha, σ, n). In the dialogue box, the value of alpha is 1 − confidence level. The Excel output shows the value of E for Example 2.

=	=CONFIDENCE(0.05,1.8,90)	
C	D	E
0.371876		

An alternate approach incorporating raw data (using the Student's t distribution presented in the next section) uses menu choices **Tools ➤ Data Analysis ➤ Describe Statistics**. Again, the value of E for the interval is given.

Minitab Raw data are required. Use the menu choice **Stat ➤ Basic Statistics ➤ 1-SampleZ.**

VIEWPOINT   *Music and Techno Theft*

Performing rights organizations ASCAP (American Society of Composers, Authors, and Publishers) and BMI (Broadcast Music, Inc.) collect royalties for song writers and music publishers. Radio, television, cable, nightclubs, restaurants, elevators, even beauty parlors play music that is copyrighted by a composer or publisher. The royalty payment for this music turns out to be more than a billion dollars a year (*Source: The Wall Street Journal*). How do ASCAP and BMI know who is playing what music? The answer is, they don't know! Instead of tracking *exactly* what gets played, they use random sampling and *confidence intervals*. For example, each radio station (there are more than 10,000 in the U.S.) has randomly chosen days of programming analyzed every year. The results are used to assess royalty fees. In fact, Deloitte & Touche (a financial services company) administers the sampling process.

Although the system is not perfect, it helps bring order into an otherwise chaotic accounting system. Such methods of "copyright policing" help prevent techno theft, ensuring that many song writers get a reasonable return for their creative work.

SECTION 9.1   PROBLEMS

Answers may vary slightly due to rounding.

1. *Medical: Vitamins and Cancer* In an article exploring blood serum levels of vitamins and lung cancer risks (*The New England Journal of Medicine*), the mean serum level of vitamin E in the control group was 11.9 mg/liter with standard deviation 4.30 mg/liter. There were 196 patients in the control group. (These patients were free of all cancer, except possibly skin cancer, in the subsequent 8 years.) Using this information, find a 95% confidence interval for the mean serum level of vitamin E in all persons similar to the control group.

2. *Medical: Vitamins and Cancer* In the same article as cited in Problem 1, 99 patients who were cancer-free at the time the blood was drawn later developed lung cancer. For these patients, the mean blood serum level of vitamin E was 10.5 mg/liter with

standard deviation 3.2 mg/liter. Using this information, find a 95% confidence interval for the mean serum level of vitamin E in the population of all persons with similar lung cancer risks.

3. *Marketing: Motel* Irv and Nancy are thinking about buying the Rockwood Motel located on Interstate 70. Before they make up their mind, they want to estimate the average number of vehicles that go by the motel each day in the summer. Fortunately, the highway department has been counting vehicles on I-70 near the motel. A random sample of 36 summer days shows an average of 16,000 cars per day with a standard deviation of 2400 cars. Find a 90% confidence interval for the mean number of cars per summer day going past the Rockwood Motel.

4. *Ballooning: Air Temperature* How hot is the air in the top (crown) of a hot air balloon? Information from *Ballooning: The Complete Guide to Riding the Winds*, by Wirth and Young (Random House), claims that the air in the crown should be an average of 100°C for a balloon to be in a state of equilibrium. However, the temperature does not need to be exactly 100°C. What is a reasonable and safe range of temperatures? This may vary with the size and (decorative) shape of the balloon. All balloons have a temperature gauge in the crown. Suppose that 56 readings (for a balloon in equilibrium) gave a mean temperature $\bar{x} = 97$°C with sample standard deviation $s = 17$°C.
   (a) Compute a 95% confidence interval for the average temperature for which this balloon will be in a steady-state equilibrium.
   (b) If the average temperature in the crown of the balloon goes above the high end of your confidence interval, do you expect the balloon will go up or down? Explain.

5. *Coyotes: Howls* A special feature of camping in the western United States is listening to coyotes howl at night. Just how long is one coyote howl? Since coyotes tend to hunt (mice, rabbits, etc.) in groups, it can seem that howls go on for a long time. This is the result of a collection of coyotes howling together. Individual coyotes tend to howl for a shorter period of time. The howls may help individual members of the pack locate each other in the dark. The following data are based on information from *Coyotes: Biology, Behavior, and Management*, edited by M. Bekoff, University of Colorado (Academic Press).
   (a) Based on a random sample of 102 coyotes, it was found that the mean duration of the howl was $\bar{x} = 1.2$ seconds with sample standard deviation $s = 0.4$ second. Compute a 99% confidence interval for the population mean duration μ of all coyote howls.
   (b) For the 102 coyotes in the sample, the mean frequency of howls was $\bar{x} = 609$ Hz (hertz) with sample standard deviation $s = 248$ Hz. Compute a 95% confidence interval for the population mean frequency μ of all coyote howls.

6. *Botany: Iris* Dr. Edgar Anderson was a botanist who collected vast amounts of data for several species of wild iris (see Data Highlights in Chapter 8). *Iris virginica*, a lovely wildflower spread over most of the American continent and much of Europe, is one of the species Dr. Anderson studied. His friend R. A. Fisher published the data in a paper entitled "The Use of Multiple Measurements in Taxonomic Problems," in *Annals of Eugenics*, 7 (part II):179–188, 1936. For a sample of 50 *Iris virginica*,
   (a) the sample mean petal length was $\bar{x} = 5.55$ cm with sample standard deviation $s = 0.57$ cm. Compute an 85% confidence interval for the population mean petal length.
   (b) the sample mean petal width was $\bar{x} = 2.03$ cm with sample standard deviation $s = 0.27$ cm. Compute a 90% confidence interval for the population mean petal width.

7. *Profits: Banks* Jobs and productivity! How do banks rate? One way to answer this question is to examine annual profits per employee. *Forbes Top Companies*, edited by J. T. Davis (John Wiley & Sons), gave the following data about annual profits per employee (in units of one thousand dollars per employee) for representative companies in financial services. Companies such as Wells Fargo, First Bank System, Key Banks, and Norwest Banks were included.

42.9	43.8	48.2	60.6	54.9	55.1	52.9	54.9	42.5	33.0	33.6
36.9	27.0	47.1	33.8	28.1	28.5	29.1	36.5	36.1	26.9	27.8
28.8	29.3	31.5	31.7	31.1	38.0	32.0	31.7	32.9	23.1	54.9
43.8	36.9	31.9	25.5	23.2	29.8	22.3	26.5	26.7		

(a) Use a calculator with mean and sample standard deviation keys to verify that, for the preceding data, $\bar{x} \approx 36.0$ and $s \approx 10.2$.

(b) Let us say that the preceding data are representative of the entire sector of (successful) financial services corporations. Find a 75% confidence interval for μ, the average annual profit per employee for all successful banks.

(c) Let us say that you are the manager for a local bank with a large number of employees. Suppose the annual profits per employee are less than 30 thousand dollars per employee. Do you think that this might be somewhat low compared with other successful financial institutions? Explain by referring to the confidence interval you computed in part (b).

(d) Suppose the annual profits are more than 40 thousand dollars per employee. As manager of the bank, would you feel somewhat better? Explain by referring to the confidence interval you computed in part (b).

(e) Repeat parts (b), (c), and (d) for a 90% confidence level.

8. *Profits: Retail* Jobs and productivity! How do retail stores rate? One way to answer this question is to examine annual profits per employee. The following data give annual profits per employee (in units of one thousand dollars per employee) for companies in retail sales. (*See* reference in Problem 7.) Companies such as Gap, Nordstrom, Circuit City, Dillards, JCPenney, Sears, Wal-Mart, Office Depot, and Toys 'Я' Us are included.

4.4	6.5	4.2	8.9	8.7	8.1	6.1	6.0	2.6	2.9	8.1	−1.9
11.9	8.2	6.4	4.7	5.5	4.8	3.0	4.3	−6.0	1.5	2.9	4.8
−1.7	9.4	5.5	5.8	4.7	6.2	15.0	4.1	3.7	5.1	4.2	

(a) Use a calculator with mean and sample standard deviation keys to verify that, for the preceding data, $\bar{x} \approx 5.1$ and $s \approx 3.8$.

(b) Let us say that the preceding data are representative of the entire sector of retail sales companies. Find an 80% confidence interval for μ, the average annual profit per employee for retail sales.

(c) Let us say that you are the manager of a retail store with a large number of employees. Suppose the annual profits per employee are less than 3 thousand dollars per employee. Do you think that this might be low compared with other retail stores? Explain by referring to the confidence interval you computed in part (b).

(d) Suppose the annual profits are more than 6.5 thousand dollars per employee. As store manager, would you feel somewhat better? Explain by referring to the confidence interval you computed in part (b).

(e) Repeat parts (b), (c), and (d) for a 95% confidence interval.

9. *Chocolate Chip Cookies: Calories* *Consumer Reports* gave the following data about calories in a 30-gram serving of chocolate chip cookies. Both fresh-baked—such as Duncan Hines and Pillsbury—and packaged cookies—such as Pepperidge Farm and Nabisco—were included.

153	152	146	138	130	146	149	138	168
147	140	156	155	163	153	155	160	145
138	150	135	155	156	150	146	129	127
171	148	155	132	155	127	150	110	

(a) Use a calculator with mean and standard deviation keys to verify that the sample mean number of calories is $\bar{x} = 146.5$ with sample standard deviation $s = 12.7$ calories.

(b) We take the point of view that the preceding data are representative of the population of all chocolate chip cookies. Find an 80% confidence interval for the mean calories μ in a 30-gram serving of all chocolate chip cookies. Find the length of this interval.

(c) Repeat part (b) using a 90% confidence interval.

(d) Repeat part (b) using a 99% confidence interval.

(e) Compare the lengths from parts (b), (c), and (d). Comment on how these lengths change as c, the confidence level, increases.

10. *Aspen Trees: Base Circumference* In Roosevelt National Forest, the rangers took random samples of live aspen trees and measured the base circumference of each tree.

(a) The first sample had 30 trees with a mean circumference of $\bar{x} = 15.71$ inches and a sample standard deviation of $s = 4.63$ inches. Find a 95% confidence interval for the mean circumference of aspen trees from these data.

(b) The next sample had 90 trees with a mean of $\bar{x} = 15.58$ inches and a sample standard deviation of $s = 4.61$ inches. Again, find a 95% confidence interval from these data.

(c) The last sample had 300 trees with a mean of $\bar{x} = 15.59$ inches and a sample standard deviation of $s = 4.62$ inches. Again, find a 95% confidence interval from these data.

(d) Find the length of each interval of parts (a), (b), and (c). Comment on how these lengths change as the sample size increases.

11. *Medical: Sleep* How long does it take to fall asleep at night? This depends on what happened the night before. Alexander Borbely is a professor at the University of Zurich Medical School and director of the Sleep Laboratory. The following is adapted from the book, *Secrets of Sleep*, by Professor Borbely.

(a) Suppose that a random sample of 38 college students was kept awake all night and the next day (24 hours total). The mean time for this group to go to sleep the next night was $\bar{x} = 2.5$ minutes with standard deviation $s = 0.7$ minute. Compute a 90% confidence interval for the mean time of all such (sleep-deprived) students to fall asleep. What is the length of this interval?

(b) Suppose that a random sample of 38 college students had a normal (8-hour) sleep and a normal (16-hour) day. The mean time for this group to go to sleep the next night was $\bar{x} = 15.2$ minutes with sample standard deviation $s = 4.8$ minutes. Compute a 90% confidence interval for the mean time to go to sleep for all people in this (normal) group. What is the length of this interval?

(c) Suppose that a random sample of 38 college students stayed in bed at least 12 hours and then after another 12 hours went back to bed. The mean time for this group to fall asleep was $\bar{x} = 25.7$ minutes with sample standard deviation

$s = 8.3$ minutes. Compute a 90% confidence interval for all people in this group to fall asleep. What is the length of this interval?

(d) Compare the lengths of the intervals in parts (a), (b), and (c). As the sample standard deviations got larger, did the intervals get longer? Why would you expect this from the method of calculation? Explain.

12. *Pro Football Players: Age* How old are professional football players? Consider the following positions: defensive end, defensive guard, defensive tackle, offensive guard, offensive end, offensive tackle. A random sample of professional players in the given positions was taken from the San Diego Chargers, Kansas City Chiefs, Denver Broncos, Oakland Raiders, and Seattle Seahawks. The following information gives age in years and was obtained from *The Sports Encyclopedia, Pro Football* (11th Edition).

26	24	25	36	26	32	31	34
32	27	32	23	24	29	30	30
29	33	25	32	24	25	28	22
22	24	23	33	32	31	26	28
32	25	22	28	25	29	25	27

(a) Use a calculator with mean and sample standard deviation keys to verify that, for the preceding data, $\bar{x} \approx 27.8$ years and $s \approx 3.8$ years.

(b) Let us say that the preceding data are representative of the entire population of professional football players holding the designated positions. Compute an 80% confidence interval for μ, the population mean age of all professional football players in the designated positions.

(c) Let us say that you are the head coach for a professional football team. Suppose that a rookie end, tackle, or guard wants to join your team. This person is 33 years old. Do you think that this age might be somewhat old for a rookie in a starting position? Explain by referring to the confidence interval you computed in part (b).

(d) Repeat parts (b) and (c) for a 99% confidence interval.

9.2
Estimating μ with Small Samples

FOCUS POINTS

✓ Learn about degrees of freedom and Student's *t* distribution.

✓ Find critical values using the degrees of freedom and confidence level.

✓ Compute confidence intervals for μ using small samples. Interpret the results.

For samples of size 30 or larger, we can approximate the population standard deviation σ by s, the sample standard deviation. Then we can use the central limit theorem to find bounds on the error of estimate and confidence intervals for μ.

There are many practical and important situations, however, where large samples are simply not available. Suppose an archaeologist discovers only seven fossil skeletons from a previously unknown species of miniature horse. Reconstructions of the skeletons of these seven miniature horses show their mean shoulder heights to be $\bar{x} = 46.1$ cm. Let μ be the mean shoulder height for this entire species of miniature horse. How can we find the maximal error of estimate $|\bar{x} - \mu|$? How can we find a confidence interval for μ? We will return to this problem later in this section.

Student's *t* distribution

To avoid the error involved in replacing σ by s—that is, approximating σ by s—when dealing with *small samples* ($n < 30$), we introduce a new variable called *Student's t variable*. The *t* variable and its corresponding distribution, called *Student's t distribution*, were discovered in 1908 by W. S. Gosset. He was employed as a statistician by Guinness brewing company, a company that frowned on the publication of research by its employees, so Gosset published his research under the pseudonym Student. Gosset was the first to recognize the importance of developing statistical methods for obtaining reliable information from small samples. It might be more fitting to call this *Gosset's t distribution;* however, in the literature of mathematical statistics it is known as *Student's t distribution.*

The *t* variable is defined by the following formula:

$$t = \frac{\bar{x} - \mu}{\frac{s}{\sqrt{n}}} \tag{5}$$

where $\bar{x}$ is the mean of a random sample of *n* measurements, μ is the population mean of the *x* distribution, and *s* is the sample standard deviation.

◇ **COMMENT** You should note that our *t* variable is just like

$$z = \frac{\bar{x} - \mu}{\frac{\sigma}{\sqrt{n}}}$$

except that we replace σ with *s*. Unlike our methods for large samples, σ cannot be approximated by *s* when the sample size is less than 30 and we cannot use the normal distribution. However, we will be using the same methods as in Section 9.1 to find the maximal error of estimate and to find confidence intervals, but we use the Student's *t* distribution. ◇

If many random samples of size *n* are drawn, then we get many *t* values from Equation (5). These *t* values can be organized into a frequency table, and a histogram can be drawn, thereby giving us an idea of the shape of the *t* distribution (for a given *n*).

Fortunately, all this work is not necessary because mathematical theorems can be used to obtain a formula for the *t* distribution. However, it is important to observe that these theorems say that the shape of the *t* distribution depends only on *n*, provided the basic variable *x* has a normal distribution. So *when we use the t distribution, we will assume that the x distribution is normal.*

Degrees of freedom

Table 5 in the Appendix gives values of the variable *t* corresponding to what we call the number of *degrees of freedom*, abbreviated *d.f.* For the methods used in this section, the number of degrees of freedom is given by the formula

$$d.f. = n - 1 \tag{6}$$

where *d.f.* stands for the degrees of freedom and *n* is the sample size being used.

Each choice for *d.f.* gives a different *t* distribution. However, for *d.f.* larger than about 30, the *t* distribution and the standard normal *z* distribution are almost the same.

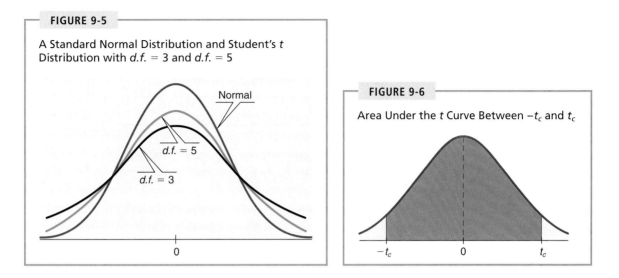

FIGURE 9-5

A Standard Normal Distribution and Student's t Distribution with $d.f. = 3$ and $d.f. = 5$

Normal

$d.f. = 5$

$d.f. = 3$

0

FIGURE 9-6

Area Under the t Curve Between $-t_c$ and t_c

$-t_c$ 0 t_c

The graph of a t distribution is always symmetrical about its mean, which (as for the z distribution) is 0. The main observable difference between a t distribution and the standard normal z distribution is that a t distribution has somewhat thicker tails.

Figure 9-5 shows a standard normal z distribution and Student's t distribution with $d.f. = 3$ and $d.f. = 5$.

Appendix Table 5 gives various t values for different degrees of freedom $d.f.$ We will use this table to find *critical values* t_c for a c confidence level. In other words, we want to find t_c so that an area equal to c under the t distribution for a given number of degrees of freedom falls between $-t_c$ and t_c. In the language of probability, we want to find t_c so that

$$P(-t_c < t < t_c) = c$$

This probability corresponds to the area shaded in Figure 9-6.

Table 5 in the Appendix has been arranged so that c is one of the column headings, and the degrees of freedom $d.f.$ are the row headings. To find t_c for any specific c, we find the column headed by that c value and read down until we reach the row headed by the appropriate number of degrees of freedom $d.f.$ (You will notice two other column headings: α' and α''. We will use these later, but for the time being, ignore them.)

Using Table 5 to find critical values for confidence intervals

EXAMPLE 3

Student's t distribution

Use Table 9-3 on the next page (an excerpt from Appendix Table 5) to find the critical value t_c for a 0.99 confidence level for a t distribution with sample size $n = 5$.

SOLUTION:

(a) First, we find the column with c heading 0.99. This is the last column.

(b) Next, we compute the number of degrees of freedom:
$d.f. = n - 1 = 5 - 1 = 4$.

(c) We read down the column under the heading $c = 0.99$ until we reach the row headed by 4 (under $d.f.$). The entry is 4.604. Therefore, $t_{0.99} = 4.604$.   ◇

TABLE 9-3   Student's t Distribution
(Excerpt from Appendix Table 5)

c	. . . 0.90	0.95	0.98	0.99
α'	—	—	—	—
α''	—	—	—	—
d.f.				
⋮				
3	. . . 2.353	3.182	4.541	5.841
4	. . . 2.132	2.776	3.747	4.604
⋮				
7	. . . 1.895	2.365	2.998	3.499
8	. . . 1.860	2.306	2.896	3.355

Source: Table 5, Appendix, was generated by Minitab.

GUIDED EXERCISE 3

Student's t distribution table

Use Table 5 in the Appendix or Table 9-3 to find t_c for a 0.90 confidence level of a t distribution with sample size $n = 9$.

(a) We find the column headed by $c = $ _____.
This is the _____ (first, second, third, fourth, fifth, sixth) column.

➡ $c = 0.90$. This is the fourth column.

(b) The degrees of freedom are given by $d.f. = n - 1 = $ _____.

➡ $d.f. = n - 1 = 9 - 1 = 8$

(c) Read down the column found in part (a) until you reach the entry in the row headed by $d.f. = 8$. The value of $t_{0.90}$ is _____ for a sample size of 9.

➡ $t_{0.90} = 1.860$ for a sample size $n = 9$.

(d) Find t_c for a 0.95 confidence level of a t distribution with sample size $n = 9$.

➡ $t_{0.95} = 2.306$ for a sample of size $n = 9$.

Maximal error of estimate

In Section 9.1, we found bounds $\pm E$ on the error of estimate for a c confidence level. Using the same basic approach, we arrive at the conclusion that

$$E = t_c \frac{s}{\sqrt{n}}$$

is the maximal error of estimate for a c confidence level with small samples (i.e., $|\bar{x} - \mu| < E$ with probability c). The analogue of Equation (1) in Section 9.1 is

$$P\left(-t_c \frac{s}{\sqrt{n}} < \bar{x} - \mu < t_c \frac{s}{\sqrt{n}}\right) = c \tag{7}$$

◇ **COMMENT** Comparing Equation (7) with Equation (1) in Section 9.1, it becomes evident that we are using the same basic method on the t distribution that we did on the z distribution.   ◇

Likewise, for small samples from normal populations, Equation (3) of Section 9.1 becomes

$$P(\bar{x} - E < \mu < \bar{x} + E) = c \tag{8}$$

where $E = t_c(s/\sqrt{n})$. Let us organize what we have been doing in a convenient summary.

Confidence interval for μ (small sample)

◇ **SUMMARY** For small samples ($n < 30$) taken from a normal population where σ is unknown, a c confidence interval for the population mean μ is as follows:

---

c Confidence interval for μ (small sample)

$$\bar{x} - E < \mu < \bar{x} + E \tag{9}$$

where $\bar{x}$ = sample mean

$$E = t_c \frac{s}{\sqrt{n}}$$

c = confidence level ($0 < c < 1$)

t_c = critical value for confidence level c, and degrees of freedom $d.f. = n - 1$ taken from t distribution

n = sample size (small samples, $n < 30$)

s = sample standard deviation

◇

◇ **COMMENT** In our applications of Student's t distribution, we have made the basic assumption that x has a normal distribution. However, the same methods apply even if x is only approximately normal. In fact, the main requirement for using the Student's t distribution is that the distribution of x values be reasonably symmetrical and mound-shaped. If this is the case, then the methods we employ with the t distribution can be considered valid for most practical applications.   ◇

EXAMPLE 4

Confidence interval for μ, small sample

Let's return to our archaeologist and the newly discovered (but extinct) species of miniature horse discussed at the beginning of this section. There are only seven known existing skeletons with shoulder heights (in centimeters) 45.3, 47.1, 44.2, 46.8, 46.5, 45.5, and 47.6. For this sample data, the mean is $\bar{x} = 46.14$ and the sample standard deviation is $s = 1.19$. Let μ be the mean shoulder height (in centimeters) for this entire species of miniature horse, and assume that the population of shoulder heights is approximately normal.

Find a 99% confidence interval for μ, the mean shoulder height of the entire population of such horses.

SOLUTION: In this case, $n = 7$, so $d.f. = n - 1 = 7 - 1 = 6$. For $c = 0.99$, Table 5 in the Appendix gives $t_{0.99} = 3.707$ (for $d.f. = 6$). The sample standard deviation is $s = 1.19$.

$$E = t_c \frac{s}{\sqrt{n}} = (3.707) \frac{1.19}{\sqrt{7}} = 1.67$$

The 99% confidence interval is

$$\bar{x} - E < \mu < \bar{x} + E$$
$$46.14 - 1.67 < \mu < 46.14 + 1.67$$
$$44.5 < \mu < 47.8$$

◇

GUIDED EXERCISE 4

Confidence interval for μ, small sample

A company has a new process for manufacturing large artificial sapphires. The production of each gem is expensive, so the number available for examination is limited. In a trial run, 12 sapphires are produced. The mean weight for these 12 gems is $\bar{x} = 6.75$ carats, and the sample standard deviation is $s = 0.33$ carat. Let μ be the mean weight for the distribution of all sapphires produced by the new process.

(a) What is $d.f.$ for this setting?

⟹ $d.f. = n - 1$ where n is the sample size. Since $n = 12$, $d.f. = 12 - 1 = 11$.

(b) Use Table 5 in the Appendix to find $t_{0.95}$.

⟹ Using Table 5 with $d.f. = 11$ and $c = 0.95$, we find $t_{0.95} = 2.201$.

(c) Find E.

⟹ $E = t_{0.95} \frac{s}{\sqrt{n}} = (2.201) \frac{0.33}{\sqrt{12}} \approx 0.21$

(d) Find a 95% confidence interval for μ.

⟹ $\bar{x} - E < \mu < \bar{x} + E$
$6.75 - 0.21 < \mu < 6.75 + 0.21$
$6.54 < \mu < 6.96$

(e) What assumption about the distribution of all sapphires had to be made to obtain these answers?

⟹ The population of artificial sapphire weights is approximately normal.

We have several formulas for confidence intervals for the population mean μ. How do we choose an appropriate one? We need to look at the sample size, the distribution of the original population, and whether or not the population standard deviation σ is known. There are essentially four cases for which we have the tools to find $c\%$ confidence intervals for the mean μ.

Summary: confidence intervals for the mean

Large-Sample Cases

$n \geq 30$

1. If σ is not known, then a $c\%$ confidence interval for μ is

$$\bar{x} - z_c \frac{s}{\sqrt{n}} < \mu < \bar{x} + z_c \frac{s}{\sqrt{n}}$$

2. If σ is known, then a $c\%$ confidence interval for μ is

$$\bar{x} - z_c \frac{\sigma}{\sqrt{n}} < \mu < \bar{x} + z_c \frac{\sigma}{\sqrt{n}}$$

Small-Sample Case

$n < 30$

If the population is approximately normal and σ is not known, then a $c\%$ confidence interval for μ is

$$\bar{x} - t_c \frac{s}{\sqrt{n}} < \mu < \bar{x} + t_c \frac{s}{\sqrt{n}} \qquad \text{Use } d.f. = n - 1.$$

For Any Sample Size

If the population *is normal* and σ is known, then for any sample size (large or small) a $c\%$ confidence interval for μ is

$$\bar{x} - z_c \frac{\sigma}{\sqrt{n}} < \mu < \bar{x} + z_c \frac{\sigma}{\sqrt{n}}$$

 TECH NOTE The TI-83Plus, Excel, and Minitab support confidence intervals using the Student's t distribution.

TI-83Plus Press the **STAT** key, select **TESTS** and choose option **8:TInterval.** You may use either raw data in a list or summary statistics.

Excel Excel gives only the value of the maximal error of estimate E. You can easily construct the confidence interval by computing $\bar{x} - E$ and $\bar{x} + E$. Use the menu choices **Tools ➤ Data Analysis ➤ Describe Statistics.** In the dialogue box, check summary statistics and check confidence level for mean. Then set the desired confidence level. Under these menu choices, Excel uses the Student's t distribution for confidence intervals regardless of sample size.

Minitab Use the menu choices **Stat ➤ Basic Statistics ➤ 1-Sample t.** In the dialogue box, indicate the column that contains the raw data. The Minitab output on the next page shows the confidence interval for Example 4.

```
T Confidence Intervals
Variable   N      Mean     StDev    SE Mean        99.0 % CI
C1         7      46.143   1.190    0.450      (44.475, 47.810)
```

VIEWPOINT

Earthquakes!

California, Washington, Nevada, and even Yellowstone National Park all have earthquakes. Some earthquakes are severe! All earthquakes bring fear and anxiety to people living near the quake. Is San Francisco due for a really big quake like the 1906 major earthquake? How big are the sizes of recent earthquakes compared with really big earthquakes? What is the duration of an earthquake? How long is the time span between major earthquakes? One way to answer questions such as these is to use existing data to estimate confidence intervals on the average size, duration, and time interval between quakes. Recent data sets for computing such confidence intervals can be found at the National Earthquake Information Service of the U.S. Geological Survey Web site. To access the site, visit the Brase/Brase *Understanding Basic Statistics* site at http://math.college.hmco.com/students and find the link to National Earthquake Information Service.

SECTION 9.2   PROBLEMS

In all the following problems, assume that the population of x values has an approximately normal distribution. Answers may vary slightly due to rounding.

1. Use Table 5 in the Appendix to find t_c for a 0.95 confidence level when the sample size is 18.

2. Use Table 5 in the Appendix to find t_c for a 0.99 confidence level when the sample size is 4.

3. Use Table 5 in the Appendix to find t_c for a 0.90 confidence level when the sample size is 22.

4. Use Table 5 in the Appendix to find the value of t_c for a 0.95 confidence level when the sample size is 12.

5. *Archaeology: Tree Rings* At Burnt Mesa Pueblo, the method of tree ring dating gave the following dates A.D. for an archaeological excavation site (*Bandelier Archaeological Excavation Project: Summer 1990 Excavations at Burnt Mesa Pueblo*, edited by Kohler, Washington State University):

    1189   1271   1267   1272   1268   1316   1275   1317   1275

    (a) Use a calculator with mean and standard deviation keys to verify that the sample mean date is $\bar{x} = 1272$ with sample standard deviation $s = 37$ years.

    (b) Find a 90% confidence interval for the mean of all tree ring dates from this archaeological site.

6. *Wildlife: Mountain Lions* How much do wild mountain lions weigh? *The 77th Annual Report of the New Mexico Department of Game and Fish,* edited by Bill Montoya, gave the following information. Adult wild mountain lions (18 months or older) captured and released for the first time in the San Andres Mountains gave the following weights (lb):

| 68 | 104 | 128 | 122 | 60 | 64 |

(a) Use a calculator with mean and sample standard deviation keys to verify that $\bar{x} \approx 91.0$ lb and $s \approx 30.7$ lb.
(b) Find a 75% confidence interval for the population average weight μ of all adult mountain lions in the specified region.

7. *Restaurant: Tips* André is head waiter at a famous gourmet restaurant in San Francisco. The Internal Revenue Service is doing an audit of his tax return this year. In particular, the IRS wants to know the average amount André gets for a tip. In an effort to satisfy the IRS, André took a random sample of eight credit card receipts, each of which indicated his tip. The results were

| $10.00 | $12.75 | $11.93 | $11.15 | $15.70 | $14.50 | $9.10 | $13.65 |

(a) Use a calculator to verify that the sample mean is $12.35 and the sample standard deviation is $2.25.
(b) Find a 90% confidence interval for the population mean of tips received by André.

8. *Franchise: Candy Store* Do you want to own your own candy store? Wow! With some interest in running your own business and a decent credit rating, you can probably get a bank loan on startup costs for franchises such as Candy Express, The Fudge Company, Karmel Corn, and Rocky Mountain Chocolate Factory. Startup costs (units in $1000) for a random sample of candy stores are given below (*Source: Entrepreneur Magazine,* Vol. 23, No. 10).

| 95 | 173 | 129 | 95 | 75 | 94 | 116 | 100 | 85 |

Use a calculator with mean and sample standard deviation keys to verify that $\bar{x} \approx 106.9$ thousand dollars and $s \approx 29.4$ thousand dollars. Find a 90% confidence interval for the population average startup costs μ for candy store franchises.

9. *Lincoln County Jail: Billy the Kid* Suppose that you are an anthropologist/historian doing research at an excavation site. How deep do you need to dig to locate artifacts of historical significance? The answer may depend on how long ago the events you hope to study occurred. If the events are relatively recent, you may not need to dig too deep. In 1881, the Lincoln County Cattle Baron Wars were in progress in the New Mexico Territory, and one infamous outlaw, Billy the Kid, killed two deputies in his escape from the Lincoln County Jail. In 1985, the anthropologist/historian Yvonne Oakes was commissioned by the Museum of New Mexico to study the now famous Lincoln Country Courthouse and Jail area (*Source: Archaeological Testing at Three Historic Sites at Lincoln State Monument,* by Y. R. Oakes, Museum of New Mexico).
(a) The depths (in inches) below grade at which significant artifacts were found in a random sample of trenches are

| 10 | 10.5 | 14 | 14 | 4 | 4.5 | 5 | 12 | 12 | 16 |
| 9 | 6 | 8 | 9 | 10 | 9 | 11 | 11 | 12 | |

Use a calculator with mean and sample standard deviation keys to verify that $\bar{x} \approx 9.8$ inches and $s \approx 3.3$ inches. Compute a 90% confidence interval for the population mean depth μ at which significant artifacts will be found.

(b) The depth (in inches) to sterile ground (no more artifacts found) for seven excavation trenches are

| 17 | 19 | 21.5 | 16 | 14 | 16 | 16 |

Use a calculator with mean and sample standard deviation keys to verify that $\bar{x} \approx 17.1$ inches and $s \approx 2.5$ inches. Compute an 80% confidence interval for the population mean depth μ to sterile ground.

10. *College Tuition: Professors' Salaries* How much has college tuition gone up? How much have professors' salaries gone up? The National Education Association's *Almanac of Higher Education*, published by the NEA, gave the following information.
   (a) A random sample of 12 East Coast colleges and universities (public and private) gave the following percentage tuition increases for a 2-year period:

| 3.9 | 2.3 | 3.7 | 9.4 | 4.8 | 3.9 |
| 6.7 | 6.6 | 4.6 | 4.3 | 3.1 | 3.1 |

Use a calculator with mean and sample standard deviation keys to verify that $\bar{x} \approx 4.70\%$ and $s \approx 1.98\%$. Compute a 90% confidence interval for the population mean percentage tuition increase μ of all East Coast colleges and universities.
   (b) A random sample of 14 East Coast colleges and universities gave the following percentage increases in salaries for professors during the same 2-year period:

| 2.8 | 4.3 | 2.4 | 3.4 | 5.5 | 3.8 | 3.4 |
| 5.1 | 2.4 | 3.0 | 2.9 | 0.4 | 1.4 | 3.7 |

Use a calculator with mean and sample standard deviation keys to verify that $\bar{x} \approx 3.18\%$ and $s \approx 1.34\%$. Compute a 90% confidence interval for the population mean percentage salary increase μ of all East Coast college and university professors.

11. *Jobs: Annual Pay* What is the average annual pay for different jobs? The following problem is based on information taken from *State and Metropolitan Area Data Book*, 4th Edition (U.S. Department of Commerce).
   (a) What if you have a factory job (manufacturing)? The following data are inflation adjusted and are from a random sample of reports from western and midwestern states. (Annual pay in units of $1000.)

| 37.0 | 34.6 | 36.1 | 44.6 | 32.3 | 24.8 | 27.1 | 28.5 |

Use a calculator with mean and sample standard deviation keys to verify that $\bar{x} \approx 33.1$ thousand dollars and $s \approx 6.4$ thousand dollars. Compute an 85% confidence interval for the population average μ of such income.
   (b) What if you are a salesperson (retail trade)? The following data are inflation adjusted and are from a random sample of reports from eastern states. (Annual pay in units of $1000.)

| 21.8 | 19.1 | 17.3 | 18.6 | 24.6 | 22.5 | 23.8 | 18.9 | 20.8 |

Use a calculator with mean and sample standard deviation keys to verify that $\bar{x} \approx 20.8$ thousand dollars and $s \approx 2.5$ thousand dollars. Compute a 99% confidence interval for the population average μ of such income.

12. *Hospitals: Length of Stay* How long do patients stay in a hospital? What percentage of hospitals provide at least some charity care? The following problem is based

on information taken from *State Health Care Data: Utilization, Spending, and Characteristics* (American Medical Association).

(a) How long do patients stay? Based on a random sample of hospital reports from eastern states, the following information was obtained (units in days):

7.2	7.9	7.0	6.9	9.9	7.5	6.6	7.3
7.3	6.8	7.1	6.8	7.3	6.9	6.8	

Use a calculator with mean and sample standard deviation keys to verify that $\bar{x} \approx 7.3$ days and $s \approx 0.8$ days. Find a 99% confidence interval for the population average number of days μ patients spend in the hospital.

(b) What percentage of the hospitals provide (at least some) charity care? Based on a random sample of hospital reports from eastern states, the following information was obtained (units in percentage of hospitals providing at least some charity care):

57.1	56.2	53.0	66.1	59.0	64.7	70.1	64.7	53.5	78.2

Use a calculator with mean and sample standard deviation keys to verify that $\bar{x} \approx 62.3\%$ and $s \approx 8.0\%$. Find a 90% confidence interval for the population average μ of the percentage of hospitals providing at least some charity care.

13. *Vital Statistics: Heights* The distribution of the heights of 18-year-old men in the United States is approximately normal with mean 68 inches and standard deviation 3 inches (U.S. Census Bureau). In Minitab we can simulate the drawing of random samples of size 20 from this population (➤ **Calc** ➤ **Random Data** ➤ **Normal** with 20 rows from a distribution with mean 68 and standard deviation 3). Then we can have Minitab compute a 95% confidence interval and draw a boxplot of the data (➤ **Stat** ➤ **Basic Statistics** ➤ **1—Sample t**, with boxplot selected in the graphs). The boxplots and confidence intervals for four different samples are shown in the accompanying figures. The four confidence intervals are

VARIABLE	N	MEAN	STDEV	SEMEAN	95.0 % CI
Sample 1	20	68.050	2.901	0.649	(66.692 , 69.407)
Sample 2	20	67.958	3.137	0.702	(66.490 , 69.426)
Sample 3	20	67.976	2.639	0.590	(66.741 , 69.211)
Sample 4	20	66.908	2.440	0.546	(65.766 , 68.050)

(a) Examine the figure [parts (a) to (d)]. How do the boxplots for the four samples differ? Why should you expect the boxplots to differ?

95% Confidence Intervals for Mean Height of 18-Year-Old Men (Sample size 20)

(a) Boxplot of Sample 1
(with 95% *t*-confidence interval for the mean)

(b) Boxplot of Sample 2
(with 95% *t*-confidence interval for the mean)

(c) Boxplot of Sample 3
(with 95% *t*-confidence interval for the mean)

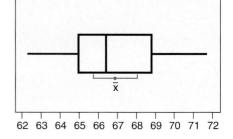

(d) Boxplot of Sample 4
(with 95% *t*-confidence interval for the mean)

(b) Examine the 95% confidence intervals for the four samples shown in the above printout. Do the intervals differ in length? Do the intervals all contain the expected population mean of 68 inches? If we draw more samples, do you expect all of the resulting 95% confidence intervals to contain $\mu = 68$? Why or why not?

9.3
Estimating *p* in the Binomial Distribution

FOCUS POINTS

✓ Compute the maximal error tolerance for proportions using a given level of confidence.

✓ Compute confidence intervals for *p* and interpret the results.

✓ Interpret poll results and the margin of error.

The binomial distribution is completely determined by the number of trials *n* and the probability *p* of success in a single trial. For most experiments, the number of trials is chosen in advance. Then the distribution is completely determined by *p*. In this section, we will consider the problem of estimating *p* under the assumption that *n* has already been selected.

Again, we are employing what are called *large-sample methods*. We will assume that the normal curve is a good approximation to the binomial distribution, and when necessary, we will use sample estimates for the standard deviation. Empirical studies have shown that these methods are quite good provided that *both*

$$np > 5 \quad \text{and} \quad nq > 5 \quad \text{where } q = 1 - p$$

Basic criteria

Let *r* be the number of successes out of *n* trials in a binomial experiment. We will take the sample proportion of successes $\hat{p}$ (read "*p* hat") $= r/n$ as our *point estimate for p*, the population proportion of successes.

Point estimate for *p*	**Point estimate for *q***
$\hat{p} = \dfrac{r}{n}$	$\hat{q} = 1 - \hat{p}$

For example, suppose that 800 students are selected at random from a student body of 20,000 and that they are each given shots to prevent a certain type of flu. These 800 students are then exposed to the flu, and 600 of them do not get the flu. What is the probability *p* that the shot will be successful for any single student selected at random from the entire population of 20,000 students? We estimate *p* for the entire student body by computing *r/n* from the sample of 800 students. The value $\hat{p} = r/n$ is 600/800, or 0.75. The value $\hat{p} = 0.75$ is then the *point estimate* for *p*.

The difference between the actual value of *p* and the estimate $\hat{p}$ is the size of our error caused by using $\hat{p}$ as a point estimate for *p*. The magnitude of $\hat{p} - p$ is called the *error of estimate* for using $\hat{p} = r/n$ as a point estimate for *p*. In absolute value notation, the error of estimate is $|\hat{p} - p|$.

To compute the bounds for the error of estimate, we need some information about the distribution of $\hat{p} = r/n$ values for different samples of the same size *n*. It turns out that, for large samples, the distribution of $\hat{p}$ values is well approximated by a *normal curve* with

$$\text{mean } \mu = p \qquad \text{and} \qquad \text{standard error } \sigma = \sqrt{pq/n}$$

Since the distribution of $\hat{p} = r/n$ is approximately normal, we use features of the standard normal distribution to find the bounds for the difference $\hat{p} - p$. Recall that z_c is the number such that an area equal to *c* under the standard normal curve falls between $-z_c$ and z_c. Then, in terms of the language of probability,

$$P\left(-z_c\sqrt{\frac{pq}{n}} < \hat{p} - p < z_c\sqrt{\frac{pq}{n}}\right) = c \tag{10}*$$

Equation (10) says that the chance is *c* that the numerical difference between $\hat{p}$ and *p* is between $-z_c\sqrt{pq/n}$ and $z_c\sqrt{pq/n}$. With the *c* confidence level, our estimate $\hat{p}$ differs from *p* by no more than

$$E = z_c\sqrt{pq/n}$$

As in Section 9.1, we call *E* the *maximal error tolerance* of the error of estimate $|\hat{p} - p|$ for a confidence level *c*.

To find a *c* confidence interval for *p*, we will use *E* in place of the expression $z_c\sqrt{pq/n}$ in Equation (10). Then we get

$$P(-E < \hat{p} - p < E) = c \tag{11}$$

Some algebraic manipulation produces the mathematically equivalent statement

$$P(\hat{p} - E < p < \hat{p} + E) = c \tag{12}$$

Equation (12) says that the probability is *c* that *p* lies in the interval from $\hat{p} - E$ to $\hat{p} + E$. Therefore, the interval from $\hat{p} - E$ to $\hat{p} + E$ is the *c* confidence interval for *p* that we wanted to find.

There is one technical difficulty in computing the *c* confidence interval for *p*. The expression $E = z_c\sqrt{pq/n}$ requires that we know the values of *p* and *q*. In most

---

*Recall from Section 7.4 that when *n* is large, the binomial distribution of the number of successes *r* is approximately normal with mean $\mu = np$ and standard deviation $\sigma = \sqrt{npq}$. Therefore, $z = (r - np)/\sqrt{npq}$. Dividing the numerator and denominator by *n* shows that $\hat{p} = r/n$ has a normal distribution with mean $\mu = p$ and standard deviation $\sigma = \sqrt{pq/n}$. Beginning with the equation $P(-z_c < z < z_c) = c$, replacing *z* by $(\hat{p} - p)/\sqrt{pq/n}$, and multiplying all parts of the inequality by $\sqrt{pq/n}$, we obtain Equation 10.

situations, we will not know the actual values of p or q, so we will use our point estimates

$$p \approx \hat{p} \quad \text{and} \quad q = 1 - p \approx 1 - \hat{p}$$

to estimate E. These estimates are safe for most practical purposes, since we are dealing with large-sample theory ($np > 5$ and $nq > 5$).

For convenient reference, we'll summarize the information about c confidence intervals for p, the probability of success in a binomial distribution.

◇ **SUMMARY** Consider a binomial distribution where n = number of trials, r = number of successes out of the n trials, p = probability of success on each trial, and q = probability of failure on each trial.

If n, p, and q are such that

$$np > 5 \quad \text{and} \quad nq > 5$$

then a c confidence interval for p is

$$\hat{p} - E < p < \hat{p} + E$$

where $\hat{p} = \dfrac{r}{n}$

$$E \approx z_c \sqrt{\frac{\hat{p}(1 - \hat{p})}{n}} \quad \text{or} \quad z_c \sqrt{\frac{\hat{p}\hat{q}}{n}}$$

z_c = critical value for confidence level c taken from a normal distribution (*see* Table 9-2 or Table 4(b) of the Appendix).

◇

EXAMPLE 5

Confidence interval for p

Let's return to our flu shot experiment described at the beginning of this section. Suppose that 800 students were selected at random from a student body of 20,000 and given shots to prevent a certain type of flu. All 800 students were exposed to the flu, and 600 of them did not get the flu. Let p represent the probability that the shot will be successful for any single student selected at random from the entire population of 20,000. Let q be the probability that the shot is not successful.

(a) What is the number of trials n? What is the value of r?

   SOLUTION: Since each of the 800 students receiving the shot may be thought of as a trial, then $n = 800$, and $r = 600$ is the number of successful trials.

(b) What are the point estimates for p and q?

   SOLUTION: We estimate p by the sample point estimate

$$\hat{p} = \frac{r}{n} = \frac{600}{800} = 0.75$$

   We estimate q by

$$\hat{q} = 1 - \hat{p} = 1 - 0.75 = 0.25$$

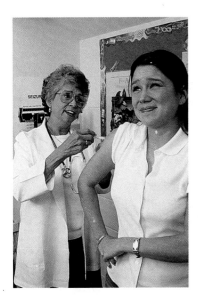

(c) Would it seem that the number of trials is large enough to justify a normal approximation to the binomial?

SOLUTION: Since $n = 800$, $p \approx 0.75$, and $q \approx 0.25$, then

$$np \approx (800)(0.75) = 600 > 5$$

and

$$nq \approx (800)(0.25) = 200 > 5$$

A normal approximation is certainly justified.

(d) Find a 99% confidence interval for p.

SOLUTION:

$$z_{0.99} = 2.58 \text{ (see Table 9-2)}$$

$$E \approx z_{0.99}\sqrt{\frac{\hat{p}(1 - \hat{p})}{n}} \approx 2.58\sqrt{\frac{(0.75)(0.25)}{800}} \approx 0.0395$$

The 99% confidence interval is then

$$\hat{p} - E < p < \hat{p} + E$$
$$0.75 - 0.0395 < p < 0.75 + 0.0395$$
$$0.71 < p < 0.79$$

GUIDED EXERCISE **5**

Confidence interval for p

A random sample of 188 books purchased at a local bookstore showed that 66 of the books were murder mysteries. Let p represent the proportion of books sold by this store that are murder mysteries.

(a) What is a point estimate for p?

$$\hat{p} = \frac{r}{n} = \frac{66}{188} \approx 0.35$$

(b) Find a 90% confidence interval for p.

$$E = z_c\sqrt{\frac{\hat{p}(1 - \hat{p})}{n}}$$

$$\approx 1.645\sqrt{\frac{(0.35)(1 - 0.35)}{188}} \approx 0.0572$$

The confidence interval is

$$\hat{p} - E < p < \hat{p} + E$$
$$0.35 - 0.0572 < p < 0.35 + 0.0572$$
$$0.29 < p < 0.41$$

Continued

(c) What is the meaning of the confidence interval you just computed?

If we had computed the interval for many different sets of 188 books, we would find that about 90% of the intervals actually contained p, the population proportion of mysteries. Consequently, we can be 90% confident that our interval is one of the ones that contains the unknown value p.

(d) To compute the confidence interval, we used a normal approximation. Does this seem justified?

$n = 188$;

$p \approx 0.35; q \approx 0.65$

Since $np \approx 65.8 > 5$ and $nq \approx 122.2 > 5$, the approximation is justified.

It is interesting to note that our sample point estimate $\hat{p} = r/n$ and the confidence interval for the population proportion p do not depend on the size of the population. In our bookstore example, it made no difference how many books the store sold. On the other hand, the size of the sample does affect the accuracy of a statistical estimate. In the next section, we will study the effect of sample size on the reliability of our estimate.

 TECH NOTE   Both the TI-83Plus and Minitab provide confidence intervals for proportions.

TI-83Plus Press the **STAT** key, select **TESTS,** and choose option **A:1-PropZInt.** The letter x represents the number of successes r. The TI-83Plus output shows the results for Guided Exercise 5.

```
1-PropZInt
(.29381,.40832)
p̂=.3510638298
n=188
```

Minitab Use menu selections **Stat ➤ Basic Statistics ➤ 1 Proportion.** In the dialogue box, select summarized data and fill in the number of trials and the number of successes. Under Options, select a confidence interval. Minitab uses the binomial distribution directly unless normal is checked. The Minitab output shows the results for Guided Exercise 5. Information from Chapter 10 material is also shown.

```
Test and Confidence Interval for One Proportion (Using Binomial)
Test of p = 0.5 vs p not = 0.5
                                                          Exact
Sample    X     N    Sample p       90.0 % CI           P-Value
   1      66   188    0.351064   (0.293222, 0.412466)    0.000
```

```
Test and Confidence Interval for One Proportion (Using Normal)
Test of p = 0.5 vs p not = 0.5
Sample   X    N   Sample p        90.0 % CI        Z-Value   P-Value
   1     66  188  0.351064  (0.293805, 0.408323)   -4.08     0.000
```

Margin of error

Newspapers frequently report the results of an opinion poll. In articles that give more information, a statement about the margin of error accompanies the poll results. The *margin of error* is the maximal error of estimate E for a confidence interval. Usually, a 95% confidence interval is assumed.

General interpretation of poll results

1. When a poll states the results of a survey, the proportion reported to respond in the designated manner is $\hat{p}$, the sample estimate of the population proportion.

2. The *margin of error* is the maximal error E of a 95% confidence interval for p.

3. A 95% confidence interval for the population proportion p is

poll report $\hat{p}$ − margin of error $E < p <$ poll report $\hat{p}$ + margin of error E.

◇ **COMMENT:** Leslie Kish, a statistician at the University of Michigan, was the first to apply the term *margin of error*. He was a pioneer in the study of population sampling techniques. His book *Survey Sampling* is still widely used all around the world.   ◇

Some articles clarify the meaning of the margin of error further by saying that it is an error due to sampling. For instance, the following comments accompany results of a political poll reported in the October 29, 1993, issue of *The Wall Street Journal.*

How Poll Was Conducted

The Wall Street Journal/NBC news poll was based on nationwide telephone interviews of 1508 adults conducted last Friday through Tuesday by the polling organizations of Peter Hart and Robert Teeter.

The sample was drawn from 315 randomly selected geographic points in the continental U.S. Each region was represented in proportion to its population. Households were selected by a method that gave all telephone numbers . . . an equal chance of being included.

One adult, 18 years or older, was selected from each household by a procedure to provide the correct number of male and female respondents.

Chances are 19 of 20 that if all adults with telephones in the U.S. had been surveyed, the findings would differ from these poll results by no more than 2.6 percentage points in either direction.

GUIDED EXERCISE 6

Margin of error

Read the last paragraph of the article on the previous page, "How Poll Was Conducted."

(a) What confidence level corresponds to the phrase "chances are 19 of 20 that if . . ."

➡ $\frac{19}{20} = 0.95$

A 95% confidence interval is being discussed.

(b) The article indicates that everyone in the sample was asked the question, "Which party, the Democratic Party or the Republican Party, do you think would do a better job handling . . . education?" Possible responses were Democrats, neither, both, or Republicans. The poll reported that 32% of the respondents said "Democrats." Does 32% represent the sample statistic $\hat{p}$ or the population parameter p for the proportion of adults responding "Democrat"?

➡ 32% represents a sample statistic $\hat{p}$ because 32% represents the percentage of the adults in the *sample* who responded "Democrats."

(c) Continue reading the last paragraph of the article. It goes on to state, ". . . if all adults with telephones in the U.S. had been surveyed, the findings would differ from these poll results by no more than 2.6 percentage points in either direction." Use this information together with parts (a) and (b) to find a 95% confidence interval for the proportion p of the specified population who would respond "Democrat" to the question.

➡ The value 2.6 percentage points represents the margin of error. Since the margin of error is equivalent to E, the maximal error of estimate for a 95% confidence interval, the confidence interval is

32% − 2.6% < p < 32% + 2.6%

29.4% < p < 34.6%

The poll indicates that at the time of the poll, between 29.4% and 34.6% of the specified population think Democrats would do a better job handling education.

VIEWPOINT   *"Band-Aid Surgery"*

Faster recovery time and less pain! Sounds great. An alternate surgical technique called *laparoscopic* ("Band-Aid") *surgery* involves small incisions in which tiny video cameras and long surgical instruments are maneuvered. Instead of a 10-inch incision, surgeons might use four little stabs of about $\frac{1}{2}$-inch in length. However, not every such surgery is successful. An article in the Health Section of *The Wall Street Journal* recommends using a surgeon who has done at least 50 such surgeries. Then the prospective patient should ask about the *rate of conversion*, that is, the proportion p of times the surgeon has been forced by complications to switch in midoperation to conventional surgery. A confidence interval for the proportion p would be useful patient information!

SECTION 9.3   PROBLEMS

For all these problems, carry at least four digits after the decimal in your calculations. Answers may vary slightly due to rounding.

1. *Myers-Briggs: Actors* Isabel Myers was a pioneer in the study of personality types. The following information is taken from *A Guide to the Development and Use of the Myers-Briggs Type Indicator,* by Myers and McCaulley (Consulting Psychologists Press). In a random sample of 62 professional actors, it was found that 39 were extroverts.

   (a) Let p represent the proportion of all actors who are extroverts. Find a point estimate for p.

   (b) Find a 95% confidence interval for p. Give a brief interpretation of the meaning of the confidence interval you have found.

   (c) Do you think the conditions $np > 5$ and $nq > 5$ are satisfied in this problem? Explain why this would be an important consideration.

2. *Myers-Briggs: Judges* In a random sample of 519 judges, it was found that 285 were introverts (*see* reference of Problem 1).

   (a) Let p represent the proportion of all judges who are introverts. Find a point estimate for p.

   (b) Find a 99% confidence interval for p. Give a brief interpretation of the meaning of the confidence interval you have found.

   (c) Do you think the conditions $np > 5$ and $nq > 5$ are satisfied in this problem? Explain why this would be an important consideration.

3. *Navajo Lifestyle: Traditional Hogans* A random sample of 5222 permanent dwellings on the entire Navajo Indian Reservation showed that 1619 were traditional Navajo hogans (*Navajo Architecture: Forms, History, Distributions,* by Jett and Spencer, University of Arizona Press).

   (a) Let p be the proportion of all permanent dwellings on the entire Navajo Reservation that are traditional hogans. Find a point estimate for p.

   (b) Find a 99% confidence interval for p. Give a brief interpretation of the confidence interval.

   (c) Do you think that $np > 5$ and $nq > 5$ are satisfied for this problem? Explain why this would be an important consideration.

4. *Archaeology: Pottery* Santa Fe black on white is a type of pottery commonly found at archaeological excavations in Bandelier National Monument. At one excavation site a sample of 592 potsherds was found, of which 360 were identified as Santa Fe black on white (*Bandelier Archaeological Excavation Project: Summer 1990 Excavations at Burnt Mesa Pueblo and Casa del Rito,* edited by Kohler and Root, Washington State University).

   (a) Let p represent the population proportion of Santa Fe black on white potsherds at the excavation site. Find a point estimate for p.

   (b) Find a 95% confidence interval for p. Give a brief statement of the meaning of the confidence interval.

   (c) Do you think that the conditions $np > 5$ and $nq > 5$ are satisfied in this problem? Why would this be important?

5. *Health Care: Colorado Physicians* A random sample of 5792 physicians in Colorado showed that 3139 provided at least some charity care (i.e., treated poor people at no cost). These data are based on information from *State Health Care Data: Utilization, Spending, and Characteristics* (American Medical Association).

(a) Let p represent the proportion of all Colorado physicians who provide some charity care. Find a point estimate for p.

(b) Find a 99% confidence interval for p. Give a brief explanation of the meaning of your answer in the context of this problem.

(c) Is the normal approximation to the binomial justified in this problem? Explain.

6. *Law Enforcement: Escaped Convicts* Case studies showed that out of 10,351 convicts who escaped from U.S. prisons, only 7867 were recaptured (*The Book of Odds*, by Shook and Shook, Signet).

(a) Let p represent the proportion of all escaped convicts who will eventually be recaptured. Find a point estimate for p.

(b) Find a 99% confidence interval for p. Give a brief statement of the meaning of the confidence interval.

(c) Is use of the normal approximation to the binomial justified in this problem? Explain.

7. *Fishing: Barbless Hooks* In a combined study of northern pike, cutthroat trout, rainbow trout, and lake trout, it was found that 26 out of 855 fish died when caught and released using barbless hooks on flies or lures. All hooks were removed from the fish (*Source: A National Symposium on Catch and Release Fishing*, Humboldt State University Press).

(a) Let p represent the proportion of all pike and trout that die (i.e., p is the mortality rate) when caught and released using barbless hooks. Find a point estimate for p.

(b) Find a 99% confidence interval for p, and give a brief explanation of the meaning of the interval.

(c) Is the normal approximation of the binomial justified in this problem? Explain.

8. *Pharmacy: Errors* Does a doctor's illegible handwriting cause problems for pharmacists? According to a report in *USA Today*, the percentage of pharmacists who say a doctor's illegible handwriting causes problems in errors or safety is 85%. Let's assume that this report is based on sample data.

(a) Compute a 95% confidence interval for the population proportion of pharmacists who say a doctor's illegible handwriting causes errors or safety problems under the assumption that the sample estimate $\hat{p}$ is based on a random sample of 100 pharmacists.

(b) Repeat part (a) under the assumption that the sample consists of 1000 pharmacists.

(c) Compare the confidence intervals computed in parts (a) and (b). What effect does increasing the sample size have on the length of the confidence interval?

9. *Surgeons: Knees* According to the American Academy of Orthopaedic Surgeons, 26.6% of all orthopaedic treatments are to the knee. Let's assume that this report is based on sample data.

(a) Compute a 95% confidence interval for the actual proportion p of orthopaedic treatments that are to the knee under the assumption that the sample estimate $\hat{p}$ is based on a random sample of 1000 general orthopaedic case treatments.

(b) Repeat part (a) under the assumption that the sample consists of 10,000 general orthopaedic case treatments.

(c) Compare the confidence intervals computed in parts (a) and (b). What effect does increasing the sample size have on the length of the confidence interval?

10. *Newspaper Reporting: Margin of Error* If you were writing a newspaper report on the results of the poll of pharmacists regarding the percentage that feel that a doctor's illegible handwriting can cause errors or safety problems [*see* Problem 8, part

(a)], what would you state as the margin of error for a sample size of 100? Follow the convention that the margin of error is based on a 95% confidence interval.

11. *Marketing: Customer Loyalty* In a marketing survey, a random sample of 730 women shoppers revealed that 628 remained loyal to their favorite supermarket during the past year (i.e., did not switch stores). (*Source: Trends in the United States: Consumer Attitudes and the Supermarket,* The Research Department, Food Marketing Institute.)
    (a) Let p represent the proportion of all women shoppers who remain loyal to their favorite supermarket. Find a point estimate for p.
    (b) Find a 95% confidence interval for p. Give a brief explanation of the meaning of the interval.
    (c) As a news writer, how would you report the survey results regarding the percentage of supermarket shoppers who remained loyal to their favorite supermarket during the past year? What is the margin of error based on a 95% confidence interval?

12. *Marketing: Bargain Hunters* In a marketing survey, a random sample of 1001 supermarket shoppers revealed that 273 always stock up on an item when they find that item at a real bargain price. (*See* reference in Problem 11.)
    (a) Let p represent the proportion of all supermarket shoppers who always stock up on an item when they find a real bargain. Find a point estimate for p.
    (b) Find a 95% confidence interval for p. Give a brief explanation of the meaning of the interval.
    (c) As a news writer, how would you report the survey results on the percentage of supermarket shoppers who stock up on items when they find the item is a real bargain? What is the margin of error based on a 95% confidence interval?

13. *Lifestyle: Smoking* In a survey of 1000 large corporations, 250 said that, given a choice between a job candidate who smokes and an equally qualified nonsmoker, the nonsmoker will get the job (*USA Today*).
    (a) Let p represent the proportion of all corporations preferring a nonsmoking candidate. Find a point estimate for p.
    (b) Find a 0.95 confidence interval for p.
    (c) As a news writer, how would you report the survey results regarding the proportion of corporations who would hire the equally qualified nonsmoker? What is the margin of error based on a 95% confidence interval?

14. *Focus Problem: Trick or Treat* Solve the Focus Problem at the beginning of this chapter.

9.4
Choosing the Sample Size

FOCUS POINTS

✓ Compute the sample size to be used for estimating a mean μ.

✓ Compute the sample size to be used for estimating a proportion p when we have an estimate for p.

✓ Compute the sample size to be used for estimating a proportion p when we have no estimate for p.

In the design stages of statistical research projects, it is a good idea to decide in advance on the confidence level you wish to use and to select the *maximum* error of estimate E you want for your project. How you choose to make these decisions depends on the requirements of the project and the practical nature of the problem. Whatever specifications you make, the next step is to determine the sample size. In this section, we will assume that the distribution of sample means $\bar{x}$ is approximately normal, and when necessary, we will approximate σ by the sample standard deviation s. These methods are technically justifiable since our samples will be of size at least 30.

Sample size for estimating μ

Let's say that at a confidence level of c, we want our point estimate $\bar{x}$ for μ to be in error either way by less than some quantity E. In other words, E is the maximum error of estimate we can tolerate. Using the language of probability, we want the following to be true:

$$P(-E < \bar{x} - \mu < E) = c \tag{13}$$

This is essentially the same as Equation (1) of Section 9.1. Let's compare them.

$$P(-E < \bar{x} - \mu < E) = c \tag{13}$$

$$P\left(-z_c \frac{\sigma}{\sqrt{n}} < \bar{x} - \mu < z_c \frac{\sigma}{\sqrt{n}}\right) = c \tag{1}$$

From this comparison, we see that we want E to be

$$E = z_c \frac{\sigma}{\sqrt{n}}$$

Solving this equation for n, we get

$$n = \left(\frac{z_c \sigma}{E}\right)^2 \tag{14}$$

To compute n from Equation (14), we must know the value of σ. If the value of σ is not previously known, we do a preliminary sampling to approximate it. For most practical purposes, a preliminary sample of size 30 or larger will give a sample standard deviation s, which we may use to approximate σ.

EXAMPLE 6

Sample size for estimating μ

A wildlife study is designed to find the mean weight of salmon caught by an Alaskan fishing company. As a preliminary study, a random sample of 50 freshly caught salmon is weighed. The sample standard deviation of the weights of these 50 fish is $s = 2.15$ lb. How large a sample should be taken to be 99% confident that the sample mean $\bar{x}$ is within 0.20 lb of the true mean weight μ?

SOLUTION: In this problem, $z_{0.99} = 2.58$ (see Table 9-2) and $E = 0.20$. The preliminary study of 50 fish is large enough to permit a good approximation of σ by $s = 2.15$. Therefore, Equation (14) becomes

$$n = \left(\frac{z_c \sigma}{E}\right)^2 \approx \left(\frac{(2.58)(2.15)}{0.20}\right)^2 = 769.2$$

NOTE In determining sample size, any fraction value of n is always rounded to the *next higher whole number*. We conclude that a sample size of 770 will be enough to satisfy the specifications. Of course, a sample size larger than 770 also works.   ◇

Salmon moving upstream

EXAMPLE 7

Sample size for estimating μ

A certain company makes light fixtures on an assembly line. An efficiency expert wants to determine the mean time it takes an employee to assemble the switch on one of these fixtures. A preliminary study used a random sample of 45 observations and found that the sample standard deviation was $s = 78$ seconds. How many more observations are necessary for the efficiency expert to be 95% sure that the point estimate $\bar{x}$ will be "off" from the true mean μ by at most 15 seconds?

SOLUTION: In this example, we approximate σ by $s = 78$. We use $z_{0.95} = 1.96$ (see Table 9-2). The maximum error of estimate is specified to be $E = 15$ seconds. Equation (14) gives us

$$n = \left(\frac{z_c \sigma}{E}\right)^2 = \left(\frac{(1.96)(78)}{15}\right)^2 = 103.9$$

The efficiency expert should use a sample of minimum size 104. Since the preliminary study has 45 observations, an additional $104 - 45 = 59$ observations are necessary.   ◇

GUIDED EXERCISE 7

Sample size for estimating μ

A large state university has over 1800 faculty members. The dean of faculty wants to estimate the average teaching experience (in years) of the faculty members. A preliminary random sample of 60 faculty members yields a sample standard deviation of $s = 3.4$ years. The dean wants to be 99% confident that the sample mean $\bar{x}$ does not differ from the population mean by more than half a year. How large a sample should be used? Let's answer this question in parts.

(a) What value can we use to approximate σ? Why can we do this?

➡ $s = 3.4$ years is a good approximation because a preliminary sample of 60 is fairly large.

(b) What is $z_{0.99}$? (*Hint: See* Table 9-2.)

➡ $z_{0.99} = 2.58$

(c) What is E for this problem?

➡ $E = 0.5$ year

(d) Which is the correct formula for n:

$$\left(\frac{z_c \sigma}{n}\right)^2, \quad \left(\frac{z_c \sigma}{E}\right)^2, \quad \text{or} \quad \left(\frac{z_c E}{\sigma}\right)^2$$

➡ $n = \left(\frac{z_c \sigma}{E}\right)^2$

(e) Use the formula for n to find the minimum sample size. Should your answer be rounded up or down to a whole number?

➡ $n = \left(\frac{(2.58)(3.4)}{0.5}\right)^2 = (17.544)^2 \approx 307.8$

Always round n up to the next whole number. Our final answer $n = 308$ is the minimum size.

Sample size for estimating $\hat{p} = r/n$

(If you omitted the binomial distribution, omit the rest of this section.)

Next, we will determine the minimum sample size when we use the sample proportion $\hat{p} = r/n$ as a point estimate for p in a binomial distribution. We will use the methods of normal approximation (large samples) discussed in Section 9.3. Suppose for a confidence level c we want the estimate $\hat{p} = r/n$ for p to be in error either way by less than some quantity E. Using the language of probability, we want the following to be true.

$$P(-E < \hat{p} - p < E) = c \tag{15}$$

Let's compare this with Equation (10) of Section 9.3. For convenience, they both are written together:

$$P(-E < \hat{p} - p < E) = c \tag{15}$$

$$P\left(-z_c\sqrt{\frac{pq}{n}} < \hat{p} - p < z_c\sqrt{\frac{pq}{n}}\right) = c \tag{10}$$

The comparison of the two equations gives a formula for E:

$$E = z_c\sqrt{\frac{pq}{n}}$$

Solving the last equation for n, we get

$$n = pq\left(\frac{z_c}{E}\right)^2$$

Since $q = 1 - p$, our equation for n can be written

$$n = p(1 - p)\left(\frac{z_c}{E}\right)^2 \tag{16}$$

Equation (16) cannot be used unless we already have a preliminary estimate for p. To get around this difficulty, we use a result from algebra showing that the maximum possible value of $p(1 - p)$ is $\frac{1}{4}$. Therefore, *when we have no preliminary estimate for p, we use the formula*

$$n = \frac{1}{4}\left(\frac{z_c}{E}\right)^2 \tag{17}$$

Since Equation (17) may make the sample size unnecessarily large, we can say the probability is *at least* (and possibly more than) c that the point estimate $\hat{p} = r/n$ for p will be in error either way by less than the quantity E.

EXAMPLE 8

Sample size for estimating p

A company is in the business of selling wholesale popcorn to grocery stores. The company buys directly from farmers. A buyer for the company is examining a large amount of corn from a certain farmer. Before the purchase is made, the buyer wants to estimate p, the probability that a kernel will pop.

Suppose that a random sample of n kernels is taken and r of these kernels pop. The buyer wants to be 95% sure that the point estimate $\hat{p} = r/n$ for p will be in error either way by less than 0.01.

(a) If no preliminary study is made to estimate p, how large a sample should the buyer use?

SOLUTION: In this case, we use Equation (17) with $z_{0.95} = 1.96$ (see Table 9-2) and $E = 0.01$.

$$n = \frac{1}{4}\left(\frac{z_c}{E}\right)^2 = \frac{1}{4}\left(\frac{1.96}{0.01}\right)^2 = 0.25(38{,}416) = 9604$$

We would need a sample of $n = 9604$ kernels.

(b) A preliminary study showed that p was approximately 0.86. If the buyer uses the results of the preliminary study, how large a sample should be used?

SOLUTION: In this case, we use Equation (16) with $p \approx 0.86$. Again, from Table 9-2, $z_{0.95} = 1.96$, and from the problem, $E = 0.01$.

$$n = p(1 - p)\left(\frac{z_c}{E}\right)^2 = (0.86)(0.14)\left(\frac{1.96}{0.01}\right)^2 = 4625.29$$

The sample size should be at least $n = 4626$ kernels. This sample is less than half the sample size necessary without the preliminary study.   ◇

GUIDED EXERCISE 8

Sample size for estimating p

In Indianapolis, the department of public health wants to estimate the proportion of children (grades 1–8) who require corrective lenses for their vision. A random sample of n children is taken, and r of these children are found to require corrective lenses. Let p be the true proportion of children requiring corrective lenses. The health department wants to be 99% sure that the point estimate $\hat{p} = r/n$ for p will be in error either way by less than 0.03.

(a) If no preliminary study is made to estimate p, how large a sample should the health department use? Let's answer this question in parts.

(i) Which formula shall we use: (16) or (17)?   ⟹   (i) We use Formula (17) because we do not have an estimate for p.

(ii) What is the value of E, and what is the value of z_c in this problem?   ⟹   (ii) $E = 0.03$ and $z_{0.99} = 2.58$ (*see* Table 9-2)

(iii) What is the value of n?   ⟹   (iii) $n = \dfrac{1}{4}\left(\dfrac{z_c}{E}\right)^2 = \dfrac{1}{4}\left(\dfrac{2.58}{0.03}\right)^2 = 1849$

So without a preliminary study to find p, we will need a sample size of at least $n = 1849$ children.

(b) A preliminary random sample of 100 children indicates that 23 require corrective lenses. Using the results of this preliminary study, how large a sample should the health department use? Again, let's answer this question in parts.

(i) Which formula should we use: (16) or (17)?   ⟹   We use Formula (16) because we have an estimate of p from a preliminary study.

Continued

9. *Highway Safety: Speeding* About what proportion p of drivers admit that they exceed the posted speed limits on interstate highways by 20 mph or more?
   (a) If you have no preliminary estimate for p, how many people should you include in a random sample to be 95% sure that the point estimate $\hat{p}$ will be within a distance of 0.05 from p?
   (b) Answer part (a) if you use the preliminary estimate from the National Highway Traffic Safety Administration survey that nationally 14% of drivers admit to speeding by 20 mph or more on interstate highways.

10. *Business: Phone Contact* How hard is it to reach a businessperson by phone? Let p be the proportion of calls to businesspeople for which the caller reaches the person being called on the *first* try.
    (a) If you have no preliminary estimate for p, how many business phone calls should you include in a random sample to be 80% sure that the point estimate $\hat{p}$ will be within a distance of 0.03 from p?
    (b) The *Book of Odds*, by Shook and Shook (Signet), reports that businesspeople can be reached by a single phone call approximately 17% of the time. Using this (national) estimate for p, answer part (a).

11. *Campus Life: Coeds* What percentage of the campus student body is female? Let p be the proportion of women students on your campus.
    (a) If no preliminary study is made to estimate p, how large a sample is needed to be 99% sure that a point estimate $\hat{p}$ will be within a distance of 0.05 from p?
    (b) The *Statistical Abstract of the United States*, 112th Edition, indicates that approximately 54% of college students are females. Answer part (a) using this estimate for p.

12. *Small Business: Bankruptcy* The National Council of Small Businesses is interested in the proportion of small businesses that declared Chapter 11 bankruptcy last year. Since there are so many small businesses, the National Council intends to estimate the proportion from a random sample. Let p be the proportion of small businesses that declared Chapter 11 bankruptcy last year.
    (a) If no preliminary sample is taken to estimate p, how large a sample is necessary to be 95% sure that a point estimate $\hat{p}$ will be within a distance of 0.10 from p?
    (b) In a preliminary random sample of 38 small businesses, it was found that 6 had declared Chapter 11 bankruptcy. How many *more* small businesses should be included in the sample to be 95% sure that a point estimate $\hat{p}$ will be within a distance of 0.10 from p?

13. *Election Results: Florida* A heavily populated county in Florida is told by the election commission to estimate the proportion p of popular vote count for the Democratic candidate for president.
    (a) The maximal error of estimate is to be 0.001 with a 99% confidence level. If no preliminary estimate for p can be agreed upon, how many ballots must be checked?
    (b) If $p = 0.5$ was used as a preliminary estimate for p, would the answer for part (a) change? Explain.

14. *Polling: Votes* You are working for a local polling firm. Your project involves interviewing a random sample of registered voters to determine the percentage favoring the use of state lottery proceeds for park improvements. The firm wants to ensure that the margin of error is no more than 3 percentage points either way. Assuming that there is no preliminary estimate for the percentage of registered voters favoring this use of lottery funds, what is the *minimum* number of respondents required?

(*Hint*: Convert the margin of error to decimal and use the *New York Times*/CBS convention that the margin of error is based on a 95% confidence level.)

SUMMARY

We have studied point estimates and interval estimates. For point estimates, we found E, the maximal error of estimate, and for interval estimates, we found the interval endpoints. In each case, E or endpoints were determined by four factors: the confidence level c, the sample estimate for μ or p, the sample standard deviation s, and the sample size n.

For large samples ($n \geq 30$), we used the normal distribution, the central limit theorem, and sometimes the normal approximation to the binomial distribution. For small samples ($n < 30$), we made use of Student's t distribution. In applications to choose the sample size, we found formulas for n so that, with probability c, the sample estimate for μ or p is in error by less than a preassigned number E.

IMPORTANT WORDS & SYMBOLS

Section 9.1
Large samples, $n \geq 30$
Point estimate for μ
Confidence level c
Critical values z_c
Error of estimate $|\bar{x} - \mu|$
Maximal error tolerance E
c confidence interval
Confidence interval for μ

Section 9.2
Small samples, $n < 30$

Student's t variable
Degrees of freedom (*d.f.*)
Critical values t_c

Section 9.3
Point estimate for p, $\hat{p}$
Confidence interval for p
Margin of error

Section 9.4
Sample size n

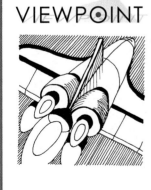

VIEWPOINT   *All Systems Go?*

On January 28, 1986, the Space Shuttle *Challenger* caught fire and blew up only seconds after launch. A great deal of good engineering went into the design of the *Challenger*. However, when a system has several confidence levels operating at once, it can happen, in rare cases, that risks will increase rather than cancel out. Diane Vaughn is a professor of sociology at Boston College and author of the book *The Challenger Launch Decision* (University of Chicago Press). Her book contains an excellent discussion of risks, the normalization of deviants, and cost/safety tradeoffs. Vaughn's book is described as "a remarkable and important analysis of how social structures can induce consequential errors in a decision process" (Robert K. Merton, Columbia University).

CHAPTER REVIEW PROBLEMS

1. In your own words, carefully explain the meaning of the following terms: point estimate, critical value, maximal error of estimate, confidence level, confidence interval, large samples, small samples.

For Problems 2–12, categorize each problem according to (a) parameter being estimated, proportion p or mean μ, and (b) large sample or small sample. Then solve the problem.

2. *Auto Insurance: Claims* Anystate Auto Insurance Company took a random sample of 370 insurance claims paid out during a 1-year period. The average claim paid was $1750 with a standard deviation of $150. Find 0.90 and 0.99 confidence intervals for the mean claim payment.

3. *Psychology: Closure* Three experiments investigating the relation between need for cognitive closure and persuasion were reported in "Motivated Resistance and Openness to Persuasion in the Presence or Absence of Prior Information," by A. W. Kruglanski (*Journal of Personality and Social Psychology*, Vol. 65, No. 5, pp. 861–874). Part of the study involved administering a "need for closure scale" to a group of students enrolled in an introductory psychology course. The "need for closure scale" has scores ranging from 101 to 201. For the 73 students in the highest quartile of the distribution, the mean score was $\bar{x} = 178.70$ with sample standard deviation $s = 7.81$. These students were all classified as high on their need for closure. Assume that the 73 students represent a random sample of all students who are classified as high on their need for closure. Find a 95% confidence interval for the population mean score μ on the "need for closure scale" for all students with a high need for closure.

4. *Psychology: Closure* How large a sample is needed in Problem 3 if we wish to be 99% confident that the sample mean score is within 2 points of the population mean score for students who are high on the need for closure?

5. *Archaeology: Excavations* The Wind Mountain archaeological site is located in southwestern New Mexico. Wind Mountain was home to an ancient culture of prehistoric Native Americans called Anasazi. A random sample of excavations at Wind Mountain gave the following depth (in centimeters) from present-day surface grade to the location of significant archaeological artifacts (*Source: Mimbres Mogollon Archaeology*, by A. Woosley and A. McIntyre, University of New Mexico Press).

85	45	120	80	75	55	65	60
65	95	90	70	75	65	68	

(a) Use a calculator with mean and sample standard deviation keys to verify that $\bar{x} \approx 74.2$ cm and $s \approx 18.3$ cm.

(b) Compute a 95% confidence interval for the mean depth μ at which archaeological artifacts from the Wind Mountain excavation site can be found.

6. *Archaeology: Pottery* Sherds of clay vessels were put together to reconstruct rim diameters of the original ceramic vessel at the Wind Mountain archaeological site (*see* source in Problem 5). A random sample of ceramic vessels gave the following rim diameters (in centimeters):

15.9	13.4	22.1	12.7	13.1	19.6	11.7	13.5	17.7	18.1

(a) Use a calculator with mean and sample standard deviation keys to verify that $\bar{x} \approx 15.8$ cm and $s \approx 3.5$ cm.

(b) Compute an 80% confidence interval for the population mean μ of rim diameters for such ceramic vessels found at the Wind Mountain archaeological site.

7. *Telephone Interviews: Survey* The National Study of the Changing Work Force conducted an extensive survey of 2958 wage and salaried workers on issues ranging from relationship with their bosses to household chores. The data were gathered through hour-long telephone interviews with a nationally representative sample (*The Wall Street Journal*). In response to the question, "What does success mean to you?" 1538 responded, "Personal satisfaction from doing a good job." Let p be the population proportion of all wage and salaried workers who would respond the same way to the stated question. Find a 90% confidence interval for p.

8. *Telephone Interviews: Survey* How large a sample is needed in Problem 7 if we wish to be 95% confident that the sample percentage of those equating success with personal satisfaction is within 1% of the population percentage? (*Hint:* Use $p \approx 0.52$ as a preliminary estimate.)

9. *Archaeology: Pottery* Three circle, red on white is one distinctive pattern painted on ceramic vessels of the Anasazi period found at the Wind Mountain archaeological site (*see* source for Problem 5). At one excavation a sample of 167 potsherds indicated that 68 were of the three circle, red on white pattern.
   (a) Find a point estimate $\hat{p}$ for the proportion of all ceramic potsherds at this site that are of the three circle, red on white pattern.
   (b) Compute a 95% confidence interval for the population proportion p of all ceramic potsherds with this distinctive pattern found at the site.

10. *Archaeology: Pottery* Consider the three circle, red on white pattern discussed in Problem 9. How many ceramic potsherds must be found and identified if we are to be 95% confident that the sample proportion $\hat{p}$ of such potsherds is within 6% of the population proportion of three circle, red on white pattern found at this excavation site? (*Hint:* Use the results of Problem 9 as a preliminary estimate.)

11. *TV: Commercials* During a television miniseries, what is the average length of time between commercial breaks? A random sample of 20 such periods was selected from miniseries that were aired on commercial television stations last year. The times between commercial breaks were (to the nearest minute)

5	7	8	14	13
10	9	8	11	12
14	11	9	10	6
8	12	5	11	8

(a) Use a calculator to verify that the sample mean of the times between commercial breaks is $\bar{x} = 9.55$ min with standard deviation $s = 2.72$ min.
(b) Find a 95% confidence interval for the mean length of time between commercial breaks.

12. *Animal Behavior: Coyotes* The book *Coyotes, Biology, Behavior and Management*, edited by Marc Bekoff, Academic Press, contains many interesting facts about coyote populations and behavior. One study involved the spring coyote density (that is, coyotes per square kilometer) in the National Elk Refuge near Grand Teton National Park in Wyoming. A sample of 5 years showed the following coyote density (coyotes per square kilometer) in the refuge.

   0.51   0.58   0.46   0.54   0.54

   Assuming the spring coyote density is approximately normal, find a 95% confidence interval for the coyote density on the National Elk Refuge.

384   Chapter 9   Estimation

Break into small groups and discuss the following topics. Organize a brief outline in which you summarize the main points of your group discussion.

Digging clams

1. Garrison Bay is a small bay in Washington state. A popular recreational activity in the bay is clam digging. For several years, this harvest has been monitored and the size distribution of clams recorded. Data for lengths and widths of little neck clams (*Protothaca staminea*) were recorded by a method of systematic sampling in a study done by S. Scherba and V. F. Gallucci ("The Application of Systematic Sampling to a Study of Infaunal Variation in a Soft Substrate Intertidal Environment," *Fishery Bulletin* 74:937–948). The data in Tables 9-4 and 9-5 give lengths and widths for 35 little neck clams.
   (a) Use a calculator to compute the sample mean and the sample standard deviation for lengths and for widths. Compute the coefficient of variation for each.
   (b) Compute a 95% confidence interval for the population mean length of all Garrison Bay little neck clams.
   (c) How many more little neck clams would be needed in a sample if you wanted to be 95% sure the sample mean length is within a maximal error of 10 mm of the population mean length?
   (d) Compute a 95% confidence interval for the population mean width of all Garrison Bay little neck clams.
   (e) How many more little neck clams would be needed in a sample if you wanted to be 95% sure the sample mean width is within a maximal error of 10 mm of the population mean width?

2. Examine Figure 9-7 on the next page, "Clocks roll back Sunday" (*USA Today*).
   (a) Of the 1024 adults surveyed, 66% were reported to favor Daylight Saving Time. How many people in the sample preferred Daylight Saving Time? Using the statistic $\hat{p} = 0.66$ and sample size $n = 1024$, find a 95% confidence interval for the proportion of people p who favor Daylight Saving Time. How could you report this information in terms of a margin of error?
   (b) Look at Figure 9-7 to find the sample statistic $\hat{p}$ for the proportion of people preferring standard time. Find a 95% confidence interval for the population proportion p of people who favor Standard Time. Report the same information in terms of a margin of error.

TABLE 9-4   Length of Little Neck Clams (mm)

530	517	505	512	487	481	485	479	452	468
459	449	472	471	455	394	475	335	508	486
474	465	420	402	410	393	389	330	305	169
91	537	519	509	511					

TABLE 9-5   Width of Little Neck Clams (mm)

494	477	471	413	407	427	408	430	395	417
394	397	402	401	385	338	422	288	464	436
414	402	383	340	349	333	356	268	264	141
77	498	456	433	447					

FIGURE 9-7

USA SNAPSHOTS®
A look at statistics that shape the nation

Clocks roll back Sunday
This Sunday clocks in most of the USA will be moved back one hour to standard time — not the first choice of most people. Those who prefer:

Daylight saving time **66%**

Standard time **28%**

No preference **6%**

Source: Hilton Time Surveys of 1,024 adults

By Sam Ward and Marcy E. Mullins, USA TODAY

Copyright © 1993 USA TODAY. Reprinted with permission.

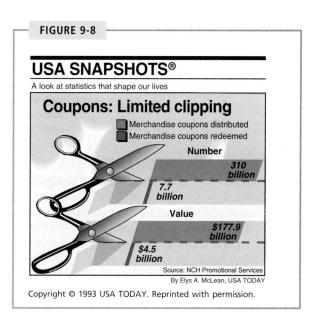

FIGURE 9-8

USA SNAPSHOTS®
A look at statistics that shape our lives

Coupons: Limited clipping
■ Merchandise coupons distributed
■ Merchandise coupons redeemed

Number
310 billion
7.7 billion

Value
$177.9 billion
$4.5 billion

Source: NCH Promotional Services

By Elys A. McLean, USA TODAY

Copyright © 1993 USA TODAY. Reprinted with permission.

3. Examine Figure 9-8, "Coupons: Limited clipping" (*USA Today*).

   (a) Use Figure 9-8 to estimate the percentage of merchandise coupons that were redeemed. Also estimate the percentage of dollar value of the coupons that were redeemed. Are these numbers approximately equal?

   (b) Suppose that you are a marketing executive working for a national chain of toy stores. You wish to estimate the percentage of coupons that will be redeemed for the toy stores. How many coupons should you check to be 95% sure that the percentage of coupons redeemed is within 1% of the population proportion of all coupons redeemed for the toy store?

   (c) Use the results of part (a) as a preliminary estimate for p, the percentage of coupons that are redeemed, and redo part (b).

   (d) Suppose that you sent out 937 coupons and found that 27 were redeemed. Explain why you could be 95% confident that the proportion of such coupons redeemed in the future would be between 1.9% and 3.9%.

   (e) Suppose that the dollar value of a collection of coupons was $10,000. Use the data in Figure 9-8 to find the expected value and standard deviation of the dollar value of the redeemed coupons. What is the probability that between $225 and $275 (out of the $10,000) is redeemed?

LINKING CONCEPTS: WRITING PROJECTS

Discuss each of the following topics in class or review the topics on your own. Then write a brief but complete essay in which you summarize the main points. Please include formulas and graphs as appropriate.

1. In this chapter, we have studied confidence intervals. Carefully read the following statements about confidence intervals.

GUIDED EXERCISE 2

Null and alternate hypotheses

A package delivery service claims it takes an average of 24 hours to send a package from New York to San Francisco. An independent consumer agency is doing a study to test the truth of this claim. Several complaints have led the agency to suspect that the delivery time is longer than 24 hours.

(a) What should be used for the null hypothesis?

⮕ The claim $\mu = 24$ hours is in question, so we take H_0: $\mu = 24$.

(b) Assuming that the delivery service does not underrate itself, what should be used for the alternate hypothesis?

⮕ If the delivery service does not underrate itself, then the only reasonable alternate hypothesis is H_1: $\mu > 24$.

Types of errors

If we *reject the null hypothesis when it is*, in fact, *true*, we have an error that is called a *type I error*. On the other hand, if we *accept the null hypothesis when it is*, in fact, *false*, we have made an error that is called a *type II error*. Table 10-1 indicates how these errors occur.

For tests of hypotheses to be well constructed, they must be designed to minimize possible errors of decision. (Usually, we do not know if an error has been made, and therefore, we can talk about only the probability of making an error.) Usually, for a given sample size, an attempt to reduce the probability of one type of error results in an increase in the probability of the other type of error. In practical applications, one type of error may be more serious than another. In such a case, careful attention is given to the more serious error. If we increase the sample size, it is possible to reduce both types of errors, but increasing the sample size may not be possible.

Level of significance

The probability with which we are willing to risk a type I error is called the *level of significance of a test*. The level of significance is denoted by the Greek letter α (pronounced "alpha"). In good statistical practice, α is specified in advance before any samples are drawn so that results will not influence the choice for the level of significance.

The *probability of making a type II error* is denoted by the Greek letter β (pronounced "beta"). Methods of hypothesis testing require us to choose α and β values to be as small as possible. In elementary statistical applications, we usually choose α first.

TABLE 10-1   Type I and Type II Errors

Truth of H_0	Our Decision	
	And if we do not reject H_0	And if we reject H_0
If H_0 is true	Correct decision; no error	Type I error
If H_0 is false	Type II error	Correct decision; no error

TABLE 10-2   Probabilities Associated with a Statistical Test

Truth of H_0	Our Decision	
	And if we accept H_0 as True	And if we reject H_0 as False
H_0 is true	Correct decision, with corresponding probability $1 - \alpha$	Type I error, with corresponding probability α called the *level of significance of the test*
H_0 is false	Type II error, with corresponding probability β	Correct decision, with corresponding probability $1 - \beta$ called the *power of the test*

Power of a test

The quantity $1 - \beta$ is called the *power of the test* and represents the probability of rejecting H_0 when it is, in fact, false. For a given level of significance, how much power can we expect from a test? The actual value of the power is usually difficult (and sometimes impossible) to obtain, since it requires us to know the H_1 distribution. However, we can make the following general comments:

1. The power of a statistical test increases as the level of significance α increases. A test performed at the $\alpha = 0.05$ level has more power than one at $\alpha = 0.01$. This means that the less stringent we make our significance level α, the more likely we will reject the null hypothesis when it is false.

2. Using a larger value of α will increase the power, but it also will increase the probability of a type I error. Despite this fact, most business executives, administrators, social scientists, and scientists use *small* α values. This choice reflects the conservative nature of administrators and scientists, who are usually more willing to make an error by failing to reject a claim (i.e., H_0) than to make an error by accepting another claim (i.e., H_1) that is false. Table 10-2 summarizes the probabilities of errors associated with a statistical test.

◇ **COMMENT** Since the calculation of the probability of a type II error is treated in advanced statistics courses, we will restrict our attention to the probability of a type I error. ◇

Meaning of accepting H_0

In most statistical applications, the level of significance is specified to be $\alpha = 0.05$ or $\alpha = 0.01$, although other values can be used. If $\alpha = 0.05$, then we say we are using a 5% level of significance. This means that in 100 similar situations, H_0 will be rejected 5 times, on average, when it should not have been rejected.

When we accept (or fail to reject) the null hypothesis, we should understand that we are *not proving the null hypothesis*. We are saying only that the sample evidence (data) is not strong enough to justify rejection of the null hypothesis. The word *accept* sometimes has a stronger meaning in common English usage than we are willing to give in our application of statistics. Therefore, we often use the expression *fail to reject H_0* instead of accept H_0. *Fail to reject* the null hypothesis simply means the evidence in favor of rejection was not strong enough (see Table 10-3 on the following page). Often, in the case H_0 cannot be rejected, a confidence interval is used to estimate the parameter in question. The confidence interval gives the statistician a range of possible values for the parameter.

TABLE 10-3   Meaning of the Terms *Fail to Reject H_0* and *Reject H_0*

Term	Meaning
Fail to reject H_0	There is not enough evidence in the data (and the test being used) to justify a rejection of H_0. This means that we retain H_0 with the understanding that we have not proved it to be true beyond all doubt.
Reject H_0	There is enough evidence in the data (and the test employed) to justify rejection of H_0. This means that we choose the alternate hypothesis H_1 with the understanding that we have not proved H_1 to be true beyond all doubt.

EXAMPLE 2

Types of errors

Let's reconsider Example 1 regarding the average gas mileage of a compact car. The hypotheses for the test are

H_0: $\mu = 47$ mpg (manufacturer's claim)

H_1: $\mu < 47$ mpg (consumer researcher's suspicion)

(a) Suppose the level of significance is $\alpha = 0.05$. Describe a type I error and its probability.

SOLUTION: If, based on sample evidence, we reject the manufacturer's claim that $\mu = 47$ mpg when, in fact, the average number of miles per gallon achieved by the compact car is 47 mpg (or higher), we have committed a type I error. The probability of making such an error is as high as 5% because the level of significance is 5%.

(b) Describe a type II error.

SOLUTION: If, based on sample evidence, we fail to reject the manufacturer's claim of $\mu = 47$ mpg when, in fact, the alternate hypothesis H_1: $\mu < 47$ mpg is true, we have made a type II error. The probability of such an error is designated by the letter β. (The computation of β is beyond the scope of this text.)   ◊

GUIDED EXERCISE 3

Types of errors

Let's reconsider Guided Exercise 1 in which we were considering the manufacturing specifications for ball bearings. The hypotheses were

H_0: $\mu = 6.0$ mm (manufacturer's specification) H_1: $\mu \neq 6.0$ mm (cause for adjusting process)

(a) Suppose the manufacturer requires a 1% level of significance. Describe a type I error, its consequence, and its probability.

A type I error is caused when sample evidence indicates that we should reject H_0 when, in fact, the average diameter of the ball bearings being produced is 6.0 mm. A type I error will cause a needless adjustment and delay of the manufacturing process. The probability of such an error is 1% because $\alpha = 0.01$.

Continued

GUIDED EXERCISE 3 continued

(b) Discuss a type II error and its consequences.

⟹ A type II error occurs if the sample evidence tells us not to reject the null hypothesis H_0: $\mu = 6.0$ mm when, in fact, the average diameter of the ball bearing is either too large or too small to meet specifications. Such an error would mean that the production process would not be adjusted when it really needed to be adjusted. This could possibly result in a large production of ball bearings that do not meet specifications.

Critical regions

We use information from a sample to determine if we should reject or not reject the null hypothesis. However, we know that there will be variability among samples. For instance, if we test drive 40 of the compact automobiles discussed in Examples 1 and 2 and compute the average miles per gallon for the 40 automobiles, we know we are likely to find that $\bar{x}$ differs from the null-hypothesis claim H_0: $\mu = 47$ mpg. A difference is likely to occur regardless of the truth of H_0. The question is, How much less than 47 mpg can $\bar{x}$ be before we begin to suspect that the population mean mileage μ is less than 47, as stated in the alternate hypothesis? In other words, how much less can $\bar{x}$ be than 47 mpg before we reject H_0: $\mu = 47$ mpg and accept the alternate hypothesis H_1: $\mu < 47$ mpg?

The answer to the question regarding the relative size of $\bar{x}$ and μ, as stated in the null hypothesis, depends on the sample size, the sampling distribution of $\bar{x}$, the alternate hypothesis H_1, and the level of significance α. If the sample test statistic $\bar{x}$ is sufficiently different from the claim about μ made in the null hypothesis, we reject the null hypothesis. The values of $\bar{x}$ for which we reject H_0 are called the *critical region* of the $\bar{x}$ distribution. Depending on the alternate hypothesis, the critical region is located on the left side, the right side, or both sides of the $\bar{x}$ distribution. If the critical region is on the left side, we call the test a *left-tailed* test. Similarly, if the critical region is on the right side, the test is called a *right-tailed* test. If the critical region is on both sides, the test is called *two-tailed*.

Left-tailed test
Right-tailed test
Two-tailed test

H_1: $\mu < k$
Left-tailed test

H_1: $\mu > k$
Right-tailed test

H_1: $\mu \neq k$
Two-tailed test

Figure 10-1 on the next page shows the relationship of the critical region to the alternate hypothesis and the level of significance α.

In the next section, we will see how to use the level of significance α and the alternate hypothesis H_1 to compute the boundary for the critical region(s). Then we will see whether or not the sample test statistic falls in the critical region. If it does, we reject H_0. If it does not, we do not reject H_0.

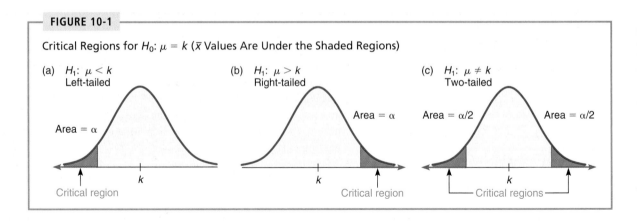

FIGURE 10-1

Critical Regions for H_0: $\mu = k$ ($\bar{x}$ Values Are Under the Shaded Regions)

(a) H_1: $\mu < k$
Left-tailed

Area = α

k

Critical region

(b) H_1: $\mu > k$
Right-tailed

Area = α

k

Critical region

(c) H_1: $\mu \neq k$
Two-tailed

Area = $\alpha/2$ Area = $\alpha/2$

k

Critical regions

Summary

In Chapter 9, we used *confidence intervals* to *estimate* the value of a parameter such as the population mean μ or probability of success p. Theory of estimation is very useful and certainly has a valuable place in statistical applications. However, confidence intervals by themselves don't answer every important type of question. For example, there are many situations in real life wherein the final decision must be a *simple yes or no*! Such a decision is based on the supposed value of a parameter, sample data, and a probability distribution involving the parameter.

Yes-or-no decisions are the essence of *hypothesis testing,* and this is the main topic of our chapter. When we make such yes-or-no decisions about the value of a statistical parameter, we realize that our decision could be wrong. Therefore, we also need to consider the types of errors and associated probabilities for such errors.

In this section, we have introduced some of the basic concepts of hypothesis testing. In particular, we have seen the null and alternate hypotheses, the level of significance, the types of errors possible, and the critical region. In the next section, we will see how to use information from a sample to conclude the test and determine whether or not to reject the null hypothesis.

◇ **COMMENT: NOTATION REGARDING THE NULL HYPOTHESIS** In statistical testing, the null hypothesis H_0 always contains the equals symbol. However, in the null hypothesis, some statistical software packages and texts also include the inequality symbol that is opposite that shown in the alternate hypothesis. For instance, if the alternate hypothesis is μ is less than 3 ($\mu < 3$), then the corresponding null hypothesis is sometimes written as μ is greater than or equal to 3 ($\mu \geq 3$). The mathematical construction of a statistical test uses the null hypothesis to assign a specific number (rather than a range of numbers) to the parameter μ in question. The null hypothesis establishes a single fixed value for μ so we are working with a single distribution having a specific mean. In this case, H_0 assigns $\mu = 3$. So when H_1: $\mu < 3$ is the alternate hypothesis, we follow the commonly used convention of writing the null hypothesis simply as H_0: $\mu = 3$. ◇

VIEWPOINT   *Lovers Take Heed!!!*

If you are going to whisper sweet nothings to your sweetheart, be sure to whisper in the *left* ear. Professor Sim of Sam Houston State University (Huntsville, Texas) found that emotionally loaded words had a higher recall rate when spoken into a person's left ear, not the right. Professor Sim presented his findings at the British Psychology Society European Congress. He told the Congress that his findings are consistent with the hypothesis that the brain's right hemisphere has more influence in the processing of emotional stimuli. The left ear is controlled by the right side of the brain. Sim's research involved statistical tests like the ones you will study in this chapter.

SECTION 10.1   PROBLEMS

1. Discuss each of the following topics in class or review the topics on your own. Then write a brief but complete essay in which you answer the following questions.
   (a) What is a null hypothesis H_0?
   (b) What is an alternate hypothesis H_1?
   (c) What is a type I error? a type II error?
   (d) What is the level of significance of a test? What is the probability of a type II error?

2. In a statistical test, we have a choice of a left-tailed critical region, right-tailed critical region, or two-tailed critical region. Is it the null hypothesis or the alternate hypothesis that determines which type of critical region is used? Explain your answer.

3. If we fail to reject (i.e., "accept") the null hypothesis, does this mean that we have *proven* it to be true beyond *all* doubt? Explain your answer.

4. If we reject the null hypothesis, does this mean that we have *proven* it to be false beyond *all* doubt? Explain your answer.

5. *Veterinary Science: Colts* The body weight of a healthy 3-month-old colt should be about $\mu = 60$ kg (*Source: The Merck Veterinary Manual,* a standard reference manual used in most veterinary colleges).
   (a) If you want to set up a statistical test to challenge the claim that $\mu = 60$ kg, what would you use for the null hypothesis H_0?
   (b) In Nevada, there are many herds of wild horses. Suppose that you want to test the claim that the average weight of a wild Nevada colt (3 months old) was less than 60 kg. What would you use for the alternate hypothesis H_1?
   (c) Suppose that you want to test the claim that the average weight of such a wild colt was greater than 60 kg. What would you use for the alternate hypothesis?
   (d) Suppose that you want to test the claim that the average weight of such a wild colt was *different* from 60 kg. What would you use for the alternate hypothesis?
   (e) For each of the tests in parts (b), (c), and (d), would the critical region be on the left, right, or both sides of the mean? Explain your answer in each case.

6. *Marketing: Shopping Time* How much customers buy is a direct result of how much time they spend in the store. A study of average shopping time in a large national

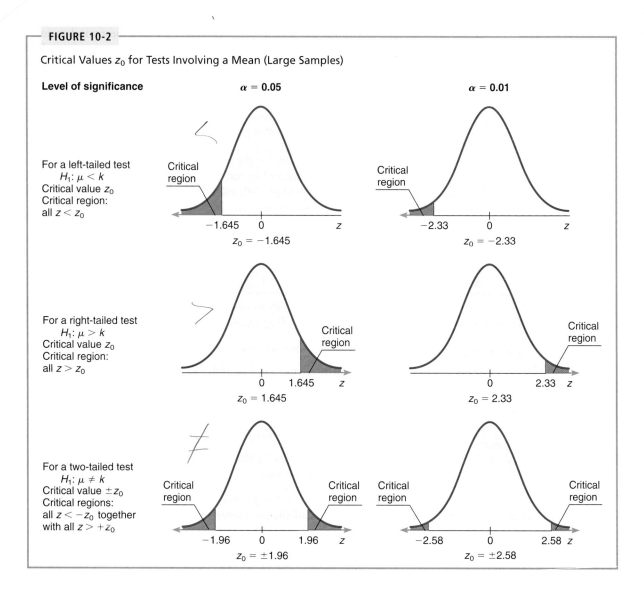

FIGURE 10-2

Critical Values z_0 for Tests Involving a Mean (Large Samples)

Level of significance $\alpha = 0.05$ $\alpha = 0.01$

For a left-tailed test
$H_1: \mu < k$
Critical value z_0
Critical region:
all $z < z_0$

Critical region -1.645 0 z $z_0 = -1.645$

Critical region -2.33 0 z $z_0 = -2.33$

For a right-tailed test
$H_1: \mu > k$
Critical value z_0
Critical region:
all $z > z_0$

0   1.645 z Critical region $z_0 = 1.645$

0   2.33 z Critical region $z_0 = 2.33$

For a two-tailed test
$H_1: \mu \neq k$
Critical value $\pm z_0$
Critical regions:
all $z < -z_0$ together
with all $z > +z_0$

Critical region -1.96 0   1.96 z Critical region $z_0 = \pm 1.96$

Critical region -2.58 0   2.58 z Critical region $z_0 = \pm 2.58$

Critical values for testing μ (large samples)

Sample test statistic

hypothesized in H_0 to merit rejecting H_0? To answer this question, we look at the $\bar{x}$ distribution *under the assumption that H_0 is true.*

According to the central limit theorem, the $\bar{x}$ distribution based on samples of size n, where n is large ($n \geq 30$), is approximately *normal.* For this reason, our critical values are z values. Figure 10-2 gives the critical values for *one-tailed* and *two-tailed tests* for two commonly used levels of significance, $\alpha = 0.05$ and $\alpha = 0.01$. The critical values given in Figure 10-2 are based on the fact that the *area* above the critical region equals the level of significance. For instance, for a left-tailed test and $\alpha = 0.05$, the area above the critical region is 0.05. Using methods of Chapter 7 and the normal distribution table (Table 4 in the Appendix), we see that $z = -1.645$ is the z value that "cuts off" 5% of the normal curve on the left.

To determine if the sample $\bar{x}$ is sufficiently far away from the mean proposed in the null hypothesis to merit rejecting H_0, we convert the sample value $\bar{x}$ to a z value and see if the sample z value falls in the critical region. If so, we reject H_0

and accept H_1. If not, we do not reject H_0. The conversion of $\bar{x}$ to z follows the formula

$$z = \frac{\bar{x} - \mu}{(\sigma/\sqrt{n})}$$

where μ = mean specified in H_0

σ = standard deviation of the x distribution

n = sample size being used

$\bar{x}$ = sample test statistic

◇ **COMMENT** *The methods used in this section will require $\bar{x}$ to be approximately normally distributed,* and we will approximate σ by the sample standard deviation s when necessary. To ensure that our methods yield accurate results, we will *require* our *sample size to be 30 or larger* ($n \geq 30$). Tests involving small samples ($n < 30$) will be presented in Section 10.4. ◇

EXAMPLE 3

Hypothesis testing procedure

Statement of Problem: Suppose the St. Louis Zoo wishes to obtain eggs of a rare Mississippi river turtle. The zoo will hatch the eggs and raise the turtles as an exhibit of a rare and endangered species. Carol Wright is the staff biologist at the zoo who has been given the job of finding the eggs to be hatched. Turtles of the area bury their eggs in a nest in sandbanks along the river. Then the nest is abandoned, and the eggs hatch by themselves. A number of different species of turtles live in the region, and eggs from each species look much alike. Past research has shown that lengths of turtle eggs are normally distributed, and lengths of the rare turtle eggs have population mean $\mu = 7.50$ cm with standard deviation $\sigma = 1.5$ cm. The rare turtle egg is the only one with mean length 7.50 cm. The mean lengths of eggs from each of the other species are longer than 7.50 cm. The eggs of all the local turtle species have the same standard deviation $\sigma = 1.5$ cm for length.

After searching for some time, Carol found a nest with 36 eggs. The way the nest was constructed makes her suspect that it was made by the rare turtle. The mean length of this collection of 36 eggs is $\bar{x} = 7.74$ cm. Since $\bar{x} = 7.74$ cm is longer than the population mean $\mu = 7.50$ cm of the rare turtle, Carol is a little worried that the eggs may come from a species that lays larger eggs.

Let's use a statistical test to help the biologist make a decision. Pay close attention to this example; it contains principal features common to most statistical tests.

I. *Summary of Known Facts:* In any testing problem, it is a good idea to make a short summary of known facts before we start to construct the test.

(a) The distribution of lengths of eggs from the rare turtle is *normal*, with population mean $\mu = 7.50$ cm and standard deviation $\sigma = 1.5$ cm.

(b) The nest contains $n = 36$ eggs with an observed mean $\bar{x} = 7.74$ cm.

(c) Since the population is *normal* and μ and σ are known, Theorem 8.1 tells us that $\bar{x}$ is also normally distributed. The mean and standard deviation of the $\bar{x}$

distribution are

$$\mu_{\bar{x}} = \mu = 7.50 \text{ cm}$$

$$\sigma_{\bar{x}} = \frac{\sigma}{\sqrt{n}} = \frac{1.5}{\sqrt{36}} = 0.25 \text{ cm}$$

II. *Establishing H_0 and H_1*: The null hypothesis H_0 is set up for the primary purpose of seeing whether or not it can be rejected. The alternate hypothesis H_1 will be chosen when the null hypothesis must be rejected.

Let μ be the mean length of the population distribution from which our sample of 36 eggs is drawn. Our biologist suspects that the nest contains the rare turtle eggs; therefore, we will use the null hypothesis

$$H_0\text{: } \mu = 7.50$$

since the rare turtle eggs are known to have mean length $\mu = 7.50$ cm. However, all other local species lay eggs with a longer population mean. Therefore, the alternate hypothesis is

$$H_1\text{: } \mu > 7.50$$

III. *Choosing the Level of Significance α*: The null hypothesis says that the eggs in the nest are from the rare turtle. A type I error means that we reject the nest as being formed by the rare turtle when it was, in fact, formed by the rare turtle. A type II error means that we accept the nest as coming from the rare turtle when it really came from a common species.

A type I error could be serious; we don't want to reject the eggs if they are from the rare turtle. A type II error is not too serious. If the eggs are from a common species, the turtles can be released after they hatch. Naturally, the zoo doesn't want to incubate the wrong eggs, but it would not be too serious if it did.

Although other levels of significance may be used, most researchers use either $\alpha = 0.05$ or $\alpha = 0.01$. Our biologist wants to use a level of significance $\alpha = 0.01$. This means that she is willing to risk a type I error with a probability of $\alpha = 0.01$.

IV. *Graphical Model for the Test*: What decision procedure shall we use to test our hypothesis? In a way, the logic of our decision process is similar to that used in a typical courtroom setting, but our methods are mathematical rather than legal. In a courtroom setting, the person charged with a crime is initially considered innocent (null hypothesis). If evidence (data) presented in court can sufficiently discredit the person's innocence, he or she is then judged to be guilty (alternate hypothesis). In a similar way, we will consider the null hypothesis to be *true* until there is enough data (mathematical evidence) to discredit it at the $\alpha = 0.01$ level of significance.

Figure 10-3 shows the graphical model for a test with hypotheses

$$H_0\text{: } \mu = 7.50 \qquad H_1\text{: } \mu > 7.50$$

and a large sample. Let us examine the details of Figure 10-3.

(a) First, we determine the *sampling distribution*. By Theorem 8.1, the *sample test statistic* $\bar{x}$ follows a *normal* distribution. We convert the $\bar{x}$ distribution to the standard normal distribution by the formula

$$z = \frac{\bar{x} - \mu}{(\sigma/\sqrt{n})}$$

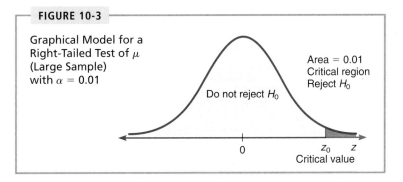

FIGURE 10-3

Graphical Model for a
Right-Tailed Test of μ
(Large Sample)
with $\alpha = 0.01$

Do not reject H_0

Area = 0.01
Critical region
Reject H_0

0   z_0 z
       Critical value

where μ is given in H_0, σ is given or estimated from the sample (large sample), and n is the sample size. In our case, $\mu = 7.50$, $\sigma = 1.5$, and $n = 36$. Figure 10-3 is the standard normal distribution.

(b) Next, we find the critical region. Since $H_1: \mu > 7.50$ claims that μ is greater than 7.50, the *critical region* for this problem is a *right tail* of the standard normal distribution. This tail has an area of $\alpha = 0.01$. If the z value corresponding to the sample statistic $\bar{x} = 7.74$ falls in the critical region, we say that there is enough evidence to discredit the null hypothesis, and we *reject H_0* at the $\alpha = 0.01$ level of significance. If our observed value falls outside the critical region, we conclude that the evidence is not strong enough to reject H_0, so in this case, we *fail to reject H_0*. Since $H_1: \mu > 7.50$ claims that μ is greater than 7.50, then an observed sample statistic $\bar{x}$ *far enough* to the *right* of $\mu = 7.50$ discredits the null hypothesis and supports the alternate hypothesis. In other examples, we will encounter left-tailed and two-tailed tests.

(c) The value z_0 is called the *critical value for the test.* It is the separation point for the critical region. If the z value of the sample statistic $\bar{x} = 7.74$ falls to the *left* of z_0, we *fail to reject H_0*. If it falls to the *right* of z_0, we *reject H_0*. Once we know z_0 and convert the sample statistic $\bar{x}$ to z, we can quickly finish the test.

V. *Finding the Critical Value, Critical Region, and z Value for $\bar{x}$*: Figure 10-2 gives critical values z_0 to use for tests with sampling distributions that are normal. We find these values, as we did in Chapter 7, by using Table 4, "Areas of a Standard Normal Distribution," in the Appendix. In this problem, we have a right-tailed test with level of significance $\alpha = 0.01$. We want to find the value z_0 so that 1% of the standard normal curve lies to the right of z_0. This is equivalent to finding the z_0 value so that 99% of the area lies to the left of z_0. Table 4 shows that $z_0 = 2.33$. The critical region is all values of z greater than or equal to $z_0 = 2.33$ (see Figure 10-4 on the next page).

The sample "evidence" we have is $\bar{x} = 7.74$ based on a nest of 36 eggs. We convert $\bar{x}$ to z using the formula in part IV:

$$z = \frac{\bar{x} - \mu}{(\sigma/\sqrt{n})} = \frac{7.74 - 7.50}{(1.5/\sqrt{36})} = 0.96$$

Now we have all the information we need to complete the test.

FIGURE 10-4

Critical Value for a Standard Normal Curve, Right-Tailed Test with $\alpha = 0.01$

Do not reject H_0

$\alpha = 0.01$

0   2.33 z
z_0
Critical value

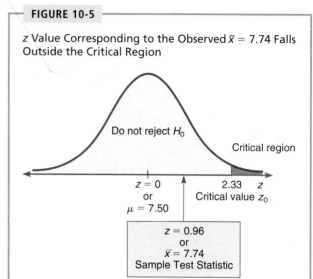

FIGURE 10-5

z Value Corresponding to the Observed $\bar{x} = 7.74$ Falls Outside the Critical Region

Do not reject H_0

Critical region

$z = 0$
or
$\mu = 7.50$

2.33 z
Critical value z_0

$z = 0.96$
or
$\bar{x} = 7.74$
Sample Test Statistic

VI. *Conclusion:* Earlier we pointed out a similarity between the logic of a statistical test and proceedings in a court of law. We are now at the point where all the sample evidence has been presented, and we are awaiting a decision. Who makes the decision? In a court, the jury or judge would make the decision, but in our statistical test, it is Mother Nature who makes the final decision. The observed value $\bar{x} = 7.74$ cm of length for turtle eggs taken from the turtle nest corresponds to the z value 0.96. This z value does not lie in the critical region (see Figure 10-5).

We conclude that there is not enough evidence to discredit the null hypothesis at the $\alpha = 0.01$ level of significance. Therefore, we do not reject $H_0: \mu = 7.50$, and we still think that the nest could have been made by the rare endangered species of turtle. The biologist should collect the eggs and take them to the zoo for incubation. It is important to remember that we do not claim to have *proven* that the eggs are from the rare turtle; all we say is that there is not enough evidence to reject H_0, and therefore, we "accept" it.

VII. *Meaning of α and β:* Remember that β is the probability of accepting H_0 when it is false. In our setting, β is the probability that the zoo incubates and hatches eggs from a common turtle species. To calculate β requires knowledge of the H_1 distribution. Since there are several species of common turtles in the area and we don't know which species made the nest, we really don't know the H_1 distribution. However, a type II error is not too serious because if the wrong eggs (from the common turtle) are hatched, there is little harm done. The baby turtles simply can be released near the place the eggs were found.

The level of significance α is the probability of rejecting H_0 when it is true. This is the probability of concluding that the eggs are not from the rare turtle species when they, in fact, are. It would be a serious mistake to reject the nest of the rare turtle. To guard against such a mistake, we have taken a relatively small α value of 0.01.   ◇

GUIDED EXERCISE 4

Testing μ (large sample)

A research meteorologist has been studying wind patterns over the Pacific Ocean. Based on these studies, a new route is proposed for commercial airlines going from San Francisco to Honolulu. The new route is intended to take advantage of existing wind patterns to reduce flying time. It is known that for the old route the distribution of flying times for a large four-engine jet has mean $\mu = 5.25$ hours with standard deviation $\sigma = 0.6$ hour. Thirty-six flights on the new route have yielded a mean flying time of $\bar{x} = 4.90$ hours. Does this indicate that the average flying time for the new route is less than 5.25 hours? Use a 5% level of significance.

(a) What is H_0?

⇒ The average time for the old route is $\mu = 5.25$ hours, so we will set up the null hypothesis $H_0: \mu = 5.25$. In words, we are saying that the mean flying time on the new route is the same as that on the old route.

(b) What is H_1?

⇒ We want to see if the average flying time for the new route is less than 5.25 hours, so the alternate hypothesis is $H_1: \mu < 5.25$.

(c) What type of critical region must be used?

⇒ Since the < symbol is used in the alternate hypothesis, Figure 10-1 tells us to use a left-tailed test.

(d) What is the critical value?

⇒ By Figure 10-2 (or using Table 4 in the Appendix), the critical value for a left-tailed test with $\alpha = 0.05$ is $z_0 = -1.645$.

(e) What is the z value corresponding to the sample test statistic $\bar{x} = 4.90$?

⇒ We use the formula from the central limit theorem with $\mu = 5.25$ (the value specified in H_0), $\sigma = 0.6$, and $n = 36$. Therefore,

$$z = \frac{\bar{x} - \mu}{(\sigma/\sqrt{n})} = \frac{4.90 - 5.25}{(0.6/\sqrt{36})} = -3.50$$

(f) Do we reject or fail to reject H_0?

⇒ Look at the critical region of Figure 10-6. The sample z value corresponding to the sample test statistic $\bar{x} = 4.90$ falls in the critical region. We reject H_0 and choose H_1 at the $\alpha = 0.05$ level. This means that we conclude that the flying time on the new route is less than the flying time on the old route.

FIGURE 10-6 Critical Region, $\alpha = 0.05$

Critical region Do not reject H_0

−1.645 $z = 0$ z
Critical value   or
              $\mu = 5.25$

$z = -3.50$
or
$\bar{x} = 4.90$
Sample Test Statistic

GUIDED EXERCISE 5 continued

(e) Show the critical region and the z value of the sample test statistic on the normal curve. Does the sample test statistic fall in the critical region or not? Do we reject or fail to reject H_0? At the 1% level of significance, can we say that the machine needs adjustment?

FIGURE 10-8 Critical Region, $\alpha = 0.01$

We see that the z value corresponding to the sample test statistic $\bar{x} = 1.87$ falls in the critical region. This means that we reject H_0 and conclude that at the 1% level of significance the machine needs adjustment.

Statistical significance

In a sense, the sample estimate of the parameter represents the evidence favoring rejection of H_0. For this reason, statisticians say the results of a random sample are *statistically significant* if the estimated parameter falls in the critical region of a test. In Guided Exercise 5, the random sample of 64 caps gave a sample test statistic $\bar{x} = 1.87$. Since the corresponding z value was in the critical region, we say that the results of the sample were statistically significant—that is, there was sufficient evidence to reject H_0 at the specified level of significance.

Statistical significance

If we reject H_0, we say that the data are *statistically significant*.
If we do not reject H_0, we say that the data are *not statistically significant*.

◇ **COMMENT: Using the Student's t Distribution to Test μ Even with Large Samples** When samples are large ($n \geq 30$), the central limit theorem assures us that the $\bar{x}$ distribution is approximately normal. In addition, we use the common convention that when we have large samples ($n \geq 30$), we can estimate σ by the sample standard deviation s. This convention allows us to use the normal distribution to test μ for large samples even when we don't know σ for the x distribution. However, when n is large and we don't know σ, we can use the Student's t distribution with $n - 1$ degrees of freedom for the $\bar{x}$ distribution. Using the Student's t distribution avoids the errors inherent in estimating σ by s.

With more widespread access to technology, such as the TI-83Plus calculator or statistical software packages, it is becoming more common to use the Student's t distribution to test μ when we have large samples, but don't know σ. Using the Student's t distribution moves the critical region farther into the tail(s) of the $\bar{x}$ distribution, and generates slightly larger P values (discussed in Section 10.3). Generally speaking, test conclusions for large sample tests of μ are the same whether we use the normal distribution with σ approximated by s or use the Student's t distribution with $n-1$ degrees of freedom. $\Diamond$

TECH NOTE   In conducting hypothesis tests for μ using the normal distribution, the TI-83Plus calculator and Minitab both provide the z value for the sample test statistic $\bar{x}$. To conclude the test, compare the z value of the test statistic with the critical value(s) for the specified level of significance and type of test.

TI-83Plus   You can select to enter raw data or summary statistics. Enter the value of $\boldsymbol{\mu_0}$ used in the null hypothesis H_0: $\mu = \boldsymbol{\mu_0}$. Select the symbol used in the alternate hypothesis ($\neq\boldsymbol{\mu_0}$, $<\boldsymbol{\mu_0}$, $>\boldsymbol{\mu_0}$). The value for sigma is also required. Using data from Guided Exercise 5 regarding the diameter of twist-off caps, we have the following displays. Press **Stat**, select **Tests**, use option **1:z-Test**.

Minitab   Enter the raw data from a sample. Use the menu selections **Stat ➤ Basic Stat ➤ 1-Sample Z**. The z value of the sample mean $\bar{x}$ is included in the display. We will see such a display in the next section.

VIEW●POINT   *Money, Money, Money!*

What is the statistical profile of U.S. millionaires? Are they flashy spenders? Are they stingy or generous with friends? Do they drive big cars and wear expensive clothes? Do millionaires work or inherit their wealth? How many millionaires are there in the U.S.? What's the buying power of a million dollars today compared with 10 or 20 years ago? For a good statistical profile of U.S. millionaires, see *The Millionaire Next Door,* by T. J. Stanley and W. D. Danko. The book points out that most U.S. millionaires *earned* what they have, behave like ordinary people, and are not flashy or arrogant. A million dollars has a great deal less buying power than it used to have. Anyone who works hard and is willing to take some risks has a good chance of becoming a millionaire. A good understanding of statistics helps.

For Problems 1 through 12, please provide the requested information.

(a) What is the null hypothesis? What is the alternate hypothesis? Will we use a left-tailed, right-tailed, or two-tailed test? What is the level of significance?

(b) What sampling distribution will we use? What is the critical value z_0 (or critical values $\pm z_0$)?

(c) Sketch the critical region and show the critical value (or critical values).

(d) Calculate the z value corresponding to the sample statistic $\bar{x}$ and show its location on the sketch of part (c).

(e) Based on your answers for parts (a) to (d), shall we reject or fail to reject (i.e., "accept") the null hypothesis at the given level of significance α? Explain your conclusion in the context of the problem.

(f) Are the data statistically significant?

1. *Meteorology: Storms* *Weatherwise* is a magazine published in association with the American Meteorological Society. In one issue, there is a rating system to classify Nor'easter storms that frequently hit New England states and can cause much damage near the ocean coasts. A *severe* storm has an average peak wave height of 16.4 feet for waves hitting the shore. Suppose that a Nor'easter is in progress at the severe storm class rating. Peak wave heights are usually measured from land (using binoculars) off fixed cement peers. Suppose that a reading of 36 peak waves showed an average wave height of $\bar{x} = 15.1$ feet with sample standard deviation $s = 3.2$ feet. Does this information indicate that the storm is (perhaps temporarily) retreating from its severe rating? Use $\alpha = 0.01$.

2. *Ford Taurus: Assembly Time* Let x be a random variable that represents assembly time for the Ford Taurus. *The Wall Street Journal* reported that the average assembly time for the Ford Taurus is $\mu = 38$ hours. A modification to the assembly procedure has been made. It is thought that the average assembly time may be reduced because of this modification. A random sample of 47 new Ford Taurus automobiles coming off the assembly line showed the average assembly time to be $\bar{x} = 37.5$ hours with sample standard deviation $s = 1.2$ hours. Does this indicate that the average assembly time has been reduced? Use $\alpha = 0.01$.

3. *Phone Calls: Priority List* Message mania! According to *The Wall Street Journal*, a professional employee working in a large company receives an average of $\mu = 31.8$ calls per day. Most of the calls are from other employees in the company. Because of the large number of calls, employees find themselves distracted and are unable to concentrate when they return to their tasks. In an effort to reduce distraction caused by such interruptions, one company established a "priority list" that all employees were to use before making phone calls. One month after the new priorities were put into use, a random sample of 63 employees showed they were receiving an average of $\bar{x} = 28.5$ calls per day with sample standard deviation $s = 10.7$ calls per day. Use a 1% level of significance to test the claim that there has been a change (either way) in the average number of calls per day received per employee.

4. *Fashion Design: Display Windows* Judy Povich is a fashion design artist who designs the display windows in front of a large clothing store in New York City. Electronic counters at the entrances total the number of people entering the store each business day. Before Judy was hired by the store, the mean number of people entering the store each day was 3218. However, since Judy has started working, it is thought that this number has increased. A random sample of 42 business days after Judy began work gave an average $\bar{x} = 3392$ people entering the store each

day. The sample standard deviation was $s = 287$ people. Does this indicate that the average number of people entering the store each day has increased? Use a 1% level of significance.

5. *Cocktail Hostess: Tips* Maureen is a cocktail hostess in a very exclusive private club. The Internal Revenue Service is auditing her tax return this year. Maureen claims that her average tip last year was $4.75. To support this claim, she sent the IRS a random sample of 52 credit card receipts showing her bar tips. When the IRS got the receipts, they computed the sample average and found it to be $\bar{x} = \$5.25$ with sample standard deviation $s = \$1.15$. Do these receipts indicate that the average tip Maureen received last year was more than $4.75? Use a 1% level of significance.

6. *Farmer's Co-op: Alfalfa* Let x be a random variable representing the percentage of protein content for early bloom alfalfa hay. The average percentage protein content of such early bloom alfalfa should be $\mu = 17.2\%$ (*Source: The Merck Veterinary Manual,* a manual commonly used in veterinary colleges). A farmer's co-op is thinking of buying a large amount of baled hay but suspects that the hay is from a later summer cutting with lower protein content. A small amount of hay was removed from each bale of a random sample of 50 bales. The average protein content from the samples was determined by the local agriculture college to be $\bar{x} = 15.8\%$ with sample standard deviation $s = 5.3\%$. Use a 5% level of significance to test the claim that this hay has lower average protein content than early bloom alfalfa.

7. *Dodge Intrepid: Acceleration* *Consumer Reports* indicated that the mean acceleration time (0 to 60 miles per hour) for the Dodge Intrepid was 10.2 seconds. In most tests of this type, regular unleaded gasoline is used. Suppose that 41 such tests were made using premium unleaded gasoline (octane over 91), and the sample mean acceleration time (0 to 60 miles per hour) was $\bar{x} = 9.7$ seconds with sample standard deviation $s = 2.1$ seconds. Does this indicate that premium gasoline tends to reduce average acceleration time (0 to 60 miles per hour)? Use $\alpha = 0.05$.

8. *Mercury Sable: Braking Distance* *Consumer Reports* also indicated that the mean braking distance (from 60 miles per hour) on wet pavement for the Mercury Sable was 159 feet. Suppose that Sables equipped with tires having a new tread designed to grip the road better on wet pavement were used in 45 tests (braking from 60 miles per hour). The sample mean braking distance was $\bar{x} = 148$ feet with sample standard deviation $s = 23.5$ feet. Does this information indicate that the population mean braking distance for Mercury Sables on wet pavement is reduced for the new tire tread? Use $\alpha = 0.01$.

9. *Vitamins: Oxygen* Let x be a random variable that represents milliliters of oxygen per deciliter of whole blood. For healthy adults, the population mean of x is $\mu = 19.0$ milliliters of oxygen per deciliter (*Source: The Merck Manual,* a commonly used reference in most medical schools and nursing programs). A company that sells vitamins claims that its multivitamin complex will increase the oxygen capacity of the blood. A random sample of 48 adults took the vitamins for 6 months. After blood tests, it was found that the sample mean was $\bar{x} = 20.7$ milliliters of oxygen per deciliter with sample standard deviation $s = 9.9$. Use a 1% level of significance to test the claim that the average oxygen capacity has been increased.

10. *Major League Baseball: Salary* The average annual salary of major league baseball players is now $1,789,556 (*Source: USA Today*). Suppose this year a random sample of 35 major league players in Florida had an average annual salary of $\bar{x} = \$1,621,726$ with sample standard deviation $s = \$591,218$. Use a 1% level of

significance to test the claim that the average salary for all Florida players is different from the national average.

11. *Blood Plasma: pH* Let x be a random variable that represents the pH of arterial plasma (i.e., the acidity of the blood). For healthy adults, the mean of the x distribution is $\mu = 7.4$ pH. (*See* source for Problem 9.) A new drug for arthritis has been developed. However, it is thought this drug might change blood pH. A random sample of 33 patients with arthritis took the drug for 3 months. Blood tests showed $\bar{x} = 8.1$ pH with sample standard deviation $s = 1.9$ pH. Use a 5% level of significance to test the claim that the drug has changed (either way) the mean pH of the blood.

12. *Air Force: Mess Hall* An Air Force base mess hall has received a shipment of 10,000 gallon-size cans of cherries. The supplier claims that the average amount of liquid is 0.25 gallon per can. A government inspector took a random sample of 100 cans and found the average liquid content to be 0.28 gallon per can with a standard deviation of 0.10. Does this indicate that the supplier's claim is too low? (Use a 5% level of significance.)

13. *Focus Problem: Pizza Franchise* Solve the Chapter 10 Focus Problem. Use $\alpha = 0.05$.

14. Suppose that you did some statistical research on a given topic.
    (a) If you had a null hypothesis that you hoped to *reject* based on your sample data, would you be better off using a one-tailed test (either left or right tail as appropriate) or a two-tailed test? Explain your answer.
    (b) Answer part (a) if you had a null hypothesis you hoped *not to reject* based on your sample data. Explain your answer. *Hint:* Examine Figure 10-2 and the size of the critical values for each test.
    (c) If a report states that certain data were used to reject (or fail to reject) a given hypothesis, would it be a good idea to know what type of test (one-tailed or two-tailed) was used? Explain your answer.

15. *U.S. Court System Analogy* Compare similarities of statistical testing with legal methods used in a U.S. court setting. Then discuss the following topics in class or consider the topics on your own. Please write a brief but complete essay in which you answer the following questions.
    (a) In a court setting, the person charged with a crime is initially considered to be innocent. The claim of innocence is maintained until the jury returns with a decision. Explain how the claim of innocence could be taken to be the null hypothesis. Do we assume that the null hypothesis is true throughout the testing procedure? What would the alternate hypothesis be in a court setting?
    (b) The court claims that a person is innocent if the evidence against the person is not adequate to find him or her guilty. This does not mean, however, that the court has necessarily *proven* the person to be innocent. It simply means that the evidence against the person was not adequate for the jury to find him or her guilty. How does this situation compare with a statistical test for which the conclusion is "do not reject" (i.e., accept) the null hypothesis? What would be a type II error in this context?
    (c) If the evidence against a person is adequate for the jury to find him or her guilty, then the court claims that the person is guilty. Remember, this docs not mean that the court has necessarily *proven* the person to be guilty. It simply means that the evidence against the person was strong enough to find him or her guilty. How does this situation compare with a statistical test for which the conclusion is "reject" the null hypothesis? What would be a type I error in this context?
    (d) In a court setting, the final decision as to whether the person charged is innocent or guilty is made at the end of the trial, usually by a jury of impartial people. In hypothesis testing, the final decision to reject or not reject the null

hypothesis is made at the end of the test by using information or data from an (impartial) random sample. Discuss these similarities between statistical hypothesis testing and a court setting.

(e) We hope that you are able to use this discussion to increase your understanding of statistical testing by comparing it with something that is a well-known part of our American way of life. However, all analogies have weak points. It is important not to take the analogy between statistical hypothesis testing and legal court methods too far. For instance, the judge does not set a level of significance and tell the jury to determine a verdict that is wrong only 5% or 1% of the time. Discuss some of these weak points in the analogy between the court setting and hypothesis testing.

16. Is there a relationship between confidence intervals and two-tailed hypothesis tests? Let *c* be the level of confidence used to construct a confidence interval from sample data. Let α be the level of significance for a two-tailed hypothesis test. The following statement applies to hypothesis tests of the mean (for both large and small samples).

> For a two-tailed hypothesis test with level of significance α and null hypothesis H_0: $\mu = k$, we *reject* H_0 whenever *k* falls *outside* the $c = 1 - \alpha$ confidence interval for μ based on the sample data. When *k* falls within the $c = 1 - \alpha$ confidence interval, we do not reject H_0.

(A corresponding relationship between confidence intervals and two-tailed hypothesis tests also is valid for other parameters such as *p*, which we will study in Section 10.5.) Whenever the value of *k* given in the null hypothesis falls *outside* the $c = 1 - \alpha$ confidence interval for the parameter, we *reject* H_0. For example, consider a two-tailed hypothesis test with $\alpha = 0.01$ and

$$H_0: \mu = 20 \qquad H_1: \mu \neq 20$$

A random sample of size 36 has a sample mean $\bar{x} = 22$ with sample standard deviation $s = 4$.

(a) What is the value of $c = 1 - \alpha$? Using the methods of Chapter 9, construct a $1 - \alpha$ confidence interval for μ from the sample data. What is the value of μ given in the null hypothesis (i.e., what is *k*)? Is this value in the confidence interval? Do we reject or fail to reject H_0 based on this information?

(b) Using methods of this chapter, find the critical region for the hypothesis test. Do we reject or fail to reject H_0? Compare your result to that of part (a).

17. Change the hypotheses of Problem 16 to H_0: $\mu = 21$ and H_1: $\mu \neq 21$. Repeat parts (a) and (b).

10.3
The *P* Value in Hypothesis Testing

FOCUS POINTS

✓ Identify the meaning of *P* value for a statistical test.

✓ Compute *P* values for large sample tests of μ.

✓ Draw test conclusions based on the *P* value and the level of significance.

The level of significance α in a test of hypothesis is the probability of making a type I error; that is, α is the probability of rejecting the null hypothesis when it is true.

The experimenter sets α before beginning the statistical test. Then this level of significance is used to determine whether or not H_0 is rejected. However, as we shall see in the next example, the same sample data may lead to two different test conclusions, depending on the level of significance used.

Summary

The P value (probability of chance) is the area of the sampling distribution that lies beyond the observed sample statistic. It tells us the probability that a sample statistic will be more extreme than the observed sample statistic. We compute the P values as follows:

1. When H_1 indicates a right-tailed test,

P value = area to the right of the observed sample statistic

P value = P(sample statistic > observed sample statistic)

2. When H_1 indicates a left-tailed test,

P value = area to the left of the observed sample statistic

P value = P(sample statistic < observed sample statistic)

3. When H_1 indicates a two-tailed test,

P value = sum of the areas in the two tails

  (a) In the case that the observed sample statistic falls in the right half of a symmetric curve, then

P value = $2P$(sample statistic > observed sample statistic)

  (b) In the case that the observed sample statistic falls in the left half of a symmetric curve, then

P value = $2P$(sample statistic < observed sample statistic)

To conclude the test, we compare the level of significance α with the P value.

1. If the P value is less than or equal to α, reject H_0.

2. If the P value is greater than α, do not reject H_0.

TECH NOTE   When doing hypothesis testing using statistical software or a calculator similar to the TI-83Plus, the P value corresponding to the sample statistic is generally provided. The user compares it with a specified level of significance α to conclude the test.

TI-83Plus The displays use the data of Guided Exercise 6 regarding the average age of students attending Fremont College. Press **STAT**, select **TESTS**, option **1:Z-Test**, then **Draw**.

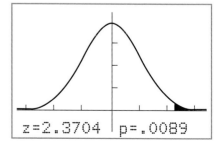

Excel The **ZTEST** function claims to return the *P* value for a two-tailed test. However, it seems to give the area in the right tail bounded by the sample statistic $\bar{x}$. With some care, you can figure out the appropriate *P* value from the given area.

Minitab Enter the raw data in a column. Use menu choices **Stat ➤ Basic Statistics ➤ 1-Sample Z**. Section Problem 10 shows a Minitab display.

VIEWP⊙INT   *Predator or Prey?*

Consider animals such as the arctic fox, gray wolf, desert lion, and South American jaguar. Each animal is a predator. What are the total sleep (hours per day), maximum life span (years), and overall danger index from other animals? Now consider prey such as rabbits, deer, wild horses, and the Brazilian tapir (a wild pig). Is there a statistically significant difference in average sleep, life span, and danger index? What about other variables such as the ratio of brain weight to body weight or the sleep exposure index (sleeping in a well-protected den or out in the open)? How did prehistoric humans fit into this picture? Scientists have collected a lot of data, and a great deal of statistical work has been done regarding such questions. For more information, visit the Brase/Brase *Understanding Basic Statistics* site at http://math.college.hmco.com/students and find the link to the StatLib site hosted by the Department of Statistics at Carnegie Mellon University. Follow links to Sleep in the Datasets Archive.

FOCUS

✓ Identify th
  sample tes

✓ Use degre
  t table to 1

✓ Compute t

✓ Estimate th
  sample tes

✓ Draw conc
  significance

Critical values
t distribution

SECTION 10.3   PROBLEMS

For Problems 1 through 9, please do the following:
(a) Convert the sample mean $\bar{x}$ to a sample *z* value. (*Hint:* Due to the large sample size, approximate σ with *s*.)
(b) Sketch a graph of the *z* distribution. Show the location of the area corresponding to the *P* value for this test.
(c) Compute the *P* value for this test.
(d) Is the data significant at the specified level of significance α?

1. Given $H_0: \mu = 5$ and $H_1: \mu > 5$ with observed sample mean $\bar{x} = 6.1$, sample standard deviation $s = 2.5$, and sample size $n = 40$. Use $\alpha = 0.01$.

2. Given $H_0: \mu = 53.1$ and $H_1: \mu < 53.1$ with observed sample mean $\bar{x} = 52.7$, sample standard deviation $s = 4.5$, and sample size $n = 41$. Use $\alpha = 0.01$.

3. Given $H_0: \mu = 21.7$ and $H_1: \mu \neq 21.7$ with observed sample mean $\bar{x} = 20.5$, sample standard deviation $s = 6.8$, and sample size $n = 45$. Use $\alpha = 0.05$.

4. Given $H_0: \mu = 18.7$ and $H_1: \mu \neq 18.7$ with observed sample mean $\bar{x} = 19.1$, sample standard deviation $s = 5.2$, and sample size $n = 32$. Use $\alpha = 0.05$.

5. *Wildlife: Coyotes* A random sample of 68 adult coyotes in a region of northern Minnesota showed the average age to be $\bar{x} = 2.05$ years with sample standard deviation $s = 0.82$ years (based on information from the book *Coyotes: Biology, Behavior and Management,* by M. Bekoff, Academic Press). However, it is thought that

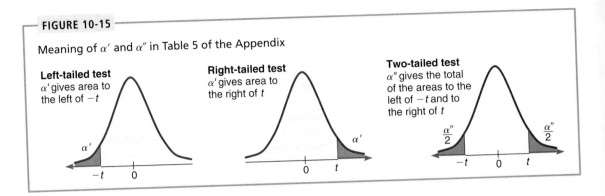

FIGURE 10-15

Meaning of α' and α'' in Table 5 of the Appendix

Left-tailed test
α' gives area to the left of $-t$

Right-tailed test
α' gives area to the right of t

Two-tailed test
α'' gives the total of the areas to the left of $-t$ and to the right of t

one-tailed test (α' is the area to the right of t or, equivalently, to the left of $-t$), and α'' is the significance level for a two-tailed test (α'' is the area beyond $-t$ and t, so, in fact, $\alpha'' = 2\alpha'$). Figure 10-15 illustrates these different values.

To find the critical value(s) t_0, we look in the column headed by the level of significance α' for a *one-tailed test* or α'' for a *two-tailed test*. The critical value t_0 is located in the row headed by degrees of freedom $d.f. = n - 1$.

EXAMPLE 8

Using the Student's t distribution table

Use Table 5 in the Appendix to find the critical value(s) t_0 and critical region for the described tests of μ with sample size $n = 11$. (Assume that the sample came from a population with a distribution that is mound-shaped and symmetric.)

(a) Suppose that we have a *right-tailed* test of μ with level of significance 0.01.

SOLUTION: Since the test is a one-tailed test, we use the column headed by $\alpha' = 0.01$. The degrees of freedom are $d.f. = 11 - 1 = 10$, so we use the row headed by 10. The critical value is $t_0 = 2.764$ (see Figure 10-16).

(b) We have a *two-tailed* test of μ with level of significance 0.01.

SOLUTION: The test is a *two-tailed* test, so we look in the column under $\alpha'' = 0.01$ and the row headed by $d.f. = 11 - 1 = 10$. The critical values are $\pm t_0 = \pm 3.169$ (see Figure 10-17). ◊

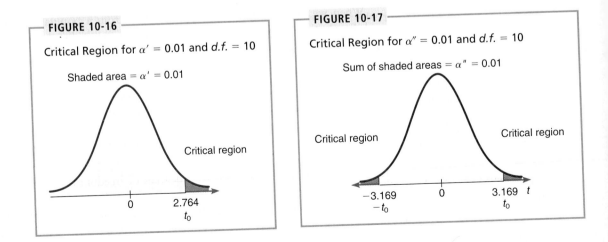

FIGURE 10-16

Critical Region for $\alpha' = 0.01$ and $d.f. = 10$

Shaded area $= \alpha' = 0.01$

Critical region

0   2.764
     t_0

FIGURE 10-17

Critical Region for $\alpha'' = 0.01$ and $d.f. = 10$

Sum of shaded areas $= \alpha'' = 0.01$

Critical region   Critical region

-3.169 0   3.169 t
$-t_0$ t_0

GUIDED EXERCISE 7

Using the Student's t distribution table

Use Table 5 of the Appendix to find the critical value(s) t_0 and critical region for the test described based on a sample of size $n = 8$.

(a) A *left-tailed* test of μ with level of significance 0.05.

⟹ In this case, we use the column headed by $\alpha' = 0.05$ and the row headed by $d.f. = n - 1 = 8 - 1 = 7$. This gives us $t = 1.895$. For a left-tailed test, we use the symmetry of the distribution to get $t_0 = -1.895$ (see Figure 10-18).

(b) A *two-tailed* test of μ with level of significance 0.05.

⟹ Use the column headed by $\alpha'' = 0.05$ and the row headed by $d.f. = n - 1 = 7$. By the symmetry of the curve, the critical values are $\pm t_0 = \pm 2.365$ (see Figure 10-19).

FIGURE 10-18 Critical Region for $d.f. = 7$ and $\alpha' = 0.05$

FIGURE 10-19 Critical Region for $d.f. = 7$ and $\alpha'' = 0.05$

Sample test statistic

Once we have the critical value(s) t_0 and critical region, we convert the sample test statistic $\bar{x}$ to a t value using the formula

$$t = \frac{\bar{x} - \mu}{(s/\sqrt{n})} \text{ with } d.f. = n - 1$$

where $\bar{x}$ = sample mean and μ is specified in H_0

n = sample size

s = sample standard deviation

To conclude the test, we locate the t value of the sample statistic on a diagram showing the critical region. If the sample t value falls in the critical region, we reject H_0. If the sample t value falls outside the critical region, we fail to reject H_0.

The next example demonstrates the traditional method of using critical regions to conclude a test of a mean using small samples.

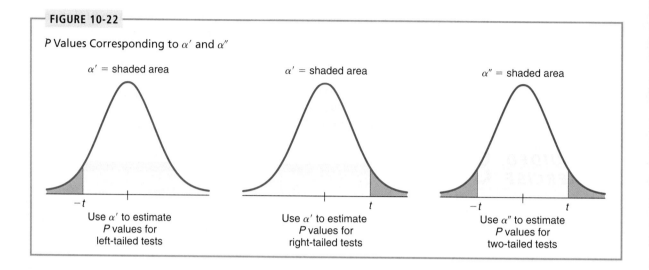

FIGURE 10-22

P Values Corresponding to α' and α''

Use α' to estimate *P* values for left-tailed tests

Use α' to estimate *P* values for right-tailed tests

Use α'' to estimate *P* values for two-tailed tests

Using *P* values for tests of μ, small samples

Recall that the *P* value for a statistical test is the probability of getting a sample statistic as far (or even farther) into the tails of the sampling distribution as the observed sample statistic. The smaller the *P* value, the stronger the evidence is to reject H_0.

In the case of small sample tests of μ, the sample distribution $\bar{x}$ follows a Student's *t* distribution. Therefore, we will use Table 5 of the Appendix to estimate the *P* values. However, the areas given in Table 5 are limited. Consequently, when we use Table 5 to estimate *P* values for $\bar{x}$, we usually find an *interval containing the P value* rather than a single number for the *P* value.

◇ **NOTE REGARDING TABLE 5 OF THE APPENDIX** In Table 5, α' represents the area in *one tail* beyond *t*. The α'' values represent the area in the *two tails*. Notice that in each column $\alpha'' = 2\alpha'$. Consequently, we use α' values as endpoints of the *P*-value intervals for *one-tailed tests* and α'' values as endpoints of the *P*-value intervals for *two-tailed tests* (see Figure 10-22).   ◇

Let's find the *P* values associated with the sample test statistic of Example 9 and of Guided Exercise 8.

EXAMPLE 10

P-value interval for two-tailed test

In Example 9, we tested the rocket manufacturer's claim that its rockets consumed an average of 5500 pounds of fuel in the first 15 seconds of operation. The test was a two-tailed test with sample statistic $\bar{x} = 5690$, $s = 250$, and $n = 6$. Find the *P* value associated with $\bar{x} = 5690$. What does the *P* value tell us?

SOLUTION:

(a) First, we convert $\bar{x} = 5690$ to a *t* value. This was done in Example 9. The corresponding *t* value is 1.862.

(b) The next step is to find an interval containing the *P* value corresponding to the sample test statistic $t = 1.862$ using a two-tailed test. Our test is two-tailed, so we will use α'' values. To find the associated *P*-value interval, we look in the row headed by *d.f.* $= n - 1 = 6 - 1 = 5$. We find that the sample *t* value 1.862 falls between 1.699 and 2.015. The corresponding *P* value then lies between corresponding α'' values 0.150 and 0.100 (see Table 10-5).

TABLE 10-5   Excerpt from Student's *t* Distribution (Table 5, Appendix)

One-tailed test *P* value	α'	···		
✓ Two-tailed test *P* value	α''	···	0.150	0.100 ···
d.f. 5		···	1.699	2.015 ···

↑
Sample *t* = 1.862

Put the adjacent α'' values in increasing order to form the interval

$$0.100 < P \text{ value} < 0.150$$

(c) We see that the *P* value may be almost as large as 0.150. For $\alpha = 0.05$, we see that the *P* value is greater than α, so we do not reject H_0. This result is consistent with the conclusion of Example 9.   ◇

The next example shows how to find an interval containing the *P* value for a one-tailed test using the Student's *t* distribution.

EXAMPLE 11

P-value interval for one-tailed test

In Guided Exercise 8, we again looked at the fuel consumption of rockets. This time the sample mean consumption for the first 15 seconds of flight was $\bar{x} = 5880$ pounds with $s = 250$ pounds and $n = 8$ rockets. We used a right-tailed test with H_0: $\mu = 5500$ and H_1: $\mu > 5500$. Find the *P* value associated with the sample statistic $\bar{x}$. What does the *P* value tell you?

SOLUTION:

(a) Again, our first step is to convert the sample statistic $\bar{x} = 5880$ to a *t* value. We did this in Guided Exercise 8 and got $t = 4.299$.

(b) This test is a one-tailed test, so we use α' values to create the *P*-value interval. To find the associated *P* value, we look in the row headed by $d.f. = n - 1 = 8 - 1 = 7$ and find that the sample *t* statistic $t = 4.299$ falls to the *right* of 3.499 (Table 10-6).

Notice that reading from left to right, the *P* values decrease. Therefore, the *P* value corresponding to the sample *t* value is smaller than that corresponding to $t = 3.499$. Therefore,

$$P \text{ value} < 0.005$$

This relation is shown in Figure 10-23 on the following page.

(c) Since the *P* value is less than the level of significance $\alpha = 0.01$, it is significant, and we reject H_0. This result is consistent with the conclusion of Guided Exercise 8.   ◇

TABLE 10-6   Excerpt from Student's *t* Distribution (Table 5, Appendix)

✓ One-tailed test *P* value	α'	···	0.010	0.005
Two-tailed test *P* value	α''	···		
d.f. 7		···	2.998	3.499

↑
Sample *t* = 4.299

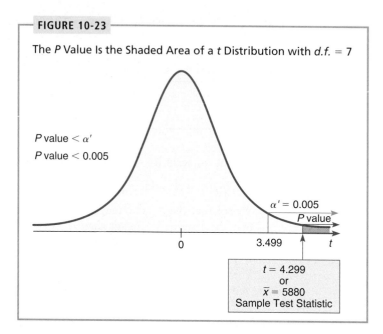

FIGURE 10-23

The *P* Value Is the Shaded Area of a *t* Distribution with *d.f.* = 7

P value $< \alpha'$
P value < 0.005

$\alpha' = 0.005$

P value

0   3.499   *t*

t = 4.299
or
$\bar{x}$ = 5880
Sample Test Statistic

TECH NOTE   Both the TI-83Plus and Minitab support testing a single mean using the Student's *t* distribution. The output for each technology is similar to the output for testing a single mean using the normal distribution. The technologies give the specific *P* value for the sample test statistic rather than an interval of *P* values.

TI-83Plus Press **STAT**, select **TESTS**, use option **2:T-Test.**

Minitab Menu selection: **Stats ➤ Basic Statistics ➤ 1-Sample t.**

 VIEWPOINT

A merry heart doeth good like a medicine.
—Proverbs 17:22

This Old Testament quote summarizes much of the research done by the non-profit group Rx Laughter. To find a link to the Rx Laughter Web site, visit the Brase/Brase *Understanding Basic Statistics* site at http://math.college.hmco.com/students and look up Rx Laughter. Lonnie Zeltzer is a professor of pediatrics at the UCLA Medical School who uses laughter to help children endure painful treatments like chemotherapy and dialysis. Professor Zeltzer works with descendants of famous comedians such as Charlie Chaplin, Harpo Marx, and Lou Costello in an unusual partnership of show business and medicine. Chris Costello, daughter of Lou Costello and a board member of Rx Laughter, claims the research done by the group is a wonderful way to bring the medical community up to speed with children's need for pain management.

Many scientists are exploring the relation between humor and health. The *Journal of the American Medical Association* contains numerous articles about laughter and its effect on stress hormones and the immune system. In all studies of this type, statistics turns out to be an extremely important research tool.

SECTION 10.4   PROBLEMS

In Problems 1–6, use Table 5 of the Appendix to find the critical value(s) t_0 for the described test of the mean (small sample).

1. Sample size $n = 9$, left-tailed test, level of significance 5%.

2. Sample size $n = 13$, right-tailed test, level of significance 1%.

3. Sample size $n = 24$, two-tailed test, level of significance 1%.

4. Sample size $n = 18$, left-tailed test, level of significance 5%.

5. Sample size $n = 12$, two-tailed test, level of significance 5%.

6. Sample size $n = 29$, right-tailed test, level of significance 1%.

For each of the following problems, please provide the requested information.
(a) What is the null hypothesis? What is the alternate hypothesis? Will we use a left-tailed, right-tailed, or two-tailed test? What is the level of significance?
(b) What sampling distribution will we use? What is the critical value t_0 (or critical values $\pm t_0$)?
(c) Sketch the critical region and show the critical value (or critical values).
(d) Calculate the t value corresponding to the sample statistic $\bar{x}$ and show its location on the sketch of part (c).
(e) Find the P value (or interval containing the P value) for your test and explain what the P value means in this context.
(f) Based on your answers for parts (a) to (e), shall we reject or fail to reject (i.e., "accept") the null hypothesis based on the given level of significance? Explain your conclusion in the context of the problem.

7. *Medical: Red Blood Cell Count* Let x be a random variable that represents red blood cell (RBC) count in millions per cubic millimeter of whole blood. Then x has a distribution that is approximately normal, and for the population of healthy female adults, the mean of the x distribution is about 4.8 (based on information from *Diagnostic Tests with Nursing Implications*, Springhouse Corporation). Suppose that a female patient has taken six laboratory blood tests over the past several months and the RBC count data sent to the patient's doctor were

3.5	4.2	4.5	4.6	3.7	3.9

   (a) Use a calculator with sample mean and sample standard deviation keys to verify that $\bar{x} = 4.07$ and $s = 0.44$.
   (b) Do the given data indicate that the population mean RBC count for this patient is lower than 4.8? Use $\alpha = 0.05$.

8. *Medical: Hemoglobin Count* Let x be a random variable that represents hemoglobin count (HC) in grams per 100 milliliters of whole blood. Then x has a distribution that is approximately normal with population mean about 14 for healthy adult women (*see* reference in Problem 7). Suppose that a female patient has taken 12 laboratory blood tests during the past year. The HC data sent to the patient's doctor were

19	23	15	21	18	16
14	20	19	16	18	21

   (a) Use a calculator with sample mean and sample standard deviation keys to verify that $\bar{x} = 18.33$ and $s = 2.71$.

(b) Does this information indicate that the population average HC for this patient is higher than 14? Use $\alpha = 0.01$.

9. *Hummingbirds: Incubation* Selasphorus platycerus is the scientific name of the broad-tailed hummingbird that is commonly found throughout the western United States. It is known that the mean incubation time for eggs of this bird is approximately 16.5 days (based on information from *The Hummingbird Book,* by D. Stokes and L. Stokes, Little, Brown and Company). Assume that the incubation times are approximately normally distributed. However, at higher elevations (above 8000 feet), it is thought that the average incubation time might be different from 16.5 days. A number of broad-tailed hummingbird nests were located in mountain country above 8000 feet, and incubation times for 18 eggs gave the following information (in days):

15	19	16	16	18	17	21	18	17
16	17	18	19	16	15	21	23	19

(a) Use a calculator with mean and standard deviation keys to verify that $\bar{x} = 17.83$ days and $s = 2.20$ days.

(b) Does this information indicate that the population average incubation time above 8000 feet elevation is different (either more or less) from 16.5 days? Use $\alpha = 0.05$.

10. *Fishing: Atlantic Salmon* Homser Lake, Oregon, has an Atlantic salmon catch and release program that has been very successful. The average fisherman's catch has been $\mu = 8.8$ Atlantic salmon per day (*Source: National Symposium on Catch and Release Fishing,* Humboldt State University). Suppose that a new quota system restricting the number of fishermen has been put into effect this season. A random sample of fishermen gave the following catch per day:

12	6	11	12	5	0	2
7	8	7	6	3	12	12

(a) Use a calculator with mean and sample standard deviation keys to verify $\bar{x} = 7.36$ and sample standard deviation $s = 4.03$.

(b) Assuming that the catch per day has an approximately normal distribution, use a 5% level of significance to test the claim that the population average catch per day is now different from 8.8.

11. *Archaeology: Tree Rings* Tree-ring dating from archaeological excavation sites is used in conjunction with other chronologic evidence to estimate occupation dates of prehistoric Indian ruins in southwestern United States. It is thought that Burnt Mesa Pueblo was occupied around 1300 A.D. (based on evidence from pot sherds and stone tools). The following data give tree-ring dates (A.D.) from adjacent archaeological sites (*Bandelier Archaeological Excavation Project: Summer 1990 Excavations at Burnt Mesa Pueblo,* edited by T. Kohler, Washington State University Department of Anthropology, 1992):

1189	1267	1268	1275	1275
1271	1272	1316	1317	1230

(a) Use a calculator with mean and standard deviation keys to verify that $\bar{x} = 1268$ and $s = 37.29$ years.

(b) Assuming that the tree-ring dates in this excavation area follow a distribution that is approximately normal, does this information indicate the population mean of tree-ring dates in the area is different (either high or low) from 1300 A.D.? Use a 1% level of significance.

12. *Ski Patrol: Avalanches* Snow avalanches can be a real problem for travelers in western United States and Canada. A very common type of avalanche is called the slab avalanche. These have been studied extensively by David McClung, a professor of civil engineering at the University of British Columbia. Slab avalanches studied in Canada had an average thickness of $\mu = 67$ cm (*Source: Avalanche Handbook*, by D. McClung and P. Schaerer). The ski patrol at Vail, Colorado, is studying slab avalanches in their region. A random sample of avalanches in spring gave the following thicknesses (cm):

59	51	76	38	65	54	49	62
68	55	64	67	63	74	65	79

    (a) Use a calculator with mean and standard deviation keys to verify that $\bar{x} = 61.8$ cm and $s = 10.6$ cm.
    (b) Assume the slab thickness has an approximately normal distribution. Use a 1% level of significance to test the claim that the mean slab thickness in the Vail region is different from that in Canada.

13. *Longevity: Honolulu* *USA Today* reported that the state with the longest mean life span is Hawaii, where the population mean life span is 77 years. A random sample of 20 obituary notices in the *Honolulu Advertizer* gave the following information about life span (in years) of Honolulu residents:

72	68	81	93	56	19	78	94	83	84
77	69	85	97	75	71	86	47	66	88

    (a) Use a calculator with mean and standard deviation keys to verify that $\bar{x} = 74.45$ years and $s = 18.09$ years.
    (b) Assuming that life span in Honolulu is approximately normally distributed, does this information indicate that the population mean life span for Honolulu residents is less than 77 years? Use a 5% level of significance.

14. *Veterinary Science: Lions* The heart rate of a healthy lion is approximately normally distributed with mean $\mu = 40$ beats per minute (*Source: The Merck Veterinary Manual*, a reference used in most veterinary colleges). A heart rate that is too slow or too fast can indicate a health problem. A veterinarian has removed an abscessed tooth from a young, healthy zoo lion. As the animal slowly starts to come out of the anesthetic, the heart rate is taken and recorded for half an hour (beats per minute):

30	37	43	38	35	36

    (a) Use a calculator with mean and standard deviation keys to verify $\bar{x} = 36.5$ and $s = 4.2$.
    (b) Use a 5% level of significance to test the claim that the population average heart rate of the lion is different (either way) from 40 beats per minute.

15. *Shopping Time: Housewares* How much customers buy is a direct result of how much time they spend in the store. The mean shopping time for a woman accompanied by children in national houseware stores is 7.3 minutes (*Source: Why We Buy: The Science of Shopping* by P. Underhill). A retail research team is studying shopping habits in the Cherry Creek Mall (Denver). A random sample of women shoppers with children in a large houseware store gave the following shopping times (minutes):

7.7	8.1	8.2	9.0	5.8	9.3	8.4	6.9	12.1	9.4
8.1	6.2	7.3	7.9	8.2	8.5	7.2	6.3	9.1	8.8

(a) Use a calculator with mean and standard deviation keys to verify that $\bar{x} = 8.1$ min and $s = 1.4$ min.
(b) Assume shopping time follows an approximately normal distribution. Use a 5% level of significance to test the claim that the average shopping time for women with children in the Cherry Creek Mall is higher than the national average for this type of store.

10.5
Tests Involving a Proportion

FOCUS POINTS
✓ Identify the components needed for testing a proportion.
✓ Find critical values for the test.
✓ Compute the sample test statistic.
✓ Find the *P* value and conclude the test.

Tests for a single proportion

Many situations arise that call for tests of proportions or percentages rather than means. For instance, a college registrar may want to determine if the proportion of students wanting 3-week intensive courses has increased.

How can we make such a test? In this section, we will study tests involving proportions (i.e., percentages or proportions). Such tests are similar to those in Sections 10.2 and 10.4. The main difference is that we are working with a distribution of proportions.

Throughout this section, we will assume that the situations we are dealing with satisfy the conditions underlying the binomial distribution. In particular, we will let r be a binomial random variable. This means that r is the number of successes out of n independent binomial trials (for the definition of binomial trial, see Section 6.2). We will use $\hat{p} = r/n$ as our estimate for p, the probability of success on each trial. The letter q again represents the probability of failure on each trial, and so $q = 1 - p$. We also assume that the samples are large (i.e., $np > 5$ and $nq > 5$).

For large samples, the distribution of $\hat{p} = r/n$ values is well approximated by a *normal curve* with mean μ and standard deviation σ as follows:

$$\mu = p$$
$$\sigma = \sqrt{\frac{pq}{n}}$$

The null and alternate hypotheses for tests of proportions are

| $H_0: p = k$ | $H_0: p = k$ | $H_0: p = k$ |
| $H_1: p < k$ | $H_1: p > k$ | $H_1: p \neq k$ |

depending on what is asked for in the problem. Notice that since p is a probability, the value k must be between 0 and 1.

Critical values for testing p ($np > 5$ and $nq > 5$)

Since the $\hat{p}$ distribution is approximately normal when n is sufficiently large, we will use the same *critical values* z_0 for our tests as those we used for testing the mean with large samples (Section 10.2). Table 10-7 shows these values.

Sample test statistic

For tests of proportions, we need to convert our sample test statistic $\hat{p}$ to a z value. Then we can compare the sample z value with a specified critical value z_0 to determine

TABLE 10-7 Critical Values for Testing a Proportion

Level of Significance	$\alpha = 0.05$	$\alpha = 0.01$
Critical value z_0 for a left-tailed test	−1.645	−2.33
Critical value z_0 for a right-tailed test	1.645	2.33
Critical values $\pm z_0$ for a two-tailed test	±1.96	±2.58

whether or not to reject H_0. The $\hat{p}$ distribution is approximately normal with mean p and standard deviation $\sqrt{pq/n}$. Therefore, the conversion of $\hat{p}$ to z follows the formula

$$z = \frac{\hat{p} - p}{\sqrt{\dfrac{pq}{n}}}$$

where $\hat{p} = r/n$ is the sample test statistic

$\quad\quad n$ = number of trials

$\quad\quad p$ = proportion specified in H_0

$\quad\quad q = 1 - p$

EXAMPLE 12

Testing p

A team of eye surgeons has developed a new technique for a risky eye operation to restore the sight of people blinded from a certain disease. Under the old method, it is known that only 30% of the patients who undergo this operation recover their eyesight.

Suppose that surgeons in various hospitals have performed a total of 225 operations using the new method and that 88 have been successful (the patients fully recovered their sight). Can we justify the claim that the new method is better than the old one? (Use a 1% level of significance.)

SOLUTION:

(a) Let p be the probability that a patient fully recovers his or her eyesight. The null hypothesis is that p is still 0.30, even for the new method of operation. Therefore,

$\quad\quad H_0$: $p = 0.30$

(b) The alternate hypothesis is that the new method has improved the patient's chances for eyesight recovery. Therefore,

$\quad\quad H_1$: $p > 0.30$

(c) Since the alternate hypothesis is H_1: $p > 0.30$, we use a right-tailed test. For $\alpha = 0.01$, Table 10-7 shows a critical value of $z_0 = 2.33$.

(d) Next, we find the sample test statistic $\hat{p}$ and convert it to a z value:

$$\hat{p} = \frac{r}{n} = \frac{88}{225} \approx 0.39$$

The z value corresponding to $\hat{p}$ is

$$z = \frac{\hat{p} - p}{\sqrt{\dfrac{pq}{n}}} = \frac{0.39 - 0.30}{\sqrt{\dfrac{0.30(0.70)}{225}}} = 2.95$$

In the formula, the value for p came from the null hypothesis. H_0 specified that $p = 0.30$, so $q = 1 - 0.30 = 0.70$.

● **CALCULATOR NOTE**   If you evaluate the denominator separately, carry at least four digits after the decimal. Since your calculator automatically carries at least eight or nine digits, it is best to do the entire calculation internally on your calculator.

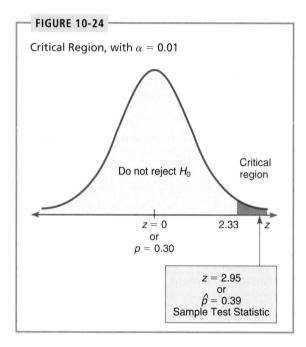

FIGURE 10-24

Critical Region, with $\alpha = 0.01$

Do not reject H_0

Critical region

$z = 0$
or
$p = 0.30$

2.33

z

$z = 2.95$
or
$\hat{p} = 0.39$
Sample Test Statistic

(e) Finally, we draw a normal distribution showing the critical region and the location of the z value corresponding to the sample test statistic $\hat{p}$ (Figure 10-24). Since the sample z value falls inside the critical region, we reject H_0 and conclude that the new method seems to have improved a patient's chances for eyesight recovery.   ◊

GUIDED EXERCISE 9

Testing p

A botanist has produced a new variety of hybrid wheat that is better able to withstand drought than other varieties. The botanist knows that for the parent plants the proportion of seeds germinating is 80%. The proportion of seeds germinating for the hybrid variety is unknown, but the botanist claims that it is 80%. To test this claim, 400 seeds from the hybrid plant are tested, and it is found that 312 germinated. Use a 5% level of significance to test the claim that the proportion germinating for the hybrid is 80%.

(a) Let p be the proportion of hybrid seeds that will germinate. What is the null hypothesis?

⟹ H_0: $p = 0.80$

(b) Because we have no prior knowledge about germination proportion for the hybrid plant, what would be a good choice for the alternate hypothesis?

⟹ H_1: $p \neq 0.80$

Continued

GUIDED EXERCISE 9 continued

(c) Should our test be a one-tailed or a two-tailed test? For $\alpha = 0.05$, what is (are) the critical value(s)?

➡ The test is a two-tailed test. By Table 10-7, the critical values are $z_0 = \pm 1.96$.

(d) Calculate the sample test statistic $\hat{p}$.

➡ The number of trials is $n = 400$, and the number of successes is $r = 312$. Thus,

$$\hat{p} = \frac{r}{n} = \frac{312}{400} = 0.78$$

(e) Next, we convert the sample test statistic $\hat{p} = 0.78$ to a z value. Based on our choice for H_0, what value should we use for p in our formula? Since $q = 1 - p$, what value should we use for q? Using these values for p and q, convert $\hat{p}$ to a z value.

➡ According to H_0, $p = 0.80$. Then $q = 1 - p = 0.20$. Using these values in the following formula gives

$$z = \frac{\hat{p} - p}{\sqrt{\dfrac{pq}{n}}}$$

$$= \frac{0.78 - 0.80}{\sqrt{\dfrac{0.80(0.20)}{400}}}$$

$$= -1.00$$

● **CALCULATOR NOTE** If you evaluate the denominator separately, be sure to carry at least 4 digits after the decimal.

(f) On the normal curve, show the critical regions and the z value corresponding to the sample test statistic $\hat{p}$. Do we reject or fail to reject H_0?

➡ The critical region is shown in Figure 10-25. Since the sample statistic does not fall in the critical region, we fail to reject H_0. At the 5% level of significance, there is not enough evidence to say the botanist is wrong.

FIGURE 10-25 Critical Region, $\alpha = 0.05$

Critical region

Do not reject H_0

Critical region

-1.96 $z = 0$ 1.96 z
or
$p = 0.80$

$z = -1.00$
or
$\hat{p} = 0.78$
Sample Test Statistic

its guests catch northern pike over 20 pounds. Suppose that last summer 64 out of a random sample of 83 guests did, in fact, catch northern pikes weighing over 20 pounds. Does this indicate that the population proportion of guests who catch pikes over 20 pounds is different from 75% (either higher or lower)? Use $\alpha = 0.05$.

7. *Plato's* Republic: *Syllable Patterns* Prose rhythm is characterized as the occurrence of five-syllable sequences in long passages of text. This characterization may be used to assess the similarity among passages of text and sometimes the identity of authors. The following information is based on an article by D. Wishart and S. V. Leach appearing in *Computer Studies of the Humanities and Verbal Behavior* (Vol. 3, pp. 90–99). Syllables were categorized as long or short. On analyzing Plato's *Republic*, Wishart and Leach found that about 26.1% of the five-syllable sequences are of the type in which two are short and three are long. Suppose that Greek archaeologists have found an ancient manuscript dating back to Plato's time (about 427–347 B.C.). A random sample of 317 five-syllable sequences from the newly discovered manuscript showed that 61 are of the type two short and three long combination. Do the data indicate that the population proportion of this type of five-syllable sequence is different (either way) from the text of Plato's *Republic*? Use $\alpha = 0.01$.

8. *Plato's* Dialogues: *Prose Rhythm Symposium* is part of a larger work referred to as Plato's *Dialogues*. Wishart and Leach (*see* source in Problem 7) found that about 21.4% of five-syllable sequences in *Symposium* are of the type in which four are short and one is long. Suppose that an antiquities store in Athens has a very old manuscript that the owner claims to be part of Plato's *Dialogues*. A random sample of 493 five-syllable sequences from this manuscript showed that 136 were of the type four short and one long. Do the data indicate that the population proportion of this type five-syllable sequence is higher than that found in Plato's *Symposium*? Use $\alpha = 0.01$.

9. *Consumers: Product Loyalty* USA Today reported that about 47% of the general consumer population in the United States is loyal to the automobile manufacturer of their choice. Suppose that Chevrolet did a study of a random sample of 1006 Chevrolet owners and found that 490 said that they would buy another Chevrolet. Does this indicate that the population proportion of consumers loyal to Chevrolet is more than 47%? Use $\alpha = 0.01$.

10. *Supermarket: Prices* Harper's Index reported that 80% of all supermarket prices end in the digit 9 or 5. Suppose that you check a random sample of 115 items in a supermarket and find that 88 have prices that end in 9 or 5. Does this indicate that less than 80% of the prices in the store end in the digits 9 or 5? Use $\alpha = 0.05$.

11. *Medical: Hypertension* This problem is based on information taken from *The Merck Manual* (a reference manual used in most medical and nursing schools). Hypertension is defined as a blood pressure over 140 mm Hg systolic and/or over 90 mm Hg diastolic. Hypertension, if not corrected, can cause long-term health problems. In the college-age population (18–24 years), about 9.2% have hypertension. Suppose that a blood donor program is occurring in a college dormitory this week (final exams week). Before each student gives blood, the nurse takes a blood pressure reading. Of 196 donors, it was found that 29 have hypertension. Do these data indicate that the population proportion of students with hypertension during final exams week is higher than 9.2%? Use a 5% level of significance.

12. *Medical: Hypertension* Diltiazem is a commonly prescribed drug for hypertension (*see* source in Problem 11). However, diltiazem causes headaches in about

12% of patients using the drug. It is hypothesized that regular exercise might help reduce the headaches. If a random sample of 209 patients using diltiazem exercised regularly and only 16 had headaches, would this indicate a reduction in the population proportion of patients having headaches? Use a 1% level of significance.

13. *Myers-Briggs: Extroverts* Are most student government leaders extroverts? According to Myers-Briggs estimates, about 82% of college student government leaders are extroverts (*Source: Myers-Briggs Type Indicator Atlas of Type Tables*). Suppose that a Myers-Briggs personality preference test was given to a random sample of 73 student government leaders attending a large national leadership conference and that 56 were found to be extroverts. Does this indicate that the population proportion of extroverts among college student government leaders is different (either way) from 82%? Use $\alpha = 0.01$.

14. *American Attitudes: NAFTA* Generally speaking, would you say that America benefits from being a member of NAFTA? Nationally, about 28% of the U.S. population believes NAFTA benefits America (*Source: American Attitudes* by S. Mitchell, Sociology Department, Ithaca College). A random sample of 48 interstate truck drivers showed that 19 believe NAFTA benefits America. Does this indicate the population proportion of interstate truckers who believe NAFTA benefits America is higher than 28%? Use a 5% level of significance.

15. *Careers: College Professors* If you could start your career all over again, would you still choose to be a college professor? About 76% of all U.S. college professors respond yes, they would choose to be a college professor again (*Source: The Chronicle of Higher Education*). A random sample of 59 college professors in Colorado showed that 47 claim they would choose college teaching again. Does this indicate the proportion of professors in Colorado who would choose the career again is different from the national rate of 76%? Use a 1% level of significance.

SUMMARY

In this chapter, we studied statistical inference methods called *hypothesis testing*. We establish an initial claim about the value of a parameter. This claim is the null hypothesis H_0, which claims the parameter in question equals a certain value. Then we propose an alternate hypothesis H_1, which indicates the parameter is less than, greater than, or different from the value in the null hypothesis. To determine whether to reject the null hypothesis or not, we use the evidence of the sample data and the predetermined level of significance α.

The basic steps we follow in the procedure of hypothesis testing are

1. State the null and alternate hypotheses H_0 and H_1.

2. Choose α, the level of significance.

3. Find the critical value(s) z_0 or t_0 and the corresponding critical region.

4. Convert the sample test statistic to a z value or t value.

5. If the sample test statistic falls in the critical region, reject H_0. Otherwise, do not reject H_0.

An alternate way to conclude a test of hypotheses is to compare the P value of the sample test statistic with α. The P value of the sample test statistic is the smallest level of significance for which we can reject H_0. Therefore, if P value $\leq \alpha$, we reject H_0, and if P value $> \alpha$, we do not reject H_0. Because statistical computer software packages generate the P value associated with the sample test statistic and the null hypothesis, the method of using P values to conclude tests of hypotheses is widely used and very popular.

In this chapter, we used hypothesis testing to conduct tests involving a single mean μ (large or small sample) and a single proportion p. The methods of hypothesis testing are very general, and we will see them used again in later chapters with other parameters and other probability distributions.

IMPORTANT WORDS & SYMBOLS

Section 10.1
Hypothesis testing
Hypothesis
Null hypothesis H_0
Alternate hypothesis H_1
Type I error
Type II error
α, the level of significance of a test and the probability of a type I error
β, the probability of a type II error
Power of a test $(1 - \beta)$
Critical region

Section 10.2
One-tailed tests
Two-tailed tests
Sample test statistic
Critical value or values for testing means
Statistical significance

Section 10.3
P value

Section 10.4
$d.f.$ for testing μ with small samples
α' and α'' with respect to the t distribution table

VIEWPOINT *Will It Rain?*

Do cloud seed experiments ever work? If you seed the clouds, will it rain? If it did rain, who would benefit? Who would be displeased by the rain? If you seed the clouds and nothing happens, would taxpayers (who support the effort) complain or rejoice? Maybe this should be studied over a remote island—such as Tasmania (near Australia). Using what you already know about statistical testing, you can conduct your own tests, given the appropriate data. Remember, there are sociological questions (pleased/displeased with result) as well as technical questions (number of inches of rain produced). For data regarding cloud-seeding experiments over Tasmania, visit the Brase/Brase *Understanding Basic Statistics* site at http://math.college.hmco.com/students and find the link to DASL, the Carnegie Mellon University Data and Story Library. From the DASL site, look under Datasets for Cloud.

CHAPTER REVIEW PROBLEMS

Before you solve each problem, first categorize it by answering the following questions:
(a) Are we testing a single mean or a single proportion?
(b) Are we using a large sample or a small sample?
(c) Then solve each problem by
  (i) stating the null and alternate hypotheses.
  (ii) finding the critical value(s) and critical region.
  (iii) sketching the critical region.
  (iv) converting the sample test statistic to a z value or a t value as appropriate and showing its location on the sketch in part (iii).
  (v) finding the P value of the sample test statistic.
  (vi) making your decision to reject or fail to reject the null hypothesis.
  (vii) relating your conclusion to the specific question asked in the problem.

1. *Environment: Miles per Vehicle* According to the Energy Information Association, the average number of miles driven annually per vehicle in the United States is 11.3 thousand miles. Suppose that a random sample of 36 vehicles owned by residents of Chicago showed that the average mileage driven last year was 11.0 thousand miles with standard deviation 600 miles. Does this indicate that the average miles driven per vehicle in Chicago is different from (higher or lower than) the national average? Use a 5% level of significance.

2. *Business: Leased Cars* According to *USA Today,* 25% of cars were leased last year instead of sold. To test if the percentage of leased cars is still the same this year, a random sample of 50 new car deals was examined. Of these, 18 involved car leases instead of car purchases. Does this indicate that the percentage of leased cars is different (either higher or lower) this year? Use a 1% level of significance.

3. *Medical: Waiting Time* New patients often have to wait a number of days to get an appointment with a physician. According to *Practice Patterns of General Internal Medicine* (published by the American Medical Association), a new patient must wait an average of 10.8 days for an appointment. Rockwood Clinic did a study of its new patients and found that for a random sample of 20 new patients, the average waiting time for an appointment was 9.2 days with standard deviation 3 days. At the 1% level of significance, do new patients at Rockwood Clinic have an average waiting time for an appointment that is less than 10.8 days?

4. *Lifestyle: Students with Jobs* Professor Jennings claims that only 35% of the students at Flora College work while attending school. Dean Renata thinks the professor has underestimated the number of students with part-time or full-time jobs. A random sample of 81 students shows that 39 have jobs. Do the data indicate that more than 35% of the students have jobs? Use a 5% level of significance.

5. *Business: Sales per Employee* In the retail business, the average annual sales per employee varies from business to business. According to an article in the magazine *American Demographics,* the average value of annual sales per employee in apparel stores is 88.9 thousand dollars. In one large chain of apparel stores it was found that for a random sample of 40 employees, the average value of sales per employee last year was 89.7 thousand dollars with standard deviation 5 thousand dollars. Is this evidence that employees in this chain have higher average annual sales than the apparel industry in general? Use a 5% level of significance.

6. *Business: Sales per Employee* For grocery stores, the average annual sales per employee is 131.4 thousand dollars (*American Demographics* magazine). Great Foods chain did a study of 50 employees from different stores and found the average

annual sales per employee to be 135.8 thousand dollars with standard deviation 8 thousand dollars. Does it appear that the average annual sales per Great Foods employee is different from 131.4 thousand? Use a 1% level of significance.

7. *Lifestyle: The Perfect Match* Is there a Mr. or Ms. Right? According to a survey conducted by CJ Olson Market Research (reported in *USA Today*), 84.6% of adults between 18 and 24 years of age believe every person has a perfect match. A random sample of 50 people in the 25-to-34 age group were asked the same question. Thirty-six in the older age group said that they believed every person has a perfect match. Use a 1% level of significance to test the hypothesis that the proportion of people in the 25-to-34 age group who believe everyone has a perfect match is less than the proportion in the 18-to-24 age group.

8. *Medical: Plasma Compress* A hospital reported that the normal death rate for patients with extensive burns (more than 40% of skin area) has been significantly reduced by the use of new fluid plasma compresses. Before the new treatment, the mortality rate for extensive burn patients was about 60%. Using the new compresses, the hospital found that only 40 out of 90 patients with extensive burns died. Use a 1% level of significance to test the claim that the mortality rate has dropped.

9. *Civil Service: College Degrees* The Congressional Budget Office reports that 36% of the federal civilian employees have a bachelor's degree or higher (*The Wall Street Journal*). A random sample of 120 employees in the private sector showed that 33 have a bachelor's degree or higher. Does this indicate that the percentage of employees who hold bachelor's or higher degrees in the private sector is less than in the federal civilian sector? Use $\alpha = 0.05$.

10. *Vending Machine: Coffee* A machine in the student lounge dispenses coffee. The average cup of coffee is supposed to contain 7.0 oz. Eight cups of coffee from this machine show the average content to be 7.3 oz with a standard deviation of 0.5 oz. Do you think the machine has slipped out of adjustment and the average amount of coffee per cup is different from 7 oz? Use a 5% level of significance.

11. *Quality Control: Fuel Injection* The manufacturer of a sports car claims that the fuel injection system lasts 48 months before it needs to be replaced. A consumer group tests this claim by surveying a random sample of 10 owners who had fuel injection systems replaced. The ages of the cars at the time of replacement were (in months):

    29 42 49 58 53
    46 30 51 42 62

    (a) Use your calculator to verify that the mean age of a car when the fuel injection system fails is $\bar{x} = 46.2$ months with standard deviation $s = 10.85$ months.
    (b) Test the claim that the fuel injection system lasts less than 48 months before needing replacement. Use a 5% level of significance.

DATA HIGHLIGHTS: GROUP PROJECTS

Break into small groups and discuss the following topics. Organize a brief outline in which you summarize the main points of your group discussion.

"With Sampling, There Is Too a Free Lunch" was a headline that appeared in *The Wall Street Journal*. The article is about food product samples available at grocery stores. Giving out food samples is expensive and labor-intensive. It clogs supermarket aisles. It is risky. What if a customer tries an item and spits it out on the floor or says the product is awful? It creates litter. Some customers drop toothpicks or small paper cups on the floor or spill the product. However, the budget that companies are willing to spend to have their products sampled is growing. The director of communications for Bigg's "hyper-market" (a combination grocery and general-merchandise store) says that more than 60% of customers sample products and about 37% of those who sample a product buy it.

(a) Let's test the hypothesis that 60% of customers sample a particular product. What is the null hypothesis? Do you believe that the percentage of customers who sample products is less than, more than, or just different from 60%? What will you use for the alternate hypothesis?
(b) Choose a level of significance α.
(c) Go to a grocery store when special products are being sampled (not just the usual in-house store samples often available at the deli or bakery). Count the number of customers going by the display when a sample is available and the number of customers who try the sample. Be sure the number of customers n is large enough to use the normal distribution to approximate the binomial.
(d) Using your sample data, conclude the hypothesis test. What is your conclusion?
(e) Do you think different food products might have higher or lower percentages of customers trying them? For instance, does a higher percentage of customers try samples of pizza than samples of yogurt? How could you use statistics to justify your answer? (*See* Chapter 11.)
(f) Do you want to include young children in your sample? Do they pick up items to include in the customer's basket, or do they just munch the samples?

LINKING CONCEPTS: WRITING PROJECTS

Discuss each of the following topics in class or review the topics on your own. Then write a brief but complete essay in which you summarize the main points. Please include formulas and graphs as appropriate.

The most important questions in life usually cannot be answered with absolute certainty. Many important questions are answered by giving an estimate and a measure of confidence in the estimate. This was the focus of Chapter 9. However, sometimes important questions must be answered in a more straightforward manner by a simple *yes* or *no*. Hypothesis testing is the statistical process of answering questions with a straightforward yes or no *and* providing an estimate of the risk in accepting the answer.

1. Review and discuss type I and type II errors associated with hypothesis testing.

2. Review and discuss the level of significance and power of a statistical test.

3. The following statements are very important. Give them some careful thought and discuss them.
   (a) When we fail to reject the null hypothesis, we do not claim it is absolutely true. We simply claim that at the given level of significance the data were not sufficient to reject the null hypothesis.
   (b) When we accept the alternate hypothesis, we do not claim the null hypothesis is absolutely false. We do claim that at the given level of significance the data presented enough evidence to reject the null hypothesis.

4. As access to computers becomes more and more prevalent, we see P values reported in hypothesis testing more frequently. Review the use of P values in hypothesis testing. What is the difference between the level of significance of a test and the P value? Considering both the P value and level of significance, under what conditions do we reject or fail to reject the null hypothesis?

SIMULATION

Recall that the level of significance α is the probability of mistakenly rejecting a true null hypothesis. If $\alpha = 0.05$, then we expect to mistakenly reject a true null hypothesis about 5% of the time. The following simulation conducted with Minitab demonstrates this phenomenon.

We draw 40 random samples of size 50 from a population that is normally distributed with mean $\mu = 30$ and $\sigma = 2.5$. The display shows the results of a hypothesis test with

$$H_0: \mu = 30 \qquad H_1: \mu > 30$$

for each of the 40 samples labeled C1 through C40. Because each of the 40 samples is drawn from a population with mean $\mu = 30$, the null hypothesis $H_0: \mu = 30$ is true for the test based on each sample. However, as the display shows, for some samples we reject the true null hypothesis.

(a) How many of the 40 samples have a sample mean $\bar{x}$ above $\mu = 30$? below $\mu = 30$?

(b) Look at the P value of the sample statistic $\bar{x}$ in each of the 40 samples. How many P values are less than or equal to $\alpha = 0.05$? What percent of the P values are less than or equal to α? What percent of the samples have us reject H_0 when, in fact, each of the samples was drawn from a normal distribution with $\mu = 30$ as hypothesized in the null hypothesis?

(c) If you have access to computer or calculator technology that creates random samples from a normal distribution with a specified mean and standard deviation, repeat this simulation. Do you expect to get the same results? Why or why not?

Minitab Display: Random samples of size 50 from a Normal Population with $\mu = 30$ and $\sigma = 2.5$

```
Z-Test
Test of mu = 30.000 vs. mu > 30.000
The assumed sigma = 2.50
Variable   N    Mean   StDev  SE Mean     Z     P
C1        50  30.002   2.776   0.354   0.01  0.50
C2        50  30.120   2.511   0.354   0.34  0.37
C3        50  30.032   2.721   0.354   0.09  0.46
C4        50  30.504   2.138   0.354   1.43  0.077
C5        50  29.901   2.496   0.354  -0.28  0.61
C6        50  30.059   2.836   0.354   0.17  0.43
C7        50  30.443   2.519   0.354   1.25  0.11
C8        50  29.775   2.530   0.354  -0.64  0.74
C9        50  30.188   2.204   0.354   0.53  0.30
C10       50  29.907   2.302   0.354  -0.26  0.60
C11       50  30.036   2.762   0.354   0.10  0.46
C12       50  30.656   2.399   0.354   1.86  0.032
C13       50  30.158   2.884   0.354   0.45  0.33
C14       50  29.830   3.129   0.354  -0.48  0.68
C15       50  30.308   2.241   0.354   0.87  0.19
C16       50  29.751   2.165   0.354  -0.70  0.76
C17       50  29.833   2.358   0.354  -0.47  0.68
C18       50  29.741   2.836   0.354  -0.73  0.77
C19       50  30.441   2.194   0.354   1.25  0.11
C20       50  29.820   2.156   0.354  -0.51  0.69
C21       50  29.611   2.360   0.354  -1.10  0.86
C22       50  30.569   2.659   0.354   1.61  0.054
C23       50  30.294   2.302   0.354   0.83  0.20
C24       50  29.978   2.298   0.354  -0.06  0.53
C25       50  29.836   2.438   0.354  -0.46  0.68
C26       50  30.102   2.322   0.354   0.29  0.39
C27       50  30.066   2.266   0.354   0.19  0.43
C28       50  29.071   2.219   0.354  -2.63  1.00
C29       50  30.597   2.426   0.354   1.69  0.046
C30       50  30.092   2.296   0.354   0.26  0.40
C31       50  29.803   2.495   0.354  -0.56  0.71
C32       50  29.546   2.335   0.354  -1.28  0.90
C33       50  29.702   1.902   0.354  -0.84  0.80
C34       50  29.233   2.657   0.354  -2.17  0.98
C35       50  30.097   2.472   0.354   0.28  0.39
C36       50  29.733   2.588   0.354  -0.76  0.78
C37       50  30.379   2.976   0.354   1.07  0.14
C38       50  29.424   2.827   0.354  -1.63  0.95
C39       50  30.288   2.396   0.354   0.81  0.21
C40       50  30.195   3.051   0.354   0.55  0.29
```

Technology Hints

TI-83Plus

Press **STAT**, select **EDIT**. Highlight the list name such as L1. Then press **MATH**, select **PRB**, highlight **6:randNorm(μ, σ, sample size)**. Press enter. Fill in the values of $\mu = 30$, $\sigma = 2.5$, and sample size $= 50$. Press enter. Now list L1 contains a random sample from the normal distribution specified.

To test the hypothesis H_0: $\mu = 30$ against H_1: $\mu > 30$, press **STAT**, select **TESTS**, and use option **1:Z-Test**. Fill in the value 30 for μ_0, 2.5 for σ, and $> \mu_0$. The output provides the value of the sample statistic $\bar{x}$, its corresponding z value, and the P value of the sample statistic.

Excel

To draw random samples from a normal distribution with $\mu = 30$ and $\sigma = 2.5$, use the menu choice **Tools ➤ Data Analysis ➤ Random Number Generator**. In the dialogue box, the number of variables is the number of samples. Fill in the rest of the dialogue box as shown.

To conduct a hypothesis test of H_0: $\mu = 30$ against H_1: $\mu > 30$, use the command **ZTEST(data range, 30, 2.5)**. This command is described as returning the two-tailed P value of the z test. However, it appears to give the P value of a right-tailed test. Use the ZTEST command with caution and check your results against the table results.

Minitab

To generate random samples from a normal distribution, use the menu choices **Calc ➤ Random Data ➤ Normal**. In the dialogue box, the number of rows refers to the sample size. Use 50 rows. Then designate the columns for the samples. Using C1–C40 will generate 40 random samples and put the samples in columns C1 through C40.

To test the hypothesis H_0: $\mu = 30$ against H_1: $\mu > 30$, use the menu choices **Stat ➤ Basic Statistics ➤ 1-SampleZ**. Use columns C1–C40 as the variables. Fill in 30 for the test mean, use greater than for the alternate hypothesis, and use 2.5 for sigma.

11

Inferences About Differences

"So what!"

—Anonymous

"Girl with Black Eye" by Norman Rockwell (1894–1978)

Norman Rockwell painted everyday people and situations. In this cover for the *Saturday Evening Post* (May 23, 1953), a young lady is about to have a conference with her school principal. So what!

We have all heard the exclamation, "So what!" Philologists (people who study cultural linguistics) tell us that this expression is a shortened version of "So what is the difference!" They also tell us that there are similar popular or slang expressions about differences in all languages and cultures. It is human nature to challenge the claim that something is better, worse, or just simply different. In this chapter, we will focus on this very human theme by studying a variety of topics regarding questions of whether or not differences exist between two populations.

For on-line student resources, visit **math.college.hmco.com/students** and follow the Statistics links to the Brase/Brase, *Understanding Basic Statistics,* 3rd edition Web site.

◇ What are the statistical advantages of paired data values? How do we construct statistical tests? (SECTION 11.1)

◇ How do we compare means from two independent populations using large samples? (SECTION 11.2)

◇ What if we want to compare means from two independent populations, but our samples are small? (SECTION 11.3)

◇ How do we use sample data to compare proportions from two independent populations? (SECTION 11.4)

FOCUS PROBLEM

The Trouble with Wood Ducks

The National Wildlife Federation published an article entitled "The Trouble with Wood Ducks" (*National Wildlife*, Vol. 31, No. 5). In this article, wood ducks are described as beautiful birds living in forested areas such as the Pacific Northwest and Southeast United States. Because of overhunting and habitat destruction, these birds were in danger of extinction. A federal ban on hunting wood ducks in 1918 helped save the species from extinction. Wood ducks like to nest in tree cavities. However, many such trees were disappearing due to heavy timber cutting. For a period of time it seemed that nesting boxes were the solution to disappearing trees. At first, the wood duck population grew, but after a few seasons, the population declined sharply. Good biology research combined with good statistics provided an answer to this disturbing phenomenon.

Cornell University professors of ecology Paul Sherman and Brad Semel found that the nesting boxes were placed too close to each other. Female wood ducks prefer a secluded nest that is a considerable distance from the next wood duck nest. In fact, female wood duck behavior changed when the nests were too close to each other. Some females would lay their eggs in another female's nest. The result was too many eggs in one nest. The biologists found that if there

451

were too many eggs in a nest, the proportion of eggs that hatched was considerably reduced. In the long run, this meant a decline in the population of wood ducks.

In their study, Sherman and Semel used two placements of nesting boxes. Group I had boxes that were well separated from each other and well hidden by available brush. Group II had boxes that were highly visible and grouped closely together.

In group I boxes, there were a total of 474 eggs, of which a field count showed that about 270 hatched. In group II boxes, there were a total of 805 eggs, of which a field count showed that, again, about 270 hatched.

(a) Find a 95% confidence interval for $p_1 - p_2$. Does the interval indicate that the proportion of eggs hatched from group I nest box placements is higher than, lower than, or not different from the proportion of eggs hatched from group II nest boxes?

(b) Use a 5% level of significance to test the hypothesis that the proportion of hatches from group I placements is greater than that from group II placements. Do we reject or fail to reject H_0? Is this result consistent with the information from the 95% confidence interval?

(c) What conclusions about placement of nest boxes can be drawn? In the article, additional concerns are raised about the higher cost of placing and maintaining group I nest box placement. At issue is also the cost efficiency per successful wood duck hatch. Data in the article do not include information that would help us answer questions of *cost* efficiency. However, the data presented do help us answer questions about proportion of successful hatches in the two nest box configurations.

11.1
Tests Involving Paired Differences (Dependent Samples)

FOCUS POINTS
✓ Identify paired data and dependent samples.
✓ Explain the advantages of paired data tests.
✓ Compute differences and the sample test statistic.
✓ Use degrees of freedom and the *t* table to find critical values.
✓ Estimate the *P* value and conclude the test.

Creating data pairs

Many statistical applications use *paired data* samples to draw conclusions about the difference between two population means. *Data pairs* occur very naturally in "before and after" situations, where the *same* object or item is measured both before and after a treatment. Applied problems in social science, natural science, and business administration frequently involve a study of matching pairs. Psychological studies of identical twins; biological studies of plant growth on plots of land matched for soil type, moisture, and sun; and business studies on sales of matched inventories are examples of paired data studies.

When working with paired data, it is very important to have a definite and uniform method of creating data pairs that clearly utilizes a natural matching of characteristics. The next example and exercise demonstrate this feature.

EXAMPLE 1

Paired data

A shoe manufacturer claims that among the general population of adults in the United States, the average length of the left foot is longer than that of the right. To compare the average length of the left foot with that of the right, we can take a random sample of 15 U.S. adults and measure the length of the left foot and then the length of the right foot for each person in the sample. Is there a natural way of pairing the measurements? How many pairs will we have?

SOLUTION: In this case, we can pair each left foot measurement with the same person's right foot measurement. The person serves as the "matching link" between the two distributions. We will have 15 pairs of measurements.   ◇

GUIDED EXERCISE 1

Paired data

A psychologist has developed a series of exercises called the Instrumental Enrichment (IE) Program, which he claims is useful in overcoming cognitive deficiencies in mentally retarded children. To test the program, extensive statistical tests are being conducted. In one experiment, a random sample of 10-year-old students with IQ scores below 80 was selected. An IQ test was given to these students before they spent 2 years in an IE Program, and an IQ test was given to the same students after the program.

(a) On what basis can you pair the IQ scores?   ⇒   Take the "before and after" IQ scores of each individual student.

(b) If there were 20 students in the sample, how many data pairs would you have?   ⇒   Twenty data pairs. Note that there would be 40 IQ scores, but only 20 pairs.

◇ **COMMENT** To compare two populations, we cannot always employ paired data tests, but when we can, what are the advantages? Using matched or paired data often can reduce the danger of introducing extraneous or uncontrollable factors into our sample measurements because the matched or paired data have essentially the *same* characteristics except for the *one* characteristic that is being measured. Furthermore, it can be shown that pairing data has the theoretical effect of reducing measurement variability (i.e., variance), which increases the accuracy of statistical conclusions.   ◇

When we wish to compare the means of two samples, the first item to be determined is whether or not there is a natural pairing between the data in the two samples. Again, data pairs are created from "before and after" situations, or from matching data by using studies of the same object, or by a process of taking measurements of closely matched items.

Testing the differences d

When testing *paired* data, we take the difference d of the data pairs *first* and look at the mean difference $\bar{d}$. Then we use a test on $\bar{d}$. Theorem 11.1 provides the basis for our work with paired data.

◇ **THEOREM 11.1** Consider a random sample of n data pairs. Suppose the differences d between the first and second members of each data pair are (approximately) normally distributed with population mean μ_d. Then the t values

$$t = \frac{\bar{d} - \mu_d}{s_d / \sqrt{n}}$$

where $\bar{d}$ is the sample mean of the d values, n is the number of data pairs, and

$$s_d = \sqrt{\frac{\Sigma(d - \bar{d})^2}{n - 1}}$$

is the sample standard deviation of the d values, follow a Student's t distribution with degrees of freedom $d.f. = n - 1$. ◊

Hypotheses for testing the mean of the differences

When testing the mean of the differences of paired data values, the null hypothesis is that there is no difference among the pairs. That is, the mean of the differences μ_d is zero.

$$H_0: \mu_d = 0$$

The alternate hypothesis depends on the problem and can be

$H_1: \mu_d < 0$	$H_1: \mu_d > 0$	$H_1: \mu_d \neq 0$
(left-tailed)	(right-tailed)	(two-tailed)

Critical values

Since the $\bar{d}$ distribution follows a Student's t distribution with degrees of freedom $d.f. = n - 1$, our critical values will be found in Table 5, "Student's t Distribution," in the Appendix. The methods to find the critical values are the same as those of Section 10.4 when we used the Student's t distribution to find critical values for tests of the mean with small samples. If the test is a *one-tailed* test, we find the critical t value in the column headed by $\alpha' =$ level of significance. If the test is a *two-tailed* test, we find the critical value in the column headed by $\alpha'' =$ level of significance. In each case, we use the row containing the degrees of freedom $n - 1$, where n is the *number of data pairs*.

Sample test statistic

For paired difference tests, we make our decision regarding H_0 according to the evidence of the sample mean $\bar{d}$ of the differences of measurements. By Theorem 11.1, we convert the sample test statistic $\bar{d}$ to a t value using the formula

$$t = \frac{\bar{d} - \mu_d}{(s_d/\sqrt{n})} \text{ with } d.f. = n - 1$$

where $s_d =$ sample standard deviation of the differences d

$n =$ number of data pairs

$\mu_d = 0$ as specified in H_0

EXAMPLE 2

Paired difference test

A team of heart surgeons at Saint Ann's Hospital knows that many patients who undergo corrective heart surgery have a dangerous buildup of anxiety before their scheduled operations. The staff psychiatrist at the hospital has started a new counseling program intended to reduce this anxiety. A test of anxiety is given to patients who know they must undergo heart surgery. Then each patient participates in a series of counseling sessions with the staff psychiatrist. At the end of the counseling sessions, each patient is retested to determine anxiety level. Table 11-1 indicates the results for a random sample of nine patients. Higher scores mean higher levels of anxiety.

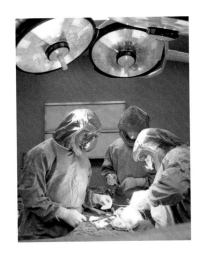

From the given data, can we conclude that the counseling sessions reduce anxiety? Use a 0.01 level of significance.

SOLUTION: Before we answer this question, let us notice two important points: (1) we have a *random sample* of nine patients, and (2) we have a *pair* of measurements taken on the same patient before and after counseling sessions. In our problem, the sample size is $n = 9$ pairs (i.e., patients), and the d values are found in the fourth column of Table 11-1.

(a) First, we need to find the sample mean $\overline{d}$ and sample standard deviation s_d of the d values. Using the formulas for $\overline{d}$ and for s_d or a calculator with mean and standard deviation keys, we find

$$\overline{d} = 33.33$$

and

$$s_d = 22.92$$

(b) Next, we set the hypotheses. In our problem, we want to test the claim that the counseling sessions reduce anxiety. This means that the anxiety level before counseling is expected to be higher than the anxiety level after counseling. In symbols, $d = B - A$ should tend to be positive, and the population mean of differences μ_d also should be positive. Therefore, we have the hypotheses

$$H_0: \mu_d = 0$$
$$H_1: \mu_d > 0$$

(c) Find the critical values. Critical values are found using the Student's t distribution. We have a right-tailed test, with level of significance 0.01, so we look in the column headed by $\alpha' = 0.01$. The degrees of freedom are $d.f. = n - 1 = 9 - 1 = 8$. In Table 5 in the Appendix, we find the critical value is $t_0 = 2.896$.

TABLE 11-1

Patient	B Score Before Counseling	A Score After Counseling	$d = B - A$ Difference
Jan	121	76	45
Tom	93	93	0
Diane	105	64	41
Barbara	115	117	−2
Mike	130	82	48
Bill	98	80	18
Frank	142	79	63
Carol	118	67	51
Alice	125	89	36

VIEWPOINT   *DUI*

DUI usually means "driving under the influence" of *alcohol,* but driving under the influence of *sleep loss* can be just as dangerous. Researchers in Australia have found that after staying awake for 24 hours straight, a person will be about as impaired as if he had enough alcohol to be legally drunk in most U.S. states (*Source: Rocky Mountain News*). Using driver simulation exams and statistical tests (paired difference tests) found in this section, it is possible to show that the null hypothesis $H_0: \mu_d = 0$ cannot be rejected. Or put another way, the average level of impairment for a given individual from alcohol (at the DUI level) is about the same as the average level of impairment for sleep loss (24 hours without sleep).

SECTION 11.1   PROBLEMS

For each of the following problems, please provide the requested information.
(a) What is the null hypothesis? What is the alternate hypothesis? Will we use a left-tailed, right-tailed, or two-tailed test? What is the level of significance?
(b) What sampling distribution will we use? What is the critical value t_0 (or critical values $\pm t_0$)?
(c) Sketch the critical region and show the critical value (or critical values).
(d) Calculate the t value corresponding to the sample statistic $\bar{d}$ and show its location on the sketch of part (c).
(e) Find the P value (or interval containing the P value) for your test and explain what the P value means in this context.
(f) Based on your answers for parts (a) to (e), shall we reject or fail to reject (i.e., "accept") the null hypothesis for the given level of significance α? Explain your conclusion in the context of the problem.

1. *Business: CEO Raises* Are America's top chief executive officers (CEOs) really worth all that money? One way to answer this question is to look at row *B,* the annual company percentage increase in revenue, versus row *A,* the CEO's annual percentage salary increase in that same company (*Source: Forbes,* Vol. 159, No. 10). A random sample of companies such as John Deere & Co., General Electric, Union Carbide, and Dow Chemical yields the following data:

B: Percent for company	24	23	25	18	6	4	21	37
A: Percent for CEO	21	25	20	14	−4	19	15	30

Do these data indicate that the population mean percentage increase in corporate revenue (row *B*) is different from the population mean percentage increase in CEO's salary? Use a 5% level of significance.

2. *Fishing: Shore or Boat?* Is fishing better from a boat or from the shore? Pyramid Lake is on the Paiute Indian Reservation in Nevada. Presidents, movie stars, and

Area on Navajo Reservation	Number of Inhabited Houses	Number of Inhabited Hogans
Bitter Springs	18	13
Rainbow Lodge	16	14
Kayenta	68	46
Red Mesa	9	32
Black Mesa	11	15
Canyon de Chelly	28	47
Cedar Point	50	17
Burnt Water	50	18

Does this information indicate that the population mean number of inhabited houses is greater than that of hogans on the Navajo Reservation? Use a 5% level of significance.

8. *Archaeology: Stone Tools* The following is based on information taken from *Bandelier Archaeological Excavation Project: Summer 1990 Excavations at Burnt Mesa Pueblo and Casa del Rito*, edited by T. A. Kohler (Washington State University, Department of Anthropology). The artifact frequency for an excavation of a kiva in Bandelier National Monument gave the following information:

Stratum	Flaked Stone Tools	Nonflaked Stone Tools
1	7	3
2	3	2
3	10	12
4	1	8
5	4	18
6	38	33
7	51	38

Does this information indicate that there tend to be more flaked stone tools than nonflaked stone tools at this excavation site? Use a 5% level of significance.

9. *Archaeology: Pot Sherds* On the same stratum of an excavated block of rooms at Bandelier National Monument, the following information was obtained about sherds of service ware in two different subareas (*see* reference in Problem 8):

Service Ware	Subarea 1	Subarea 2
Socorro black on white	10	4
Santa Fe black on white	42	39
Galisteo black on white	15	21
Puerco black on red	6	9
Wingate black on red	11	6

Does this information indicate that there is a difference (either higher or lower) for the population mean number of service ware sherds in subarea 1 as compared with subarea 2? Use a 5% level of significance.

10. *Psychology: Babies* Many mothers say that they can recognize the cries of their own babies and that they can distinguish between a cry of pain and a hunger cry. A psychologist studied this phenomenon using the following experiment. A random sample of mothers listened to a tape-recorded set of five cries from different babies, one of which was their own. They had to decide which was their baby. Each mother heard 20 such sets, in which 10 were cries of hungry babies and 10 were cries produced by a slight pin prick on a foot. The results are shown in the following table, where row *B* is the correct number of identifications (out of 10) for a hunger cry, and row *A* is the correct number of identifications (out of 10) for a pain cry. The psychologist claims that the mothers are more successful in picking out their own babies when a hunger cry is involved, since the mothers have more experience with that situation. Test this claim at the 5% level of significance.

Mother	1	2	3	4	5	6	7	8
B: Hunger cry	6	6	6	5	3	7	9	2
A: Pain cry	5	4	7	3	2	6	4	3

11. *Medical: Heart Rate* The following problem is based on information taken from *Diagnostic Tests with Nursing Implications* (Springhouse). A patient who had undergone a triple coronary artery bypass was tested in the following way. The patient took his pulse at rest and then on a treadmill exercised at a 10% grade at 1.7 miles per hour for 2 minutes. This procedure was repeated on six different days. The pulse rate 6 minutes after the test and the pulse rate before the test are shown in the following table (beats per minute):

Test	1	2	3	4	5	6
Pulse before	69	72	75	73	70	74
Pulse after	85	79	83	84	87	78

Do these data indicate that the population mean heart rate 6 minutes after the test is higher than before the test? Use a 5% level of significance.

12. *Medical: Blood Pressure* The patient described in Problem 11 also had his systolic blood pressure measured before and after the treadmill tests. The results for the tests are shown in the following table:

Test	1	2	3	4	5	6
Before	135	120	138	127	122	133
After	146	132	144	122	121	130

Does this indicate that the population mean systolic blood pressure is different (either higher or lower) before and 6 minutes after the treadmill test? Use a 5% level of significance.

13. *Golf: Tournaments* Do professional golfers play better in their first round? Let row *B* represent the score in the fourth (and final) round, and let row *A* represent the score in the first round of a professional golf tournament. A random sample of

finalists in the British Open gave the following data for their first and last rounds in the tournament (*Source: Golf Almanac*).

B: Last	73	68	73	71	71	72	68	68	74
A: First	66	70	64	71	65	71	71	71	71

Do the data indicate that the population mean score on the last round is higher than that on the first? Use a 5% level of significance.

14. *Academic: Professors* The following Excel output shows data based on information taken from *Academe: Bulletin of the American Association of University Professors* (Vol. 79, No. 2). The data are from a random sample of nine small- and medium-sized colleges and universities in the western United States and give the numbers of male and female assistant professors at the selected colleges. The Excel menu choices ➤ **Tools** ➤ **Data Analysis** ➤ *t*-Test: Paired Two-Sample for Means produced the following results.

Male	Female
17	16
59	55
7	8
41	36
22	30
25	34
20	15
23	22
14	11

B	C	D
t-Test: Paired Two Sample for Means		
	Male	*Female*
Mean	25.33333333	25.22222222
Variance	244.75	225.1944444
Observations	9	9
Pearson Correlation	0.941529561	
Hypothesized Mean Difference	0	
df	8	
t Stat	0.063150898	
P(T<=t) one-tail	0.475597855	
t Critical one-tail	1.85954832	
P(T<=t) two-tail	0.95119571	
t Critical two-tail	2.306005626	

In the dialogue box for t-Test, we used Male as array 1 and Female as array 2, so the results are based on differences $d = $ #Male $-$ #Female professors.

(a) Looking at the output, what is the null hypothesis H_0?

(b) The entry t Stat gives the t value of the sample test statistic $\bar{d}$. The sign of t Stat is positive. Does this indicate that the value of $\bar{d}$ is positive or negative?

(c) For a one-tailed test of $H_1: \mu_d > 0$, the P value is given by P(T<=t). Does this P value mean that we reject or fail to reject H_0 for a one-tailed test?

(d) In the dialogue box for the t-Test, we specified a level of significance $\alpha = 0.05$. Compare the t Stat value with the critical value for a one-tailed test. Based on critical values, do we reject H_0 or not? Is this result consistent with the result from part (c)?

(e) State your conclusion from part (d) in the context of the question: Do we have statistical evidence that in small and medium-sized colleges in the western United States the population mean number of male assistant professors is greater than that of female assistant professors?

(f) The output P(T<=t) two-tail gives the *P* value of the sample test statistic for a two-tailed test while t Critical two-tail gives the positive critical value for a 5% level of significance. State H_1 for a two-tailed test. Then complete parts (c) through (e) for a two-tailed test.

11.2   Inferences About the Difference of Two Means (Large, Independent Samples)

Independent Samples

Many practical applications of statistics involve a comparison of two population means or two population proportions. With dependent samples, we could pair the data and then consider the difference of data measurements *d*. In this section, we will turn our attention to tests of differences of means from *independent samples*. We will see new techniques for testing the difference of means from *independent samples*.

First, let's consider independent samples.

> **Definition** We say that two sampling distributions are *independent* if there is no relation whatsoever between specific values of the two distributions.

EXAMPLE 3

Independent sample

A teacher wishes to compare the effectiveness of two teaching methods. Students are randomly divided into two groups: The first group is taught by method 1; the second group, by method 2. At the end of the course, a comprehensive exam is given to all students, and the mean score $\overline{x}_1$ for group 1 is compared with the mean score $\overline{x}_2$ for group 2. Are the samples independent or dependent?

SOLUTION: Because the students were *randomly* divided into two groups, it is reasonable to say that the $\overline{x}_1$ distribution is independent of the $\overline{x}_2$ distribution.   ◇

EXAMPLE 4

Dependent sample

In Section 11.1, we considered a situation in which a shoe manufacturer claims that for the general population of adult U.S. citizens, the average length of the left foot is longer than the average length of the right foot. To study this claim, the manufacturer gathers data in this fashion: Sixty adult U.S. citizens are drawn at random, and for these 60 people, both their left and right feet are measured. Let $\overline{x}_1$ be the mean length of the left feet and $\overline{x}_2$ be the mean length of the right feet.

Are the $\overline{x}_1$ and $\overline{x}_2$ distributions independent for this method of collecting data?

SOLUTION: In this method, there is only *one* random sample of people drawn, and both the left and right feet are measured from this sample. The length of a person's left foot is usually related to the length of the right foot, so in this case the $\overline{x}_1$ and $\overline{x}_2$ distributions are *not* independent. In fact, we could pair the data and consider the distribution of the differences, left foot length minus right foot length. Then we would use the techniques of paired difference tests as found in Section 11.1.   ◇

GUIDED EXERCISE **3**

Independent sample

Suppose the shoe manufacturer of Example 4 gathers data in the following way: Sixty adult U.S. citizens are drawn at random and their left feet are measured; then another 60 adult U.S. citizens are drawn at random and their right feet are measured. Again, $\bar{x}_1$ is the mean of the left foot measurements and $\bar{x}_2$ is the mean of the right foot measurements.

Are the $\bar{x}_1$ and $\bar{x}_2$ distributions independent for this method of collecting data?	⟹	For this method of gathering data, two random samples are drawn: one for the left foot measurements and one for the right foot measurements. The first sample is not related to the second sample. The $\bar{x}_1$ and $\bar{x}_2$ distributions are independent.

Testing Difference of Means for Large, Independent Samples

Properties of $\bar{x}_1 - \bar{x}_2$ distribution, large sample size

In this portion, we will use distributions that arise from a difference of means from independent samples. How do we obtain such distributions? If we have two statistical variables x_1 and x_2, each with its own distribution, we take independent random samples of size n_1 from the x_1 distribution and size n_2 from the x_2 distribution. Then we can compute the respective means $\bar{x}_1$ and $\bar{x}_2$. Consider the difference $\bar{x}_1 - \bar{x}_2$. This represents a difference of means. If we repeat the sampling process over and over, we will create lots of $\bar{x}_1 - \bar{x}_2$ values. Figure 11-3 illustrates the sampling distribution of $\bar{x}_1 - \bar{x}_2$. These values can be arranged in a frequency table, and we can make a histogram for the distribution of $\bar{x}_1 - \bar{x}_2$ values. This would give us an experimental idea of the theoretical distribution of $\bar{x}_1 - \bar{x}_2$.

FIGURE 11-3

Sampling Distribution of $\bar{x}_1 - \bar{x}_2$

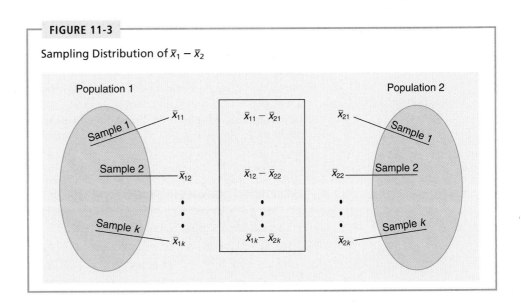

Fortunately, it is not necessary to carry out this lengthy process for each example. The results have already been worked out mathematically. The next theorem presents the main results for large, independent samples.

◇ **Theorem 11.2** Let x_1 have a normal distribution with mean μ_1 and standard deviation σ_1. Let x_2 have a normal distribution with mean μ_2 and standard deviation σ_2. If we take independent random samples of size n_1 from the x_1 distribution and of size n_2 from the x_2 distribution, then the variable $\bar{x}_1 - \bar{x}_2$ has

1. A normal distribution

2. Mean $\mu_1 - \mu_2$

3. Standard deviation

$$\sqrt{\frac{\sigma_1^2}{n_1} + \frac{\sigma_2^2}{n_2}} \quad ◇$$

◇ **COMMENT** The theorem requires that x_1 and x_2 have normal distributions. However, if both n_1 and n_2 are 30 or larger, then for most practical applications, the central limit theorem assures us that $\bar{x}_1$ and $\bar{x}_2$ are approximately normally distributed. In this case, the conclusions of the theorem are again valid even if the original x_1 and x_2 distributions were not normal.   ◇

Hypotheses for testing difference of means

When testing the difference of means, it is customary to use the null hypothesis

$$H_0: \mu_1 - \mu_2 = 0 \text{ or equivalently, } H_0: \mu_1 = \mu_2$$

As mentioned in Section 10.1, the null hypothesis is set up to see if it can be rejected. When testing the difference of means, we first set up the hypothesis H_0 that there is no difference. The alternate hypothesis could then be any of the ones listed in Table 11-5. The alternate hypothesis and consequent type of test used depend on the particular problem. Note that μ_1 is always listed first.

Critical values

Since the $\bar{x}_1 - \bar{x}_2$ distribution is *normal*, we use the usual critical values z_0 for normal distributions given in Figure 10-2 of Section 10.2. For convenience, we list the values again in Table 11-6 on the following page.

Sample test statistic

For tests of difference of means $\mu_1 - \mu_2$, independent samples, we make a decision regarding H_0 based on the sample evidence $\bar{x}_1 - \bar{x}_2$. For large samples, Theorem 11.2 tells us that the $\bar{x}_1 - \bar{x}_2$ distribution is normal with mean $\mu_1 - \mu_2$ and

TABLE 11-5   Alternate Hypotheses and Type of Test: Difference of Two Means

H_1			Type of Test
$H_1: \mu_1 - \mu_2 < 0$	or equivalently	$H_1: \mu_1 < \mu_2$	Left-tailed test
$H_1: \mu_1 - \mu_2 > 0$	or equivalently	$H_1: \mu_1 > \mu_2$	Right-tailed test
$H_1: \mu_1 - \mu_2 \neq 0$	or equivalently	$H_1: \mu_1 \neq \mu_2$	Two-tailed test

TABLE 11-7   Some Levels of Confidence and Their
Corresponding Critical Values

Level of Confidence c	Critical Value z_c
0.70, or 70%	1.04
0.75, or 75%	1.15
0.80, or 80%	1.28
0.85, or 85%	1.44
0.90, or 90%	1.645
0.95, or 95%	1.96
0.98, or 98%	2.33
0.99, or 99%	2.58

EXAMPLE 6

Confidence interval for
$\mu_1 - \mu_2$, *large samples*

Yellowstone National Park

In the summer of 1988, Yellowstone National Park had some major fires that destroyed large tracts of old timber near many famous trout streams. Fishermen were concerned about the long-term effect of the fire on these streams. However, biologists claimed that the new meadows that would spring up under dead trees would produce a lot more insects, which would in turn mean better fishing in the years ahead. Guide services registered with the park provided data about the daily catch for fishermen over many years. Ranger checks on the streams also provided data about the daily number of fish caught by fishermen. *Yellowstone Today* (a national park publication) indicated that the biologists' claim is basically correct and that Yellowstone anglers are delighted by their average increased catch.

Suppose that you are a biologist studying data from Yellowstone streams. A random sample of $n_1 = 167$ fishing reports in the years 1983 to 1988 (before the fire) showed that the average catch per day was $\bar{x}_1 = 5.2$ trout with sample standard deviation $s_1 = 1.9$. Then another random sample of $n_2 = 125$ fishing reports in the years 1990 to 1993 (after the fire) showed that the average catch per day was $\bar{x}_2 = 6.8$ trout with sample standard deviation $s_2 = 2.3$.

(a) For each sample, what is the population? Are the samples dependent or independent? Explain.

SOLUTION: The population for the first sample is the number of trout caught per day by fishermen before the fire. The population of the second sample is the number of trout caught per day after the fire. Both samples were random samples taken in their respective time periods. There was no effort to pair individual data values. Therefore, the samples can be thought of as independent samples.

(b) Compute a 95% confidence interval for $\mu_1 - \mu_2$, the difference of population means.

SOLUTION: Since $n_1 = 167$, $\bar{x}_1 = 5.2$, and $s_1 = 1.9$ and $n_2 = 125$, $\bar{x}_2 = 6.8$, $s_2 = 2.3$, and $z_{0.95} = 1.96$ (see Table 11-7), then

$$E \approx z_c \sqrt{\frac{s_1^2}{n_1} + \frac{s_2^2}{n_2}}$$

$$= 1.96\sqrt{\frac{(1.9)^2}{167} + \frac{(2.3)^2}{125}} \approx 1.96\sqrt{0.0639} = 0.4955 \approx 0.50$$

The 95% confidence interval is

$$(\bar{x}_1 - \bar{x}_2) - E < \mu_1 - \mu_2 < (\bar{x}_1 - \bar{x}_2) + E$$
$$(5.2 - 6.8) - 0.50 < \mu_1 - \mu_2 < (5.2 - 6.8) + 0.50$$
$$-2.10 < \mu_1 - \mu_2 < -1.10$$

(c) Explain the meaning of the confidence interval you computed in part (b).

SOLUTION: We are 95% confident that the interval -2.10 to -1.10 fish per day is one of the intervals containing the population difference $\mu_1 - \mu_2$, where μ_1 represents the population average daily catch before the fire and μ_2 represents the population average daily catch after the fire. Put another way, since the confidence interval contains only *negative values*, we can be 95% sure that $\mu_1 - \mu_2 < 0$. This means that we are 95% sure that $\mu_1 < \mu_2$. In words, we are 95% sure that the average catch before the fire is less than the average catch after the fire. ◇

Meaning of confidence intervals for $\mu_1 - \mu_2$

In Example 6, we saw that all the numbers contained in the confidence interval for $\mu_1 - \mu_2$ were negative. Two other cases can occur. The next display summarizes the interpretations for all three cases.

Interpretation of confidence intervals for $\mu_1 - \mu_2$

Suppose that we construct a $c\%$ confidence interval for $\mu_1 - \mu_2$. Then three cases arise:

1. The $c\%$ confidence interval contains only *negative values*. In this case, we conclude that $\mu_1 - \mu_2 < 0$, and we are therefore $c\%$ confident that $\mu_1 < \mu_2$.

2. The $c\%$ confidence interval contains only *positive values*. In this case, we conclude that $\mu_1 - \mu_2 > 0$, and we can be $c\%$ confident that $\mu_1 > \mu_2$.

3. The $c\%$ confidence interval contains *both positive and negative values*. In this case, we cannot at the $c\%$ confidence level conclude that either μ_1 or μ_2 is larger. However, if we *reduce* the confidence level c to a *smaller value*, then the confidence interval will, in general, be shorter (explain why). A shorter confidence interval *might* put us back into case 1 or case 2 above (again, explain why).

GUIDED EXERCISE 5

Interpret a confidence interval

(a) A study reported a 90% confidence interval for the difference of means to be

$$10 < \mu_1 - \mu_2 < 20$$

⟹ At a 90% level of confidence, we can say that the difference $\mu_1 - \mu_2$ is positive, so $\mu_1 - \mu_2 > 0$ and $\mu_1 > \mu_2$.

Continued

GUIDED EXERCISE 5 continued

For this interval, what can you conclude about the respective values of μ_1 and μ_2?

(b) A study reported a 95% confidence interval for the difference of means to be

$$-5.32 < \mu_1 - \mu_2 < 3.16$$

From this interval, what can you conclude about the respective values of μ_1 and μ_2?

⇒ At the 95% confidence level, we see that the difference of means ranges from negative to positive values. We cannot tell from this interval whether μ_1 is greater than μ_2 or μ_1 is less than μ_2.

TECH NOTE *TI-83Plus* Use the **STAT** key and highlight **TESTS**. The choice **3:2-SampZTest** performs a test of hypothesis for the difference of means using the normal distribution. The choice **9:2-SampZInt** generates confidence intervals for the difference of means. Both hypothesis tests and confidence intervals for $\mu_1 - \mu_2$ require values for σ_1 and σ_2 or estimates for the values.

Excel Enter the data in two columns. The menu selection **Tools ➤ Data Analysis ➤ z-Test Two Sample for Means** conducts a test of difference of means for large samples. Again, values for σ_1 and σ_2 or estimates for these values are required. *See* Problem 7 for an Excel display of results.

Minitab Enter the data in two columns. Use menu choices **Stat ➤ Basic Statistics.** Then the choice **2 sample t** can be used for difference of means test and confidence intervals. In the dialogue box, do not select equal variances. Minitab always uses the Student's t distribution and sample standard deviations for hypothesis tests and confidence intervals of $\mu_1 - \mu_2$ regardless of sample size. Consequently, the final results for hypothesis tests and confidence intervals will differ slightly from those obtained by using the normal distribution. Confidence intervals generated using the Student's t distribution are slightly wider than those from the normal distribution, and P values are somewhat larger. *See* Problem 8 for a Minitab display of results.

VIEWPOINT   *What's the Difference?*

Will two 15-minute piano lessons a week significantly improve a child's analytical reasoning skills? Why piano? Why not computer keyboard instruction or maybe voice lessons? Professor Frances Rauscher, University of Wisconsin, and Professor Gordon Shaw, University of California at Irvine, claim that there is a difference! How could this be measured? A large number of piano students were given complicated tests of mental ability. Independent control groups of other students were given the same tests. Techniques involving the study of differences of means were used to draw the conclusion that students taking piano lessons did better on tests measuring analytical reasoning skills. (Reported in *The Denver Post*.)

SECTION 11.2   PROBLEMS

For Problems 1–6, please provide the requested information.
(a) What is the null hypothesis? What is the alternate hypothesis? Will we use a left-tailed, right-tailed, or two-tailed test? What is the level of significance?
(b) What sampling distribution will we use? Find the appropriate critical value z_0 (or critical values $\pm z_0$).
(c) Sketch the critical region and show the critical value (or critical values).
(d) Calculate the z value as appropriate that corresponds to the sample statistic. Show its location on the sketch in part (c).
(e) Find the P value for your test and explain what the P value means in the context of the problem.
(f) Based on your answers for parts (a) to (e), shall we reject or fail to reject (i.e., "accept") the null hypothesis? Explain your conclusion in simple, nontechnical terms.

1. *Medical: Sleep* This problem is based on information from the book *Secrets of Sleep,* by Alexander Borbely. Professor Borbely is director of the Sleep Laboratory at the University of Zurich Medical School. REM sleep is rapid eye movement sleep, during which most dreams occur. Each night a person has both REM and non-REM sleep. However, it is thought that children have more REM sleep than older adults. A sample of $n_1 = 33$ children 10 years old showed they had an average REM sleep of $\bar{x}_1 = 2.6$ hours per night with sample standard deviation $s_1 = 0.5$ hour. Another random sample of $n_2 = 32$ people 35 years old showed they had an average of $\bar{x}_2 = 1.9$ hours of REM sleep per night with sample standard deviation $s_2 = 0.8$ hour. Does this indicate that, on average, 10-year-old children tend to have more REM sleep than 35-year-old adults? Use a 1% level of significance.

2. *Franchise: Startup Cost* *Franchise and Business Opportunities Annual Report* lists over 1500 business startup opportunities. Based on information from this source, a random sample of $n_1 = 62$ small clothing stores had a sample mean startup cost $\bar{x}_1 = \$83,000$ with sample standard deviation $s_1 = \$17,000$. Another random sample of $n_2 = 51$ small bakeries had a sample mean startup cost of $\bar{x}_2 = \$91,000$ with sample standard deviation $s_2 = \$22,000$. Does this indicate that there is a difference (either higher or lower) in the mean startup costs of small clothing stores compared with small bakeries? Use a 1% level of significance.

3. *Environment: Air Pollution* Based on information from the *Rocky Mountain News*, a random sample of $n_1 = 45$ winter days in Denver gave a sample mean pollution index $\bar{x}_1 = 43$ with sample standard deviation $s_1 = 22$. For Englewood (a suburb of Denver), a random sample of $n_2 = 47$ winter days gave a sample mean pollution index $\bar{x}_2 = 30$ with sample standard deviation $s_2 = 12$. Does this indicate that the mean pollution index for Englewood is less than that for Denver in the winter? Use a 1% level of significance.

4. *Leisure: Outdoor Activities* A Michigan study concerning preferences for outdoor activities used a questionnaire with a 6-point Likert-type response in which 1 designated "not important" and 6 designated "extremely important." A random sample of $n_1 = 201$ adults were asked about fishing as an outdoor activity. The mean response was $\bar{x}_1 = 4.7$ with sample standard deviation $s_1 = 1.1$. Another random sample of $n_2 = 135$ adults were asked about camping as an outdoor activity. For this group, the mean response was $\bar{x}_2 = 4.2$ with $s_2 = 1.4$. (Based on information taken from *National Symposium on Catch and Release Fishing*, Humboldt State University.) Does this indicate a difference (either way) regarding preference for camping or preference for fishing as an outdoor activity? Use a 1% level of significance.

5. *Psychology: Peer Tutoring* In the journal *Mental Retardation* (April 1985), an article reported the results of a peer tutoring program to help mildly mentally retarded children learn to read. In the experiment to assess the value of peer tutoring, the mildly retarded children were divided into two groups: One, the experimental group, received peer tutoring along with regular instruction, and the other, the control group, received regular instruction with no peer tutoring. There were 30 children in each group. The Gates-MacGintie Reading Test was given to both groups before instruction began. For the experimental group, the mean score on the vocabulary portion of the test was 344.5 with standard deviation 49.1. For the control group, the mean score on the same test was 345.9 with standard deviation 50.9. Use a 5% level of significance to test the hypothesis that there was no difference in the vocabulary scores of the two groups before the instruction began.

6. *Psychology: Peer Tutoring* In the same article cited in Problem 5, the results of the following experiment were reported. Form 2 of the Gates-MacGintie Reading Test was administered to both a control group and an experimental group after 6 weeks of instruction during which the experimental group received peer tutoring and the control group did not. The average score for the 30 subjects in the control group on the vocabulary portion of the test was 349.2 with standard deviation 56.6. For the experimental group of 30 children, the mean score on the same portion of the test was 368.4 with standard deviation 39.5. Use a 1% level of significance to test the claim that the experimental group performed better than the control group.

7. *Pro Athletes: Heights* Is there a difference in average heights of professional football players and professional basketball players? The following Excel output utilizes a random sample of 45 pro football players and a random sample of 40 pro basketball players (*Sources: Sports Encyclopedia Pro Football* and *The Official NBA Basketball Encyclopedia*, respectively). Excel requires the user to input estimates for the variance of each population. We used the menu choice ➤ **Paste Function** $\boxed{f_x}$ ➤ **VAR** to compute the sample variance for each data set. Then we used the menu choices ➤ **Tools** ➤ **Data Analysis** ➤ *z*-**Test Two Sample for Means** to generate the Excel output. A 5% level of significance was chosen.

B	C	D
Variance for Football Player Heights	0.134	
Variance for Basketball Player Heights	0.0987	
z-Test: Two Sample for Means		
	HtFt	*HtBk*
Mean	6.178888872	6.453249979
Known Variance	0.134	0.0987
Observations	45	40
Hypothesized Mean Difference	0	
z	−3.718026896	
P(Z<=z) one-tail	0.000100422	
z Critical one-tail	1.644853476	
P(Z<=z) two-tail	0.000200843	
z Critical two-tail	1.959962787	

(a) The hypothesized Mean Difference 0 implies that the null hypothesis is H_0: $\mu_1 - \mu_2 = 0$. Look at the printout. What is the z value of the sample test statistic?

(b) What is the alternate hypothesis for a two-tailed test? Look at the printout. What is the P value of the sample test statistic for a two-tailed test [see P(Z< = z) two-tail]? For a 5% level of significance, what is the test conclusion regarding football player heights (HtFt) versus basketball player heights (HtBk)?

8. *Botany: Iris* Species of iris differ in many ways. A study by E. Anderson in the *Bulletin of the American Iris Society* shows a random sample of petal lengths for *Iris virginica* and a random sample of petal lengths for *Iris setosa*. Minitab generates confidence intervals for the difference of means $\mu_1 - \mu_2$. The Student's t distribution is used. The following printout shows the results.

```
Minitab Display: 2 sample t
          N     Mean    StDev    SE Mean
VIRGI    35    5.477    0.553     0.093
SETOSA   38    1.489    0.209     0.034

95% CI for mu VIRGI - mu SETOSA: (3.787, 4.188)  T-Test mu
VIRGI = mu SETOSA (vs not =): T = 40.13  P = 0.0000  DF = 42
```

(a) Based on the printout, what level of confidence is used for the confidence interval?
(b) Examine the confidence interval and explain what it means in this context. The confidence interval contains all positive numbers. What does this tell you about the relationship between the average petal length μ_1 for *Iris virginica* and the average petal length μ_2 for *Iris setosa*?

9. *Nursing: Parent Sensitivity* "Parental Sensitivity to Infant Cues: Similarities and Differences Between Mothers and Fathers," by M. V. Graham (*Journal of Pediatric Nursing*, Vol. 8, No. 6), reports a study of parental empathy for sensitivity cues and baby temperament (higher scores mean more empathy). Let x_1 be a random variable that represents the score of a mother on an empathy test (as regards her baby). Let x_2 be the empathy score of a father. A random sample of 32 mothers gave a sample mean $\bar{x}_1 = 69.44$ with sample standard deviation $s_1 = 11.69$. A random sample of 32 fathers gave $\bar{x}_2 = 59$ with $s_2 = 11.60$.
(a) Let μ_1 be the population mean of x_1 and μ_2 be the population mean of x_2. Find a 99% confidence interval for $\mu_1 - \mu_2$.
(b) Examine the confidence interval and explain what it means in this context. Does the confidence interval contain all positive, all negative, or both positive and negative numbers? What does this tell you about the relationship between average empathy scores for mothers and those for fathers at the 99% confidence level?

10. *U.S. Geological Survey: Old Faithful Geyser* The U.S. Geological Survey compiled historical data about Old Faithful Geyser (Yellowstone National Park) from 1870 to 1987. Some of these data are published in the book *The Story of Old Faithful*, by G. D. Marler (Yellowstone Association Press). Let x_1 be a random variable that represents the time interval (minutes) between Old Faithful eruptions in the years 1948 to 1952. Based on 9340 observations, the sample mean interval was $\bar{x}_1 = 63.3$ minutes with sample standard deviation $s_1 = 9.17$ minutes. Let x_2 be a random variable that represents the time interval in minutes between Old Faithful eruptions in the years 1983 to 1987. Based on 25,111 observations, the sample mean time interval was $\bar{x}_2 = 72.1$ minutes with sample standard deviation $s_2 = 12.67$ minutes. Let μ_1 be the population mean of x_1 and μ_2 be the population mean of x_2.

(a) Compute a 99% confidence interval for $\mu_1 - \mu_2$.
(b) Comment on the meaning of the confidence interval in the context of this problem. Does the interval consist of positive numbers only, negative numbers only, or a mix of positive and negative numbers? Does it appear (at the 99% confidence level) that a change in the interval length between eruptions has occurred? Many geologic experts believe that the distribution of eruption times of Old Faithful changed after the major earthquake that occurred in 1959.

11. *Wildlife: Deer* Mule deer in Colorado have been studied extensively by G. W. Mierau and J. L. Schmidt. In their book, *The Mule Deer of Mesa Verde National Park* (Mesa Verde Museum Association), they give the following information about weights of adult male mule deer in two Colorado regions. In the Cache la Poudre Region, a random sample of $n_1 = 51$ deer weighed an average of $\bar{x}_1 = 74.04$ kg with sample standard deviation $s_1 = 17.19$ kg. In the Mesa Verde Region, a random sample of $n_2 = 36$ deer gave a mean weight of $\bar{x}_2 = 94.53$ kg with sample standard deviation $s_2 = 19.66$ kg.
(a) Let μ_1 be the population mean weight of all bucks in the Cache la Poudre Region. Let μ_2 be the population mean weight of all bucks in the Mesa Verde Region. Find a 95% confidence interval for $\mu_1 - \mu_2$.
(b) Examine the confidence interval and comment on its meaning. Does it include numbers that are all positive, all negative, or mixed? What conclusion can you draw (at the 95% level) about the average weight of bucks in the Cache la Poudre Region compared with those in the Mesa Verde Region? The bucks in Mesa Verde National Park are not hunted (and thus tend to be older), and the browse is more abundant. How might these conditions account for the results shown by the confidence interval?

12. *Life Insurance: Male vs. Female* What does life insurance cost? Does it matter if you are male or female? This problem is based on information from *Consumer Reports*. For similar benefits (male and female), the annual premiums paid by a person 45 years old for a $250,000 annual renewable term life insurance policy were as follows:

> *Males:* $x_1 =$ annual premium for described life insurance; sample size
> $n_1 = 47$; sample mean $\bar{x}_1 = \$483.43$; sample standard deviation
> $s_1 = \$126.62$

> *Females:* $x_2 =$ annual premium for described life insurance; sample size
> $n_2 = 51$; sample mean $\bar{x}_2 = \$414.43$; sample standard deviation
> $s_2 = \$105.99$

(a) Assume that the preceding values are representative of the premiums paid for the described life insurance by all males and females (45 years old). Let μ_1 be the population mean annual premium for the described insurance for a male who is 45 years old. Let μ_2 be the population mean annual premium for the described insurance for a female who is 45 years old. Find a 90% confidence interval for $\mu_1 - \mu_2$.
(b) Explain the meaning of the confidence interval found in part (a) in the context of the problem. Does the interval contain numbers that are all positive, all negative, or both positive and negative? Can you conclude that the population mean premium for males is greater than the population mean premium for females at the 90% confidence level?

13. *General: Interpretation* Suppose that a 95% confidence interval for the difference of means (large sample) contains both positive and negative numbers. Will a 99% confidence interval based on the same data necessarily contain both positive and negative numbers? Explain. What about a 90% confidence interval? Explain.

11.3   Inferences About the Difference of Two Means (Small, Independent Samples)

FOCUS POINTS

✓ Compute a pooled variance for two independent samples.
✓ Compute the sample test statistic, critical value, and P value for $\mu_1 - \mu_2$ (small samples).
✓ Use the level of significance α and the P value to conclude the test.
✓ Use the Student's t distribution to find confidence intervals for $\mu_1 - \mu_2$ (small samples).
✓ Interpret the meaning of a confidence interval for $\mu_1 - \mu_2$.

Testing Difference of Means for Small Samples

Statistical methods involving small samples and the difference between two means are much like the methods for large samples. However, for small samples, we will use the Student's t distribution for critical values instead of the normal distribution.

Independent random samples of sizes n_1 and n_2, respectively, are drawn from two populations that have means μ_1 and μ_2. *We assume that the parent populations have normal distributions, and we also assume that the standard deviations σ_1 and σ_2 for the two populations are equal.* The condition $\sigma_1 = \sigma_2$ may seem quite restrictive. However, in a great many practical applications this condition is satisfied. Furthermore, our methods still apply even if the standard deviations are known to be only approximately equal.

Suppose we draw two independent random samples, one from the x_1 population and one from the x_2 population. Say the sample from the x_1 population is of size n_1 and has sample standard deviation s_1. Likewise for the x_2 population, the sample size is n_2 and the sample standard deviation is s_2. We estimate the common standard deviation for the two populations by using a *pooled variance* of the s_1^2 and s_2^2 values. Research shows that the best estimate of the common variance of the x_1 and x_2 populations is given by the formula

$$s^2 = \frac{(n_1 - 1)s_1^2 + (n_2 - 1)s_2^2}{n_1 + n_2 - 2}$$

Pooled standard deviation

The best estimate of the common or pooled standard deviation is then

$$s = \sqrt{\frac{(n_1 - 1)s_1^2 + (n_2 - 1)s_2^2}{n_1 + n_2 - 2}}$$

Null hypothesis

Just as with tests of difference of means $\mu_1 - \mu_2$ for large independent samples, the null hypothesis we use for testing the difference of means for small samples is that the population means do not differ. In symbols, we have

$$H_0\!: \mu_1 - \mu_2 = 0 \quad \text{or, equivalently,} \quad H_0\!: \mu_1 = \mu_2$$

Alternate hypotheses

For tests of difference of means $\mu_1 - \mu_2$ for small independent samples, we use the same alternate hypotheses as we did for tests of difference of means for large samples. Table 11-8 lists the possibilities for the alternate hypothesis. This table is simply a repetition of Table 11-5 in Section 11.2.

TABLE 11-8   Alternate Hypotheses and Type of Test: Difference of Two Means

H_1			Type of Test
$H_1\!: \mu_1 - \mu_2 < 0$	or equivalently	$H_1\!: \mu_1 < \mu_2$	Left-tailed test
$H_1\!: \mu_1 - \mu_2 > 0$	or equivalently	$H_1\!: \mu_1 > \mu_2$	Right-tailed test
$H_1\!: \mu_1 - \mu_2 \neq 0$	or equivalently	$H_1\!: \mu_1 \neq \mu_2$	Two-tailed test

The next theorem presents the main results necessary to conduct tests of hypotheses regarding $\mu_1 - \mu_2$ for small samples.

◇ **THEOREM 11.3** Let x_1 and x_2 have normal (or approximately normal) distributions with means μ_1 and μ_2 and standard deviations σ_1 and σ_2, respectively. We assume that σ_1 and σ_2 are (approximately) equal. Suppose that we take independent random samples of sizes n_1 and n_2 from the x_1 and x_2 populations. Let $\bar{x}_1$ and $\bar{x}_2$ and s_1 and s_2 be the sample means and standard deviations. Let s be the sample pooled standard deviation. If $n_1 < 30$ and/or $n_2 < 30$, then

$$t = \frac{(\bar{x}_1 - \bar{x}_2) - (\mu_1 - \mu_2)}{s\sqrt{\dfrac{1}{n_1} + \dfrac{1}{n_2}}}$$

has a Student's t distribution with degrees of freedom

$$d.f. = n_1 + n_2 - 2$$
◇

Critical values

Since the sample statistic $\bar{x}_1 - \bar{x}_2$ follows a Student's t distribution, we find the critical value(s) t_0 and critical regions by using Table 5 in the Appendix. The critical value(s) t_0 are found in the row headed by $d.f. = n_1 + n_2 - 2$. For *one-tailed tests*, we use the column headed by $\alpha' = $ level of significance. For *two-tailed tests*, we use the column headed by $\alpha'' = $ level of significance.

Sample test statistic

The null hypothesis states that $\mu_1 - \mu_2 = 0$. Consequently, we use the following two formulas to convert the sample test statistic $\bar{x}_1 - \bar{x}_2$ to a t value:

$$t = \frac{\bar{x}_1 - \bar{x}_2}{s\sqrt{\dfrac{1}{n_1} + \dfrac{1}{n_2}}} \text{ with } d.f. = n_1 + n_2 - 2$$

where the pooled standard deviation s is

$$s = \sqrt{\frac{(n_1 - 1)s_1^2 + (n_2 - 1)s_2^2}{n_1 + n_2 - 2}}$$

Then we compare the sample t statistic with the critical value(s) t_0 to conclude the test.

EXAMPLE 7
Testing the difference of means (small samples)

Two competing headache remedies claim to give fast-acting relief. An experiment was performed to compare the mean lengths of time required for bodily absorption of brand A and brand B headache remedies.

Twelve people were randomly selected and given an oral dosage of brand A. Another 12 were randomly selected and given an equal dosage of brand B. The length of time in minutes for the drugs to reach a specified level in the blood was recorded. The means, standard deviations, and sizes of the two samples follow:

Brand A: $\bar{x}_1 = 20.1$; $s_1 = 8.7$; $n_1 = 12$
Brand B: $\bar{x}_2 = 18.9$; $s_2 = 7.5$; $n_2 = 12$

Past experience with the drug composition of the two remedies permits researchers to assume that the standard deviations of the two time distributions are approximately equal. Let us use a 5% level of significance to test the claim that there is no difference in the mean time required for bodily absorption. Also, find the P value of the sample test statistic.

SOLUTION:

(a) The null hypothesis is

$$H_0: \mu_1 = \mu_2 \qquad \text{or} \qquad H_0: \mu_1 - \mu_2 = 0$$

Since we have no prior knowledge about which brand is faster, the alternate hypothesis will be simply

$$H_1: \mu_1 \neq \mu_2 \qquad \text{or} \qquad H_0: \mu_1 - \mu_2 \neq 0$$

(b) Since the alternate hypothesis is $H_1: \mu_1 \neq \mu_2$, we use a two-tailed test. The critical values are found in Table 5 in the Appendix using the row headed by $d.f. = n_1 + n_2 - 2 = 12 + 12 - 2 = 22$ and the column headed by $\alpha'' = 0.05$. The critical values are $\pm t_0 = \pm 2.074$.

(c) Next, we compute the sample test statistic $\bar{x}_1 - \bar{x}_2$ and convert it to a t value. Since $\bar{x}_1 = 20.1$ and $\bar{x}_2 = 18.9$, their difference is

$$\bar{x}_1 - \bar{x}_2 = 20.1 - 18.9 = 1.2$$

To find the t value corresponding to 1.2, we first find the *pooled standard deviation s*.

$$s = \sqrt{\frac{(n_1 - 1)s_1^2 + (n_2 - 1)s_2^2}{n_1 + n_2 - 2}}$$

$$= \sqrt{\frac{(12 - 1)(8.7)^2 + (12 - 1)(7.5)^2}{12 + 12 - 2}} = \sqrt{65.97} \approx 8.12$$

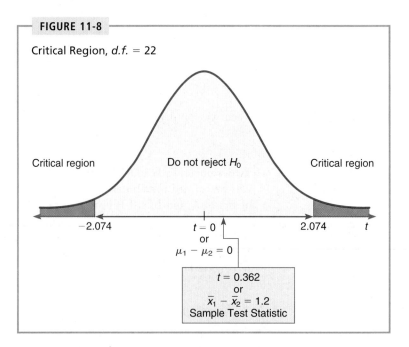

FIGURE 11-8

Critical Region, *d.f.* = 22

The c confidence interval is

$$(\bar{x}_1 - \bar{x}_2) - 1.19 < \mu_1 - \mu_2 < (\bar{x}_1 - \bar{x}_2) + 1.19$$
$$(19.65 - 6.59) - 1.19 < \mu_1 - \mu_2 < (19.65 - 6.59) + 1.19$$
$$11.87 < \mu_1 - \mu_2 < 14.25$$

Therefore, after further rounding, we have

$$11.9 \text{ hertz} < \mu_1 - \mu_2 < 14.3 \text{ hertz}$$

(d) Explain the meaning of the confidence interval you computed in part (c).

SOLUTION: μ_1 represents the population average brain wave activity for people who drink 1/2 liter of wine before sleeping. μ_2 represents the population average brain wave activity for people who take no alcohol before sleeping. Both periods of measurement are from 4 to 6 A.M. We are 90% confident the difference $\mu_1 - \mu_2$ is between 11.9 and 14.3 hertz (rounded values). It would seem reasonable to conclude that people who drink before sleeping might wake up in the early morning and have trouble going back to sleep. Since the confidence interval from 11.9 to 14.3 contains only *positive values*, we could express this by saying that we are 90% confident that $\mu_1 - \mu_2$ is *positive*. This means that $\mu_1 - \mu_2 > 0$. Thus we are 90% confident that $\mu_1 > \mu_2$ (that is, average brain wave activity from 4 to 6 A.M. for the group drinking wine is more than average brain wave activity for the group not drinking). ◊

TECH NOTE The TI-83Plus, Excel, and Minitab all support testing the difference of means for independent small samples. When testing the difference of means for small independent samples, the technologies give the option of using the pooled standard deviation or not. When the standard deviations of the two populations are approximately equal, it is appropriate to use the pooled standard deviation. The technologies take care of the more complicated calculations required when the two populations do not have approximately equal standard deviations. In such case, do not use the pooled standard deviation.

TI-83Plus Use the **STAT** key and highlight **TESTS**. The choice **4:2-SampTTest** performs a test of hypothesis for the difference of means using the Student's t distribution. The choice **0:2-SampTInt** generates confidence intervals for the difference of means. Use **Pooled: Yes** when the assumption of equal variances or standard deviations is valid. Notice that the pooled variance is given in the output as S_{XP}.

Excel Enter the data in columns. The menu selection **Tools ➤ Data Analysis ➤ t-Test: Two-Sample Assuming Equal Variances** conducts a test of difference of means for small samples when the variances are equal. If the variances are not equal, then use the option ***t*-Test: Two-Sample Assuming Unequal Variances.** The output gives both the critical value and P value for a one-tailed test and for a two-tailed test.

Minitab Enter the data in two columns. Use menu choices **Stat ➤ Basic Statistics.** Then the choice **2 sample *t*** can be used for difference of means test and confidence intervals. In the dialogue box, select equal variances. *See* Problem 11 for a Minitab display of results.

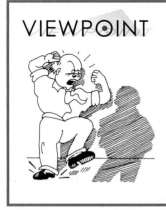

VIEWPOINT   *Temper! Temper!*

In her book *Red Ink Behaviors*, Jean Hollands discusses inappropriate, problem behaviors of professional employees in the corporate business world. Temper tantrums, flaming e-mails, omitting essential information, sabotaging fellow workers, and the arrogant opinion that others are "dumb and dispensable" create personnel problems that cost companies a lot in the form of wasted time, reduced productivity, and lost revenues. A study of major industries in the Silicon Valley area gave Hollands data for estimating just how much time and money are wasted by such "red ink behaviors." For more information, see Problems 3 and 4 in this section.

SECTION 11.3   PROBLEMS

For Problems 1–6, please provide the requested information.

(a) What is the null hypothesis? What is the alternate hypothesis? Will we use a left-tailed, right-tailed, or two-tailed test? What is the level of significance?

(b) What sampling distribution will we use? Find the appropriate critical value t_0 (or critical values $\pm t_0$).

(c) Sketch the critical region and show the critical value (or critical values).

(d) Calculate the t value that corresponds to the sample statistic. Show its location on the sketch in part (c).

(e) Find the P value for your test and explain what the P value means in the context of the problem.

(f) Based on your answers for parts (a) to (e), shall we reject or fail to reject (i.e., "accept") the null hypothesis for the given level of significance α? Explain your conclusion in the context of the problem.

1. *Wildlife: Fox Rabies* A study of fox rabies in southern Germany gave the following information about different regions and the occurrence of rabies in each region (B. Sayers et al., "A Pattern Analysis Study of a Wild Life Rabies Epizootic," *Medical Informatics* 2:11–34). Based on information from this article, a random sample of $n_1 = 16$ locations in region I gave the following information about the number of cases of fox rabies near that location.

    x_1 data: 1   8   8   8   7   8   8   1
              3   3   3   2   5   1   4   6

    A second random sample of $n_2 = 15$ locations in region II gave the following information about the number of cases of fox rabies near that location.

    x_2 data: 1   1   3   1   4   8   5   4
              4   4   2   2   5   6   9

    (a) Use a calculator with sample mean and sample standard deviation keys to verify that $\bar{x}_1 = 4.75$ and $s_1 = 2.82$ in region I and that $\bar{x}_2 = 3.93$ and $s_2 = 2.43$ in region II.

    (b) Does this information indicate that there is a difference in the mean number of cases of fox rabies between the two regions? Use a 5% level of significance.

2. *Agriculture: Bell Peppers* The pathogen *Phytophthora capsici* causes bell peppers to wilt and die. Because bell peppers are an important commercial crop, this disease has undergone a great deal of agricultural research. It is thought that too much water aids the spread of the pathogen. Two fields are under study. The first step in the research project is to compare the mean soil water content for the two fields (*Source: Journal of Agricultural, Biological, and Environmental Statistics*, Vol. 2, Number 2).

Soil Water Content (% by volume)

Field A samples x_1:	10.2	10.7	15.5	10.4	9.9	10.0	16.6	15.1
	15.2	13.8	14.1	11.4	11.5	11.0		
Field B samples x_2:	8.1	8.5	8.4	7.3	8.0	7.1	13.9	12.2
	13.4	11.3	12.6	12.6	12.7	12.4	11.3	

(a) Use a calculator with mean and standard deviation keys to verify that $\bar{x}_1 = 12.53$, $s_1 = 2.39$, $\bar{x}_2 = 10.65$, and $s_2 = 2.44$.

(b) Use a 5% level of significance to test the claim that field A has on average higher soil water content than field B.

3. *Management: Lost Time* In her book *Red Ink Behaviors,* Jean Hollands reports the assessment of leading Silicon Valley companies regarding a manager's lost time due to inappropriate behavior of employees. Consider the following independent random variables—manager's hours per week lost due to hot tempers, flaming e-mails, and general unproductive tensions, x_1:

1	5	8	4	2	4	10

Manager's hours per week lost due to disputes regarding technical worker's superior attitudes that their colleagues are "dumb and dispensable," x_2:

10	5	4	0	0	4	10

(a) Use a calculator with sample mean and standard deviation keys to verify that $\bar{x}_1 = 4.86$, $s_1 = 3.18$, $\bar{x}_2 = 4.71$, and $s_2 = 4.11$.

(b) Does this indicate that the population mean time lost for hot tempers is different from population mean time lost due to disputes arising from technical worker's superior attitudes? Use $\alpha = 0.05$.

4. *Management: Intimidators and Stressors* This problem is based on information regarding productivity in leading Silicon Valley companies. (*See* reference in Problem 3.) In large corporations, an "intimidator" is an employee who tries to stop communication and sometimes sabotages others but, above all, likes to listen to himself or herself talk. Let x_1 be a random variable representing productive hours per week lost by peer employees of an intimidator.

8	3	6	2	2	5	2

A "stressor" is an employee with a hot temper that leads to unproductive tantrums in corporate society. Let x_2 be a random variable representing productive hours per week lost by peer employees of a stressor.

3	3	10	3	4	2	5

(a) Use a calculator with mean and standard deviation keys to verify that $\bar{x}_1 = 4.00$, $s_1 = 2.38$, $\bar{x}_2 = 4.29$, and $s_2 = 2.69$.

(b) Assuming that the variables x_1 and x_2 are independent, do the data indicate that the population mean time lost due to stressors is greater than the population mean time lost due to intimidators? Use a 5% level of significance.

5. *Environment: Power Plants* A large electric power plant uses ocean water for its cooling system (and returns the water to the ocean). A random sample of 10 temperature readings showed the changes in water temperature to be (in °F):

   6   8   4   5   10   3   9   11   7   9

   (a) For these temperatures, verify that the mean is $\bar{x}_1 = 7.2$°F and the sample standard deviation is $s_1 = 2.7$°F.

   A new generator was added to the plant, and environmentalists fear that the average change in water temperature has increased. A random sample of 12 temperature readings showed the changes in water temperatures now to be (in °F):

9	11	15	12	7	12
10	13	8	11	14	8

   (b) For these temperatures, verify that the sample mean is $\bar{x}_2 = 10.8$°F with sample standard deviation $s_2 = 2.5$°F.
   (c) Use a 5% level of significance to test the claim that the mean change in water temperature has increased since the addition of the new generator.

6. *Psychology: Perceptual Illusions* Irv is doing a psychology study of susceptibility of people to perceptual illusions. He claims airplane pilots are less susceptible than the general population. To test the claim, a random sample of 11 airplane pilots judged the length of a line in a large illusionary figure. For each judgment, the magnitude of the deviation from the actual length was recorded. The mean of these magnitudes was $\bar{x}_1 = 76$ mm with sample standard deviation $s_1 = 9$ mm. A random sample of 15 nonpilots were then asked to judge the length, and the magnitudes of their deviations were recorded. For the nonpilots the mean of the magnitudes was $\bar{x}_2 = 82$ mm with standard deviation $s_2 = 7$ mm. Do the data support the claim at the 1% level of significance?

7. *Business: Profit Sectors* Many companies take in a lot of money. How much of that money can they keep as profit? Do some industry groups have a better performance than others? One way to answer such questions is to study profit as a percentage of revenue. Annual profit as a percentage of revenue is shown below for a group of companies that manufacture electronic equipment. Names such as General Electric, Motorola, Whirlpool, Maytag, and so on are included.

8.1	5.1	5.3	6.2	6.8	4.8	5.5	5.4
6.0	4.9	4.5	5.1	5.2	4.7	6.3	

   For a group of food and drug stores, annual profit as a percentage of revenue is shown next. Names such as Kroger, Safeway, Albertson's, Walgreen, and so on are included.

1.5	2.7	3.6	2.0	3.2	2.5
2.9	1.9	1.5	2.1	2.0	4.5

   (*Source: Fortune 500*, Vol. 135, No. 8.)

   (a) Assume that the preceding data are representative of each given industry group. Let x_1 represent annual profit as a percentage of revenue for the electronic companies, and let x_2 represent profit as a percentage of revenue for food and drug

(b) Because $H_1: p_1 < p_2$, we use a left-tailed test. For $\alpha = 0.05$, Table 11-11 gives the critical value $z_0 = -1.645$.

(c) Next, we compute the sample statistic $\hat{p}_1 - \hat{p}_2$ and convert it to a z value.

● **CALCULATOR NOTE**   Carry the values for $\hat{p}_1, \hat{p}_2, \hat{q}_1, \hat{q}_2$, and the pooled estimates $\hat{p}$ and $\hat{q}$ out to at least three places after the decimal. Then round the z value of the corresponding test statistic to two places after the decimal.

For the first group, the number of successes is $r_1 = 295$ out of $n_1 = 625$ trials. For the second group, there are $r_2 = 350$ successes out of $n_2 = 625$ trials. Since

$$\hat{p}_1 = \frac{r_1}{n_1} = \frac{295}{625} = 0.472 \quad \text{and} \quad \hat{p}_2 = \frac{r_2}{n_2} = \frac{350}{625} = 0.560$$

then

$$\hat{p}_1 - \hat{p}_2 = 0.472 - 0.560 = -0.088$$

To convert this $\hat{p}_1 - \hat{p}_2$ value to a z value, we need to find the *pooled estimate* $\hat{p}$ for the common values of p_1 and p_2 and the corresponding value for $\hat{q}$:

$$\hat{p} = \frac{r_1 + r_2}{n_1 + n_2} = \frac{295 + 350}{625 + 625} = 0.516 \quad \text{and} \quad \hat{q} = 1 - \hat{p} = 0.484$$

Using these values, we find

$$z = \frac{\hat{p}_1 - \hat{p}_2}{\sqrt{\dfrac{\hat{p}\hat{q}}{n_1} + \dfrac{\hat{p}\hat{q}}{n_2}}} = \frac{-0.088}{\sqrt{\dfrac{(0.516)(0.484)}{625} + \dfrac{(0.516)(0.484)}{625}}} \approx -3.11$$

(d) Next, we show the critical region and the location of the sample test statistic with respect to the critical region. From this information, we can conclude the test (see Figure 11-10). We see that the z value corresponding to the sample test statistic $\hat{p}_1 - \hat{p}_2 = -0.088$ falls in the critical region. Therefore, we reject H_0 and conclude that the county clerk's claim is valid at the 0.05 level of significance.

FIGURE 11-10

Critical Region

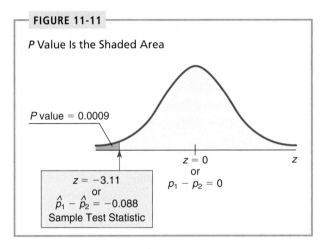

FIGURE 11-11

P Value Is the Shaded Area

P value = 0.0009

z = −3.11
or
$\hat{p}_1 - \hat{p}_2 = -0.088$
Sample Test Statistic

z = 0
or
$p_1 - p_2 = 0$

z

(e) Since this is a *left-tailed* test, the P value corresponding to the test statistic is the area to the *left* of $z = -3.11$ (see Figure 11-11). By Table 4 in the Appendix, we see that

$$P \text{ value} = 0.0009$$

This means that we reject H_0 for any $\alpha \geq 0.0009$. In particular, we reject H_0 for $\alpha = 0.05$ (and also for $\alpha = 0.01$). This is consistent with the conclusion reached in part (d).   ◇

GUIDED EXERCISE 7

Testing the difference of proportions

In Example 9 about voter registration, suppose that a random sample of 1100 potential voters was randomly divided into two groups.

    Group 1: 500 potential voters; no registration reminders sent; 248 registered to vote

    Group 2: 600 potential voters; registration reminders sent; 332 registered to vote

Do these data support the claim that the proportion of voters who register was greater in the group that received reminders than in the group that did not? Use a 1% level of significance.

(a) What should we use for H_0 and H_1? Should we use a left-, right-, or two-tailed test?

⟹ As before, H_0: $p_1 = p_2$ and H_1: $p_1 < p_2$. We use a left-tailed test.

(b) What is $\hat{p}$ for this problem? What is $\hat{q}$?

⟹ $n_1 = 500$ and $r_1 = 248$; $n_2 = 600$ and $r_2 = 332$. Thus,

$$\hat{p} = \frac{r_1 + r_2}{n_1 + n_1} = \frac{248 + 332}{500 + 600} \approx 0.527$$

$$\hat{q} = 1 - \hat{p} \approx 1 - 0.527 \approx 0.473$$

Continued

GUIDED EXERCISE 7 continued

(c) Find the critical value z_0.

⟹ Since this is a left-tailed test with $\alpha = 0.01$, the critical value is $z_0 = -2.33$.

(d) What is the value of the sample test statistic $\hat{p}_1 - \hat{p}_2$?

⟹ $\hat{p}_1 = \dfrac{r_1}{n_1} = \dfrac{248}{500} = 0.496$

$\hat{p}_2 = \dfrac{r_2}{n_2} = \dfrac{332}{600} \approx 0.553$

$\hat{p}_1 - \hat{p}_2 \approx -0.057$

(e) Convert the sample test statistic $\hat{p}_1 - \hat{p}_2 = -0.057$ to a z value.

⟹ $z = \dfrac{\hat{p}_1 - \hat{p}_2}{\sqrt{\dfrac{\hat{p}\hat{q}}{n_1} + \dfrac{\hat{p}\hat{q}}{n_2}}} \approx \dfrac{-0.057}{\sqrt{\dfrac{(0.527)(0.473)}{500} + \dfrac{(0.527)(0.473)}{600}}} \approx -1.89$

(f) Show the critical region and the sample test statistic on a diagram. Do we reject or fail to reject H_0?

⟹ Since the sample test statistic falls outside the critical region, we fail to reject H_0 (see Figure 11-12). At the 1% level of significance, the data do not support the claim that reminders increase the proportion of registered voters.

(g) Find the P value for the sample test statistic. What does this tell you?

⟹ The z value corresponding to the sample test statistic is -1.89. We are using a *left-tailed* test, so the P value is the area to the left of $z = -1.89$ (see Figure 11-13). By Table 4 of the Appendix, the P value = 0.0294. This tells us to reject H_0 for all $\alpha \geq 0.0294$. We cannot reject H_0 for $\alpha = 0.01$.

FIGURE 11-12 Critical Region

FIGURE 11-13 P Value Is the Shaded Area

Confidence Intervals for $p_1 - p_2$

Estimating the difference of proportions $p_1 - p_2$

We conclude this section with a discussion of confidence intervals for $p_1 - p_2$, the difference of two proportions from binomial probability distributions. Based on information similar to that used for hypothesis testing, we can use the following results.

c Confidence interval for $p_1 - p_2$ (large samples)

$$(\hat{p}_1 - \hat{p}_2) - E < p_1 - p_2 < (\hat{p}_1 - \hat{p}_2) + E$$

where

$$E \approx z_c \hat{\sigma} = z_c \sqrt{\frac{\hat{p}_1 \hat{q}_1}{n_1} + \frac{\hat{p}_2 \hat{q}_2}{n_2}}$$

c = confidence level, $0 < c < 1$

z_c = critical value for confidence level c (See Table 11-7 for frequently used values.)

n_1 = number of trials in binomial experiment 1

r_1 = number of successes in binomial experiment 1

n_2 = number of trials in binomial experiment 2

r_2 = number of successes in binomial experiment 2

$\hat{p}_1 = \dfrac{r_1}{n_1}$ and $\hat{q}_1 = 1 - \dfrac{r_1}{n_1}$

$\hat{p}_2 = \dfrac{r_2}{n_2}$ and $\hat{q}_2 = 1 - \dfrac{r_2}{n_2}$

$\hat{\sigma} \approx$ standard deviation of the $\hat{p}_1 - \hat{p}_2$ distribution

We assume that all four quantities

$$n_1 \hat{p}_1 \qquad n_1 \hat{q}_1 \qquad n_2 \hat{p}_2 \qquad n_2 \hat{q}_2$$

are greater than 5.

EXAMPLE 10

Confidence interval for $p_1 - p_2$

In his book, *Secrets of Sleep*, Professor Borbely describes research on dreams in the sleep laboratory at the University of Zurich Medical School. During normal sleep, there is a phase known as *REM* (rapid eye movement). For most people, REM sleep occurs about every 90 minutes or so, and it is thought that dreams occur just before or during the REM phase. Using electronic equipment in the sleep laboratory, it is possible to detect the REM phase in a sleeping person. If a person is wakened immediately after the REM phase, he or she usually can describe a dream that has just taken place. Based on a study of over 650 people in the Zurich sleep laboratory, it was found that about one-third of all dream reports contain feelings of fear, anxiety, or aggression. There is a conjecture that if a person is in a good mood when going to sleep, the proportion of "bad" dreams (fear, anxiety, aggression) might be reduced.

Suppose that two groups of subjects were randomly chosen for a sleep study. In group I, before going to sleep, the subjects spent 1 hour watching a comedy movie. In this group, there were a total of $n_1 = 175$ dreams recorded, of which $r_1 = 49$ were dreams with feelings of anxiety, fear, or aggression. In group II, the subjects did not watch a movie but simply went to sleep. In this group, there were a total of $n_2 = 180$ dreams recorded, of which $r_2 = 63$ were dreams with feelings of anxiety, fear, or aggression.

(a) Why could groups I and II be considered independent binomial distributions? Why do we have a "large-sample" situation?

married couples who have two or more personality preferences in common. Let p_2 be the population proportion of all married couples who have no personality preferences in common.

(a) Find a 99% confidence interval for $p_1 - p_2$.

(b) Explain the meaning of the confidence interval in part (a) in the context of this problem. Does the confidence interval contain all positive, all negative, or both positive and negative numbers? What does this tell you (at the 99% confidence level) about the proportion of married couples with two or more personality preferences in common compared with the proportion of married couples sharing no personality preferences in common?

8. *Myers-Briggs: Married Couples* Most married couples have two or three personality preferences in common (*see* reference in Problem 7). Myers used a random sample of 375 married couples and found that 132 had three preferences in common. Another random sample of 571 couples showed that 217 had two personality preferences in common. Let p_1 be the population proportion of all married couples who have three personality preferences in common. Let p_2 be the population proportion of all married couples who have two personality preferences in common.

(a) Find a 90% confidence interval for $p_1 - p_2$.

(b) Examine the confidence interval in part (a) and explain what it means in this context. Does the confidence interval contain all positive, all negative, or both positive and negative numbers? What does this tell you about the proportion of married couples with three personality preferences in common compared with the proportion of couples with two preferences in common (at the 90% confidence level)?

9. *Navajo Reservation: Permanent Dwellings* S. C. Jett is a professor of geography at the University of California, Davis. He and a colleague, V. E. Spencer, are experts on modern Navajo culture and geography. The following information is taken from their book, *Navajo Architecture: Forms, History, Distribution* (University of Arizona Press). On the Navajo Reservation, a random sample of 210 permanent dwellings in the Fort Defiance Region showed that 65 were traditional Navajo hogans. In the Indian Wells Region, a random sample of 152 permanent dwellings showed that 18 were traditional hogans. Let p_1 be the population proportion of all traditional hogans in the Fort Defiance Region, and let p_2 be the population proportion of all traditional hogans in the Indian Wells Region.

(a) Find a 99% confidence interval for $p_1 - p_2$.

(b) Examine the confidence interval and comment on its meaning. Does it include numbers that are all positive, all negative, or mixed? What if it is hypothesized that Navajo who follow the traditional culture of their people tend to live in hogans? Comment on the confidence interval for $p_1 - p_2$ in this context.

10. *Anthropology: Cultural Affiliation* "Unknown cultural affiliations and loss of identity at high elevations." These are words used to propose the hypothesis that archaeological sites tend to lose their identity as altitude extremes are reached. This idea is based on the notion that prehistoric people tended not to take trade wares to temporary settings and/or isolated areas (*Reference: Prehistoric New Mexico: Background for Survey*, by D. E. Stuart and R. P. Gauthier, University of New Mexico Press). As elevation zones of prehistoric people (in what is now the state of New Mexico) increased, there seemed to be a loss of artifact identification. Consider the following information.

Elevation Zone	Number of Artifacts	Number Unidentified
7000–7500 ft	112	69
5000–5500 ft	140	26

Let p_1 be the population proportion of unidentified archaeological artifacts at the elevation zone 7000–7500 ft in the given archaeological area. Let p_2 be the population proportion of unidentified archaeological artifacts at the elevation zone 5000–5500 ft in the given archaeological area.

(a) Find a 99% confidence interval for $p_1 - p_2$.

(b) Explain the meaning of the confidence interval in part (a) in the context of this problem. Does the confidence interval contain all positive numbers, all negative numbers, or both positive and negative numbers? What does this tell you (at the 99% confidence level) about the comparison of the population proportion of unidentified artifacts at high elevations (7000–7500 ft) with the population proportion of unidentified artifacts at lower elevations (5000–5500 ft)? How does this relate to the stated hypothesis?

11. *Focus Problem: Wood Ducks* Use the following Minitab printout to solve the Focus Problem at the beginning of this chapter. In Minitab, the symbol x is used for the number of successes. Sample p refers to $\hat{p}$.

```
Test and Confidence Interval for Two Proportions
Sample      X        N       Sample p
1          270      474      0.569620
2          270      805      0.335404
Estimate for p(1) - p(2): 0.234217
95% CI for p(1) - p(2): (0.178985, 0.289448)
Test for p(1) - p(2) = 0 (vs > 0) : Z = 8.19 P-Value = 0.000
```

SUMMARY

In this chapter, we continued our discussion of hypothesis testing. We applied the techniques to the difference of means μ_d formed from paired data of dependent samples. For large and small, independent samples we considered the difference of means $\mu_1 - \mu_2$, where μ_1 is the mean of the first population and μ_2 is the mean of the second population. We used hypothesis testing to determine if the sample difference $\bar{x}_1 - \bar{x}_2$ was significant. We also constructed confidence intervals for $\mu_1 - \mu_2$.

Finally, we looked at the difference of proportions $p_1 - p_2$ from large, independent samples. The sample statistic $\hat{p}_1 - \hat{p}_2$ was used to determine if the difference in proportions was significant. The same sample statistic $\hat{p}_1 - \hat{p}_2$ was used as a point estimate for $p_1 - p_2$ in the construction of confidence intervals.

IMPORTANT WORDS & SYMBOLS

Section 11.1
Dependent samples
Data pairs
μ_d, difference of means from data pairs
$d.f.$ for testing μ_d

Section 11.2
Independent samples

Section 11.3
Pooled variance

Section 11.4
Pooled estimate of proportions, $\hat{p}$

VIEWPOINT *Who Watches Cable TV?*

Consider the following claim: Average cable TV viewers are, generally speaking, as affluent as newspaper readers and better off than radio or magazine audiences. How do we know that this claim is true? One way to answer such a question is to construct several individual tests of difference of means. One test would compare cable TV viewers with newspaper readers, a second test would compare cable TV viewers with radio audiences, and a third test would compare cable TV viewers with magazine audiences. Another way to handle all three tests at once would be to use techniques of ANOVA (analysis of variance)—a subject often introduced in a second course in statistics. For more information and data, see *American Demographics* (Vol. 17, No. 6).

CHAPTER REVIEW PROBLEMS

For each of these problems, please do the following:
(a) Determine whether the samples are independent or dependent.
(b) Determine whether the sampling distribution for the difference follows a Student's t distribution or a normal distribution.
(c) In hypothesis testing problems, please provide the following information:
   (i)   The null and alternate hypotheses.
   (ii) The appropriate critical value t_0 to z_0 (or critical values $\pm t_0$ or $\pm z_0$).
   (iii) A sketch of the sampling distribution showing the critical region(s).
   (iv) Calculate the t or z value as appropriate that corresponds to the sample statistic. Show this value on the sketch in part (iii).
   (v) Find the P value for the test.
   (vi) Based on your answers for parts (i–v), shall we reject or fail to reject the null hypothesis? Explain your conclusion in the context of the problem.

1. *Super Bowl Football: Super Ads* The broadcast of the Super Bowl on commercial television offers a stage for "super ads." *USA Today* asks readers to watch and rate the ads aired during the Super Bowl on a scale of 1 to 10 (with 10 the highest score). How do ratings on 30-second ads compare with ratings on 60-second ads? Ratings for samples of ads aired during past Super Bowls follow. Assume that these data are representative of all 30- and 60-second ads.

   30-Second Ads: x_1 = rating; sample size $n_1 = 142$; $\bar{x}_1 = 6.03$; estimated sample standard deviation $s_1 = 1.2$

   60-Second Ads: x_2 = rating; sample size $n_2 = 47$; $\bar{x}_2 = 7.09$; estimated sample standard deviation $s_2 = 1.3$

   Let μ_1 represent the population of x_1 values and let μ_2 represent the population of x_2 values.
   (a) Use a 5% level of significance to test the claim that there is no difference in the average ratings for 30-second ads and 60-second ads.
   (b) Find a 95% confidence interval. Comment on the meaning of the confidence interval. Does the confidence interval contain values that are all positive, all negative, or mixed? At the 95% confidence level, what does the interval tell you about the average rating of 30-second ads compared with the average rating of 60-second ads?

2. *Psychology: Rats in a Maze* The following data are based on information from the Regis University Psychology Department. In an effort to determine if rats perform certain tasks more quickly when offered larger rewards, the following experiment was performed. On day 1, a group of 3 rats were given a reward of one food pellet each time they ran a maze. A second group of 3 rats were given a reward of five food pellets each time they ran the maze. On day 2, the groups were reversed, so the first group now got five food pellets for running the maze and the second group got only one pellet for running the same maze. The average times in seconds for each rat to run the maze 30 times are shown in the following table.

Rat	A	B	C	D	E	F
Time with one food pellet	3.6	4.2	2.9	3.1	3.5	3.9
Time with five food pellets	3.0	3.7	3.0	3.3	2.8	3.0

Do these data indicate that rats receiving larger rewards tend to run the maze in less time? Use a 5% level of significance.

3. *Psychology: Rats Climbing Ladders* The same experimental design discussed in Problem 2 also was used to test rats trained to climb a sequence of short ladders. Times in seconds for eight rats to do this task are shown in the following table.

Rat	A	B	C	D	E	F	G	H
Time 1 pellet	12.5	13.7	11.4	12.1	11.0	10.4	14.6	12.3
Time 5 pellets	11.1	12.0	12.2	10.6	11.5	10.5	12.9	11.0

Do these data indicate that rats receiving larger rewards tend to perform the ladder climb in less time? Use a 5% level of significance.

4. *Transportation: Two Bus Lines* A comparison is made between two bus lines to determine if arrival times of their regular buses from Denver to Durango are off schedule by the same amount of time. For 81 randomly selected runs, bus line A was observed to be off schedule an average time of 53 min with standard deviation 19 min. For 100 randomly selected runs, bus line B was observed to be off schedule an average of 62 min with standard deviation 15 min. Do the data indicate a significant difference in off-schedule times? Use a 5% level of significance.

5. *Surveys: Telephone vs. Face to Face* The book *Survey Responses: An Evaluation of Their Validity*, by E. J. Wentland and K. Smith (Academic Press, 1993), includes studies reporting accuracy of answers to questions from surveys. A study by Locander et al. (1976) considered the question, "Are you a registered voter?" Accuracy of response was confirmed by a check of city voting records. Two methods of survey were used: a face-to-face interview and a telephone interview. A random sample of 93 people were asked the voter registration question face to face. Seventy-nine respondents gave accurate answers (as verified by city records). Another random sample of 83 people were asked the same question during telephone interviews. Seventy-four respondents gave accurate answers. Assume that the samples are representative of the general population. Let p_1 be the population proportion of all people who answer the voter registration question accurately during face-to-face interviews. Let p_2 be the population proportion of all people who answer the question accurately during telephone interviews.

   (a) Use a 1% level of significance to test the hypothesis that the proportion of respondents who answer accurately during face-to-face interviews is different from the proportion who answer accurately during telephone interviews.

   (b) Find a 99% confidence interval for $p_1 - p_2$. Comment on the meaning of the confidence interval in the context of this problem. At the 99% level, do you detect any difference in the proportion of accurate responses from the face-to-face interviews compared with the proportion of accurate responses from the phone interviews?

6. *Surveys: Sensitive Material* Locander et al. (1976) also studied the accuracy of responses on questions involving material more sensitive than voter registration. From public records, individuals were identified as having been charged with drunken driving not less than 6 months or more than 12 months from the starting date of the study. Two random samples from this group were studied. In the first sample of 30 individuals, the respondents were asked in face-to-face interviews if they had been charged with drunken driving in the last 12 months. Of these 30 people interviewed face to face, 16 answered the question accurately. The second random sample consisted of 46 people who had been charged with drunken driving. During telephone interviews, 25 of these responded accurately to the question asking if they had been charged with drunken driving during the past 12 months. Assume that the samples are representative of all people recently charged with drunken driving. Let p_1 represent the population proportion of all people with recent charges of drunken driving who respond accurately to face-to-face interviews asking if they have been charged with drunken driving during the past 12 months. Let p_2 represent the population proportion of people who respond accurately to the question when it is asked in telephone interviews.
   (a) Use a 5% level of significance to test the hypothesis that the proportion of respondents who answer accurately during face-to-face interviews is different from the proportion who answer accurately during telephone interviews.
   (b) Find a 95% confidence interval for $p_1 - p_2$. Comment on the meaning of the confidence interval in the context of this problem. At the 95% level, do you detect any difference in the proportion of accurate responses from the face-to-face interviews compared with the proportion of accurate responses from the phone interviews?

7. *Hamburger Stands: Service* An independent rating service is trying to determine which of two hamburger stands has quicker service. Over a period of 16 randomly selected times, the average waiting period at Burger Queen is 4.8 min with standard deviation 2.0 min. The average waiting period at McGregor's over a period of 14 randomly selected times is 5.2 min with standard deviation 1.8 min.
   (a) Find a 90% confidence interval for the difference of population mean waiting times.
   (b) Does the confidence interval contain all positive numbers, all negative numbers, or some positive and some negative numbers? What conclusion can you draw regarding the difference in mean waiting time for these two hamburger stands?

8. *Highway Maintenance: Reflecting Paint* The highway department tested two types of reflecting paint for concrete bridge end pillars. The two kinds of paint were alike in every respect except that one was red and the other was yellow. The red paint was applied to 12 bridges, and the yellow paint was applied to 12 bridges. After a period of 1 year, reflectometer readings were made on all these bridge end pillars. (A higher reading meant better visibility.) For the red paint, the mean reflectometer reading was $\bar{x}_1 = 9.4$ with standard deviation $s_1 = 2.1$. For the yellow paint, the mean was $\bar{x}_2 = 6.8$ with standard deviation $s_2 = 2.0$. Based on these data, can we conclude that the yellow paint had less visibility after 1 year? (Use a 1% level of significance.)

9. *Archaeology: Projectile Points* The Wind Mountain archaeological site is in southwestern New Mexico. Prehistoric Native Americans called Anasazi once lived and hunted small game in this region. A stemmed projectile point is an arrowhead that has a notch on each side of the base. Both stemmed and stemless projectile points

were found at the Wind Mountain site. A random sample of $n_1 = 55$ stemmed projectile points showed the mean length to be $\bar{x}_1 = 3.0$ cm with sample standard deviation $s_1 = 0.8$ cm. Another random sample of $n_2 = 52$ stemless projectile points showed the mean length to be $\bar{x}_2 = 2.7$ cm with $s_2 = 0.9$ cm (*Source: Mimbres Mogollon Archaeology*, by A. I. Woosley and A. J. McIntyre, University of New Mexico Press). Do these data indicate a difference (either way) for the population mean length of the two types of projectile points? Use a 5% level of significance.

DATA HIGHLIGHTS: GROUP PROJECT

Break into small groups and discuss the following topic. Organize a brief outline in which you summarize the main points of your group discussion.

"Sweets May Not Be Culprit in Hyper Kids" was a *USA Today* (February 3, 1994) headline reporting results of a study that appeared in the *New England Journal of Medicine*. In this study, the subjects were 25 normal preschoolers aged 3 to 5, and 23 kids aged 6 to 10, who had been described as "sensitive to sugar." The kids and their families were put on three different diets for 3 weeks each. One diet was high in sugar, one was low in sugar and contained aspartame, and one was low in sugar and contained saccharin. The diets were all free of additives, artificial food coloring, preservatives, and chocolate. All food in the household was removed, and then meals were delivered to the families. Researchers gathered information about the kids' behavior from parents, babysitters, and teachers. In addition, researchers tested the kids for memory, concentration, reading, and math skills. The result: "We couldn't find any difference in terms of their behavior or their learning on any of the three diets," said Mark Wolraich, professor of pediatrics at Vanderbilt University Medical Center, who oversaw the project. In another interview, Dr. Wolraich was quoted as saying, "Our study would say there is no evidence sugar has an adverse effect on children's behavior."

(a) This research involved comparison of several means, not just two. However, let us take a simplified view of the problem and consider the difference of behavior when children consumed the diet with sugar compared with their behavior when they consumed the diet with aspartame and low sugar. List some variables that might be measured to reflect the behavior of the children.

(b) Let's assume that the general null hypothesis was that there is no difference in children's behavior when they have a diet high in sugar. Was the evidence sufficient to allow the researchers to reject the null hypothesis and conclude that there are differences in children's behavior when they have a diet high in sugar? When we cannot reject H_0, have we *proved* that H_0 is true? In your own words, paraphrase the comments made by Dr. Wolraich.

LINKING CONCEPTS: WRITING PROJECTS

Discuss each of the following topics in class or review the topics on your own. Then write a brief but complete essay in which you summarize the main points. Please include formulas and graphs as appropriate.

Is there a relationship between confidence intervals and two-tailed hypothesis tests? The answer is yes. Let c be the level of confidence used to construct a confidence interval from sample data. Let α be the level of significance for a two-tailed hypothesis test. The following statement applies to hypothesis tests of the mean.

12.1
Chi-Square: Tests of Independence

FOCUS POINTS

✓ Set up a test to investigate independence of random variables.

✓ Use contingency tables to compute the sample χ^2 statistic.

✓ Complete the test and estimate the P value.

Innovative Machines Incorporated has developed two new letter arrangements for computer keyboards. The company wishes to see if there is any relationship between the arrangement of letters on the keyboard and the number of hours it takes a new typing student to learn to type at 20 words per minute. Or, from another point of view, is the time it takes a student to learn to type *independent* of the arrangement of the letters on a keyboard?

To answer questions of this type, we test the hypotheses

H_0: Keyboard arrangement and learning times *are independent.*

H_1: Keyboard arrangement and learning times *are not independent.*

Chi-square distribution

In problems of this sort, we are testing the *independence* of two factors. The probability distribution we use to make the decision is the *chi-square distribution*. Recall from the overview of the chi-square distribution that *chi* is pronounced like the first two letters of the word *kite* and is a Greek letter denoted by the symbol χ, so chi-square is denoted by χ^2.

The first task for Innovative Machines is to gather data. Suppose that the company took a random sample of 300 beginning typing students and randomly assigned them to learn to type on one of three keyboards. The learning times for this sample are shown in Table 12-2 on the next page.

Contingency table

Table 12-2 is called a *contingency table*. The *shaded boxes* that contain observed frequencies are called *cells*. The row and column totals are not considered to be cells. This contingency table is of size 3×3 (read, "three-by-three") because there are three rows of cells and three columns. When giving the size of a contingency table, we always list the number of *rows first*.

GUIDED EXERCISE 1

Size of contingency table

Give the size of the contingency tables in Figures 12-2 and 12-3. Also, count the number of cells in each table. (Remember, each cell is a shaded box.)

(a) **FIGURE 12-2** Contingency Table

⟹ There are two rows and four columns, so this is a 2×4 table. There are eight cells.

Continued

GUIDED EXERCISE 1 continued

(b) **FIGURE 12-3** Contingency Table

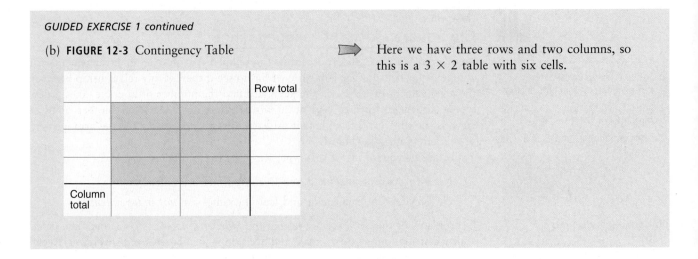

Here we have three rows and two columns, so this is a 3 × 2 table with six cells.

We are testing the null hypothesis that the keyboard arrangement and the time it takes a student to learn to type are *independent*. We use this hypothesis to determine the *expected frequency* of each cell.

Expected frequency

For instance, to compute the expected frequency of cell 1 in Table 12-2, we observe that cell 1 consists of all the students in the sample who learned to type on keyboard A and who mastered the skill at the 20-words-per-minute level in 21 to 40 hours. By the assumption (null hypothesis) that the two events are independent, we use the multiplication law to obtain the probability that a student is in cell 1.

$$P(\text{cell 1}) = P(\text{keyboard A } and \text{ skill in 21–40 h})$$
$$= P(\text{keyboard A}) \cdot P(\text{skill in 21–40 h})$$

Because there are 300 students in the sample and 80 used keyboard A,

$$P(\text{keyboard A}) = \frac{80}{300}$$

Also, 90 of the 300 students learned to type in 21–40 hours, so

$$P(\text{skill in 21–40 h}) = \frac{90}{300}$$

TABLE 12-2   Keyboard versus Time to Learn to Type at 20 wpm

Keyboard	21–40 h	41–60 h	61–80 h	Row Total
A	#1 25	#2 30	#3 25	80
B	#4 30	#5 71	#6 19	120
Standard	#7 35	#8 49	#9 16	100
Column total	90	150	60	300 Sample size

Using these two probabilities and the assumption of independence,

$$P(\text{keyboard A } and \text{ skill in 21--40 h}) = \frac{80}{300} \cdot \frac{90}{300}$$

Finally, because there are 300 students in the sample, we have the *expected frequency E* for cell 1.

$E = P(\text{student in cell 1}) \cdot (\text{no. of students in sample})$

$$= \frac{80}{300} \cdot \frac{90}{300} \cdot 300$$

$$= \frac{80 \cdot 90}{300}$$

$$= 24$$

We can repeat this process for each cell. However, the last step yields an easier formula for the expected frequency E.

Formula for expected frequency *E*

$$E = \frac{(\text{Row total})(\text{Column total})}{\text{Sample size}}$$

Note: If the expected value is not a whole number, do *not* round it to the nearest whole number.

Let's use this formula in Example 1 to find the expected frequency for cell 2.

EXAMPLE 1

Expected frequency

Find the expected frequency for cell 2 of contingency Table 12-2.

SOLUTION: Cell 2 is in row 1 and column 2. The *row total* is 80, and the *column total* is 150. The size of the sample is still 300.

$$E = \frac{(\text{Row total})(\text{Column total})}{\text{Sample size}}$$

$$= \frac{(80)(150)}{300} = 40 \qquad \qquad \diamond$$

GUIDED EXERCISE 2

Expected frequency

Table 12-3 on the following page contains the *observed frequencies O* and *expected frequencies E* for the contingency table giving keyboard arrangement and number of hours it takes a student to learn to type at 20 words per minute. Fill in the missing expected frequencies.

Continued

GUIDED EXERCISE 2 continued

TABLE 12-3 Complete Contingency Table of Keyboard Arrangement and Time to Learn to Type

Keyboard	21–40 h	41–60 h	61–80 h	Row Total
A	#1 $O = 25$ $E = 24$	#2 $O = 30$ $E = 40$	#3 $O = 25$ $E =$ ——	80
B	#4 $O = 30$ $E = 36$	#5 $O = 71$ $E =$ ——	#6 $O = 19$ $E =$ ——	120
Standard	#7 $O = 35$ $E =$ ——	#8 $O = 49$ $E = 50$	#9 $O = 16$ $E = 20$	100
Column Total	90	150	60	300 Sample Size

For cell 3, we have

$$E = \frac{(80)(60)}{300} = 16$$

For cell 5, we have

$$E = \frac{(120)(150)}{300} = 60$$

For cell 6, we have

$$E = \frac{(120)(60)}{300} = 24$$

For cell 7, we have

$$E = \frac{(100)(90)}{300} = 30$$

Computing the sample test statistic χ^2

Now we are ready to compute the sample statistic χ^2 for the typing students. The χ^2 value is a measure of the sum of differences between observed frequency O and expected frequency E in each cell. These differences are listed in Table 12-4.

As we see, if we sum the differences between the observed frequencies and the expected frequencies of the cells, we get the value zero. This total certainly does not reflect the fact that there were differences between the observed and expected frequencies. To obtain a measure whose sum does reflect the magnitude of the differences, we square the differences and work with the quantities $(O - E)^2$. But instead of using the terms $(O - E)^2$, we use the values $(O - E)^2/E$. The reason we use this expression is that a small difference between the observed and expected frequency is not nearly as important if the expected frequency is large as it is if the

TABLE 12-4 Difference Between the Observed and Expected Frequencies

Cell	Observed O	Expected E	Difference $(O - E)$
1	25	24	1
2	30	40	−10
3	25	16	9
4	30	36	−6
5	71	60	11
6	19	24	−5
7	35	30	5
8	49	50	−1
9	16	20	−4
			$\Sigma(O - E) = 0$

2. *Census: Type of Household* The type of household for the U.S. population and a random sample of 411 households of the community of Dove Creek, Montana, are shown (based on *Statistical Abstract of the United States*).

Type of Household	Percent of U.S. Households	Observed Number of Households in Dove Creek
Married, with children	26%	102
Married, no children	29%	112
Single parent	9%	33
One person	25%	96
Other (e.g., roommates, siblings)	11%	68

Use a 5% level of significance to test the claim that the distribution of U.S. households fits the Dove Creek distribution.

3. *Archaeology: Stone Tools* The type of raw material used to construct stone tools found at the archaeological site Casa del Rito is shown below (*Bandelier Archaeological Excavation Project*, edited by Kohler and Root). A random sample of 1486 stone tools was obtained from a current excavation site.

Raw Material	Regional Percent of Stone Tools	Observed Number of Tools at Current Excavation Site
Basalt	61.3%	906
Obsidian	10.6%	162
Welded tuff	11.4%	168
Pedernal chert	13.1%	197
Other	3.6%	53

Use a 1% level of significance to test the claim that the regional distribution of raw materials fits the distribution at the current excavation site.

4. *Ecology: Deer* The type of browse favored by deer is shown in the following table (*The Mule Deer of Mesa Verde National Park*, edited by Mierau and Schmidt). Using binoculars, volunteers observed feeding habits of a random sample of 320 deer.

Type of Browse	Plant Composition in Study Area	Observed Number of Deer Feeding on This Plant
Sage brush	32%	102
Rabbit brush	38.7%	125
Salt brush	12%	43
Service berry	9.3%	27
Other	8%	23

Use a 5% level of significance to test the claim that the natural distribution of browse fits the deer feeding pattern.

5. *Meteorology: Colorado* The following problem is based on information from the *National Oceanic and Atmospheric Administration (NOAA) Environmental Data Service*. Let x be a random variable that represents the average daily temperature (degrees

Fahrenheit) in July in the town of Kit Carson, Colorado. The x distribution has a mean μ approximately 75°F and standard deviation σ approximately 8°F. A 20-year study (620 July days) gave the entries in the right column of the following table.

I	II	III	IV
Region under Normal Curve	x°F	Expected % from Normal Curve	Observed Number of Days in 20 Years
$\mu - 3\sigma \le x < \mu - 2\sigma$	$51 \le x < 59$	2.35%	16
$\mu - 2\sigma \le x < \mu - \sigma$	$59 \le x < 67$	13.5%	78
$\mu - \sigma \le x < \mu$	$67 \le x < 75$	34%	212
$\mu \le x < \mu + \sigma$	$75 \le x < 83$	34%	221
$\mu + \sigma \le x < \mu + 2\sigma$	$83 \le x < 91$	13.5%	81
$\mu + 2\sigma \le x < \mu + 3\sigma$	$91 \le x < 99$	2.35%	12

(i) Remember that $\mu = 75$ and $\sigma = 8$. Examine Figure 7-5 in Chapter 7. Write a brief explanation for columns I, II, and III in the context of this problem.

(ii) Use a 1% level of significance to test the claim that the average daily July temperature follows a normal distribution with $\mu = 75$ and $\sigma = 8$.

6. *Meteorology: Hawaii* Let x be a random variable that represents the average daily temperature (degrees Fahrenheit) in January at the town of Hana, Maui. The x variable has a mean μ approximately 68°F and standard deviation σ approximately 4°F (*see* reference in Problem 5). A 20-year study (620 January days) gave the entries in the right column of the following table.

I	II	III	IV
Region under Normal Curve	x°F	Expected % from Normal Curve	Observed Number of Days in 20 Years
$\mu - 3\sigma \le x < \mu - 2\sigma$	$56 \le x < 60$	2.35%	14
$\mu - 2\sigma \le x < \mu - \sigma$	$60 \le x < 64$	13.5%	86
$\mu - \sigma \le x < \mu$	$64 \le x < 68$	34%	207
$\mu \le x < \mu + \sigma$	$68 \le x < 72$	34%	215
$\mu + \sigma \le x < \mu + 2\sigma$	$72 \le x < 76$	13.5%	83
$\mu + 2\sigma \le x < \mu + 3\sigma$	$76 \le x < 80$	2.35%	15

(i) Remember that $\mu = 68$ and $\sigma = 4$. Examine Figure 7-5 in Chapter 7. Write a brief explanation for columns I, II, and III in the context of this problem.

(ii) Use a 1% level of significance to test the claim that the average daily January temperature follows a normal distribution with $\mu = 68$ and $\sigma = 4$.

7. *Ecology: Fish* The Fish and Game Department stocked Lake Lulu with fish in the following proportions: 30% catfish, 15% bass, 40% bluegill, and 15% pike. Five years later they sampled the lake to see if the distribution of fish had changed. They found the 500 fish in the sample were distributed as follows:

Catfish	Bass	Bluegill	Pike
120	85	220	75

In the 5-year interval, did the distribution of fish change at the 0.05 level?

Complete parts (a) through (c).
(d) If a team had $x = 4$ fouls over and above the opposing team, what does the least-squares equation forecast for y?
(e) Find an 80% confidence interval for the forecast in part (d).

5. *Sociology: College vs. Income* The following data are based on information from the book *Life in America's Small Cities* (by G. S. Thomas, Prometheus Books). Let x = percentage of those 25 years or older with 4 or more years of college. Let y = per capita income in thousands of dollars. Five small cities in South Carolina (Greenwood, Hilton Head Island, Myrtle Beach, Orangeburg, and Sumpter) reported the following information regarding the x and y variables.

x	13.8	21.9	12.5	12.7	11.5
y	9.0	10.8	8.8	6.9	7.2

$SS_{xy} \approx 22.854$ $SS_x \approx 71.488$ $b \approx 0.320$ $\bar{x} \approx 14.48$ $\bar{y} \approx 8.54$
$y = 3.91 + 0.320x$

Complete parts (a) through (c).
(d) For a small city in South Carolina where $x = 20$ percent of the population 25 years or older who have had 4 or more years of college, what would the least-squares equation forecast for y = per capita income (in thousands of dollars)?
(e) Find an 80% confidence interval for your forecast y value in part (d).

6. *Sociology: High School Dropouts vs. Income* Five small cities in California (El Centro, Eureka, Hanford, Madera, and San Luis Obispo–Atascadero) reported the following information. Let x = percentage of 16- to 19-year-olds not in school and not high school graduates. Let y = per capita income in thousands of dollars. The following information was obtained (*see* reference in Problem 5).

x	16.2	9.9	19.5	19.7	9.8
y	7.2	8.8	7.9	8.1	10.3

$SS_{xy} \approx -17.026$ $SS_x \approx 96.828$ $b \approx -0.176$ $\bar{x} \approx 15.02$ $\bar{y} \approx 8.46$
$y = 11.1 - 0.176x$

Complete parts (a) through (c).
(d) For a small city in California where $x = 17$, what would the least-squares equation forecast for y = per capita income (in thousands of dollars)?
(e) Find a 75% confidence interval for the forecast y value in part (d).

7. *Archaeology: Cultural Affiliation* Data for this problem are based on information taken from *Prehistoric New Mexico: Background for Survey* (by D. E. Stuart and R. P. Gauthier, University of New Mexico Press). It is thought that prehistoric Indians did not take their best tools, pottery, and household items when they visited higher elevations for their summer camps. It is hypothesized that archaeological sites tend to lose their cultural identity and specific cultural affiliation as the elevation of the site increases. Let x = elevation (in thousands of feet) for an archaeological site in the southwestern United States. Let y = percentage of unidentified artifacts (no specific cultural affiliation) at a given elevation. The following data were obtained for a collection of archaeological sites in New Mexico.

x	5.25	5.75	6.25	6.75	7.25
y	19	13	33	37	62

$SS_{xy} \approx 55$ $SS_x \approx 2.5$ $b \approx 22$ $\bar{x} \approx 6.25$ $\bar{y} \approx 32.8$
$y = -104.7 + 22x$

Complete parts (a) through (c).
(d) At an archaeological site with elevation $x = 6.5$ (thousand feet), what does the least-squares equation forecast for y = percentage of culturally unidentified artifacts?
(e) Find a 75% confidence interval for the forecast y value in part (d).

8. *Psychology: Irrelevant Responses* A child psychiatrist is studying the mental development of children. A random sample of nine children were asked a standard set of questions appropriate to the age of each child. The number of irrelevant responses to the questions was recorded for each child. In the following data, x = age of child in years and y = number of irrelevant responses.

x	2	3	4	5	7	9	10	11	12
y	15	15	12	13	11	10	8	6	5

$SS_{xy} \approx -104$ $SS_x \approx 108$ $b \approx -0.963$ $\bar{x} \approx 7$ $\bar{y} \approx 10.556$
$y = 17.296 - 0.963x$

Complete parts (a) through (c).
(d) If a child is 9.5 years old, what does the least-squares line predict for the number of irrelevant responses?
(e) Find a 99% confidence interval for the number of irrelevant responses for a child who is 9.5 years old.

9. *General Discussion: Residual Plots* There are several ways to assess how well a least-squares line serves as a model for the data. One method is a graphic tool called a *residual plot*. Let x be the given explanatory variable and y the corresponding response variable of the data. Let the symbol y_p represent the predicted value for x determined by the least-squares line. To make a residual plot, we put the x values in order on the horizontal axis and plot the corresponding residuals $y - y_p$ in the vertical direction. Since for a least-squares model the mean of the residuals is always zero, we dash in a horizontal line at zero. The accompanying figure shows a residual plot for the data of Guided Exercise 3 of Section 4.2, in which the relationship between the number of ads run per week and the number of cars sold that week was explored. To make the residual plot, first compute all the residuals. Remember that x and y are the given data values and y_p is computed from the least-squares line $y_p \approx 6.56 + 1.01x$.

Residual

x	y	y_p	$y - y_p$
6	15	12.6	2.4
20	31	26.8	4.2
0	10	6.6	3.4
14	16	20.7	-4.7
25	28	31.8	-3.8

Residual

x	y	y_p	$y - y_p$
16	20	22.7	-2.7
28	40	34.8	5.2
18	25	24.7	0.3
10	12	16.7	-4.7
8	15	14.6	0.4

Residual Plot (Produced by Minitab)

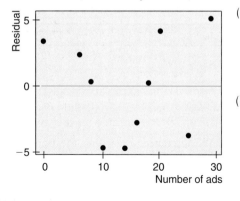

(a) If the least-squares line provides a reasonable model for the data, the pattern of points in the plot will seem random and unstructured around the horizontal line at 0. Is this the case for the residual plot?

(b) If a point on the residual plot seems far outside the pattern of other points, it might reflect an unusual data point (x, y), called an *outlier*. Such points may have quite an influence on the least-squares model. Do there appear to be any outliers in the data for the residual plot?

10. Consider the data in Problem 3.
    (a) Make a residual plot for the least-squares model.
    (b) Use the residual plot to comment about the appropriateness of the least-squares model for these data. *See* Problem 9.

12.5
Testing the Correlation Coefficient and the Slope

Learn more, earn more! We have probably all heard this platitude. The question is whether or not there is some truth in the statement. Do college graduates have an improved chance at a better income? Is there a trend in the general population to support the "learn more, earn more" statement?

Consider the following variables: x = percentage of the population 25 or older with at least four years of college and y = percentage *growth* in per capita income over the past seven years. A random sample of six communities in Ohio gave the following information (based on *Life in America's Small Cities,* by G. S. Thomas):

If we use what we learned in Sections 4.2 and 4.3, we can compute the correlation coefficient r and the least-squares line $y = a + bx$ using the data of Table 12-15.

TABLE 12-15   Education and Income Growth Percentages

x	9.9	11.4	8.1	14.7	8.5	12.6
y	37.1	43.0	33.4	47.1	26.5	40.2

However, r is only a *sample* correlation coefficient and $y = a + bx$ is only a "sample-based" least-squares line. What if we used *all* possible data pairs (x, y) from *all* U.S. cities, not just the six towns in Ohio? If you did this seemingly impossible task, you would have the *population* of all (x, y) pairs.

From this population of (x, y) pairs you could (in theory) compute the *population correlation coefficient,* which we call ρ (Greek letter rho that is pronounced like "row"). You could also compute the least-squares line for the entire population, which we denote as $y = \alpha + \beta x$ using more Greek letters α (alpha) and β (beta) (see Table 12-16 on the next page).

TABLE 12-16 Sample Statistics and Corresponding Population Parameters

Sample		Population
r	$\rightarrow$	ρ
$y = a + bx$	$\rightarrow$	$y = \alpha + \beta x$

Assumptions for inferences concerning linear regression

To make inferences regarding the population correlation coefficient ρ and the slope β of the population least-squares line, we need to be sure that

(a) The set (x, y) of ordered pairs is a *random sample* from the population of all possible such (x, y) pairs.

(b) For each fixed value of x, the y values have a normal distribution.

We *assume these conditions are met for all inferences presented in this section.*

Testing the Correlation Coefficient

The first topic we want to study is the statistical significance of the sample correlation coefficient r. To do this, we construct a statistical test of ρ, the population correlation coefficient. The test will be based on the following theorem.

◇ **THEOREM 12.2** Let r be the sample correlation coefficient computed using data pairs (x, y). We use the null hypothesis

H_0: x and y have no linear correlation, so $\rho = 0$

The alternate hypothesis may be

H_1: $\rho > 0$ or H_1: $\rho < 0$ or H_1: $\rho \neq 0$

The conversion of r to a Student's t distribution is

$$t = \frac{r\sqrt{n-2}}{\sqrt{1-r^2}} \quad \text{with } d.f. = n - 2$$

where n is the number of sample (x, y) data pairs $(n \geq 3)$. ◇

EXAMPLE 6

Testing ρ

Let's return to our data from Ohio regarding the percentage of the population with at least four years of college and the percentage of growth in per capita income (Table 12-15). We'll develop a test for the population correlation coefficient ρ.

SOLUTION: First, we compute the sample correlation coefficient r. Using a calculator, statistical software, or "by-hand" calculation from Section 4.3, we find

$r \approx 0.887$

Now we test the correlation coefficient ρ. Remember x represents percentage college "graduates" and y represents percentage salary increases in the general pop-

GUIDED EXERCISE 10 continued

(c) Use a 1% level of significance to test the claim that $\beta > 0$.

⟹ Convert $b \approx 0.712$ to a t value. Since the t value for b equals the t value for r, we use the result of part (b) and conclude $t \approx 7.56$.

$H_0: \beta = 0$; $H_1: \beta > 0$; $d.f. = 5$; $t_0 = 3.365$. Reject H_0 and conclude the slope of the least-squares line is positive.

TECH NOTE   It turns out that the t value corresponding to the correlation coefficient r is the same as the t value corresponding to b, the slope of the least-squares line. Consequently, the two tests $H_0: \rho = 0$ and $H_0: \beta = 0$ (with similar corresponding alternate hypotheses) have the same conclusions. The TI-83Plus uses this fact explicitly. Minitab and Excel show the two-tailed P value for the slope b of the least-squares line. Excel also shows confidence intervals for β. The displays show data from Guided Exercise 10 regarding the age and weight of wolf pups.

TI-83Plus Under **STAT**, select **TEST** and use option **E:LinRegTTest.**

```
LinRegTTest
y=a+bx
β>0 and ρ>0
t=7.5632
p=3.2037E-4
df=5.0000
↓a=5.9317
```

```
LinRegTTest
y=a+bx
β>0 and ρ>0
↑b=.7120
s=3.3676
r²=.9196
r=.9590
```

Note that the value of S_e is given as s.

Excel Use menu selection **Tools ➤ Data Analysis ➤ Regression.**

Regression Statistics	
Multiple R	0.958966516
R Square	0.919616778
Adjusted R Square	0.903540133
Standard Error	3.367628886	← Value of S_e
Observations	7

	Coefficients	Standard Error	t Stat	P-value	Lower 95%	Upper 95%
Intercept	5.931681179	2.558126184	2.318760198	0.068158803	-0.644180779	12.50754314
X Variable 1	0.711989283	0.094138596	7.563202697	0.000640746	0.469998714	0.953979853

b for two-tailed test

Minitab Use menu selection **Stat** ➤ **Regression** ➤ **Regression**. The value of S_e is S; P is the P value of a two-tailed test. For a one-tailed test, divide the P value by 2.

```
Regression Analysis
The regression equation is
y = 5.93 + 0.712 x
Predictor       Coef       StDev        T          P
Constant        5.932      2.558       2.32       0.068
x               0.71199    0.09414     7.56       0.001
S = 3.368       R-Sq = 92.0%        R-Sq(adj) = 90.4%
```

VIEWPOINT   *Hazardous to Your Health!*

The Federal Trade Commission (FTC) evaluates domestic cigarettes according to their tar, nicotine, and carbon monoxide contents. The United States Surgeon General considers each of these substances to be a health hazard. If a cigarette has a "richer" taste, with more tar or nicotine, does it also tend to produce more carbon monoxide? Is there a significant positive correlation? For more information, visit the Brase/Brase *Understanding Basic Statistics* site at http://math.college.hmco.com/students and find the link to the *Journal of Statistics Education*. Follow the links to the cigarette data.

SECTION 12.5   PROBLEMS

Answers to these problems may vary slightly due to rounding.

1. *Salary: Tuition* Is tuition correlated to professors' salaries? If you pay more tuition, does your professor get a higher salary? Let x = average annual tuition and fees for 2-year public colleges (units $100). Let y = average annual full-time faculty salary (units $1,000). A random sample of 8 states gave the following information (*Source: The Chronicle of Higher Education*, Vol. XLVI, Number 1):

x	13.4	8.2	9.4	3.8	14.5	12.5	23.9	8.2
y	37.9	50.5	33.3	56.5	37.9	42.5	43.9	41.6

(a) Please verify that $r \approx -0.377$; $b \approx -0.468$. (Note that $SS_x = 252.40$; $SS_y = 387.63$; $SS_{xy} = -118.06$; $S_e \approx 7.443$.)
(b) Use a 5% level of significance to test the claim that $\rho < 0$. Include an estimate for the P value.
(c) Use a 5% level of significance to test the claim that $\beta \neq 0$. Include an estimate for the P value.

2. *Salary: Tuition* Is tuition correlated to professors' salaries? If you pay more tuition, does your professor get a higher salary? Let x = average annual tuition and fees for 4-year public colleges (units $100). Let y = average annual full-time faculty salary (units $1,000). A random sample of 9 states gave the following information (*Source: The Chronicle of Higher Education*, Vol. XLVI, Number 1):

x	24.9	20.6	24.5	27.1	26.2	19.1	40.1	22.7	33.9
y	42.7	47.8	42.3	62.7	48.3	50.7	50.3	45.0	49.1

(a) Please verify that $r \approx 0.187$; $b \approx 0.172$. (Note that $SS_x = 350.70$; $SS_y = 294.30$; $SS_{xy} = 60.17$; $S_e \approx 6.369$.)
(b) Use a 5% level of significance to test the claim that $\rho \neq 0$. Include an estimate for the P value.
(c) Use a 5% level of significance to test the claim that $\beta \neq 0$. Include an estimate for the P value.

3. *Scuba Diving: Depth* What is the optimal time for a scuba diver to be on the bottom? That depends on the depth of the dive. The U.S. Navy has done a lot of research on this topic. The Navy defines the "optimal time" to be the time at each depth for the best balance between length of work period and decompression time after surfacing. Let x = depth of dive in meters, and let y = optimal time in hours. A random sample of divers gave the following data (based on information taken from *Medical Physiology* by A. C. Guyton, M.D.)

x	14.1	24.3	30.2	38.3	51.3	20.5	22.7
y	2.58	2.08	1.58	1.03	0.75	2.38	2.20

(a) Please verify that $r \approx -0.976$; $b \approx -0.054$. (Note that $SS_x = 940.89$; $SS_y = 2.93$; $SS_{xy} = -51.23$; $S_e \approx 0.166$.)
(b) Use a 1% level of significance to test the claim that $\rho < 0$. Include an estimate for the P value.
(c) Use a 1% level of significance to test the claim that $\beta < 0$. Include an estimate for the P value.

4. *Physiology: Oxygen* Aviation and high-altitude physiology is a specialty in the study of medicine. Let x = partial pressure of oxygen in the alveoli (air cells in the lungs) when breathing naturally available air. Let y = partial pressure when breathing pure oxygen. The (x, y) data pairs correspond to elevations from 10,000 feet up to 30,000 feet in 5,000-foot intervals for a random sample of volunteers. Although the medical data were collected using airplanes, it will apply equally well to Mt. Everest climbers (summit 29,028 feet).

x	6.7	5.1	4.2	3.3	2.1 (units: mm Hg/10)
y	43.6	32.9	26.2	16.2	13.9 (units: mm Hg/10)

(based on information taken from *Medical Physiology* by A. C. Guyton, M.D.)

(a) Please verify that $r \approx 0.984$; $b \approx 6.876$. (Note that $SS_x = 12.25$; $SS_y = 598.29$; $SS_{xy} = 84.22$; $S_e \approx 2.532$.)

(b) Use a 1% level of significance to test the claim that the population correlation coefficient is positive. Include an estimate for the P value.

(c) Use a 1% level of significance to test the claim that the slope of the population least-squares line is positive. Include an estimate for the P value.

5. *New Car: Negotiating Price* Suppose you are interested in buying a new Toyota Corolla. You are standing on the sales lot looking at a model with different options. The list price is on the vehicle. As a salesperson approaches, you wonder what the dealer invoice price is for this model with its options. The following data are based on information taken from *Consumer Guide* (vol. 677). Let x be the list price (in thousands of dollars) for a random selection of Toyota Corollas of different models and options. Let y be the dealer invoice (in thousands of dollars) for the given vehicle.

x	12.6	13.0	12.8	13.6	13.4	14.2
y	11.6	12.0	11.5	12.2	12.0	12.8

(a) Please verify that $r \approx 0.956$; $b \approx 0.758$. (Note that $SS_x = 1.733$; $SS_y = 1.088$; $SS_{xy} = 1.313$; $S_e \approx 0.1527$.)

(b) Use a 1% level of significance to test the claim that $\rho > 0$. Include an estimate for the P value.

(c) Use a 1% level of significance to test the claim that $\beta > 0$. Include an estimate for the P value.

6. *New Car: Negotiating Price* Suppose you are interested in buying a new Lincoln Navigator or Town Car. You are standing on the sales lot looking at a model with different options. The list price is on the vehicle. As a salesperson approaches, you wonder what the dealer invoice price is for this model with its options. The following data are based on information taken from *Consumer Guide* (vol. 677). Let x be the list price (in thousands of dollars) for a random selection of these cars of different models and options. Let y be the dealer invoice (in thousands of dollars) for the given vehicle.

x	32.1	33.5	36.1	44.0	47.8
y	29.8	31.1	32.0	42.1	42.2

(a) Please verify that $r \approx 0.977$; $b \approx 0.879$. (Note that $SS_x = 188.26$; $SS_y = 152.53$; $SS_{xy} = 165.55$; $S_e \approx 1.522$.)

(b) Use a 1% level of significance to test the claim that $\rho > 0$. Include an estimate for the P value.

(c) Use a 1% level of significance to test the claim that $\beta > 0$. Include an estimate for the P value.

7. (a) Suppose $n = 6$ and the sample correlation coefficient is $r = 0.90$. Is r significant at the 1% level of significance (based on a two-tailed test)?

(b) Suppose $n = 10$ and the sample correlation coefficient is $r = 0.90$. Is r significant at the 1% level of significance (based on a two-tailed test)?

(c) Explain why the test results of parts (a) and (b) are different even though the sample correlation coefficient $r = 0.90$ in both parts. Does it appear that sample size plays an important role in determining the significance of a correlation coefficient? Explain.

Distribution of Drunk Driver Arrests by Age

Age	National Percentage	Number in Freemont County
16–17	3.7	8
18–24	18.9	35
25–29	12.9	23
30–34	10.3	19
35–39	8.5	12
40–44	7.9	14
45–49	8.0	16
50–54	7.9	13
55–59	6.8	10
60–64	5.7	9
65 and over	9.4	15
	100%	174

Use a chi-square test with 5% level of significance to test the claim that the age distribution of drunk drivers arrested in Freemont County is the same as the national age distribution of drunk drivers arrested.

(a) State the null and alternate hypotheses.
(b) Find the value of the chi-square statistic from the sample.
(c) Find the degrees of freedom and the appropriate chi-square critical value.
(d) Sketch the critical region and locate your sample chi-square value and critical chi-square value on the sketch.
(e) Decide whether you should reject or not reject the null hypothesis. State your conclusion in the context of the problem.

LINKING CONCEPTS: WRITING PROJECTS

Discuss each of the following topics in class or review the topics on your own. Then write a brief but complete essay in which you summarize the main points. Please include formulas and graphs as appropriate.

Consider the results of a study conducted by the makers of Advil. A random sample of patients were assigned to one of two groups: One received ibuprofen and the other received a placebo. Then patients were asked if they experienced stomach upset (yes or no). The results of the study follow.

Group	Stomach Upset		Row Total
	Yes	No	
Ibuprofen	8	664	672
Placebo	6	645	651
Column total	14	1309	1323

1. In Chapter 5 on probability (Section 5.2), we considered contingency tables and the probabilities of events described by different combinations of the rows and columns. Use these techniques to compute the probability that a patient had stomach upset, *given* that the patient was in the ibuprofen group. Compute the probability that a patient had stomach upset, *given* that the patient was in the placebo group.

2. Are the proportions of patients having stomach upset in the two different treatment groups the same or not? We treated questions of this type in Section 11.4 where we applied the normal distribution to test the difference of proportions. Use such a test to determine whether or not the proportions are different at the 5% level of significance.

3. Now consider the 2×2 contingency table. We can apply techniques of tests of independence to see if the proportions of patients experiencing stomach upset in the two treatment groups are the same or not. When we apply these techniques to tests of proportions, we assume that the sample size for each treatment group is *assigned* ahead of time and is not itself a random variable. In such a case, we apply a test of independence, but we change the hypotheses to H_0: The proportions are equal and H_1: The proportions are not equal. Perform such a test.

4. Compare the results of Problems 2 and 3. Do the test conclusions agree? When testing two proportions, you can use either the normal distribution test of Section 11.4 or a chi-square test. Can you think of any advantages of one choice over the other?

Using TECHNOLOGY

TI-83PLUS • EXCEL • MINITAB • COMPUTERSTAT

APPLICATION

Tests of Independence

A study involving people who are food processors gave the following information about work shift and number of sick days.

Shift	Number of Sick Days				
	0	1	2	3	4 or more
Day	134	44	24	10	61
Aft/ev	90	39	23	18	99
Night	107	37	21	20	82
Rotating	56	20	14	17	92

Source: United States Department of Health, Education, and Welfare, NIOSH Technical Report. Tasto, Colligan, *et al. Health Consequences of Shift Work.* Washington: GPO, 1978, 29. (*Note:* This table was adapted from Table 7 on page 25 of the source.)

Use a 1% level of significance to test the null hypothesis that work shift and number of sick days are independent against the alternate hypothesis that they are not independent. What is the sample chi-square value? What is the test conclusion?

Technology Hints

TI-83Plus

Press the **Matrix** key and enter the observed values in a matrix. Name the matrix [A]. Then dimension matrix [B] to the same dimensions. Matrix [B] will hold the expected values. Press the **STAT** key and under the **TESTS** menu, select option **C:χ^2-Test. Calculate** will give the sample χ^2 value and its P value. **Draw** will show the sample test statistic on the χ^2 distribution with the P value shaded.

Excel

There are no built-in commands for executing a chi-square test. The *Technology Guide* shows a multistep process for conducting a chi-square test. The Tech Note of Section 12.1 gives a brief outline of the process.

Minitab

In Minitab, enter data for tests of independence in columns. Be sure to enter data only and not row or column totals. Then use the menu choices ➤ **Stat** ➤ **Tables** ➤ **Chisquare Test.** An example of a typical Minitab display for tests of independence is given in Problem 9 in Section 12.1.

ComputerStat

Under the main menu item Hypothesis Testing, select Chi-Square Test for Independence. Follow the instructions on the screen to enter data and generate the test results.

Appendix: Tables

TABLE 1   Random Numbers

92630	78240	19267	95457	53497	23894	37708	79862	76471	66418
79445	78735	71549	44843	26104	67318	00701	34986	66751	99723
59654	71966	27386	50004	05358	94031	29281	18544	52429	06080
31524	49587	76612	39789	13537	48086	59483	60680	84675	53014
06348	76938	90379	51392	55887	71015	09209	79157	24440	30244
28703	51709	94456	48396	73780	06436	86641	69239	57662	80181
68108	89266	94730	95761	75023	48464	65544	96583	18911	16391
99938	90704	93621	66330	33393	95261	95349	51769	91616	33238
91543	73196	34449	63513	83834	99411	58826	40456	69268	48562
42103	02781	73920	56297	72678	12249	25270	36678	21313	75767
17138	27584	25296	28387	51350	61664	37893	05363	44143	42677
28297	14280	54524	21618	95320	38174	60579	08089	94999	78460
09331	56712	51333	06289	75345	08811	82711	57392	25252	30333
31295	04204	93712	51287	05754	79396	87399	51773	33075	97061
36146	15560	27592	42089	99281	59640	15221	96079	09961	05371
29553	18432	13630	05529	02791	81017	49027	79031	50912	09399
23501	22642	63081	08191	89420	67800	55137	54707	32945	64522
57888	85846	67967	07835	11314	01545	48535	17142	08552	67457
55336	71264	88472	04334	63919	36394	11196	92470	70543	29776
10087	10072	55980	64688	68239	20461	89381	93809	00796	95945
34101	81277	66090	88872	37818	72142	67140	50785	21380	16703
53362	44940	60430	22834	14130	96593	23298	56203	92671	15925
82975	66158	84731	19436	55790	69229	28661	13675	99318	76873
54827	84673	22898	08094	14326	87038	42892	21127	30712	48489
25464	59098	27436	89421	80754	89924	19097	67737	80368	08795
67609	60214	41475	84950	40133	02546	09570	45682	50165	15609
44921	70924	61295	51137	47596	86735	35561	76649	18217	63446
33170	30972	98130	95828	49786	13301	36081	80761	33985	68621
84687	85445	06208	17654	51333	02878	35010	67578	61574	20749
71886	56450	36567	09395	96951	35507	17555	35212	69106	01679
00475	02224	74722	14721	40215	21351	08596	45625	83981	63748
25993	38881	68361	59560	41274	69742	40703	37993	03435	18873

TABLE 1 *continued*

92882	53178	99195	93803	56985	53089	15305	50522	55900	43026
25138	26810	07093	15677	60688	04410	24505	37890	67186	62829
84631	71882	12991	83028	82484	90339	91950	74579	03539	90122
34003	92326	12793	61453	48121	74271	28363	66561	75220	35908
53775	45749	05734	86169	42762	70175	97310	73894	88606	19994
59316	97885	72807	54966	60859	11932	35265	71601	55577	67715
20479	66557	50705	26999	09854	52591	14063	30214	19890	19292
86180	84931	25455	26044	02227	52015	21820	50599	51671	65411
21451	68001	72710	40261	61281	13172	63819	48970	51732	54113
98062	68375	80089	24135	72355	95428	11808	29740	81644	86610
01788	64429	14430	94575	75153	94576	61393	96192	03227	32258
62465	04841	43272	68702	01274	05437	22953	18946	99053	41690
94324	31089	84159	92933	99989	89500	91586	02802	69471	68274
05797	43984	21575	09908	70221	19791	51578	36432	33494	79888
10395	14289	52185	09721	25789	38562	54794	04897	59012	89251
35177	56986	25549	59730	64718	52630	31100	62384	49483	11409
25633	89619	75882	98256	02126	72099	57183	55887	09320	73463
16464	48280	94254	45777	45150	68865	11382	11782	22695	41988

Source: Reprinted from *A Million Random Digits with 100,000 Normal Deviates* by the Rand Corporation (New York: The Free Press, 1955). Copyright 1955 and 1983 by the Rand Corporation. Used by permission.

TABLE 2 Binomial Coefficients $C_{n,r}$

n \ r	0	1	2	3	4	5	6	7	8	9	10
1	1	1									
2	1	2	1								
3	1	3	3	1							
4	1	4	6	4	1						
5	1	5	10	10	5	1					
6	1	6	15	20	15	6	1				
7	1	7	21	35	35	21	7	1			
8	1	8	28	56	70	56	28	8	1		
9	1	9	36	84	126	126	84	36	9	1	
10	1	10	45	120	210	252	210	120	45	10	1
11	1	11	55	165	330	462	462	330	165	55	11
12	1	12	66	220	495	792	924	792	495	220	66
13	1	13	78	286	715	1,287	1,716	1,716	1,287	715	286
14	1	14	91	364	1,001	2,002	3,003	3,432	3,003	2,002	1,001
15	1	15	105	455	1,365	3,003	5,005	6,435	6,435	5,005	3,003
16	1	16	120	560	1,820	4,368	8,008	11,440	12,870	11,440	8,008
17	1	17	136	680	2,380	6,188	12,376	19,448	24,310	24,310	19,448
18	1	18	153	816	3,060	8,568	18,564	31,824	43,758	48,620	43,758
19	1	19	171	969	3,876	11,628	27,132	50,388	75,582	92,378	92,378
20	1	20	190	1,140	4,845	15,504	38,760	77,520	125,970	167,960	184,756

TABLE 3 Binomial Probability Distribution $C_{n,r}p^r q^{n-r}$

This table shows the probability of r successes in n independent trials, each with probability of success p.

n	r	.01	.05	.10	.15	.20	.25	.30	.35	.40	.45	.50	.55	.60	.65	.70	.75	.80	.85	.90	.95
2	0	.980	.902	.810	.723	.640	.563	.490	.423	.360	.303	.250	.203	.160	.123	.090	.063	.040	.023	.010	.002
	1	.020	.095	.180	.255	.320	.375	.420	.455	.480	.495	.500	.495	.480	.455	.420	.375	.320	.255	.180	.095
	2	.000	.002	.010	.023	.040	.063	.090	.123	.160	.203	.250	.303	.360	.423	.490	.563	.640	.723	.810	.902
3	0	.970	.857	.729	.614	.512	.422	.343	.275	.216	.166	.125	.091	.064	.043	.027	.016	.008	.003	.001	.000
	1	.029	.135	.243	.325	.384	.422	.441	.444	.432	.408	.375	.334	.288	.239	.189	.141	.096	.057	.027	.007
	2	.000	.007	.027	.057	.096	.141	.189	.239	.288	.334	.375	.408	.432	.444	.441	.422	.384	.325	.243	.135
	3	.000	.000	.001	.003	.008	.016	.027	.043	.064	.091	.125	.166	.216	.275	.343	.422	.512	.614	.729	.857
4	0	.961	.815	.656	.522	.410	.316	.240	.179	.130	.092	.062	.041	.026	.015	.008	.004	.002	.001	.000	.000
	1	.039	.171	.292	.368	.410	.422	.412	.384	.346	.300	.250	.200	.154	.112	.076	.047	.026	.011	.004	.000
	2	.001	.014	.049	.098	.154	.211	.265	.311	.346	.368	.375	.368	.346	.311	.265	.211	.154	.098	.049	.014
	3	.000	.000	.004	.011	.026	.047	.076	.112	.154	.200	.250	.300	.346	.384	.412	.422	.410	.368	.292	.171
	4	.000	.000	.000	.001	.002	.004	.008	.015	.026	.041	.062	.092	.130	.179	.240	.316	.410	.522	.656	.815
5	0	.951	.774	.590	.444	.328	.237	.168	.116	.078	.050	.031	.019	.010	.005	.002	.001	.000	.000	.000	.000
	1	.048	.204	.328	.392	.410	.396	.360	.312	.259	.206	.156	.113	.077	.049	.028	.015	.006	.002	.000	.000
	2	.001	.021	.073	.138	.205	.264	.309	.336	.346	.337	.312	.276	.230	.181	.132	.088	.051	.024	.008	.001
	3	.000	.001	.008	.024	.051	.088	.132	.181	.230	.276	.312	.337	.346	.336	.309	.264	.205	.138	.073	.021
	4	.000	.000	.000	.002	.006	.015	.028	.049	.077	.113	.156	.206	.259	.312	.360	.396	.410	.392	.328	.204
	5	.000	.000	.000	.000	.000	.001	.002	.005	.010	.019	.031	.050	.078	.116	.168	.237	.328	.444	.590	.774
6	0	.941	.735	.531	.377	.262	.178	.118	.075	.047	.028	.016	.008	.004	.002	.001	.000	.000	.000	.000	.000
	1	.057	.232	.354	.399	.393	.356	.303	.244	.187	.136	.094	.061	.037	.020	.010	.004	.002	.000	.000	.000
	2	.001	.031	.098	.176	.246	.297	.324	.328	.311	.278	.234	.186	.138	.095	.060	.033	.015	.006	.001	.000
	3	.000	.002	.015	.042	.082	.132	.185	.236	.276	.303	.312	.303	.276	.236	.185	.132	.082	.042	.015	.002
	4	.000	.000	.001	.006	.015	.033	.060	.095	.138	.186	.234	.278	.311	.328	.324	.297	.246	.176	.098	.031
	5	.000	.000	.000	.000	.002	.004	.010	.020	.037	.061	.094	.136	.187	.244	.303	.356	.393	.399	.354	.232
	6	.000	.000	.000	.000	.000	.000	.001	.002	.004	.008	.016	.028	.047	.075	.118	.178	.262	.377	.531	.735
7	0	.932	.698	.478	.321	.210	.133	.082	.049	.028	.015	.008	.004	.002	.001	.000	.000	.000	.000	.000	.000
	1	.066	.257	.372	.396	.367	.311	.247	.185	.131	.087	.055	.032	.017	.008	.004	.001	.000	.000	.000	.000
	2	.002	.041	.124	.210	.275	.311	.318	.299	.261	.214	.164	.117	.077	.047	.025	.012	.004	.001	.000	.000
	3	.000	.004	.023	.062	.115	.173	.227	.268	.290	.292	.273	.239	.194	.144	.097	.058	.029	.011	.003	.000
	4	.000	.000	.003	.011	.029	.058	.097	.144	.194	.239	.273	.292	.290	.268	.227	.173	.115	.062	.023	.004
	5	.000	.000	.000	.001	.004	.012	.025	.047	.077	.117	.164	.214	.261	.299	.318	.311	.275	.210	.124	.041
	6	.000	.000	.000	.000	.000	.001	.004	.008	.017	.032	.055	.087	.131	.185	.247	.311	.367	.396	.372	.257
	7	.000	.000	.000	.000	.000	.000	.000	.001	.002	.004	.008	.015	.028	.049	.082	.133	.210	.321	.478	.698

TABLE 3 *continued*

		p																			
n	r	.01	.05	.10	.15	.20	.25	.30	.35	.40	.45	.50	.55	.60	.65	.70	.75	.80	.85	.90	.95
8	0	.923	.663	.430	.272	.168	.100	.058	.032	.017	.008	.004	.002	.001	.000	.000	.000	.000	.000	.000	.000
	1	.075	.279	.383	.385	.336	.267	.198	.137	.090	.055	.031	.016	.008	.003	.001	.000	.000	.000	.000	.000
	2	.003	.051	.149	.238	.294	.311	.296	.259	.209	.157	.109	.070	.041	.022	.010	.004	.001	.000	.000	.000
	3	.000	.005	.033	.084	.147	.208	.254	.279	.279	.257	.219	.172	.124	.081	.047	.023	.009	.003	.000	.000
	4	.000	.000	.005	.018	.046	.087	.136	.188	.232	.263	.273	.263	.232	.188	.136	.087	.046	.018	.005	.000
	5	.000	.000	.000	.003	.009	.023	.047	.081	.124	.172	.219	.257	.279	.279	.254	.208	.147	.084	.033	.005
	6	.000	.000	.000	.000	.001	.004	.010	.022	.041	.070	.109	.157	.209	.259	.296	.311	.294	.238	.149	.051
	7	.000	.000	.000	.000	.000	.000	.001	.003	.008	.016	.031	.055	.090	.137	.198	.267	.336	.385	.383	.279
	8	.000	.000	.000	.000	.000	.000	.000	.000	.001	.002	.004	.008	.017	.032	.058	.100	.168	.272	.430	.663
9	0	.914	.630	.387	.232	.134	.075	.040	.021	.010	.005	.002	.001	.000	.000	.000	.000	.000	.000	.000	.000
	1	.083	.299	.387	.368	.302	.225	.156	.100	.060	.034	.018	.008	.004	.001	.000	.000	.000	.000	.000	.000
	2	.003	.063	.172	.260	.302	.300	.267	.216	.161	.111	.070	.041	.021	.010	.004	.001	.000	.000	.000	.000
	3	.000	.008	.045	.107	.176	.234	.267	.272	.251	.212	.164	.116	.074	.042	.021	.009	.003	.001	.000	.000
	4	.000	.001	.007	.028	.066	.117	.172	.219	.251	.260	.246	.213	.167	.118	.074	.039	.017	.005	.001	.000
	5	.000	.000	.001	.005	.017	.039	.074	.118	.167	.213	.246	.260	.251	.219	.172	.117	.066	.028	.007	.001
	6	.000	.000	.000	.001	.003	.009	.021	.042	.074	.116	.164	.212	.251	.272	.267	.234	.176	.107	.045	.008
	7	.000	.000	.000	.000	.000	.001	.004	.010	.021	.041	.070	.111	.161	.216	.267	.300	.302	.260	.172	.063
	8	.000	.000	.000	.000	.000	.000	.000	.001	.004	.008	.018	.034	.060	.100	.156	.225	.302	.368	.387	.299
	9	.000	.000	.000	.000	.000	.000	.000	.000	.000	.001	.002	.005	.010	.021	.040	.075	.134	.232	.387	.630
10	0	.904	.599	.349	.197	.107	.056	.028	.014	.006	.003	.001	.000	.000	.000	.000	.000	.000	.000	.000	.000
	1	.091	.315	.387	.347	.268	.188	.121	.072	.040	.021	.010	.004	.002	.000	.000	.000	.000	.000	.000	.000
	2	.004	.075	.194	.276	.302	.282	.233	.176	.121	.076	.044	.023	.011	.004	.001	.001	.000	.000	.000	.000
	3	.000	.010	.057	.130	.201	.250	.267	.252	.215	.166	.117	.075	.042	.021	.009	.003	.001	.000	.000	.000
	4	.000	.001	.011	.040	.088	.146	.200	.238	.251	.238	.205	.160	.111	.069	.037	.016	.006	.001	.000	.000
	5	.000	.000	.001	.008	.026	.058	.103	.154	.201	.234	.246	.234	.201	.154	.103	.058	.026	.008	.001	.000
	6	.000	.000	.000	.001	.006	.016	.037	.069	.111	.160	.205	.238	.251	.238	.200	.146	.088	.040	.011	.001
	7	.000	.000	.000	.000	.001	.003	.009	.021	.042	.075	.117	.166	.215	.252	.267	.250	.201	.130	.057	.010
	8	.000	.000	.000	.000	.000	.000	.001	.004	.011	.023	.044	.076	.121	.176	.233	.282	.302	.276	.194	.075
	9	.000	.000	.000	.000	.000	.000	.000	.000	.002	.004	.010	.021	.040	.072	.121	.188	.268	.347	.387	.315
	10	.000	.000	.000	.000	.000	.000	.000	.000	.000	.000	.001	.003	.006	.014	.028	.056	.107	.197	.349	.599
11	0	.895	.569	.314	.167	.086	.042	.020	.009	.004	.001	.000	.000	.000	.000	.000	.000	.000	.000	.000	.000
	1	.099	.329	.384	.325	.236	.155	.093	.052	.027	.013	.005	.002	.001	.000	.000	.000	.000	.000	.000	.000
	2	.005	.087	.213	.287	.295	.258	.200	.140	.089	.051	.027	.013	.005	.002	.001	.000	.000	.000	.000	.000

TABLE 3 *continued*

											p										
n	r	.01	.05	.10	.15	.20	.25	.30	.35	.40	.45	.50	.55	.60	.65	.70	.75	.80	.85	.90	.95
11	3	.000	.014	.071	.152	.221	.258	.257	.225	.177	.126	.081	.046	.023	.010	.004	.001	.000	.000	.000	.000
	4	.000	.001	.016	.054	.111	.172	.220	.243	.236	.206	.161	.113	.070	.038	.017	.006	.002	.000	.000	.000
	5	.000	.000	.002	.013	.039	.080	.132	.183	.221	.236	.226	.193	.147	.099	.057	.027	.010	.002	.000	.000
	6	.000	.000	.000	.002	.010	.027	.057	.099	.147	.193	.226	.236	.221	.183	.132	.080	.039	.013	.002	.000
	7	.000	.000	.000	.000	.002	.006	.017	.038	.070	.113	.161	.206	.236	.243	.220	.172	.111	.054	.016	.001
	8	.000	.000	.000	.000	.000	.001	.004	.010	.023	.046	.081	.126	.177	.225	.257	.258	.221	.152	.071	.014
	9	.000	.000	.000	.000	.000	.000	.001	.002	.005	.013	.027	.051	.089	.140	.200	.258	.295	.287	.213	.087
	10	.000	.000	.000	.000	.000	.000	.000	.000	.001	.002	.005	.013	.027	.052	.093	.155	.236	.325	.384	.329
	11	.000	.000	.000	.000	.000	.000	.000	.000	.000	.000	.000	.001	.004	.009	.020	.042	.086	.167	.314	.569
12	0	.886	.540	.282	.142	.069	.032	.014	.006	.002	.001	.000	.000	.000	.000	.000	.000	.000	.000	.000	.000
	1	.107	.341	.377	.301	.206	.127	.071	.037	.017	.008	.003	.001	.000	.000	.000	.000	.000	.000	.000	.000
	2	.006	.099	.230	.292	.283	.232	.168	.109	.064	.034	.016	.007	.002	.001	.000	.000	.000	.000	.000	.000
	3	.000	.017	.085	.172	.236	.258	.240	.195	.142	.092	.054	.028	.012	.005	.001	.000	.000	.000	.000	.000
	4	.000	.002	.021	.068	.133	.194	.231	.237	.213	.170	.121	.076	.042	.020	.008	.002	.001	.000	.000	.000
	5	.000	.000	.004	.019	.053	.103	.158	.204	.227	.223	.193	.149	.101	.059	.029	.011	.003	.001	.000	.000
	6	.000	.000	.000	.004	.016	.040	.079	.128	.177	.212	.226	.212	.177	.128	.079	.040	.016	.004	.000	.000
	7	.000	.000	.000	.001	.003	.011	.029	.059	.101	.149	.193	.223	.227	.204	.158	.103	.053	.019	.004	.000
	8	.000	.000	.000	.000	.001	.002	.008	.020	.042	.076	.121	.170	.213	.237	.231	.194	.133	.068	.021	.002
	9	.000	.000	.000	.000	.000	.000	.001	.005	.012	.028	.054	.092	.142	.195	.240	.258	.236	.172	.085	.017
	10	.000	.000	.000	.000	.000	.000	.000	.001	.002	.007	.016	.034	.064	.109	.168	.232	.283	.292	.230	.099
	11	.000	.000	.000	.000	.000	.000	.000	.000	.000	.001	.003	.008	.017	.037	.071	.127	.206	.301	.377	.341
	12	.000	.000	.000	.000	.000	.000	.000	.000	.000	.000	.000	.001	.002	.006	.014	.032	.069	.142	.282	.540
15	0	.860	.463	.206	.087	.035	.013	.005	.002	.000	.000	.000	.000	.000	.000	.000	.000	.000	.000	.000	.000
	1	.130	.366	.343	.231	.132	.067	.031	.013	.005	.002	.000	.000	.000	.000	.000	.000	.000	.000	.000	.000
	2	.009	.135	.267	.286	.231	.156	.092	.048	.022	.009	.003	.001	.000	.000	.000	.000	.000	.000	.000	.000
	3	.000	.031	.129	.218	.250	.225	.170	.111	.063	.032	.014	.005	.002	.000	.000	.000	.000	.000	.000	.000
	4	.000	.005	.043	.116	.188	.225	.219	.179	.127	.078	.042	.019	.007	.002	.001	.000	.000	.000	.000	.000
	5	.000	.001	.010	.045	.103	.165	.206	.212	.186	.140	.092	.051	.024	.010	.003	.001	.000	.000	.000	.000
	6	.000	.000	.002	.013	.043	.092	.147	.191	.207	.191	.153	.105	.061	.030	.012	.003	.001	.000	.000	.000
	7	.000	.000	.000	.003	.014	.039	.081	.132	.177	.201	.196	.165	.118	.071	.035	.013	.003	.001	.000	.000
	8	.000	.000	.000	.001	.003	.013	.035	.071	.118	.165	.196	.201	.177	.132	.081	.039	.014	.003	.000	.000
	9	.000	.000	.000	.000	.001	.003	.012	.030	.061	.105	.153	.191	.207	.191	.147	.092	.043	.013	.002	.000
	10	.000	.000	.000	.000	.000	.001	.003	.010	.024	.051	.092	.140	.186	.212	.206	.165	.103	.045	.010	.001

TABLE 3 *continued*

n	r											p									
		.01	.05	.10	.15	.20	.25	.30	.35	.40	.45	.50	.55	.60	.65	.70	.75	.80	.85	.90	.95
15	11	.000	.000	.000	.000	.000	.000	.001	.002	.007	.019	.042	.078	.127	.179	.219	.225	.188	.116	.043	.005
	12	.000	.000	.000	.000	.000	.000	.000	.000	.002	.005	.014	.032	.063	.111	.170	.225	.250	.218	.129	.031
	13	.000	.000	.000	.000	.000	.000	.000	.000	.000	.001	.003	.009	.022	.048	.092	.156	.231	.286	.267	.135
	14	.000	.000	.000	.000	.000	.000	.000	.000	.000	.000	.000	.002	.005	.013	.031	.067	.132	.231	.343	.366
	15	.000	.000	.000	.000	.000	.000	.000	.000	.000	.000	.000	.000	.000	.002	.005	.013	.035	.087	.206	.463
16	0	.851	.440	.185	.074	.028	.010	.003	.001	.000	.000	.000	.000	.000	.000	.000	.000	.000	.000	.000	.000
	1	.138	.371	.329	.210	.113	.053	.023	.009	.003	.001	.000	.000	.000	.000	.000	.000	.000	.000	.000	.000
	2	.010	.146	.275	.277	.211	.134	.073	.035	.015	.006	.002	.001	.000	.000	.000	.000	.000	.000	.000	.000
	3	.000	.036	.142	.229	.246	.208	.146	.089	.047	.022	.009	.003	.001	.000	.000	.000	.000	.000	.000	.000
	4	.000	.006	.051	.131	.200	.225	.204	.155	.101	.057	.028	.011	.004	.001	.000	.000	.000	.000	.000	.000
	5	.000	.001	.014	.056	.120	.180	.210	.201	.162	.112	.067	.034	.014	.005	.001	.000	.000	.000	.000	.000
	6	.000	.000	.003	.018	.055	.110	.165	.198	.198	.168	.122	.075	.039	.017	.006	.001	.000	.000	.000	.000
	7	.000	.000	.000	.005	.020	.052	.101	.152	.189	.197	.175	.132	.084	.044	.019	.006	.001	.000	.000	.000
	8	.000	.000	.000	.001	.006	.020	.049	.092	.142	.181	.196	.181	.142	.092	.049	.020	.006	.001	.000	.000
	9	.000	.000	.000	.000	.001	.006	.019	.044	.084	.132	.175	.197	.189	.152	.101	.052	.020	.005	.000	.000
	10	.000	.000	.000	.000	.000	.001	.006	.017	.039	.075	.122	.168	.198	.201	.165	.110	.055	.018	.003	.000
	11	.000	.000	.000	.000	.000	.000	.001	.005	.014	.034	.067	.112	.162	.201	.210	.180	.120	.056	.014	.001
	12	.000	.000	.000	.000	.000	.000	.000	.001	.004	.011	.028	.057	.101	.155	.204	.225	.200	.131	.051	.006
	13	.000	.000	.000	.000	.000	.000	.000	.000	.001	.003	.009	.022	.047	.089	.146	.208	.246	.229	.142	.036
	14	.000	.000	.000	.000	.000	.000	.000	.000	.000	.001	.002	.006	.015	.035	.073	.134	.211	.277	.275	.146
	15	.000	.000	.000	.000	.000	.000	.000	.000	.000	.000	.000	.001	.003	.009	.023	.053	.113	.210	.329	.371
	16	.000	.000	.000	.000	.000	.000	.000	.000	.000	.000	.000	.000	.000	.001	.003	.010	.028	.074	.185	.440
20	0	.818	.358	.122	.039	.012	.003	.001	.000	.000	.000	.000	.000	.000	.000	.000	.000	.000	.000	.000	.000
	1	.165	.377	.270	.137	.058	.021	.007	.002	.000	.000	.000	.000	.000	.000	.000	.000	.000	.000	.000	.000
	2	.016	.189	.285	.229	.137	.067	.028	.010	.003	.001	.000	.000	.000	.000	.000	.000	.000	.000	.000	.000
	3	.001	.060	.190	.243	.205	.134	.072	.032	.012	.004	.001	.000	.000	.000	.000	.000	.000	.000	.000	.000
	4	.000	.013	.090	.182	.218	.190	.130	.074	.035	.014	.005	.001	.000	.000	.000	.000	.000	.000	.000	.000
	5	.000	.002	.032	.103	.175	.202	.179	.127	.075	.036	.015	.005	.001	.000	.000	.000	.000	.000	.000	.000
	6	.000	.000	.009	.045	.109	.169	.192	.171	.124	.075	.037	.015	.005	.001	.000	.000	.000	.000	.000	.000
	7	.000	.000	.002	.016	.055	.112	.164	.184	.166	.122	.074	.037	.015	.005	.001	.000	.000	.000	.000	.000
	8	.000	.000	.000	.005	.022	.061	.114	.161	.180	.162	.120	.073	.035	.014	.004	.001	.000	.000	.000	.000
	9	.000	.000	.000	.001	.007	.027	.065	.116	.160	.177	.160	.119	.071	.034	.012	.003	.001	.000	.000	.000

Chapter 11

Sample Test Statistics for Tests of Hypotheses

for paired differences d $t = \dfrac{\bar{d} - \mu_d}{s_d/\sqrt{n}}$ with $d.f. = n - 1$

difference of means large sample

$$z = \dfrac{(\bar{x}_1 - \bar{x}_2) - (\mu_1 - \mu_2)}{\sqrt{\dfrac{\sigma_1^2}{n_1} + \dfrac{\sigma_2^2}{n_2}}}$$

difference of means small sample with $\sigma_1 \approx \sigma_2$;
$d.f. = n_1 + n_2 - 2$

$$t = \dfrac{(\bar{x}_1 - \bar{x}_2) - (\mu_1 - \mu_2)}{s\sqrt{\dfrac{1}{n_1} + \dfrac{1}{n_2}}}$$

where $s = \sqrt{\dfrac{(n_1 - 1)s_1^2 + (n_2 - 1)s_2^2}{n_1 + n_2 - 2}}$

difference of proportions

$$z = \dfrac{\hat{p}_1 - \hat{p}_2}{\sqrt{\dfrac{\hat{p}\hat{q}}{n_1} + \dfrac{\hat{p}\hat{q}}{n_2}}} \text{ where } \hat{p} = \dfrac{r_1 + r_2}{n_1 + n_2}; \hat{q} = 1 - \hat{p};$$

$$\hat{p}_1 = r_1/n_1; \hat{p}_2 = r_2/n_2$$

Confidence Intervals

for difference of means (when $n_1 \geq 30$ and $n_2 \geq 30$)

$$(\bar{x}_1 - \bar{x}_2) - z_c\sqrt{\dfrac{\sigma_1^2}{n_1} + \dfrac{\sigma_2^2}{n_2}} < \mu_1 - \mu_2 < (\bar{x}_1 - \bar{x}_2)$$

$$+ z_c\sqrt{\dfrac{\sigma_1^2}{n_1} + \dfrac{\sigma_2^2}{n_2}}$$

for difference of means ($n_1 < 30$ and/or $n_2 < 30$ and $\sigma_1 \approx \sigma_2$)
$d.f. = n_1 + n_2 - 2$

$$(\bar{x}_1 - \bar{x}_2) - t_c s\sqrt{\dfrac{1}{n_1} + \dfrac{1}{n_2}} < \mu_1 - \mu_2 < (\bar{x}_1 - \bar{x}_2)$$

$$\mp t_c s\sqrt{\dfrac{1}{n_1} + \dfrac{1}{n_2}}$$

where $s = \sqrt{\dfrac{(n_1 - 1)s_1^2 + (n_2 - 1)s_2^2}{n_1 + n_2 - 2}}$

for difference of proportions

where $\hat{p}_1 = r_1/n_1; \hat{p}_2 = r_2/n_2; \hat{q}_1 = 1 - \hat{p}_1; \hat{q}_2 = 1 - \hat{p}_2$

$$(\hat{p}_1 - \hat{p}_2) - z_c\sqrt{\dfrac{\hat{p}_1\hat{q}_1}{n_1} + \dfrac{\hat{p}_2\hat{q}_2}{n_2}} < p_1 - p_2 < (\hat{p}_1 - \hat{p}_2)$$

$$+ z_c\sqrt{\dfrac{\hat{p}_1\hat{q}_1}{n_1} + \dfrac{\hat{p}_2\hat{q}_2}{n_2}}$$

Chapter 12

$$\chi^2 = \sum\dfrac{(O - E)^2}{E} \text{ where } E = \dfrac{(\text{row total})(\text{column total})}{\text{sample size}}$$

Tests of Independence $d.f. = (R - 1)(C - 1)$
Goodness of fit $d.f. = (\text{number of entries}) - 1$
Sample test statistic for H_0: $\sigma^2 = k$; $d.f. = n - 1$

$$\chi^2 = \dfrac{(n - 1)s^2}{\sigma^2}$$

Linear Regression

Standard error of estimate $S_e = \sqrt{\dfrac{SS_y - bSS_{xy}}{n - 2}}$

where $b = \dfrac{SS_{xy}}{SS_x}$

Confidence interval for y

$y_p - E < y < y_p + E$ where y_p is the predicted y value
for x and

$$E = t_c S_e\sqrt{1 + \dfrac{1}{n} + \dfrac{(x - \bar{x})^2}{SS_x}} \text{ with } d.f. = n - 2$$

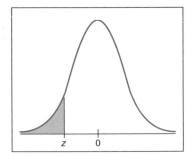

Table entry for z is the area to the left of z.

Areas of a Standard Normal Distribution

(a) Table of Areas to the Left of z									
z	.00	.01	.02	.03	.04	.05	.06	.07	.08	.09
−3.4	.0003	.0003	.0003	.0003	.0003	.0003	.0003	.0003	.0003	.0002
−3.3	.0005	.0005	.0005	.0004	.0004	.0004	.0004	.0004	.0004	.0003
−3.2	.0007	.0007	.0006	.0006	.0006	.0006	.0006	.0005	.0005	.0005
−3.1	.0010	.0009	.0009	.0009	.0008	.0008	.0008	.0008	.0007	.0007
−3.0	.0013	.0013	.0013	.0012	.0012	.0011	.0011	.0011	.0010	.0010
−2.9	.0019	.0018	.0018	.0017	.0016	.0016	.0015	.0015	.0014	.0014
−2.8	.0026	.0025	.0024	.0023	.0023	.0022	.0021	.0021	.0020	.0019
−2.7	.0035	.0034	.0033	.0032	.0031	.0030	.0029	.0028	.0027	.0026
−2.6	.0047	.0045	.0044	.0043	.0041	.0040	.0039	.0038	.0037	.0036
−2.5	.0062	.0060	.0059	.0057	.0055	.0054	.0052	.0051	.0049	.0048
−2.4	.0082	.0080	.0078	.0075	.0073	.0071	.0069	.0068	.0066	.0064
−2.3	.0107	.0104	.0102	.0099	.0096	.0094	.0091	.0089	.0087	.0084
−2.2	.0139	.0136	.0132	.0129	.0125	.0122	.0119	.0116	.0113	.0110
−2.1	.0179	.0174	.0170	.0166	.0162	.0158	.0154	.0150	.0146	.0143
−2.0	.0228	.0222	.0217	.0212	.0207	.0202	.0197	.0192	.0188	.0183
−1.9	.0287	.0281	.0274	.0268	.0262	.0256	.0250	.0244	.0239	.0233
−1.8	.0359	.0351	.0344	.0336	.0329	.0322	.0314	.0307	.0301	.0294
−1.7	.0446	.0436	.0427	.0418	.0409	.0401	.0392	.0384	.0375	.0367
−1.6	.0548	.0537	.0526	.0516	.0505	.0495	.0485	.0475	.0465	.0455
−1.5	.0668	.0655	.0643	.0630	.0618	.0606	.0594	.0582	.0571	.0559
−1.4	.0808	.0793	.0778	.0764	.0749	.0735	.0721	.0708	.0694	.0681
−1.3	.0968	.0951	.0934	.0918	.0901	.0885	.0869	.0853	.0838	.0823
−1.2	.1151	.1131	.1112	.1093	.1075	.1056	.1038	.1020	.1003	.0985
−1.1	.1357	.1335	.1314	.1292	.1271	.1251	.1230	.1210	.1190	.1170
−1.0	.1587	.1562	.1539	.1515	.1492	.1469	.1446	.1423	.1401	.1379
−0.9	.1841	.1814	.1788	.1762	.1736	.1711	.1685	.1660	.1635	.1611
−0.8	.2119	.2090	.2061	.2033	.2005	.1977	.1949	.1922	.1894	.1867
−0.7	.2420	.2389	.2358	.2327	.2296	.2266	.2236	.2206	.2177	.2148
−0.6	.2743	.2709	.2676	.2643	.2611	.2578	.2546	.2514	.2483	.2451
−0.5	.3085	.3050	.3015	.2981	.2946	.2912	.2877	.2843	.2810	.2776
−0.4	.3446	.3409	.3372	.3336	.3300	.3264	.3228	.3192	.3156	.3121
−0.3	.3821	.3783	.3745	.3707	.3669	.3632	.3594	.3557	.3520	.3483
−0.2	.4207	.4168	.4129	.4090	.4052	.4013	.3974	.3936	.3897	.3859
−0.1	.4602	.4562	.4522	.4483	.4443	.4404	.4364	.4325	.4286	.4247
−0.0	.5000	.4960	.4920	.4880	.4840	.4801	.4761	.4721	.4681	.4641

For values of z less than −3.49, use 0.000 to approximate the area.

TABLE 3 *continued*

n	r											p										
		.01	.05	.10	.15	.20	.25	.30	.35	.40	.45	.50	.55	.60	.65	.70	.75	.80	.85	.90	.95
20	10	.000	.000	.000	.000	.002	.010	.031	.069	.117	.159	.176	.159	.117	.069	.031	.010	.002	.000	.000	.000
	11	.000	.000	.000	.000	.000	.003	.012	.034	.071	.119	.160	.177	.160	.116	.065	.027	.007	.001	.000	.000
	12	.000	.000	.000	.000	.000	.001	.004	.014	.035	.073	.120	.162	.180	.161	.114	.061	.022	.005	.000	.000
	13	.000	.000	.000	.000	.000	.000	.001	.005	.015	.037	.074	.122	.166	.184	.164	.112	.055	.016	.002	.000
	14	.000	.000	.000	.000	.000	.000	.000	.001	.005	.015	.037	.075	.124	.171	.192	.169	.109	.045	.009	.000
	15	.000	.000	.000	.000	.000	.000	.000	.000	.001	.005	.015	.036	.075	.127	.179	.202	.175	.103	.032	.002
	16	.000	.000	.000	.000	.000	.000	.000	.000	.000	.001	.005	.014	.035	.074	.130	.190	.218	.182	.090	.013
	17	.000	.000	.000	.000	.000	.000	.000	.000	.000	.000	.001	.004	.012	.032	.072	.134	.205	.243	.190	.060
	18	.000	.000	.000	.000	.000	.000	.000	.000	.000	.000	.000	.001	.003	.010	.028	.067	.137	.229	.285	.189
	19	.000	.000	.000	.000	.000	.000	.000	.000	.000	.000	.000	.000	.000	.002	.007	.021	.058	.137	.270	.377
	20	.000	.000	.000	.000	.000	.000	.000	.000	.000	.000	.000	.000	.000	.000	.001	.003	.012	.039	.122	.358

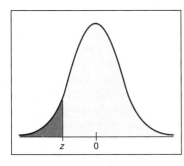

Table entry for *z* is the area to the left of *z*.

TABLE 4 Areas of a Standard Normal Distribution

(a) Table of Areas to the Left of *z*

z	.00	.01	.02	.03	.04	.05	.06	.07	.08	.09
−3.4	.0003	.0003	.0003	.0003	.0003	.0003	.0003	.0003	.0003	.0002
−3.3	.0005	.0005	.0005	.0004	.0004	.0004	.0004	.0004	.0004	.0003
−3.2	.0007	.0007	.0006	.0006	.0006	.0006	.0006	.0005	.0005	.0005
−3.1	.0010	.0009	.0009	.0009	.0008	.0008	.0008	.0008	.0007	.0007
−3.0	.0013	.0013	.0013	.0012	.0012	.0011	.0011	.0011	.0010	.0010
−2.9	.0019	.0018	.0018	.0017	.0016	.0016	.0015	.0015	.0014	.0014
−2.8	.0026	.0025	.0024	.0023	.0023	.0022	.0021	.0021	.0020	.0019
−2.7	.0035	.0034	.0033	.0032	.0031	.0030	.0029	.0028	.0027	.0026
−2.6	.0047	.0045	.0044	.0043	.0041	.0040	.0039	.0038	.0037	.0036
−2.5	.0062	.0060	.0059	.0057	.0055	.0054	.0052	.0051	.0049	.0048
−2.4	.0082	.0080	.0078	.0075	.0073	.0071	.0069	.0068	.0066	.0064
−2.3	.0107	.0104	.0102	.0099	.0096	.0094	.0091	.0089	.0087	.0084
−2.2	.0139	.0136	.0132	.0129	.0125	.0122	.0119	.0116	.0113	.0110
−2.1	.0179	.0174	.0170	.0166	.0162	.0158	.0154	.0150	.0146	.0143
−2.0	.0228	.0222	.0217	.0212	.0207	.0202	.0197	.0192	.0188	.0183
−1.9	.0287	.0281	.0274	.0268	.0262	.0256	.0250	.0244	.0239	.0233
−1.8	.0359	.0351	.0344	.0336	.0329	.0322	.0314	.0307	.0301	.0294
−1.7	.0446	.0436	.0427	.0418	.0409	.0401	.0392	.0384	.0375	.0367
−1.6	.0548	.0537	.0526	.0516	.0505	.0495	.0485	.0475	.0465	.0455
−1.5	.0668	.0655	.0643	.0630	.0618	.0606	.0594	.0582	.0571	.0559
−1.4	.0808	.0793	.0778	.0764	.0749	.0735	.0721	.0708	.0694	.0681
−1.3	.0968	.0951	.0934	.0918	.0901	.0885	.0869	.0853	.0838	.0823
−1.2	.1151	.1131	.1112	.1093	.1075	.1056	.1038	.1020	.1003	.0985
−1.1	.1357	.1335	.1314	.1292	.1271	.1251	.1230	.1210	.1190	.1170
−1.0	.1587	.1562	.1539	.1515	.1492	.1469	.1446	.1423	.1401	.1379
−0.9	.1841	.1814	.1788	.1762	.1736	.1711	.1685	.1660	.1635	.1611
−0.8	.2119	.2090	.2061	.2033	.2005	.1977	.1949	.1922	.1894	.1867
−0.7	.2420	.2389	.2358	.2327	.2296	.2266	.2236	.2206	.2177	.2148
−0.6	.2743	.2709	.2676	.2643	.2611	.2578	.2546	.2514	.2483	.2451
−0.5	.3085	.3050	.3015	.2981	.2946	.2912	.2877	.2843	.2810	.2776
−0.4	.3446	.3409	.3372	.3336	.3300	.3264	.3228	.3192	.3156	.3121
−0.3	.3821	.3783	.3745	.3707	.3669	.3632	.3594	.3557	.3520	.3483
−0.2	.4207	.4168	.4129	.4090	.4052	.4013	.3974	.3936	.3897	.3859
−0.1	.4602	.4562	.4522	.4483	.4443	.4404	.4364	.4325	.4286	.4247
−0.0	.5000	.4960	.4920	.4880	.4840	.4801	.4761	.4721	.4681	.4641

For values of *z* less than −3.49, use 0.000 to approximate the area.

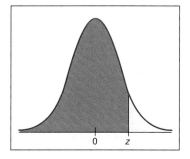

Table entry for z is the area to the left of z.

TABLE 4(a) *continued*

z	.00	.01	.02	.03	.04	.05	.06	.07	.08	.09
0.0	.5000	.5040	.5080	.5120	.5160	.5199	.5239	.5279	.5319	.5359
0.1	.5398	.5438	.5478	.5517	.5557	.5596	.5636	.5675	.5714	.5753
0.2	.5793	.5832	.5871	.5910	.5948	.5987	.6026	.6064	.6103	.6141
0.3	.6179	.6217	.6255	.6293	.6331	.6368	.6406	.6443	.6480	.6517
0.4	.6554	.6591	.6628	.6664	.6700	.6736	.6772	.6808	.6844	.6879
0.5	.6915	.6950	.6985	.7019	.7054	.7088	.7123	.7157	.7190	.7224
0.6	.7257	.7291	.7324	.7357	.7389	.7422	.7454	.7486	.7517	.7549
0.7	.7580	.7611	.7642	.7673	.7704	.7734	.7764	.7794	.7823	.7852
0.8	.7881	.7910	.7939	.7967	.7995	.8023	.8051	.8078	.8106	.8133
0.9	.8159	.8186	.8212	.8238	.8264	.8289	.8315	.8340	.8365	.8389
1.0	.8413	.8438	.8461	.8485	.8508	.8531	.8554	.8577	.8599	.8621
1.1	.8643	.8665	.8686	.8708	.8729	.8749	.8770	.8790	.8810	.8830
1.2	.8849	.8869	.8888	.8907	.8925	.8944	.8962	.8980	.8997	.9015
1.3	.9032	.9049	.9066	.9082	.9099	.9115	.9131	.9147	.9162	.9177
1.4	.9192	.9207	.9222	.9236	.9251	.9265	.9279	.9292	.9306	.9319
1.5	.9332	.9345	.9357	.9370	.9382	.9394	.9406	.9418	.9429	.9441
1.6	.9452	.9463	.9474	.9484	.9495	.9505	.9515	.9525	.9535	.9545
1.7	.9554	.9564	.9573	.9582	.9591	.9599	.9608	.9616	.9625	.9633
1.8	.9641	.9649	.9656	.9664	.9671	.9678	.9686	.9693	.9699	.9706
1.9	.9713	.9719	.9726	.9732	.9738	.9744	.9750	.9756	.9761	.9767
2.0	.9772	.9778	.9783	.9788	.9793	.9798	.9803	.9808	.9812	.9817
2.1	.9821	.9826	.9830	.9834	.9838	.9842	.9846	.9850	.9854	.9857
2.2	.9861	.9864	.9868	.9871	.9875	.9878	.9881	.9884	.9887	.9890
2.3	.9893	.9896	.9898	.9901	.9904	.9906	.9909	.9911	.9913	.9916
2.4	.9918	.9920	.9922	.9925	.9927	.9929	.9931	.9932	.9934	.9936
2.5	.9938	.9940	.9941	.9943	.9945	.9946	.9948	.9949	.9951	.9952
2.6	.9953	.9955	.9956	.9957	.9959	.9960	.9961	.9962	.9963	.9964
2.7	.9965	.9966	.9967	.9968	.9969	.9970	.9971	.9972	.9973	.9974
2.8	.9974	.9975	.9976	.9977	.9977	.9978	.9979	.9979	.9980	.9981
2.9	.9981	.9982	.9982	.9983	.9984	.9984	.9985	.9985	.9986	.9986
3.0	.9987	.9987	.9987	.9988	.9988	.9989	.9989	.9989	.9990	.9990
3.1	.9990	.9991	.9991	.9991	.9992	.9992	.9992	.9992	.9993	.9993
3.2	.9993	.9993	.9994	.9994	.9994	.9994	.9994	.9995	.9995	.9995
3.3	.9995	.9995	.9995	.9996	.9996	.9996	.9996	.9996	.9996	.9997
3.4	.9997	.9997	.9997	.9997	.9997	.9997	.9997	.9997	.9997	.9998

For z values greater than 3.49, use 1.000 to approximate the area.

TABLE 4 *continued*

(b) Confidence Interval Critical Values z_c

Level of Confidence c	Critical Value z_c
0.75, or 75%	1.15
0.80, or 80%	1.28
0.85, or 85%	1.44
0.90, or 90%	1.645
0.95, or 95%	1.96
0.98, or 98%	2.33
0.99, or 99%	2.58

TABLE 4 *continued*

(c) Hypothesis Testing, Critical Values z_0

Level of Significance	$\alpha = 0.05$	$\alpha = 0.01$
Critical value z_0 for a left-tailed test	−1.645	−2.33
Critical value z_0 for a right-tailed test	1.645	2.33
Critical values $\pm z_0$ for a two-tailed test	±1.96	±2.58

TABLE 5   Student's *t* Distribution

		Student's *t* values generated by Minitab Version 9.2						
	c	0.750	0.800	0.850	0.900	0.950	0.980	0.990
	α'	0.125	0.100	0.075	0.050	0.025	0.010	0.005
d.f.	α''	0.250	0.200	0.150	0.100	0.050	0.020	0.010
1		2.414	3.078	4.165	6.314	12.706	31.821	63.657
2		1.604	1.886	2.282	2.920	4.303	6.965	9.925
3		1.423	1.638	1.924	2.353	3.182	4.541	5.841
4		1.344	1.533	1.778	2.132	2.776	3.747	4.604
5		1.301	1.476	1.699	2.015	2.571	3.365	4.032
6		1.273	1.440	1.650	1.943	2.447	3.143	3.707
7		1.254	1.415	1.617	1.895	2.365	2.998	3.499
8		1.240	1.397	1.592	1.860	2.306	2.896	3.355
9		1.230	1.383	1.574	1.833	2.262	2.821	3.250
10		1.221	1.372	1.559	1.812	2.228	2.764	3.169
11		1.214	1.363	1.548	1.796	2.201	2.718	3.106
12		1.209	1.356	1.538	1.782	2.179	2.681	3.055
13		1.204	1.350	1.530	1.771	2.160	2.650	3.012
14		1.200	1.345	1.523	1.761	2.145	2.624	2.977
15		1.197	1.341	1.517	1.753	2.131	2.602	2.947
16		1.194	1.337	1.512	1.746	2.120	2.583	2.921
17		1.191	1.333	1.508	1.740	2.110	2.567	2.898
18		1.189	1.330	1.504	1.734	2.101	2.552	2.878
19		1.187	1.328	1.500	1.729	2.093	2.539	2.861
20		1.185	1.325	1.497	1.725	2.086	2.528	2.845
21		1.183	1.323	1.494	1.721	2.080	2.518	2.831
22		1.182	1.321	1.492	1.717	2.074	2.508	2.819
23		1.180	1.319	1.489	1.714	2.069	2.500	2.807
24		1.179	1.318	1.487	1.711	2.064	2.492	2.797
25		1.178	1.316	1.485	1.708	2.060	2.485	2.787
26		1.177	1.315	1.483	1.706	2.056	2.479	2.779
27		1.176	1.314	1.482	1.703	2.052	2.473	2.771
28		1.175	1.313	1.480	1.701	2.048	2.467	2.763
29		1.174	1.311	1.479	1.699	2.045	2.462	2.756
30		1.173	1.310	1.477	1.697	2.042	2.457	2.750
35		1.170	1.306	1.472	1.690	2.030	2.438	2.724
40		1.167	1.303	1.468	1.684	2.021	2.423	2.704
45		1.165	1.301	1.465	1.679	2.014	2.412	2.690
50		1.164	1.299	1.462	1.676	2.009	2.403	2.678
55		1.163	1.297	1.460	1.673	2.004	2.396	2.668
60		1.162	1.296	1.458	1.671	2.000	2.390	2.660
90		1.158	1.291	1.452	1.662	1.987	2.369	2.632
120		1.156	1.289	1.449	1.658	1.980	2.358	2.617
∞		1.150	1.282	1.440	1.645	1.960	2.326	2.58

c is a confidence level:

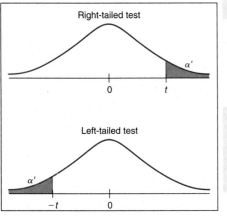

α' is the level of significance for a one-tailed test:

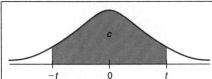

Right-tailed test

Left-tailed test

α'' is the level of significance for a two-tailed test:

(c) Tonnes of Wheat—Histogram

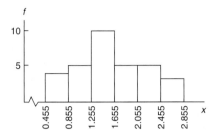

13. (a) One. (b) 5/51 or 9.8%. (c) Interval from 650 to 750.
15. Dotplot for Percentage of Children in Neighborhood

Percentage

Section 2.3

1. (a) Longevity of Cowboys

4		7 = 47 years
4		7
5		2 7 8 8
6		1 6 6 8 8
7		0 2 2 3 3 5 6 7
8		4 4 4 5 6 6 7 9
9		0 1 1 2 3 7

(b) Yes, certainly these cowboys lived long lives.
3. Average Length of Hospital Stay

5		2 = 5.2 days
5		2 3 5 5 6 7
6		0 2 4 6 6 7 7 8 8 8 8 9 9
7		0 0 0 0 0 1 1 1 2 2 2 3 3 3 3 4 4 5 5 6 6 8
8		4 5 7
9		4 6 9
10		0 3
11		1

The distribution is skewed right.

5. (a) Minutes Beyond 2 Hours (1961–1980)

0		9 = 9 minutes past 2 hours
0˙		9 9
1*		0 0 2 3 3 4
1˙		5 5 6 6 7 8 8 9
2*		0 2 3 3

(b) Minutes Beyond 2 Hours (1981–2000)

0		7 = 7 minutes past 2 hours
0˙		7 7 7 8 8 8 8 9 9 9 9 9 9 9 9 9
1*		0 0 1 1 4

(c) In more recent times, the winning times have been closer to 2 hours, with all the times between 7 and 14 minutes over two hours. In the earlier period, more than half the times were more than 2 hours and 14 minutes.
7. Milligrams of Tar per Cigarette

1		0 = 1.0 mg tar		
1		0	11	4
2			12	0 4 8
3			13	7
4		1 5	14	1 5 9
5			15	0 1 2 8
6			16	0 6
7		3 8	17	0
8		0 6 8		
9		0		
10			29	8

9. Milligrams of Nicotine per Cigarette

0		1 = 0.1 milligram
0*		1 4 4
0˙		5 6 6 6 7 7 7 8 8 9 9 9
1*		0 0 0 0 0 0 1 2
1˙		
2*		0

11. (a) $49,000 to $126,000 for California; $45,000 to
$120,000 for New York.
(b) New York; California.
(c) California has slightly higher average salaries.

Chapter 2 Review

1. Figure 2-1(a) is essentially a bar graph, but the bar
lengths do not have the same scales. Figure 2-1(b) is a
time plot. This figure gives clear, undistorted representation of fuel economy standards by year.
3. Problems with Tax Returns

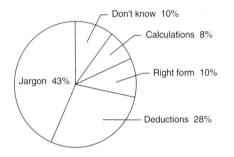

Don't know 10%
Calculations 8%
Right form 10%
Jargon 43%
Deductions 28%

5. (a) Class width = 6.

Class Limits	Boundaries	Midpoint	Frequency	Relative Frequency
59–64	58.5–64.5	61.5	1	0.02
65–70	64.5–70.5	67.5	7	0.13
71–76	70.5–76.5	73.5	6	0.11
77–82	76.5–82.5	79.5	13	0.25
83–88	82.5–88.5	85.5	18	0.34
89–94	88.5–94.5	91.5	7	0.13
95–100	94.5–100.5	97.5	1	0.02

(b, c) Glucose Level—Histogram and Relative-
Frequency Histogram

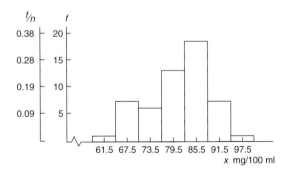

7. (a) 1240s had 40 data.   (b) 75.   (c) From 1203 to
1212. Little if any repairs or new construction.

CHAPTER 3

Section 3.1

1. Mean = 156.33; median = 157; mode = 157. A gardener in Colorado should look at seed and plant descriptions to determine if the plant can thrive and mature in
the designated number of frost-free days. The mean, median, and mode are all close. About half the locations
have 157 or fewer frost-free days.
3. $\bar{x} \approx 167.3$ °F; median = 171 °F; mode = 178 °F.
5. (a) $\bar{x} \approx 3.27$; median = 3; mode = 3.   (b) $\bar{x} \approx 4.21$;
median = 2; mode = 1.   (c) Lower canyon mean is
greater; median and mode are less.   (d) Trimmed mean
= 3.75 and is closer to upper canyon mean.
7. (a) Mean = 28.83 thousand dollars.   (b) Median =
18.5 thousand dollars. The median best describes the
salary of the majority of employees, since the mean is influenced by the high salaries of the president and vice
president.   (c) Mean = 17.3 thousand dollars; median
= 17 thousand dollars.   (d) Without the salaries for the
two executives, the mean and the median are closer, and
both reflect the salary of most of the other workers more
accurately. The mean changed quite a bit, while the median did not, a difference that indicates that the mean is
more sensitive to the absence or presence of extreme
values.
9. (a) Mode.   (b) Mean, median, mode if it exists.
(c) Mode if it exists; median, mean if 24-hour clock is
used.
11. (a) If the largest data value is *replaced* by a larger value,
the mean will increase because the sum of the data values
will increase, but the number of them will remain the
same. The median will not change. The same value will
still be in the eighth position when the data are ordered.
(b) If the largest value is replaced by a value that is
smaller (but still higher than the median), the mean will
decrease because the sum of the data values will decrease. The median will not change. The same value will
be in the eighth position in increasing order.   (c) If the
largest value is replaced by a value that is smaller than
the median, the mean will decrease because the sum of
the data values will decrease. The median also will decrease because the former value in the eighth position
will move to the ninth position in increasing order. The
median will be the new value in the eighth position.

Section 3.2

1. (a) Range = 54 deer/km^2; sample mean $\bar{x}$ = 20.9 deer/km^2; sample variance s^2 = 225.0; sample standard deviation s = 15.0 deer/km^2. (b) CV = 71.8%. Since the standard deviation is about 72% of the mean, there is considerable variation in the distribution of deer from one part of the park to another.

3. (a) Range = 77.6; $\bar{x}$ = 55.7. (b) $s^2 \approx$ 537; $s \approx$ 23.17. (c) $CV \approx$ 41.6%. Larger than for geese and so greater relative standard deviation.

5. (a) Pax $CV \approx$ 98.9%; Vanguard $CV \approx$ 222.8%; Pax seems less risky. (b) Pax: -11.43% to 34.81%; Vanguard -19.39% to 30.61%; The performance range for Pax seems better than that for Vanguard (based on these historical data).

7. (a) Range = 737; $\bar{x} \approx$ 566.9. (b) $s^2 \approx$ 71,202; $s \approx$ 266.8. (c) $CV \approx$ 47.1%. (d) 33 to 1100.

9. (a) Use calculator. (b) Wal-Mart $CV \approx$ 2%; Disney $CV \approx$ 3%; yes. (c) Wal-Mart 48.85 to 55.21; Disney 29.29 to 35.17.

11. Midpoints: 25.5, 35.5, 45.5; $\bar{x} \approx$ 35.80; $s^2 \approx$ 61.1; $s \approx$ 7.82.

13. Midpoints: 10.55, 14.55, 18.55, 22.55, 26.55; $\bar{x} \approx$ 15.6; $s^2 \approx$ 23.4; $s \approx$ 4.8.

15. Midpoints and frequencies are shown on the figure. $\bar{x}$ = 7.9 hours; s = 1.05 hours; CV = 13.29%.

Section 3.3

1. 82% or more of the scores were at or below Angela's score; 18% or fewer of the scores were above Angela's score.

3. No, the score 82 might have a percentile rank less than 70.

5. Low = 2; Nurses' Length of Employment (months)
Q_1 = 9.5;
median = 23;
Q_3 = 28.5;
high = 42;
IQR = 19.

7. (a) Low = 17; Bachelor's Degree Percentage by State
Q_1 = 22;
median = 24;
Q_3 = 27;
high = 38;
IQR = 5.
(b) 3rd quartile, since it is between the median and Q_3.

9. (a) California has the lowest premium; Pennsylvania has the highest. (b) Pennsylvania has the highest median premium. (c) California has the smallest range. Texas has the smallest interquartile range. (d) Part (a) is the five-number summary for Texas. It has the smallest IQR. Part (b) is the five-number summary for Pennsylvania. It has the largest minimum. Part (c) is the five-number summary for California. It has the lowest minimum.

11. (a) Smallest median, assistant; highest increase, associate.
(b) Instructor.
(c) Assistant.
(d) Professor; yes.
(e) Associate, 9.16; instructor, 10.23; yes, the points indicated by asterisks are above the upper limit for each rank.

Chapter 3 Review

1. (a) $\bar{x}$ = 109.5; s = 31.7; CV = 28.9%; range = 69.
(b) $\bar{x}$ = 110.125; s = 7.2; CV = 6.5%; range = 20.
(c) The first distribution is more spread than the second.

3. (a) Low = 31; Percentage of Democratic Vote by County
Q_1 = 40;
median = 45;
Q_3 = 52.5;
high = 68;
IQR = 12.5.

(b) Class width = 8.

Class	Midpoint	f
31–38	34.5	11
39–46	42.5	24
47–54	50.5	15
55–62	58.5	7
63–70	66.5	3

$\bar{x} \approx 46.1$; $s \approx 8.64$; 28.82 to 63.38.
(c) $\bar{x} \approx 46.15$; $s \approx 8.63$.
5. (a) $\bar{x} = 7.83$ lb; $s = 2.32$ lb; $CV = 29.6\%$; range = 4.8 lb.   (b) $\bar{x} = 9.95$ lb; $s = 0.29$ lb; $CV = 2.9\%$; range = 0.7 lb.   (c) Second line has more consistent performance as reflected by the smaller standard deviation, CV, and range.
7. Mean weight = 156.25 lb.

CHAPTER 4

Section 4.1

1. Moderate or low linear correlation.
3. High linear correlation.
5. High linear correlation.
7. (a) Ages and Average Weights of Shetland Ponies

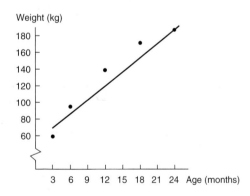

(b) Draw a line with a positive slope that you think best fits the data.   (c) High.

9. (a) Change in Wages and in Consumer Prices in Various Countries (%)

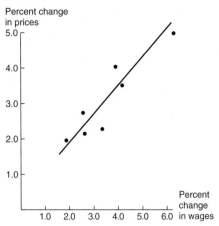

(b) Draw a line with a positive slope that you think best fits the data. In Section 4.2, you will learn how to get the equation for the line.   (c) The linear correlation appears to be moderate.

11. (a) Body Diameter and Weight of Prehistoric Pottery

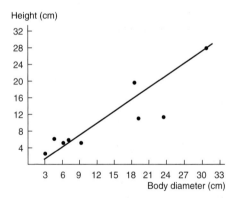

(b) Draw a line with a positive slope that you think best fits the data.   (c) Moderate.

3. (a) Outcomes for Drawing Two Balls (without replacement)

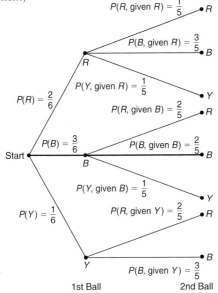

1st Ball   2nd Ball

(b) $P(R \text{ and } R) = 2/6 \cdot 1/5 = 1/15$.
$P(R \text{ 1st and } B \text{ 2nd}) = 2/6 \cdot 3/5 = 1/5$.
$P(R \text{ 1st and } Y \text{ 2nd}) = 2/6 \cdot 1/5 = 1/15$.
$P(B \text{ 1st and } R \text{ 2nd}) = 3/6 \cdot 2/5 = 1/5$.
$P(B \text{ 1st and } B \text{ 2nd}) = 3/6 \cdot 2/5 = 1/5$.
$P(B \text{ 1st and } Y \text{ 2nd}) = 3/6 \cdot 1/5 = 1/10$.
$P(Y \text{ 1st and } R \text{ 2nd}) = 1/6 \cdot 2/5 = 1/15$.
$P(Y \text{ 1st and } B \text{ 2nd}) = 1/6 \cdot 3/5 = 1/10$.
5. $4 \cdot 3 \cdot 2 \cdot 1 = 24$ ways.
7. (a) $52 \cdot 52 = 2704$ (b) $4 \cdot 4 = 16$.
(c) $16/2704 = 0.006$.
9. $P_{5,2} = (5!/3!) = 5 \cdot 4 = 20$.
11. $P_{7,7} = (7!/0!) = 7! = 5040$.
13. $C_{5,2} = (5!/(2!3!)) = 10$.
15. $C_{7,7} = (7!/(7!0!)) = 1$.
17. $P_{15,3} = 2730$.
19. $C_{15,5} = (15!/(5!10!)) = 3003$.
21. (a) $C_{12,6} = (12!/(6!6!)) = 924$.
(b) $C_{7,6} = (7!/(6!1!)) = 7$.   (c) $7/924 = 0.008$.

Chapter 5 Review

1. $P(\text{asked}) = 24\%$; $P(\text{received, given asked}) = 45\%$;
$P(\text{ask and receive}) = (0.24)(0.45) = 10.8\%$.
3. (a) If the first card is replaced, the outcomes are independent. Replacing the first card restores the original sample space. If the first card is not replaced, the outcomes are not independent, because removing the first card changes the sample space.
(b) $P(\text{heart and heart}) = (13/52)(13/52) = 0.063$.
(c) $P(\text{heart and heart}) = (13/52)(12/51) = 0.059$.

5. (a) Drop a fixed number of tacks and count how many land flat side down. Then form the ratio of the number landing flat side down to the total number dropped.
(b) Up, down.   (c) $P(\text{up}) = 160/500 = 0.32$; $P(\text{down}) = 340/500 = 0.68$.
7. (a)

Outcomes x	2	3	4	5	6
$P(x)$	0.028	0.056	0.083	0.111	0.139

(b)

x	7	8	9	10	11	12
$P(x)$	0.167	0.139	0.111	0.083	0.056	0.028

9. $C_{8,2} = (8!/(2!6!)) = (8 \cdot 7/2) = 28$.
11. $3 \cdot 2 \cdot 1 = 6$.
13. $4 \cdot 4 \cdot 4 \cdot 4 \cdot 4 = 1024$ choices; $P(\text{all correct}) = 1/1024 = 0.00098$.

CHAPTER 6

Section 6.1

1. (a) Discrete.   (b) Continuous.   (c) Continuous.
(d) Discrete.   (e) Continuous.
3. (a) Yes.   (b) No; probabilities total to more than 1.
5. (a) Yes, events are distinct and probabilities total 1.
(b) Income Distribution ($1000).

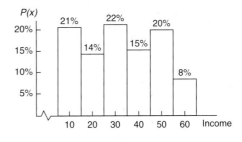

(c) 32.3 thousand dollars.   (d) 16.12 thousand dollars.
7. (a) Number of Fish Caught in a 6-Hour Period at Pyramid Lake, Nevada

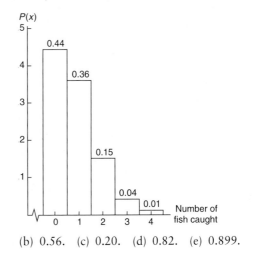

(b) 0.56.   (c) 0.20.   (d) 0.82.   (e) 0.899.

9. (a) 15/719; 704/719.   (b) $0.73; $14.27.
11. (a) 0.01191; $595.50.   (b) $646; $698; $751.50;
$806.50; $3497.50 total.   (c) $4197.50.
(d) $1502.50.

Section 6.2

1. A trial is one flip of a fair quarter. Success = coin shows
heads. Failure = coin shows tails. $n = 3$; $p = 0.5$; $q = 0.5$.   (a) $P(r = 3 \text{ heads}) = C_{3,3}p^3q^0 = 1(0.5)^3(0.5)^0 = 0.125$. To find this value in Table 3 of the Appendix, use
the group in which $n = 3$, the column headed by $p = 0.5$, and the row headed by $r = 3$.   (b) $P(r = 2 \text{ heads}) = C_{3,2}p^2q^1 = 3(0.5)^2(0.5)^1 = 0.375$. To find this
value in Table 3 of the Appendix, use the group in which
$n = 3$, the column headed by $p = 0.5$, and the row
headed by $r = 2$.   (c) $P(r \text{ is 2 or more}) = P(r = 2 \text{ heads}) + P(r = 3 \text{ heads}) = 0.375 + 0.125 = 0.500$.
(d) The probability of getting three tails when you toss a
coin three times is the same as getting no or zero heads.
Therefore, $P(3 \text{ tails}) = P(r = 0 \text{ heads}) = C_{3,0}p^0q^3 = 1(0.5)^0(0.5)^3 = 0.125$. To find this value in Table 3 of the
Appendix, use the group in which $n = 3$, the column
headed by $p = 0.5$, and the row headed by $r = 0$.
3. (a) A trial is a man's response to the question,
"Would you marry the same woman again?" Success = a
positive response. Failure = a negative response. $n = 10$;
$p = 0.80$; $q = 0.20$. Using values in Table 3 of the
Appendix, $P(r \text{ is at least } 7) = P(r = 7) + P(r = 8) + P(r = 9) + P(r = 10) = 0.201 + 0.302 + 0.268 + 0.107 = 0.878$. $P(r \text{ is less than half of } 10) = P(r < 5) = P(r = 0) + P(r = 1) + P(r = 2) + P(r = 3) + P(r = 4) = 0.000 + 0.000 + 0.000 + 0.001 + 0.006 = 0.007$.   (b) A trial
is a woman's response to the question, "Would you
marry the same man again?" Success = a positive response. Failure = a negative response. $n = 10$; $p = 0.5$;
$q = 0.5$. Using values in Table 3 of the Appendix, $P(r \text{ is}$
at least $7) = P(r = 7) + P(r = 8) + P(r = 9) + P(r = 10) = 0.117 + 0.044 + 0.010 + 0.001 = 0.172$. $P(r \text{ is less}$
than half of $10) = P(r < 5) = P(r = 0) + P(r = 1) + P(r = 2) + P(r = 3) + P(r = 4) = 0.001 + 0.010 + 0.044 + 0.117 + 0.205 = 0.377$.
5. A trial consists of a woman's response regarding her
mother-in-law. Success = dislike. Failure = like. $n = 6$;
$p = 0.90$; $q = 0.10$.   (a) $P(r = 6) = 0.531$.   (b) $P(r = 0) = 0.000$ (to 3 digits).   (c) $P(r \geq 4) = P(r = 4) + P(r = 5) + P(r = 6) = 0.098 + 0.354 + 0.531 = 0.983$.
(d) $P(r \leq 3) = 1 - P(r \geq 4) \approx 1 - 0.983 = 0.017$ or
0.016 directly from table.

7. A trial is taking a polygraph exam. Success = pass. Fail-
ure = fail. $n = 9$; $p = 0.85$; $q = 0.15$.   (a) $P(r = 9) = 0.232$.   (b) $P(r \geq 5) = P(r = 5) + P(r = 6) + P(r = 7) + P(r = 8) + P(r = 9) = 0.028 + 0.107 + 0.260 + 0.368 + 0.232 = 0.995$.   (c) $P(r \leq 4) = 1 - P(r \geq 5) \approx 1 - 0.995 = 0.005$ or 0.006 directly from table.
(d) $P(r = 0) = 0.000$ (to 3 digits).
9. A trial is catching and releasing a pike. Success = pike
dies. Failure = pike lives. $n = 16$; $p = 0.05$; $q = 0.95$.
(a) $P(r = 0) = 0.440$.   (b) $P(r < 3) = 0.957$.
(c) $P(r = 0) = 0.440$ (all live is equivalent to none die).
(d) Change success to live; $p = 0.95$; $P(r > 14) = 0.811$.
11. (a) A trial consists of using the Myers-Briggs instrument
to determine if a person in marketing is an extrovert.
Success = extrovert. Failure = not extrovert. $n = 15$;
$p = 0.75$; $q = 0.25$. $P(r \geq 10) = 0.851$; $P(r \geq 5) = 0.999$; $P(r = 15) = 0.013$.   (b) A trial consists of using
the Myers-Briggs instrument to determine if a computer
programmer is an introvert. Success = introvert.
Failure = not introvert. $n = 5$; $p = 0.60$; $q = 0.40$.
$P(r = 0) = 0.010$; $P(r \geq 3) = 0.683$; $P(r = 5) = 0.078$.
13. (a) $n = 10$; $p = 0.40$; $P(r = 0) = 0.006$.   (b) $n = 10$;
$p = 0.40$; $P(r < 5) = 0.633$.   (c) $n = 10$; $p = 0.30$;
$P(r \leq 2) = 0.382$.   (d) $n = 10$; $p = 0.70$; $P(r \geq 6) = 0.849$.
15. $n = 8$; $p = 0.53$; $q = 0.47$.   (a) 0.812515; yes,
truncated at 5 digits.   (b) 0.187486; 0.18749; yes,
rounded to 5 digits.
17. (a) They are the same.   (b) They are the same.
(c) $r = 1$.   (d) The one headed by $p = 0.80$.

Section 6.3

1. (a) Binomial Distribution
The distribution is symmetrical.

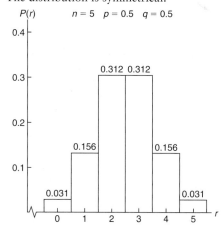

(b) Binomial Distribution
The distribution is skewed right.

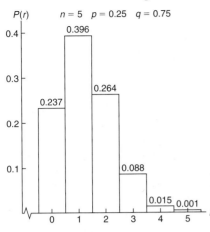

(c) Binomial Distribution
The distribution is skewed left.

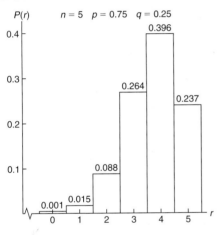

(d) The distributions are mirror images of one another.
(e) The distribution would be skewed left for $p = 0.73$ because the more likely number of successes are to the right of the middle.

3. (a) Households with Children Under 2 That Buy Film

(b) Households with No Children Under 21 That Buy Film

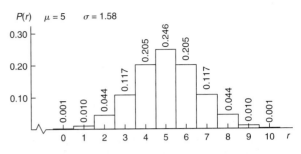

(c) Yes. Adults with children seem to buy more film.

5. (a) Binomial Distribution for Number of Gullible Consumers

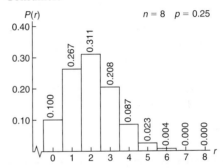

(b) $\mu = 2$; $\sigma = 1.225$.

7. (a) $P(r = 7 \text{ guilty in Japan}) = 0.698$; $P(r = 7 \text{ guilty in U.S.}) = 0.028$. (b) For guilty in Japan, $\mu = 6.65$, $\sigma \approx 0.58$; For guilty in U.S., $\mu = 4.2$; $\sigma \approx 1.30$.

9. (a) Yes, symmetric; $\mu = 5.17$; yes. (b) Yes; yes; right; $\mu = 3.96$; left. (c) 0.64; yes; left; $\mu = 7.04$; right. (d) They are the same, respectively. (e) $\sigma = 1.59$; yes, because the values for p and q are exchanged.

Chapter 6 Review

1. (a) 38; 11.6
(b) Leases in Months

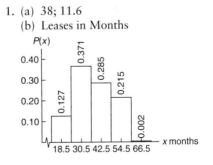

3. (a) Claimants Under 25

(b) $P(r \geq 6) = 0.504$.   (c) $\mu = 5.5$; $\sigma = 1.57$.
5. (a) 0.039.   (b) 0.403.   (c) 8.
7. (a) Number of Good Grapefruit

(b) 0.244, 0.999.   (c) 7.5.   (d) 1.37.
9. $P(r \leq 2) = 0.000$ (to 3 digits). The data seem to indicate that the percent favoring the increase in fees is less than 85%.

CHAPTER 7

Section 7.1

1. (a) No, it's skewed.   (b) No, it crosses the horizontal axis.   (c) No, it has three peaks.   (d) No, the curve is not smooth.
3. Figure 7-16 has the larger standard deviation. The mean of Figure 7-16 is $\mu = 10$. The mean of Figure 7-17 is $\mu = 4$.
5. (a) 50%.   (b) 68%.   (c) 99.7%.
7. (a) 50%.   (b) 50%.   (c) 68%.   (d) 95%.
9. (a) From 1207 to 1279.   (b) From 1171 to 1315.   (c) From 1135 to 1351.
11. (a) From 1.70 mA to 4.60 mA.   (b) From 0.25 mA to 6.05 mA.

13. (a) Tri-County Bank Monthly Loan Request—First Year (thousands of dollars)

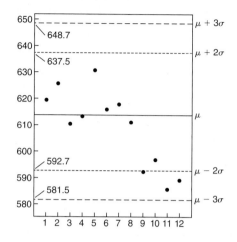

The process is out of control with type III warning signal, since 2 of 3 consecutive points are more than 2 standard deviations below the mean. The trend is down.
(b) Tri-County Bank Monthly Loan Requests—Second Year (thousands of dollars).

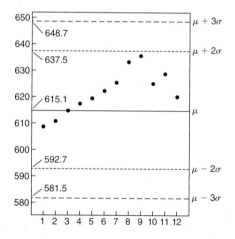

The process shows warning signal II, a run of nine consecutive points above the mean. It would say the economy is heating up.

15. Visibility Standard Index

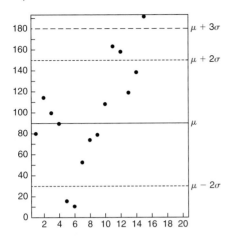

There is one point above 3σ. Thus control signal I indicates "out of control." Control signal III is present. There are two consecutive points below $\mu - 2\sigma$ and two consecutive points above $\mu + 2\sigma$. The out-of-control signals that cause the most concern are those above the mean. Special pollution regulations may be appropriate for those periods.

Section 7.2

1. (a) Robert, Jan, and Linda each scored above the mean. (b) Joel scored on the mean. (c) Susan and John scored below the mean. (d) Robert, 172; Jan, 184; Susan, 110; Joel, 150; John, 134; Linda, 182.
3. (a) $-4.00 < z < 4.00$. (b) $z < -1.6$. (c) $1.00 < z$. (d) $81.75°F < x$. (e) $x < 63.5°F$. (f) $64°F < x < 81.25°F$.
5. (a) $-1.77 < z$. (b) $z < 1.61$. (c) $-1.45 < z < 1.45$. (d) $3706 < x < 5907$. (e) $x < 5615$. (f) $6000 < x$. (g) A population of 2800 deer corresponds to a z value of -2.58. Data values this far below the mean occur less than 2.5% of the time. This would be an unusually low number. The population 6300 corresponds to a z value of 3.06. Fall deer populations are practically never that large. Such a population would be considered an unusually high population.
7. (a) $-1.00 < z$. (b) $z < -2.00$. (c) $-2.67 < z < 2.33$. (d) $x < 4.4$. (e) $5.2 < x$. (f) $4.1 < x < 4.5$. (g) A red blood cell count of 5.9 or higher corresponds to a standard z score of 3.67. Practically no data values occur this far above the mean. Such a count would be considered unusually high for a healthy female.
9. 0.5000. 11. 0.0934. 13. 0.6736. 15. 0.0643.
17. 0.8888. 19. 0.4993. 21. 0.4778. 23. 0.8953.
25. 0.3471. 27. 0.0306. 29. 0.5000. 31. 0.4483.
33. 0.8849. 35. 0.0885. 37. 0.8849. 39. 0.8808.
41. 0.3226. 43. 0.4474. 45. 0.2939. 47. 0.6704.

Section 7.3

1. $P(3 \le x \le 6) = P(-0.50 \le z \le 1.00) = 0.5328$.
3. $P(50 \le x \le 70) = P(0.67 \le z \le 2.00) = 0.2286$.
5. $P(8 \le x \le 12) = P(-2.19 \le z \le -0.94) = 0.1593$.
7. $P(x \ge 30) = P(z \ge 2.94) = 0.0016$.
9. $P(x \ge 90) = P(z \ge -0.67) = 0.7486$.
11. -1.555. 13. 0.13. 15. 1.41. 17. -0.92.
19. ±2.33.
21. (a) $P(x > 60) = P(z > -1) = 0.8413$. (b) $P(x < 110) = P(z < 1) = 0.8413$. (c) $P(60 \le x \le 110) = P(-1.00 \le z \le 1.00) = 2(0.3413) = 0.6826$. (d) $P(x > 140) = P(z > 2.20) = 0.0139$.
23. (a) $P(x > 675) = P(z > 1.75) = 0.0401$. (b) $P(x < 450) = P(z < -0.50) = 0.3085$. (c) $P(450 \le x \le 675) = P(-0.50 \le z \le 1.75) = 0.6514$. (d) $P(x > 28) = P(z > 1.67) = 0.0475$. (e) $P(x > 12) = P(z > -1.00) = 0.8413$. (f) $P(12 \le x \le 28) = P(-1.00 \le z \le 1.67) = 0.7938$.
25. (a) $P(x < 36 \text{ mo}) = P(z < -1.13) = 0.1292$. They will replace 13% of their batteries. (b) $P(z < z_0) = 10\%$ for $z_0 = -1.28$; $x = -1.28(8) + 45 = 34.76$. Guarantee the batteries for 35 months.
27. (a) According to the empirical rule, about 95% of the data lies between $\mu - 2\sigma$ and $\mu + 2\sigma$. Since this interval is 4σ wide, we have $4\sigma \approx 6$ years, so $\sigma = 1.5$ years. (b) $P(x > 5) = P(z > -2.00) = 0.9772$. (c) $P(x < 10) = P(z < 1.33) = 0.9082$. (d) $P(z < z_0) = 0.10$ for $z_0 = -1.28$; $x = -1.28(1.5) + 8 = 6.08$ years. Guarantee the TVs for about 6.1 years.
29. (a) $P(z \ge z_0) = 0.99$ for $z_0 = -2.33$; $x = -2.33(3.7) + 90 \approx 81.38$ mo. Guarantee the microchips for 81 months. (b) $P(x \le 84) = P(z \le -1.62) = 0.0526$. (c) Expected loss $= (50,000,000)(0.0526) = \$2,630,000$. (d) Profit $= \$370,000$.

Section 7.4

Note: Answers may differ slightly depending on how many digits are carried in the computation of the standard deviation and the computation of z.

1. (a) $P(r \ge 50) = P(x \ge 49.5) = P(z \ge -27.53) \approx 1$ or almost certain. (b) $P(r \ge 50) = P(x \ge 49.5) = P(z \ge 7.78) \approx 0$ or almost impossible for a random sample.
3. (a) $P(r \ge 15) = P(x \ge 14.5) = P(z \ge -1.61) = 0.9463$. (b) $P(r \ge 28) = P(x \ge 27.5) = P(z \ge 1.49) = 0.0681$. (c) $P(15 \le r \le 28) = P(14.5 \le x \le 28.5) = P(-1.61 \le z \le 1.73) = 0.9045$. (d) Since both np and nq are larger than 5, the normal approximation is appropriate.
5. (a) $P(r \ge 15) = P(x \ge 14.5) = P(z \ge -2.35) = 0.9906$. (b) $P(r \ge 30) = P(x \ge 29.5) = P(z \ge 0.62) = 0.2676$. (c) $P(25 \le r \le 35) + P(24.5 \le x \le 35.5) = P(-0.37 \le z \le 1.81) = 0.6092$. (d) $P(r > 40) = P(r \ge 41) = P(x \ge 40.5) = P(z \ge 2.80) = 0.0026$.

7. (a) $P(r \geq 47) = P(x \geq 46.5) = P(z \geq -1.94) = 0.9738$.
(b) $P(r \leq 58) = P(x \leq 58.5) = P(z \leq 1.75) = 0.9599$. In parts (c) and (d), let r be the number of products that succeed, and use $p = 1 - 0.80 = 0.20$. (c) $P(r \geq 15) = P(x \geq 14.5) = P(z \geq 0.40) = 0.3446$. (d) $P(r < 10) = P(r \leq 9) = P(x \leq 9.5) = P(z \leq -1.14) = 0.1271$.

9. (a) $P(r > 280) = P(r \geq 281) = P(x > 280.5) = P(z \geq -2.16) = 0.9846$. (b) $P(r \geq 320) = P(x \geq 319.5) = P(z \geq 1.95) = 0.0256$.
(c) $P(280 \leq r \leq 320) = P(279.5 \leq x \leq 320.5) = P(-2.26 \leq z \leq 2.05) = 0.9679$. (d) $n = 430$; $p = 0.70$; $q = 0.30$; np and nq are both greater than 5. These conditions mean that the normal approximation to the binomial is appropriate.

11. (a) $P(r \geq 540) = P(x \geq 539.5) = P(z \geq 3.81) \approx 0.000$.
(b) $P(r \leq 500) = P(x \leq 500.5) = P(z \leq 1.11) = 0.8665$.
(c) $P(485 \leq r \leq 525) = P(484.5 \leq x \leq 525.5) = P(0 \leq z \leq 2.84) = 0.4977$.

Chapter 7 Review

1. (a) 0.4599. (b) 0.4015. (c) 0.0384. (d) 0.0104.
(e) 0.0250. (f) 0.8413.
3. (a) 0.9821. (b) 0.3156. (c) 0.2977.
5. 1.645.
7. $z = \pm 1.96$.
9. (a) 0.89. (b) 0. (c) 0.2514.
11. (a) 0.0166. (b) 0.975.
13. (a) 0.9772. (b) 17.3 hr.
15. (a) From $1.81 to $3.51 is a 68% range of errors.
(b) From $0.96 to $4.36 is a 95% range of errors.
(c) Almost all errors are from $0.11 to $5.21.
17. (a) 0.5812. (b) 0.0668. (c) 0.0122.
19. (a) 0.8665. (b) 0.7330.

CHAPTER 8

Section 8.1

1. A set of measurements or counts either existing or conceptual. For example, the population of all ages of all people in Colorado; the population of weights of all students in your school; the population count of all antelope in Wyoming.
3. A numerical descriptive measure of a population, such as μ, the population mean; σ, the population standard deviation; σ^2, the population variance.
5. A statistical inference is a conclusion about the value of a population parameter. We will do both estimation and testing.
7. They help us visualize the sampling distribution by using tables and graphs that approximately represent the sampling distribution.

9. We studied the sampling distribution of mean trout lengths based on samples of size 5. Other such sampling distributions abound.

Section 8.2

Note: Answers may differ slightly depending on the number of digits carried in the standard deviation.
1. (a) $\mu_{\bar{x}} = 15$; $\sigma_{\bar{x}} = 2.0$; $P(15 \leq \bar{x} \leq 17) = P(0 \leq z \leq 1.00) = 0.3413$. (b) $\mu_{\bar{x}} = 15$; $\sigma_{\bar{x}} = 1.75$; $P(15 \leq \bar{x} \leq 17) = P(0 \leq z \leq 1.14) = 0.3729$. (c) The standard deviation is smaller in part (b) because of the larger sample size. Therefore, the distribution about $\mu_{\bar{x}}$ is narrower in part (b).
3. (a) No; the sample size is only 9 and so is too small.
(b) Yes; the $\bar{x}$ distribution also will be normal with $\mu_{\bar{x}} = 25$; $\sigma_{\bar{x}} = 3.5/3$; $P(23 \leq \bar{x} \leq 26) = P(-1.71 \leq z \leq 0.86) = 0.7615$.
5. (a) $P(x < 74.5) = P(z < -0.63) = 0.2643$.
(b) $P(\bar{x} < 74.5) = P(z < -2.79) = 0.0026$. (c) No. If the weight of only one car were less than 74.5 tons, we cannot conclude that the loader is out of adjustment. If the mean weight for a sample of 20 cars were less than 74.5 tons, we would suspect that the loader is malfunctioning. As we see in part (b), the probability of this happening is very low if the loader is correctly adjusted.
7. (a) $P(x < 40) = P(z < -1.80) = 0.0359$. (b) Since the x distribution is approximately normal, the $\bar{x}$ distribution is approximately normal with mean 85 and standard deviation 17.678. $P(\bar{x} < 40) = P(z < -2.55) = 0.0054$.
(c) $P(\bar{x} < 40) = P(z < -3.12) = 0.0009$.
(d) $P(\bar{x} < 40) = P(z < -4.02) < 0.0002$. (e) Yes; if the average value based on five tests were less than 40, the patient is almost certain to have excess insulin.
9. (a) $P(x < 54) = P(z < -1.27) = 0.1020$. (b) The expected number undernourished is 2200(0.1020), or about 224. (c) $P(\bar{x} \leq 60) = P(z \leq -2.99) = 0.0014$.
(d) $P(\bar{x} < 64.2) = P(z < 1.20) = 0.8849$. Since the sample average is above the mean, it is quite unlikely that the doe population is undernourished.
11. (a) Since x itself represents a sample mean return based on a large (random) sample of stocks, x has a distribution that is approximately normal (central limit theorem). (b) $P(1\% \leq \bar{x} \leq 2\%) = P(-1.63 \leq z \leq 1.09) = 0.8105$. (c) $P(1\% \leq \bar{x} \leq 2\%) = P(-3.27 \leq z \leq 2.18) = 0.9849$. (d) Yes. The standard deviation decreases as the sample size increases.
(e) $P(\bar{x} < 1\%) = P(z < -3.27) = 0.0005$. This is very unlikely if $\mu = 1.6\%$. One would suspect that μ has slipped below 1.6%.
13. (a) 30 or more. (b) No.

Chapter 8 Review

1. (a) A normal distribution. (b) The mean μ of the x distribution. (c) $\sigma/\sqrt{n}$, where σ is the standard deviation of the x distribution. (d) They will both be approximately normal with the same mean, but the standard deviations will be $\sigma/\sqrt{50}$ and $\sigma/\sqrt{100}$, respectively.
3. (a) $P(x \geq 40) = P(z \geq 0.71) = 0.2389$. (b) $P(\bar{x} \geq 40)$ $= P(z \geq 2.14) = 0.0162$.
5. $P(98 \leq \bar{x} \leq 102) = P(-1.33 \leq z \leq 1.33) = 0.8164$.
7. (a) $P(\bar{x} \geq 750) = P(z \geq 0) = 0.5000$.
   (b) $P(745 \leq \bar{x} \leq 755) = P(-2.00 \leq z \leq 2.00) =$ 0.9544.

CHAPTER 9

Section 9.1

1. 11.3 to 12.5 mg/l.
3. 15,342 to 16,658.
5. (a) 1.10 sec to 1.30 sec. (b) 560.87 Hz to 657.13 Hz.
7. (a) The mean and standard deviation round to the values given. (b) Using the rounded values of part (a), the 75% interval is from 34.19 thousand to 37.81 thousand. (c) Yes. 30 thousand dollars is below the mean annual profit per employee. We can say with 75% confidence that the mean lies between 34.19 thousand and 37.81 thousand. (d) Yes. 40 thousand is above the mean annual profit per employee. (e) 33.41 thousand to 38.59 thousand. We can say with 90% confidence that the mean lies between 33.4 thousand and 38.6 thousand dollars. 30 thousand is below the mean and 40 thousand is above the mean.
9. (a) The mean and standard deviation round to the values given. (b) Using the rounded values for the mean and standard deviation given in part (a), the interval is from 143.8 to 149.2. (c) Using the rounded values for the mean and standard deviation given in part (a), the interval is from 143.0 to 150.0. (d) Using the rounded values for the mean and standard deviation given in part (a), the interval is from 141.0 to 152.0. (e) The lengths increase as c increases because the values of z_c increase as c increases. If we want to be more certain that μ is in the interval based on the given sample, we have to make the interval wider.
11. (a) 2.3 to 2.7 min. (b) 13.9 to 16.5 min. (c) 23.5 to 27.9 min. (d) The intervals got longer as s increased. This is to be expected, since the endpoints are $\pm z_c s/\sqrt{n}$ from the sample mean and z_c and n were fixed. The length of each interval is $2z_c s/\sqrt{n}$.

Section 9.2

1. 2.110.
3. 1.721.
5. (a) The mean and standard deviation round to the values given. (b) Using the rounded values for the mean and standard deviation given in part (a), the interval is from 1249 to 1295.
7. (a) Use calculator. (b) $10.84 to $13.86.
9. (a) Use calculator; 8.5 to 11.1 in. (b) Use calculator; 15.7 to 18.5 in.
11. (a) Use calculator; 29.4 thousand to 36.8 thousand. (b) Use calculator; 18.0 thousand to 23.6 thousand.
13. (a) Boxplots differ in length of interquartile box, location of median, and length of whiskers. The boxplots come from different samples. (b) Yes; No; For 95% confidence intervals, we expect about 95% of the samples to generate intervals that contain the mean of the population.

Section 9.3

1. (a) $\hat{p} = 39/62 = 0.6290$. (b) 0.51 to 0.75. If this experiment were repeated many times, about 95% of the intervals would contain p. (c) Both np and nq are greater than 5. If either is less than 5, the normal curve will not necessarily give a good approximation to the binomial.
3. (a) $\hat{p} = 1619/5222 = 0.3100$. (b) 0.29 to 0.33. If we repeat the survey with many different samples of 5222 dwellings, about 99% of the intervals will contain p. (c) Both np and nq are greater than 5. If either is less than 5, the normal curve will not necessarily give a good approximation to the binomial.
5. (a) $\hat{p} = 0.5420$. (b) 0.53 to 0.56. (c) Yes. np and nq are greater than 5.
7. (a) $\hat{p} = 0.0304$. (b) 0.02 to 0.05. (c) Yes. np and nq are greater than 5.
9. (a) 0.239 to 0.293. (b) 0.257 to 0.275. (c) The confidence interval based on a larger sample is shorter or narrower.
11. (a) $\hat{p} = 0.8603$. (b) 0.84 to 0.89. (c) A recent study shows that 86% of women shoppers remained loyal to their favorite supermarket last year. The margin of error was 2.5 percentage points.
13. (a) $\hat{p} = 0.25$. (b) 0.22 to 0.28. (c) A survey of 1000 large corporations has shown that 25% will choose a nonsmoking job candidate over an equally qualified smoker. The margin of error was 2.7%.

Section 9.4

1. Estimate a mean. Use 75 plots.
3. Estimate a mean; 120 total or 64 more.
5. Estimate a mean; 117 or 34 more.
7. Estimate a mean; 385 or 218 more.

9. (a) Estimate a proportion; 385.   (b) 186.
11. (a) Estimate a proportion; 666.   (b) 662.
13. (a) 1,664,100.   (b) No. When $p = 0.5$, the formula for sample size without a preliminary estimate is the same as the formula with a preliminary estimate.

Chapter 9 Review

1. See text.
3. Interval for mean, large sample; 176.91 to 180.49.
5. Interval for mean, small sample.
   (a) Use calculator.   (b) 64.1 to 84.3.
7. Interval for proportion; 0.50 to 0.54.
9. Interval for proportion.
   (a) $\hat{p} = 0.4072$.   (b) 0.333 to 0.482.
11. (a) Mean and standard deviation round to results shown.
   (b) Confidence interval for a mean with small sample. Using the rounded values for the mean and standard deviation given in part (a), the interval is from 8.3 to 10.8.

CHAPTER 10

Section 10.1

1. See text.
3. No, if we fail to reject the null hypothesis, we have not proven it beyond all doubt. We have failed only to find sufficient evidence to reject it.
5. (a) $H_0: \mu = 60$ kg.   (b) $H_1: \mu < 60$ kg.   (c) $H_1: \mu > 60$ kg.   (d) $H_1: \mu \neq 60$ kg.   (e) For part (b), the critical region is on the left. For part (c), the critical region is on the right. For part (d), the critical region is on both sides of the mean.
7. (a) $H_0: \mu = 16.4$ feet.   (b) $H_1: \mu > 16.4$ feet.   (c) $H_1: \mu < 16.4$ feet.   (d) $H_1: \mu \neq 16.4$ feet.   (e) For part (b), the critical region is on the right. For part (c), the critical region is on the left. For part (d), the critical region is on both sides of the mean.
9. (a) $H_0: \mu = 288$ lb.   (b) Higher: $H_1: \mu > 288$ lb; lower: $H_1: \mu < 288$ lb; different: $H_1: \mu \neq 288$ lb. For higher, the critical region is the right tail; for lower, the critical region is the left tail; for different, the critical region is both tails.

Section 10.2

1. $H_0: \mu = 16.4$ feet; $H_1: \mu < 16.4$ feet; left-tailed; normal distribution; $z_0 = -2.33$; sample $z = -2.44$; Reject H_0; The storm is lessening; Results statistically significant.
3. $H_0: \mu = 31.8$ calls/day; $H_1: \mu \neq 31.8$ calls/day; two-tailed; normal distribution; $z_0 = \pm 2.58$; sample $z = -2.45$; Do not reject H_0; There is not enough evidence that the mean number of messages has decreased; Results not statistically significant.

5. $H_0: \mu = \$4.75$; $H_1: \mu > \$4.75$; right-tailed; normal distribution; $z_0 = 2.33$; sample $z = 3.14$; Reject H_0; There is evidence that her average tip is more than \$4.75; Results statistically significant.
7. $H_0: \mu = 10.2$ seconds; $H_1: \mu < 10.2$ seconds; left-tailed; normal distribution; $z_0 = -1.645$; sample $z = -1.52$; Do not reject H_0. There is not enough evidence to conclude that the mean acceleration time is less; Results not statistically significant.
9. $H_0: \mu = 19.0$; $H_1: \mu > 19.0$; right-tailed; normal distribution; $z_0 = 2.33$; sample $z = 1.19$; Do not reject H_0. There is not sufficient evidence to conclude that the average oxygen capacity has increased; Results not statistically significant.
11. $H_0: \mu = 7.4$; $H_1: \mu \neq 7.4$; two-tailed; normal distribution; $z_0 = \pm 1.96$; sample $z = 2.12$; Reject H_0; The drug has changed the pH of the blood; Results are statistically significant.
13. The mean and standard deviation round to the value given; $H_0: \mu = \$13.9$ thousand; $H_1: \mu > \$13.9$ thousand; right-tailed; normal distribution; $z_0 = 1.645$; sample $z = 2.34$; Reject H_0; The mean franchise costs for pizza businesses are significantly higher; Results are statistically significant.
15. Essay or class discussion.
17. $H_0: \mu = 21$; $H_1: \mu \neq 21$; $\alpha = 0.01$; two-tailed test.
   (a) $c = 1 - 0.01 = 0.99$ or 99% confidence interval; interval is from 20.28 to 23.72 based on the sample statistics. Since the hypothesized value $\mu = 21$ falls in the confidence interval, we do not reject H_0.   (b) $z_0 = \pm 2.58$; sample $z = 1.50$; Do not reject H_0; The conclusions are the same.

Section 10.3

1. Sample $z = 2.78$; P value $= 0.0027$; The data are significant at the 1% level.
3. Sample $z = -1.18$; P value $= 2(0.1190) = 0.238$; No, not significant at the 5% level.
5. $H_0: \mu = 1.75$; $H_1: \mu > 1.75$; sample $z = 3.02$; P value $= 0.0013$; Yes, the data are significant at the 1% level.
7. $H_0: \mu = 19$ inches; $H_1: \mu < 19$ inches; sample $z = -0.80$; P value $= 0.2119$; No, the data are not significant at the 5% level. We cannot conclude that the average length of trout caught in Pyramid Lake is different from 19 inches.
9. $H_0: \mu = \$15.35$; $H_1: \mu < 15.35$; sample $z = -3.29$; P value $= 0.0005$; Yes, the data are significant at the 5% level. There is evidence that college students' average daily ownership expenses are less than the national average.

Index

Chapter 2

Class Width = $\dfrac{\text{high} - \text{low}}{\text{number classes}}$ (increase to next integer)

Class Midpoint = $\dfrac{\text{upper limit} + \text{lower limit}}{2}$

Lower boundary = lower boundary of previous class
   + class width

Chapter 3

Sample mean $\bar{x} = \dfrac{\Sigma x}{n}$

Population mean $\mu = \dfrac{\Sigma x}{N}$

Weighted average = $\dfrac{\Sigma xw}{\Sigma w}$

Range = largest data value − smallest data value

Sample standard deviation $s = \sqrt{\dfrac{\Sigma(x - \bar{x})^2}{n - 1}}$

Computation formula $s = \sqrt{\dfrac{SS_x}{n - 1}}$ where
$SS_x = \Sigma x^2 - \dfrac{(\Sigma x)^2}{n}$

Population standard deviation $\sigma = \sqrt{\dfrac{\Sigma(x - \mu)^2}{N}}$

Sample variance s^2

Population variance σ^2

Sample Coefficient of Variation $CV = \dfrac{s}{\bar{x}} \cdot 100$

Sample mean for grouped data $\bar{x} = \dfrac{\Sigma xf}{n}$

Sample standard deviation for grouped data

$s = \sqrt{\dfrac{\Sigma(x - \bar{x})^2 f}{n - 1}} = \sqrt{\dfrac{\Sigma x^2 f - (\Sigma xf)^2/n}{n - 1}}$

Chapter 4

Regression and Correlation

$$
\begin{aligned}
&\text{In all these formulas}\\
&SS_x = \Sigma x^2 - \frac{(\Sigma x)^2}{n}\\
&SS_y = \Sigma y^2 - \frac{(\Sigma y)^2}{n}\\
&SS_{xy} = \Sigma xy - \frac{(\Sigma x)(\Sigma y)}{n}
\end{aligned}
$$

Least squares line $y = a + bx$ where $b = \dfrac{SS_{xy}}{SS_x}$ and
   $a = \bar{y} - b\bar{x}$

Pearson product-moment correlation coefficient
   $r = \dfrac{SS_{xy}}{\sqrt{SS_x SS_y}}$

Coefficient of determination = r^2

Chapter 5

Probability of the complement of event A
   $P(not\ A) = 1 - P(A)$

Multiplication rule for independent events
   $P(A\ and\ B) = P(A) \cdot P(B)$

General multiplication rules
   $P(A\ and\ B) = P(A) \cdot P(B,\ given\ A)$
   $P(A\ and\ B) = P(B) \cdot P(A,\ given\ B)$

Addition rule for mutually exclusive events
   $P(A\ or\ B) = P(A) + P(B)$

General addition rule
   $P(A\ or\ B) = P(A) + P(B) - P(A\ and\ B)$

Permutation rule $P_{n,r} = \dfrac{n!}{(n - r)!}$

Combination rule $C_{n,r} = \dfrac{n!}{r!(n - r)!}$

Chapter 6

Mean of a discrete probability distribution $\mu = \Sigma x P(x)$

Standard deviation of a discrete probability distribution
$$\sigma = \sqrt{\Sigma(x - \mu)^2 P(x)}$$

For Binomial Distributions

r = number of successes; p = probability of success;
$q = 1 - p$

Binomial probability distribution $P(r) = C_{n,r} p^r q^{n-r}$

Mean $\mu = np$

Standard deviation $\sigma = \sqrt{npq}$

Chapter 7

Raw score $x = z\sigma + \mu$

Standard score $z = \dfrac{x - \mu}{\sigma}$

Chapter 8

Mean of $\bar{x}$ distribution $\mu_{\bar{x}} = \mu$

Standard deviation of $\bar{x}$ distribution $\sigma_{\bar{x}} = \dfrac{\sigma}{\sqrt{n}}$

Standard score for $\bar{x}$ $\quad z = \dfrac{\bar{x} - \mu}{\sigma/\sqrt{n}}$

Chapter 9

Confidence Interval

for $\mu(n \geq 30)$
$$\bar{x} - z_c \frac{\sigma}{\sqrt{n}} < \mu < \bar{x} + z_c \frac{\sigma}{\sqrt{n}}$$

for μ $(n < 30)$
$$d.f. = n - 1$$
$$\bar{x} - t_c \frac{s}{\sqrt{n}} < \mu < \bar{x} + t_c \frac{s}{\sqrt{n}}$$

for p $(np > 5$ and $nq > 5)$
$$\hat{p} - z_c \sqrt{\frac{\hat{p}(1 - \hat{p})}{n}} < p < \hat{p} + z_c \sqrt{\frac{\hat{p}(1 - \hat{p})}{n}}$$
where $\hat{p} = r/n$

Sample Size n for Estimating

means $n = \left(\dfrac{z_c \sigma}{E}\right)^2$

proportions

$n = p(1 - p)\left(\dfrac{z_c}{E}\right)^2$ with preliminary estimate for p

$n = \dfrac{1}{4}\left(\dfrac{z_c}{E}\right)^2$ without preliminary estimate for p

Chapter 10

Sample Test Statistics for Tests of Hypotheses

for μ $(n \geq 30)$ $\qquad z = \dfrac{\bar{x} - \mu}{\sigma\sqrt{n}}$

for μ $(n < 30)$; $d.f. = n - 1$ $\qquad t = \dfrac{\bar{x} - \mu}{s/\sqrt{n}}$

for p $\qquad z = \dfrac{\hat{p} - p}{\sqrt{pq/n}}$ where $q = 1 - p$

Chapter 11

Sample Test Statistics for Tests of Hypotheses

for paired differences d $t = \dfrac{\bar{d} - \mu_d}{s_d/\sqrt{n}}$ with $d.f. = n - 1$

difference of means large sample

$$z = \frac{(\bar{x}_1 - \bar{x}_2) - (\mu_1 - \mu_2)}{\sqrt{\dfrac{\sigma_1^2}{n_1} + \dfrac{\sigma_2^2}{n_2}}}$$

difference of means small sample with $\sigma_1 \approx \sigma_2$; $d.f. = n_1 + n_2 - 2$

$$t = \frac{(\bar{x}_1 - \bar{x}_2) - (\mu_1 - \mu_2)}{s\sqrt{\dfrac{1}{n_1} + \dfrac{1}{n_2}}}$$

where $s = \sqrt{\dfrac{(n_1 - 1)s_1^2 + (n_2 - 1)s_2^2}{n_1 + n_2 - 2}}$

difference of proportions

$$z = \frac{\hat{p}_1 - \hat{p}_2}{\sqrt{\dfrac{\hat{p}\hat{q}}{n_1} + \dfrac{\hat{p}\hat{q}}{n_2}}}$$ where $\hat{p} = \dfrac{r_1 + r_2}{n_1 + n_2}; \hat{q} = 1 - \hat{p};$

$\hat{p}_1 = r_1/n_1; \hat{p}_2 = r_2/n_2$

Confidence Intervals

for difference of means (when $n_1 \geq 30$ and $n_2 \geq 30$)

$$(\bar{x}_1 - \bar{x}_2) - z_c\sqrt{\frac{\sigma_1^2}{n_1} + \frac{\sigma_2^2}{n_2}} < \mu_1 - \mu_2 < (\bar{x}_1 - \bar{x}_2)$$
$$+ z_c\sqrt{\frac{\sigma_1^2}{n_1} + \frac{\sigma_2^2}{n_2}}$$

for difference of means ($n_1 < 30$ and/or $n_2 < 30$ and $\sigma_1 \approx \sigma_2$) $d.f. = n_1 + n_2 - 2$

$$(\bar{x}_1 - \bar{x}_2) - t_c s\sqrt{\frac{1}{n_1} + \frac{1}{n_2}} < \mu_1 - \mu_2 < (\bar{x}_1 - \bar{x}_2)$$
$$+ t_c s\sqrt{\frac{1}{n_1} + \frac{1}{n_2}}$$

where $s = \sqrt{\dfrac{(n_1 - 1)s_1^2 + (n_2 - 1)s_2^2}{n_1 + n_2 - 2}}$

for difference of proportions

where $\hat{p}_1 = r_1/n_1; \hat{p}_2 = r_2/n_2; \hat{q}_1 = 1 - \hat{p}_1; \hat{q}_2 = 1 - \hat{p}_2$

$$(\hat{p}_1 - \hat{p}_2) - z_c\sqrt{\frac{\hat{p}_1\hat{q}_1}{n_1} + \frac{\hat{p}_2\hat{q}_2}{n_2}} < p_1 - p_2 < (\hat{p}_1 - \hat{p}_2)$$
$$+ z_c\sqrt{\frac{\hat{p}_1\hat{q}_1}{n_1} + \frac{\hat{p}_2\hat{q}_2}{n_2}}$$

Chapter 12

$$\chi^2 = \sum \frac{(O - E)^2}{E} \text{ where } E = \frac{(\text{row total})(\text{column total})}{\text{sample size}}$$

Tests of Independence $d.f. = (R - 1)(C - 1)$

Goodness of fit $d.f. = (\text{number of entries}) - 1$

Sample test statistic for $H_0: \sigma^2 = k; d.f. = n - 1$

$$\chi^2 = \frac{(n - 1)s^2}{\sigma^2}$$

Linear Regression

Standard error of estimate $S_e = \sqrt{\dfrac{SS_y - bSS_{xy}}{n - 2}}$

where $b = \dfrac{SS_{xy}}{SS_x}$

Confidence interval for y

$y_p - E < y < y_p + E$ where y_p is the predicted y value for x and

$$E = t_c S_e\sqrt{1 + \frac{1}{n} + \frac{(x - \bar{x})^2}{SS_x}} \text{ with } d.f. = n - 2$$

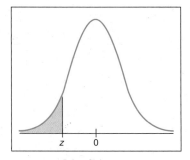

Table entry for z is the area to the left of z.

Areas of a Standard Normal Distribution

(a) Table of Areas to the Left of z

z	.00	.01	.02	.03	.04	.05	.06	.07	.08	.09
−3.4	.0003	.0003	.0003	.0003	.0003	.0003	.0003	.0003	.0003	.0002
−3.3	.0005	.0005	.0005	.0004	.0004	.0004	.0004	.0004	.0004	.0003
−3.2	.0007	.0007	.0006	.0006	.0006	.0006	.0006	.0005	.0005	.0005
−3.1	.0010	.0009	.0009	.0009	.0008	.0008	.0008	.0008	.0007	.0007
−3.0	.0013	.0013	.0013	.0012	.0012	.0011	.0011	.0011	.0010	.0010
−2.9	.0019	.0018	.0018	.0017	.0016	.0016	.0015	.0015	.0014	.0014
−2.8	.0026	.0025	.0024	.0023	.0023	.0022	.0021	.0021	.0020	.0019
−2.7	.0035	.0034	.0033	.0032	.0031	.0030	.0029	.0028	.0027	.0026
−2.6	.0047	.0045	.0044	.0043	.0041	.0040	.0039	.0038	.0037	.0036
−2.5	.0062	.0060	.0059	.0057	.0055	.0054	.0052	.0051	.0049	.0048
−2.4	.0082	.0080	.0078	.0075	.0073	.0071	.0069	.0068	.0066	.0064
−2.3	.0107	.0104	.0102	.0099	.0096	.0094	.0091	.0089	.0087	.0084
−2.2	.0139	.0136	.0132	.0129	.0125	.0122	.0119	.0116	.0113	.0110
−2.1	.0179	.0174	.0170	.0166	.0162	.0158	.0154	.0150	.0146	.0143
−2.0	.0228	.0222	.0217	.0212	.0207	.0202	.0197	.0192	.0188	.0183
−1.9	.0287	.0281	.0274	.0268	.0262	.0256	.0250	.0244	.0239	.0233
−1.8	.0359	.0351	.0344	.0336	.0329	.0322	.0314	.0307	.0301	.0294
−1.7	.0446	.0436	.0427	.0418	.0409	.0401	.0392	.0384	.0375	.0367
−1.6	.0548	.0537	.0526	.0516	.0505	.0495	.0485	.0475	.0465	.0455
−1.5	.0668	.0655	.0643	.0630	.0618	.0606	.0594	.0582	.0571	.0559
−1.4	.0808	.0793	.0778	.0764	.0749	.0735	.0721	.0708	.0694	.0681
−1.3	.0968	.0951	.0934	.0918	.0901	.0885	.0869	.0853	.0838	.0823
−1.2	.1151	.1131	.1112	.1093	.1075	.1056	.1038	.1020	.1003	.0985
−1.1	.1357	.1335	.1314	.1292	.1271	.1251	.1230	.1210	.1190	.1170
−1.0	.1587	.1562	.1539	.1515	.1492	.1469	.1446	.1423	.1401	.1379
−0.9	.1841	.1814	.1788	.1762	.1736	.1711	.1685	.1660	.1635	.1611
−0.8	.2119	.2090	.2061	.2033	.2005	.1977	.1949	.1922	.1894	.1867
−0.7	.2420	.2389	.2358	.2327	.2296	.2266	.2236	.2206	.2177	.2148
−0.6	.2743	.2709	.2676	.2643	.2611	.2578	.2546	.2514	.2483	.2451
−0.5	.3085	.3050	.3015	.2981	.2946	.2912	.2877	.2843	.2810	.2776
−0.4	.3446	.3409	.3372	.3336	.3300	.3264	.3228	.3192	.3156	.3121
−0.3	.3821	.3783	.3745	.3707	.3669	.3632	.3594	.3557	.3520	.3483
−0.2	.4207	.4168	.4129	.4090	.4052	.4013	.3974	.3936	.3897	.3859
−0.1	.4602	.4562	.4522	.4483	.4443	.4404	.4364	.4325	.4286	.4247
−0.0	.5000	.4960	.4920	.4880	.4840	.4801	.4761	.4721	.4681	.4641

For values of z less than −3.49, use 0.000 to approximate the area.